Nuclear Magnetic Resonance Spectroscopy

Nuclear Magnetic Resonance Spectroscopy

An Introduction to Principles, Applications, and Experimental Methods

Joseph B. Lambert
Trinity University
San Antonio, Texas, USA

Eugene P. Mazzola
University of Maryland
College Park, Maryland, USA

Clark D. Ridge
Center for Food Safety and Applied Nutrition
U.S. Food and Drug Administration,
College Park, Maryland, USA

Second Edition

This edition first published 2019
© 2019 John Wiley & Sons Ltd

Previous edition published by Pearson, 2004

All rights reserved. No part of this publication may be reproduced, stored in a retrieval system, or transmitted, in any form or by any means, electronic, mechanical, photocopying, recording or otherwise, except as permitted by law. Advice on how to obtain permission to reuse material from this title is available at http://www.wiley.com/go/permissions.

The right of Joseph B. Lambert, Eugene P. Mazzola, and Clark D. Ridge to be identified as the authors of this work has been asserted in accordance with law.

Registered Offices
John Wiley & Sons, Inc., 111 River Street, Hoboken, NJ 07030, USA
John Wiley & Sons Ltd, The Atrium, Southern Gate, Chichester, West Sussex, PO19 8SQ, UK

Editorial Office
The Atrium, Southern Gate, Chichester, West Sussex, PO19 8SQ, UK

For details of our global editorial offices, customer services, and more information about Wiley products visit us at www.wiley.com.

Wiley also publishes its books in a variety of electronic formats and by print-on-demand. Some content that appears in standard print versions of this book may not be available in other formats.

Limit of Liability/Disclaimer of Warranty
In view of ongoing research, equipment modifications, changes in governmental regulations, and the constant flow of information relating to the use of experimental reagents, equipment, and devices, the reader is urged to review and evaluate the information provided in the package insert or instructions for each chemical, piece of equipment, reagent, or device for, among other things, any changes in the instructions or indication of usage and for added warnings and precautions. While the publisher and authors have used their best efforts in preparing this work, they make no representations or warranties with respect to the accuracy or completeness of the contents of this work and specifically disclaim all warranties, including without limitation any implied warranties of merchantability or fitness for a particular purpose. No warranty may be created or extended by sales representatives, written sales materials or promotional statements for this work. The fact that an organization, website, or product is referred to in this work as a citation and/or potential source of further information does not mean that the publisher and authors endorse the information or services the organization, website, or product may provide or recommendations it may make. This work is sold with the understanding that the publisher is not engaged in rendering professional services. The advice and strategies contained herein may not be suitable for your situation. You should consult with a specialist where appropriate. Further, readers should be aware that websites listed in this work may have changed or disappeared between when this work was written and when it is read. Neither the publisher nor authors shall be liable for any loss of profit or any other commercial damages, including but not limited to special, incidental, consequential, or other damages.

Library of Congress Cataloging-in-Publication Data

Names: Lambert, Joseph B., author. | Mazzola, Eugene P., author. | Ridge, Clark D., author.
Title: Nuclear magnetic resonance spectroscopy : an introduction to principles, applications, and experimental methods / Joseph B. Lambert, Eugene P. Mazzola, Clark D. Ridge.
Description: Second edition. | Hoboken, NJ : John Wiley & Sons, 2019. | Includes bibliographical references and index. |
Identifiers: LCCN 2018026834 (print) | LCCN 2018036673 (ebook) | ISBN 9781119295273 (Adobe PDF) | ISBN 9781119295280 (ePub) | ISBN 9781119295235 (hardcover)
Subjects: LCSH: Nuclear magnetic resonance spectroscopy.
Classification: LCC QD96.N8 (ebook) | LCC QD96.N8 L36 2018 (print) | DDC 543/.66–dc23
LC record available at https://lccn.loc.gov/2018026834

Cover design by Wiley
Cover image: Background © imagewerks/Getty;
all other images courtesy of Clark D. Ridge

Set in 10/12pt WarnockPro by SPi Global, Chennai, India

Contents

Preface to First Edition *xiii*
Acknowledgments *xiv*
Preface to Second Edition *xv*
Acknowledgments *xvi*
Solutions *xvii*
Symbols *xix*
Abbreviations *xxi*

1	**Introduction** *1*	
1.1	Magnetic Properties of Nuclei *1*	
1.2	The Chemical Shift *6*	
1.3	Excitation and Relaxation *10*	
1.4	Pulsed Experiments *13*	
1.5	The Coupling Constant *16*	
1.6	Quantitation and Complex Splitting *23*	
1.7	Commonly Studied Nuclides *25*	
1.8	Dynamic Effects *28*	
1.9	Spectra of Solids *30*	
	Problems *33*	
	Tips on Solving NMR Problems *36*	
	References *37*	
	Further Reading *38*	
2	**Introductory Experimental Methods** *39*	
2.1	The Spectrometer *39*	
2.2	Sample Preparation *41*	
2.3	Optimizing the Signal *42*	
	2.3.1 Sample Tube Placement *42*	
	2.3.2 Probe Tuning *43*	
	2.3.3 Field/Frequency Locking *43*	
	2.3.4 Spectrometer Shimming *44*	
2.4	Determination of NMR Spectral-Acquisition Parameters *48*	
	2.4.1 Number of Data Points *50*	
	2.4.2 Spectral Width *50*	

- 2.4.3 Filter Bandwidth 52
- 2.4.4 Acquisition Time 52
- 2.4.5 Transmitter Offset 52
- 2.4.6 Flip Angle 52
- 2.4.7 Receiver Gain 54
- 2.4.8 Number of Scans 55
- 2.4.9 Steady-State Scans 55
- 2.4.10 Oversampling and Digital Filtration 56
- 2.4.11 Decoupling for X Nuclei 56
- 2.4.12 Typical NMR Experiments 57
- 2.5 Determination of NMR Spectral-Processing Parameters 58
  - 2.5.1 Exponential Weighting 59
  - 2.5.2 Zero Filling 59
  - 2.5.3 FID Truncation and Spectral Artifacts 60
  - 2.5.4 Resolution 62
- 2.6 Determination of NMR Spectra: Spectral Presentation 63
  - 2.6.1 Signal Phasing and Baseline Correction 63
  - 2.6.2 Zero Referencing 66
  - 2.6.3 Determination of Certain NMR Parameters 66
    - 2.6.3.1 Chemical Shifts and Coupling Constants 66
    - 2.6.3.2 ^{1}H Integration 68
- 2.7 Calibrations 70
  - 2.7.1 Pulse Width (Flip Angle) 70
  - 2.7.2 Decoupler Field Strength 72
  - Problems 73
  - References 74
  - Further Reading 74

3 The Chemical Shift 75

- 3.1 Factors That Influence Proton Shifts 75
  - 3.1.1 Local Fields 75
  - 3.1.2 Nonlocal Fields 77
- 3.2 Proton Chemical Shifts and Structure 85
  - 3.2.1 Saturated Aliphatics 85
    - 3.2.1.1 Alkanes 85
    - 3.2.1.2 Functionalized Alkanes 86
  - 3.2.2 Unsaturated Aliphatics 87
    - 3.2.2.1 Alkynes 87
    - 3.2.2.2 Alkenes 88
    - 3.2.2.3 Aldehydes 89
  - 3.2.3 Aromatics 89
  - 3.2.4 Protons on Oxygen and Nitrogen 90
  - 3.2.5 Programs for Empirical Calculations 91
- 3.3 Medium and Isotope Effects 92
  - 3.3.1 Medium Effects 92
  - 3.3.2 Isotope Effects 95
- 3.4 Factors That Influence Carbon Shifts 96

3.5 Carbon Chemical Shifts and Structure 98
    3.5.1 Saturated Aliphatics 98
        3.5.1.1 Acyclic Alkanes 98
        3.5.1.2 Cyclic Alkanes 101
        3.5.1.3 Functionalized Alkanes 101
    3.5.2 Unsaturated Compounds 103
        3.5.2.1 Alkenes 103
        3.5.2.2 Alkynes and Nitriles 104
        3.5.2.3 Aromatics 104
    3.5.3 Carbonyl Groups 105
    3.5.4 Programs for Empirical Calculations 105
3.6 Tables of Chemical Shifts 106
Problems 110
Further Tips on Solving NMR Problems 119
References 122
Further Reading 122

4 The Coupling Constant *125*
4.1 First- and Second-order Spectra 125
4.2 Chemical and Magnetic Equivalence 126
4.3 Signs and Mechanisms of Coupling 132
4.4 Couplings over One Bond 134
4.5 Geminal Couplings 136
4.6 Vicinal Couplings 139
4.7 Long-range Couplings 143
    4.7.1 σ–π Overlap 143
    4.7.2 Zigzag Pathways 144
    4.7.3 Through-Space Coupling 145
4.8 Spectral Analysis 146
4.9 Second-order Spectra 147
    4.9.1 Deceptive Simplicity 147
    4.9.2 Virtual Coupling 149
    4.9.3 Shift Reagents 150
    4.9.4 Isotope Satellites 150
4.10 Tables of Coupling Constants 151
Problems 157
References 169
Further Reading 170

5 Further Topics in One-Dimensional NMR Spectroscopy *173*
5.1 Spin–Lattice and Spin–Spin Relaxation 173
    5.1.1 Causes of Relaxation 173
    5.1.2 Measurement of Relaxation Time 175
    5.1.3 Transverse Relaxation 176
    5.1.4 Structural Ramifications 177
    5.1.5 Anisotropic Motion 177
    5.1.6 Segmental Motion 178
    5.1.7 Partially Relaxed Spectra 178
    5.1.8 Quadrupolar Relaxation 178

5.2 Reactions on the NMR Time Scale *180*
  5.2.1 Hindered Rotation *181*
  5.2.2 Ring Reversal *183*
  5.2.3 Atomic Inversion *183*
  5.2.4 Valence Tautomerizations and Bond Shifts *185*
  5.2.5 Quantification *187*
  5.2.6 Magnetization Transfer and Spin Locking *187*
5.3 Multiple Resonance *188*
  5.3.1 Spin Decoupling *188*
  5.3.2 Difference Decoupling *190*
  5.3.3 Classes of Multiple Resonance Experiments *190*
  5.3.4 Off-resonance Decoupling *191*
5.4 The Nuclear Overhauser Effect *194*
  5.4.1 Origin *194*
  5.4.2 Observation *195*
  5.4.3 Difference NOE *198*
  5.4.4 Applications *199*
  5.4.5 Limitations *200*
5.5 Spectral Editing *200*
  5.5.1 The Spin–Echo Experiment *201*
  5.5.2 The Attached Proton Test *201*
  5.5.3 The DEPT Sequence *204*
5.6 Sensitivity Enhancement *205*
  5.6.1 The INEPT sequence *206*
  5.6.2 Refocused INEPT *208*
  5.6.3 Spectral Editing with Refocused INEPT *208*
  5.6.4 DEPT Revisited *210*
5.7 Carbon Connectivity *212*
5.8 Phase Cycling, Composite Pulses, and Shaped Pulses *213*
  5.8.1 Phase Cycling *213*
  5.8.2 Composite Pulses *215*
  5.8.3 Shaped Pulses *215*
  Problems *217*
  References *231*
  Further Reading *231*

6 **Two-Dimensional NMR Spectroscopy** *237*
6.1 Proton–Proton Correlation Through J Coupling *237*
  6.1.1 COSY45 *247*
  6.1.2 Long-Range COSY (LRCOSY or Delayed COSY) *248*
  6.1.3 Phase-Sensitive COSY (ϕ-COSY) *249*
  6.1.4 Multiple Quantum Filtration *250*
  6.1.5 TOtal Correlation SpectroscopY (TOCSY) *252*
  6.1.6 Relayed COSY *252*
  6.1.7 J-Resolved Spectroscopy *252*
  6.1.8 COSY for Other Nuclides *254*

6.2 Proton–Heteronucleus Correlation 254
    6.2.1 HETCOR 255
    6.2.2 HMQC 257
    6.2.3 BIRD-HMQC 257
    6.2.4 HSQC 260
    6.2.5 COLOC 260
    6.2.6 HMBC 260
    6.2.7 Heteronuclear Relay Coherence Transfer 263
6.3 Proton–Proton Correlation Through Space or Chemical Exchange 264
6.4 Carbon–Carbon Correlation 268
6.5 Higher Dimensions 270
6.6 Pulsed Field Gradients 273
6.7 Diffusion-Ordered Spectroscopy 277
6.8 Summary of 2D Methods 279
Problems 280
References 305
Further Reading 306

7 Advanced Experimental Methods 309

7.1 Part A: One-Dimensional Techniques 309
    7.1.1 T_1 Measurements 309
    7.1.2 ^{13}C Spectral Editing Experiments 311
        7.1.2.1 The APT Experiment 311
        7.1.2.2 The DEPT Experiment 312
    7.1.3 NOE Experiments 313
        7.1.3.1 The NOE Difference Experiment 314
        7.1.3.2 The Double-Pulse, Field-Gradient, Spin-Echo NOE Experiment 315
7.2 Part B: Two-Dimensional Techniques 316
    7.2.1 Two-Dimensional NMR Data-Acquisition Parameters 316
        7.2.1.1 Number of Data Points 316
        7.2.1.2 Number of Time Increments 317
        7.2.1.3 Spectral Widths 317
        7.2.1.4 Acquisition Time 317
        7.2.1.5 Transmitter Offset 318
        7.2.1.6 Flip Angle 318
        7.2.1.7 Relaxation Delay 318
        7.2.1.8 Receiver Gain 318
        7.2.1.9 Number of Scans per Time Increment 319
        7.2.1.10 Steady-State Scans 319
    7.2.2 Two-Dimensional NMR Data-Processing Parameters 319
        7.2.2.1 Weighting Functions 319
        7.2.2.2 Zero Filling 321
        7.2.2.3 Digital Resolution 321
        7.2.2.4 Linear Prediction 322
    7.2.3 Two-Dimensional NMR Data Display 324
        7.2.3.1 Phasing and Zero Referencing 324

- 7.2.3.2 Symmetrization *325*
- 7.2.3.3 Use of Cross Sections in Analysis *325*

7.3 Part C: Two-Dimensional Techniques: The Experiments *325*
- 7.3.1 Homonuclear Chemical-Shift Correlation Experiments via Scalar Coupling *326*
  - 7.3.1.1 The COSY Family: COSY-90°, COSY-45°, Long-Range COSY, and DQF-COSY *326*
  - 7.3.1.2 The TOCSY Experiment *330*
- 7.3.2 Direct Heteronuclear Chemical-Shift Correlation via Scalar Coupling *331*
  - 7.3.2.1 The HMQC Experiment *331*
  - 7.3.2.2 The HSQC Experiment *332*
  - 7.3.2.3 The HETCOR Experiment *334*
- 7.3.3 Indirect Heteronuclear Chemical-Shift Correlation via Scalar Coupling *335*
  - 7.3.3.1 The HMBC Experiment *336*
  - 7.3.3.2 The FLOCK Experiment *338*
  - 7.3.3.3 The HSQC–TOCSY Experiment *340*
- 7.3.4 Homonuclear Chemical-Shift Correlation via Dipolar Coupling *342*
  - 7.3.4.1 The NOESY Experiment *342*
  - 7.3.4.2 The ROESY Experiment *343*
- 7.3.5 1D and Advanced 2D Experiments *345*
  - 7.3.5.1 The 1D TOCSY Experiment *345*
  - 7.3.5.2 The 1D NOESY and ROESY Experiments *347*
  - 7.3.5.3 The Multiplicity-Edited HSQC Experiment *347*
  - 7.3.5.4 The H2BC Experiment *348*
  - 7.3.5.5 Nonuniform Sampling *352*
  - 7.3.5.6 Pure Shift NMR *355*
  - 7.3.5.7 Covariance NMR *358*
- 7.3.6 Pure Shift-Covariance NMR *362*

References *362*

8 Structural Elucidation: Two Methods *365*

8.1 Part A: Spectral Analysis *365*
- 8.1.1 ^{1}H NMR Data *365*
- 8.1.2 ^{13}C NMR Data *366*
- 8.1.3 The DEPT Experiment *369*
- 8.1.4 The HSQC Experiment *370*
- 8.1.5 The COSY Experiment *370*
- 8.1.6 The HMBC Experiment *372*
- 8.1.7 General Molecular Assembly Strategy *372*
- 8.1.8 A Specific Molecular Assembly Procedure *374*
- 8.1.9 The NOESY Experiment *379*

8.2 Part B: Computer-Assisted Structure Elucidation *382*
- 8.2.1 CASE Procedures *383*
- 8.2.2 T-2 Toxin *384*

Appendix A   Derivation of the NMR Equation   *389*

Appendix B   The Bloch Equations   *391*
Reference   *395*

Appendix C   Quantum Mechanical Treatment of the Two-Spin System   *397*

Appendix D   Analysis of Second-Order, Three- and Four-Spin Systems by Inspection   *409*

Appendix E   Relaxation   *415*

Appendix F   Product-Operator Formalism and Coherence-Level Diagrams   *421*
Reference   *433*

Appendix G   Stereochemical Considerations   *435*
G.1   Homotopics Groups   *436*
G.2   Enantiotopic Groups   *438*
G.3   Diastereotopic Groups   *440*
References   *441*

Index   *443*

Preface to First Edition

Nuclear magnetic resonance (NMR) has become the chemist's most general structural tool. It is one of the few techniques that may be applied to all three states of matter. Some spectra may be obtained from less than a microgram of material. In the early 1960s, spectra were taken crudely on strip-chart recorders. The field has since seen one major advance after another, culminating in the Nobel prizes awarded to Richard R. Ernst in 1991 and to Kurt Wüthrich in 2002. The very richness of the field, however, has made it intimidating to many users. How can they take full advantage of the power of the method when so much of the methodology seems to be highly technical, beyond the grasp of the casual user? This text was written to answer this question. The chapters provide an essentially nonmathematical introduction to the entire field, with emphasis on structural analysis.

The early chapters introduce classical NMR spectroscopy. A thorough understanding of proton and carbon chemical shifts (Chapter 3) is required in order to initiate any analysis of spectra. The role of other nuclei is key to the examination of molecules containing various heteroatoms. An analysis of coupling constants (Chapter 4) provides information about stereochemistry and connectivity relationships between nuclei. The older concepts of chemical shifts and coupling constants are emphasized, because they provide the basis for the application of modern pulse sequences.

Chapters 5 and 6 describe the basics of modern NMR spectroscopy. The phenomena of relaxation, of chemical dynamics, and of multiple resonance are considered thoroughly. One-dimensional multipulse sequences are explored to determine the number of protons attached to carbon atoms, to enhance spectral sensitivity, and to determine connectivities among carbon atoms. Concepts that have been considered advanced, but are now moving towards the routine, are examined, including phase cycling, composite pulses, pulsed field gradients, and shaped pulses. Two-dimensional methods represent the current apex of the field. We discuss a large number of these experiments. It is our intention to describe not only what the pulse sequences do, but also how they work, so that the user has a better grasp of the techniques.

Two chapters are dedicated to experimental methodologies. Although many people are provided with spectra by expert technicians, increasing numbers of chemists must record spectra themselves. They must consider and optimize numerous experimental variables. These chapters address not only the basic parameters, such as spectral width and acquisition time, but also the parameters of more advanced techniques, such as spectral editing and two-dimensional spectra.

To summarize modern NMR spectroscopy, Chapter 8 carries out the total structural proof of a single complex natural product. This chapter illustrates the tactics and strategies of structure elucidation, from one-dimensional assignments to two-dimensional spectral correlations, culminating in stereochemical analysis based on Overhauser effects.

The theory behind NMR not only is beautiful in itself, but also offers considerable insight into the methodology. Consequently, a series of appendices presents a full treatment of this theoretical underpinning, necessary to the physical or analytical chemist, but possibly still edifying to the synthetic organic or inorganic chemist.

This text thus offers
- classical analysis of chemical shifts and coupling constants for both protons and other nuclei,
- modern multipulse and multidimensional methods, both explained and illustrated,
- experimental procedures and practical advice relative to the execution of NMR experiments,
- a chapter-long worked-out problem that illustrates the application of nearly all current methods to determine the structure and stereochemistry of a complex natural product,
- appendices containing the theoretical basis of NMR, including the most modern approach that uses product operators and coherence-level diagrams, and
- extensive problems throughout the book.

Joseph B. Lambert
Eugene P. Mazzola

Acknowledgments

The authors are indebted to numerous people for assistance in preparing this manuscript. For expert word processing, artwork, recording of spectra for figures, or general assistance, we thank Curtis N. Barton, Gwendolyn N. Chmurny, Frederick S. Fry, Jr., D. Aaron Lucas, Peggy L. Mazzola, Marcia L. Meltzer, William F. Reynolds, Carol J. Slingo, Mitchell J. Smith, Que. N. Van, and Yuyang Wang. In addition, we are grateful to the following individuals for reviewing all or part of the manuscript: Lyle D. Isaacs (University of Maryland, College Park), William F. Reynolds (University of Toronto), Que. N. Van (National Cancer Institute, Frederick, Maryland), and R. Thomas Williamson (Wyeth Research).

Preface to Second Edition

During the 15 years since the first edition of this book was published, experimental techniques have gained prominence and are changing the way that NMR is, and will continue to be, practiced. In this second edition, we introduce and explain several new techniques in a manner that should be comprehensible to advanced undergraduate and junior graduate students in chemistry.

The first new procedure is "nonuniform sampling," which is a data-processing method to enhance indirect-dimension NMR data. It has the capability of enabling either (i) NMR spectra of equal resolution in the nondetected, f_1 dimension to be acquired in less time than in the past or (ii) spectra of greater f_1 resolution to be acquired in the same time as standard two-dimensional NMR spectra.

The second new technique is "pure shift NMR," which involves both NMR data accumulation and processing. This method permits the acquisition of both one- and two-dimensional proton-decoupled, NMR spectra to be acquired. The procedure is extremely useful when heavily overlapping proton NMR spectra are encountered, an increasingly prevalent situation.

The third procedure, "covariance NMR," is also a data-processing technique, which comes in two forms: homonuclear "direct" and heteronuclear "general indirect" covariance. Direct covariance is applied to symmetrical, homonuclear 2D data, such as COSY and NOESY, to produce two-dimensional data in which the resolution is identical in both the f_1 and the f_2 domains. General indirect covariance is employed with heteronuclear data so that two, relatively short, NMR experiments, such as HSQC and TOCSY, can be combined to yield the HSQC + TOCSY spectrum in far less time than it would take to acquire the HSQC-TOCSY spectrum directly.

We consider this book to be introductory, and these topics at present are covered in no other introductory book. In addition to adding these topics, we have revised the entire book, made minor corrections throughout, and added many new problems, to bring the material up to a standard for the 2020s.

18 January 2018

Joseph B. Lambert
San Antonio, Texas

Eugene P. Mazzola
College Park, Maryland

Clark D. Ridge
College Park, Maryland

Acknowledgments

The authors thank David Rovnyak (Bucknell University) for many helpful discussions concerning nonuniform sampling, Carlos Cobas of Mestrelab Research for assistance with nonuniform sampling, pure shift, and covariance data processing, Jill Clouse of MilliporeSigma for supplying a generous sample of 2-norbornene, Advanced Chemistry Development, Inc. (ACD Labs, Toronto, Ontario) for assistance with their Structural Elucidator program, and Lilly Ridge for typing large portions of the book.

Solutions

Please visit http://booksupport.wiley.com and enter the book title, author name, or isbn to access the Solutions Manual and Powerpoint slides of the figures to accompany this text.

Symbols

B_0	main magnetic field
B_1	magnetic field due to transmitter
B_2	magnetic field in double-resonance experiments
Hz	hertz (a unit of frequency)
I	dimensionless spin
I_z	spin quantum number in the z direction
J	indirect spin–spin coupling constant
M	magnetization
p	coherence order
T	tesla (unit of magnetic flux density, commonly, the magnetic-field strength)
T_1	spin–lattice (longitudinal) relaxation time
$T_{1\rho}$	spin–lattice relaxation time (spin locked) in the rotating frame
T_2	spin–spin (transverse) relaxation time
T_2^*	effective spin–spin relaxation time (includes magnetic-field inhomogeneity effects on xy magnetization)
T_c	coalescence temperature
t_a	acquisition time
t_p	transmitter pulse duration or pulse width (in μs)
t_1	two-dimensional (2D) incremented time
t_2	two-dimensional (2D) acquisition time
W	designation for relaxation pathways, with units of rate constants
α	flip angle
α^0	optimum (Ernst) flip angle
γ	gyromagnetic or magnetogyric ratio
γB_0	resonance or Larmor frequency (ω_0)
γB_2	decoupler field strength (ω_2)
δ	chemical shift
η	nuclear Overhauser enhancement
μ	magnetic moment
ν	linear frequency
σ	magnetic shielding
τ	time delay or lifetime
τ_c	effective correlation time
τ_m	mixing time
ω	angular frequency

Abbreviations

APT	attached proton test
ASIS	aromatic solvent-induced shift
BIRD	bilinear rotation decoupling
COLOC	correlation spectroscopy via long-range coupling
COSY	correlation spectroscopy
CP	cross polarization
CW	continuous wave
CYCLOPS	cyclically ordered phase sequence
DANTE	delays alternating with nutation for tailored excitation
DEPT	distortionless enhancement by polarization transfer
DPFGSE	double pulsed field gradient spin echo
DQF	double quantum filtered
DR	digital resolution
DT	relaxation delay time
EXSY	exchange spectroscopy
FID	free-induction decay
Fn	Fourier number
FT	Fourier transform or transformation
H2BC	heteronuclear two-bond correlation
HETCOR	heteronuclear chemical-shift correlation
HMBC	heteronuclear multiple bond correlation
HMQC	heteronuclear multiple quantum correlation
HOD	monodeuterated water
HSQC	heteronuclear single quantum correlation
INADEQUATE	incredible natural abundance double-quantum transfer experiment
INEPT	insensitive nuclei enhanced by polarization transfer
LP	linear prediction
MAS	magic angle spinning
MQC	multiple quantum coherence
MRI	magnetic resonance imaging
n_i	number of time increments
NMR	nuclear magnetic resonance
NOE	nuclear Overhauser effect or enhancement
NOESY	NOE spectroscopy
np	number of data points

ns	number of scans	
ns/i	number of scans per time increment	
NUS	nonuniform sampling	
PFG	pulsed field gradient	
ppm	parts per million	
PSYCHE	pure shift yielded by chirp excitation	
RF	radio frequency	
ROESY	rotating frame nuclear Overhauser effect spectroscopy	
RT	repetition time (DT + t_a)	
S/N	signal-to-noise ratio	
SR	spectral resolution	
sw	spectral width	
TMS	tetramethylsilane	
TOCSY	total correlation spectroscopy	
WALTZ	wideband, alternating-phase, low-power technique for zero residual splitting	
WATERGATE	water suppression by gradient tailored excitation	

1

Introduction

Structure determination of almost any organic or biological molecule, as well as that of many inorganic molecules, begins with nuclear magnetic resonance (NMR) spectroscopy. During its existence of more than half a century, NMR spectroscopy has undergone several internal revolutions, repeatedly redefining itself as an increasingly complex and effective structural tool. Aside from X-ray crystallography, which can uncover the complete molecular structure of some pure crystalline materials, NMR spectroscopy is the chemist's most direct and general tool for identifying the structure of both pure compounds and mixtures, as either solids or liquids. The process often involves performing several NMR experiments to deduce the molecular structure from the magnetic properties of the atomic nuclei and the surrounding electrons.

1.1 Magnetic Properties of Nuclei

The simplest atom, hydrogen, is found in almost all organic compounds and is composed of a single proton and a single electron. The hydrogen atom is denoted as ^{1}H, in which the superscript signifies the sum of the atom's protons and neutrons, that is, the atomic mass of the element. For the purpose of NMR, the key aspect of the hydrogen nucleus is its angular momentum properties, which resemble those of a classical spinning particle. Because the spinning hydrogen nucleus is positively charged, it generates a magnetic field and possesses a *magnetic moment* **μ**, just as a charge moving in a circle creates a magnetic field (Figure 1.1). The magnetic moment **μ** is a vector, because it has both magnitude and direction, as defined by its axis of spin in the figure. In this context, *boldface* symbols connote a vectorial parameter; when only the magnitude is under consideration, the symbol is depicted without boldface, as μ. The NMR experiment exploits the magnetic properties of nuclei to provide information on the molecular structure.

The spin properties of protons and neutrons in the nuclei of heavier elements combine to define the overall spin of the nucleus. When both the atomic number (the number of protons) and the atomic mass (the sum of the protons and neutrons) are even, the nucleus has no magnetic properties, as signified by a zero value of its *spin quantum number*, I (Figure 1.2). Such nuclei are considered not to be spinning. Common nonmagnetic (nonspinning) nuclei are carbon (^{12}C) and oxygen (^{16}O), which therefore are invisible to the NMR experiment. When either the atomic number or the atomic mass is odd, or when both are odd, the nucleus has magnetic properties that correspond to spin.

Nuclear Magnetic Resonance Spectroscopy: An Introduction to Principles, Applications, and Experimental Methods, Second Edition. Joseph B. Lambert, Eugene P. Mazzola, and Clark D. Ridge.
© 2019 John Wiley & Sons Ltd. Published 2019 by John Wiley & Sons Ltd.

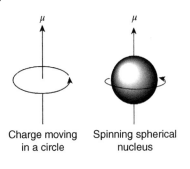

Figure 1.1 Analogy between a charge moving in a circle and a spinning nucleus.

Figure 1.2 Three classes of nuclei.

For spinning nuclei, the spin quantum number can take on only certain values, which is to say that it is quantized. Those nuclei with a spherical shape have a spin I of $½$, and those with a nonspherical, or quadrupolar, shape have a spin of 1 or more (in increments of $½$).

Common nuclei with a spin of $½$ include ^{1}H, ^{13}C, ^{15}N, ^{19}F, ^{29}Si, and ^{31}P. Thus, many of the most common elements found in organic molecules (H, C, N, P) have at least one isotope with $I = ½$ (although oxygen does not). The class of nuclei with $I = ½$ is the most easily examined by the NMR experiment. *Quadrupolar nuclei* ($I > ½$) include ^{2}H, ^{11}B, ^{14}N, ^{17}O, ^{33}S, and ^{35}Cl.

The magnitude of the magnetic moment produced by a spinning nucleus varies from atom to atom in accordance with the equation $\mu = \gamma \hbar I$ (see Appendix A for a derivation of this equation). The quantity $\hbar$ is Planck's constant h divided by 2π, and γ is a characteristic of the nucleus called the *gyromagnetic* or the *magnetogyric ratio*. The larger the gyromagnetic ratio, the larger is the magnetic moment of the nucleus. Nuclei that have the same number of protons, but different numbers of neutrons, are called *isotopes* (^{1}H/^{2}H, ^{14}N/^{15}N). The term *nuclide* generally is applied to any atomic nucleus.

To study nuclear magnetic properties, the experimentalist subjects nuclei to a strong laboratory magnetic field B_0 with units of tesla, or T (1 T = 10^4 Gauss, or G). In the absence of the laboratory field, nuclear magnets of the same isotope have the same energy. When the B_0 field is turned on along a direction designated as the z-axis, the energies of the nuclei in a sample are affected. There is a slight tendency for magnetic moments to move along the general direction of B_0 (+z) rather than the opposite direction (−z). (This motion will be more fully described presently.) Nuclei with a spin of $½$ assume only these two modes of motion. The splitting of spins into specific groups has been called the *Zeeman effect*.

The interaction is illustrated in Figure 1.3. At the left is a magnetic moment with a +z component, and at the right is one with a −z component. The nuclear magnets are

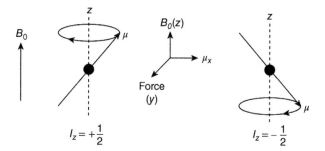

Figure 1.3 Interaction between a spinning nucleus and an external magnetic field B_0.

not actually lined up parallel to the $+z$ or $-z$ direction. Rather, the force of B_0 causes the magnetic moment to move in a circular fashion about the $+z$ direction in the first case and about the $-z$ direction in the second. In terms of vector analysis, the B_0 field in the z-direction operates on the x component of μ to create a force in the y-direction (Figure 1.3, inset in the middle). The force F is the cross, or vector, product between the magnetic moment μ and the magnetic field B (a vector with magnitude only in the z-direction at this stage with value B_0), that is, $F = \mu \times B$. The nuclear moment then begins to move toward the y-direction. Because the force of B_0 on μ, is always perpendicular to both B_0 and μ (according to the definition of a cross product), the motion of μ describes a circular orbit around the $+z$ or the $-z$-direction, in complete analogy to the forces present in a spinning top or gyroscope. This motion is termed *precession*.

As the process of quantization allows only two directions of precession for a spin-½ nucleus (Figure 1.3), two assemblages or *spin states* are created, designated as $I_z = +½$ for those precessing with the field ($+z$) and $I_z = -½$ for those precessing against the field ($-z$) (some texts refer to the quantum number I_z as m_I). The assignment of signs ($+$ or $-$) is entirely arbitrary. The designation $I_z = +½$ is given to the slightly lower energy. In the absence of B_0, the precessional motions are absent, and all nuclei have the same energy.

The relative proportions of nuclei with $+z$ and $-z$ precession in the presence of B_0 is defined by Boltzmann's law (Eq. (1.1)),

$$\frac{n\left(+\tfrac{1}{2}\right)}{n\left(-\tfrac{1}{2}\right)} = \exp\left(\frac{\Delta E}{kT}\right) \tag{1.1}$$

in which n is the population of a spin state, k is Boltzmann's constant, T is the absolute temperature in kelvin (K), and ΔE is the energy difference between the spin states. Figure 1.4a depicts the energies of the two states and the difference ΔE between them.

The precessional motion of the magnetic moment around B_0 occurs with angular frequency ω_0, called the *Larmor frequency*, whose units are radians per second (rad s^{-1}). As B_0 increases, so does the angular frequency, that is, $\omega_0 \propto B_0$, as is demonstrated in Appendix A. The constant of proportionality between ω_0 and B_0 is the gyromagnetic ratio γ, so that $\omega_0 = \gamma B_0$. The natural precession frequency can be expressed as linear frequency in Planck's relationship $\Delta E = h\nu_0$, or as angular frequency $\Delta E = \hbar \omega_0$

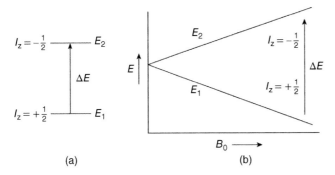

Figure 1.4 (a) The energy difference between spin states. (b) The energy difference as a function of the external field B_0.

($\omega_0 = 2\pi\nu_0$). In this way, the energy difference between the spin states is related to the Larmor frequency by the formula of Eq. (1.2).

$$\Delta E = \hbar\omega_0 = h\nu_0 = \gamma\hbar B_0 \tag{1.2}$$

Thus, as the B_0 field increases, the difference in energy between the two spin states increases, as illustrated in Figure 1.4b. Appendix A provides a complete derivation of these relationships.

The foregoing equations indicate that the natural precession frequency of a spinning nucleus ($\omega_0 = \gamma B_0$) depends only on the nuclear properties contained in the gyromagnetic ratio γ and on the laboratory-determined value of the magnetic field B_0. For a proton in a magnetic field B_0 of 7.05 T, the frequency of precession is 300 MHz, and the difference in energy between the spin states is only 0.0286 cal mol^{-1} (0.120 J mol^{-1}). This value is extremely small in comparison with the energy differences between vibrational or electronic states. At a higher field, such as 14.1 T, the frequency increases proportionately to 600 MHz in this case.

In the NMR experiment, the two states illustrated in Figure 1.4 are made to interconvert by applying a second magnetic field B_1 at radio frequency (RF) range. When the frequency of the B_1 field is the same as the Larmor frequency of the nucleus, energy can flow by absorption and emission between this newly applied field and the nuclei. Absorption of energy occurs as $+\frac{1}{2}$ nuclei become $-\frac{1}{2}$ nuclei, and emission occurs as $-\frac{1}{2}$ nuclei become $+\frac{1}{2}$ nuclei. Since there is an excess of $+\frac{1}{2}$ nuclei at the beginning of the experiment, there is a net absorption of energy. The process is called *resonance*, and the absorption may be detected electronically and displayed as a plot of frequency vs amount of energy absorbed. Because the resonance frequency ν_0 is highly dependent on the structural environment of the nucleus, NMR spectroscopy has become the structural tool of choice for chemists. Figure 1.5 illustrates the NMR spectrum for the protons in benzene. Absorption is represented by a peak directed upward from the baseline.

Because gyromagnetic ratios vary among elements and even among isotopes of a single element, resonance frequencies also vary ($\omega_0 = \gamma B_0$). There is essentially no overlap in the resonance frequencies of different nuclides, including isotopes. At the field strength at which protons resonate at 300 MHz (7.05 T), ^{13}C nuclei resonate at 75.45 MHz, ^{15}N nuclei at 30.42 MHz, and so on. At 14.1 T, the frequencies would be doubled, respectively, 600, 150.9, and 60.84 MHz.

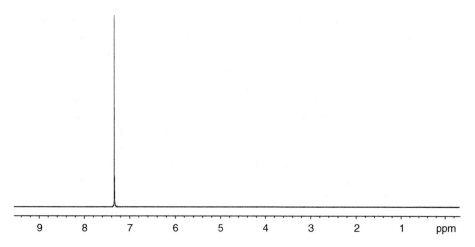

Figure 1.5 The 300 MHz ^{1}H spectrum of benzene.

The magnitude of the gyromagnetic ratio γ also has an important influence on the intensity of the resonance. The difference in energy, $\Delta E = \gamma \hbar B_0$ (Eq. (1.2)), between the two spin states is directly proportional not only to B_0, as illustrated in Figure 1.4b, but also to γ. From Boltzmann's law (Eq. (1.1)), when ΔE is larger, there is a greater population difference between the two states. A greater excess of $I_z = +\frac{1}{2}$ spins (designated the E_1 state) means that more nuclei are available to flip to the E_2 state with $I_z = -\frac{1}{2}$, so the resonance intensity is larger. The proton has one of the largest gyromagnetic ratios, so its spin states are relatively far apart, and the value of ΔE is especially large. The proton signal, consequently, is very strong. Many other important nuclei, such as ^{13}C and ^{15}N, have much smaller gyromagnetic ratios and hence have smaller differences between the energies of the two spin states (Figure 1.6). Thus, their signals are much less intense.

When spins have values greater than $\frac{1}{2}$, more than two spin states are allowed. For $I = 1$ nuclei, such as ^{2}H and ^{14}N, the magnetic moments may precess about three directions relative to B_0: parallel ($I_z = +1$), perpendicular (0), and opposite (-1). In general, there are $(2I + 1)$ spin states—for example, six for $I = 5/2$ (^{17}O has this spin). The values

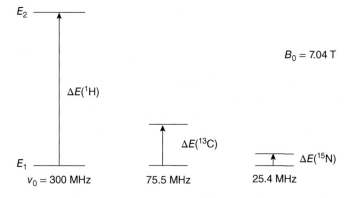

Figure 1.6 The energy difference between spin states for three nuclides with various relative magnitudes of the gyromagnetic ratio ($|\gamma|$): 26.75 for ^{1}H, 6.73 for ^{13}C, and 2.71 for ^{15}N.

of I_z extend from $+I$ to $-I$ in increments of 1 $(+I, (+I-1), (+I-2), \ldots, -I)$. For example, $I_z = +1, 0,$ and -1 for $I = 1$, and $+3/2, +\frac{1}{2}, -\frac{1}{2},$ and $-3/2$ for $I = 3/2$. Hence, the energy state picture is more complex for quadrupolar than for spherical nuclei.

In summary, the NMR experiment consists of immersing magnetic nuclei in a strong field B_0 to distinguish them according to their values of I_z ($+\frac{1}{2}$ and $-\frac{1}{2}$ for spin-$\frac{1}{2}$ nuclei), followed by the application of a B_1 field whose frequency corresponds to the Larmor frequency of the nuclei under examination ($\omega_0 = \gamma B_0$). This application of energy results in a net absorption, as the excess $+\frac{1}{2}$ nuclei are converted to $-\frac{1}{2}$ nuclei. The resonance frequency varies from nuclide to nuclide according to the value of the gyromagnetic ratio γ. The energy difference between the I_z spin states, $\Delta E = h\nu$, which determines the intensity of the absorption, depends on the value of B_0 (Figure 1.4) and on the gyromagnetic ratio of the nucleus ($\Delta E = \gamma \hbar B_0$) (Figure 1.6).

1.2 The Chemical Shift

The remaining sections in this chapter discuss the various factors that determine the content of NMR spectra. Uppermost is the location of the resonance in the spectrum, the so-called resonance frequency ν_0 (or ω_0 as angular frequency), which depends on the molecular environment as well as on γ and B_0 ($\nu_0 = \gamma B_0/2\pi$ or $\omega_0 = \gamma B_0$). This dependence of the resonance frequency on structure is the ultimate reason for the importance of NMR spectroscopy in chemistry.

The electron cloud that surrounds the nucleus also has charge, motion, and, hence, a magnetic moment. The magnetic field generated by the electrons alters the B_0 field in the microenvironment around the nucleus. The actual field present at a given nucleus thus depends on the nature of the surrounding electrons. This electronic modulation of the B_0 field is termed *shielding* and is represented quantitatively by the Greek letter sigma (σ). The actual field at the nucleus becomes B_{local} and may be expressed as $B_{local} = B_0(1 - \sigma)$, in which the electronic shielding σ is positive for protons. The variation of the resonance frequency with shielding has been termed the *chemical shift*.

By substituting B_{local} for B_0 in Eq. (1.2), the expression for the resonance frequency in terms of shielding becomes Eq. (1.3).

$$\nu_0 = \frac{\gamma B_0(1-\sigma)}{2\pi} \qquad (1.3)$$

Decreased shielding thus results in a higher resonance frequency ν_0 at constant B_0, since σ enters the equation after a negative sign. For example, the presence of an electron-withdrawing group in a molecule reduces the electron density around a proton so that there is less shielding and, consequently, a higher resonance frequency than in the case of a molecule that lacks the electron-withdrawing group. Hence, protons in fluoromethane (CH_3F) resonate at a higher frequency than those in methane (CH_4), because the fluorine atom withdraws electrons from around the hydrogen nuclei.

Figure 1.7 separately shows the NMR spectra of the protons and the carbons of methyl acetate ($CH_3CO_2CH_3$). Although 98.9% of naturally occurring carbon is the nonmagnetic ^{12}C, the carbon NMR experiment is carried out on the 1.1% of ^{13}C, which has an I of $\frac{1}{2}$. Because of differential electronic shielding, the ^1H spectrum contains separate resonances for the two types of protons (O—CH_3 and C—CH_3), and the ^{13}C spectrum

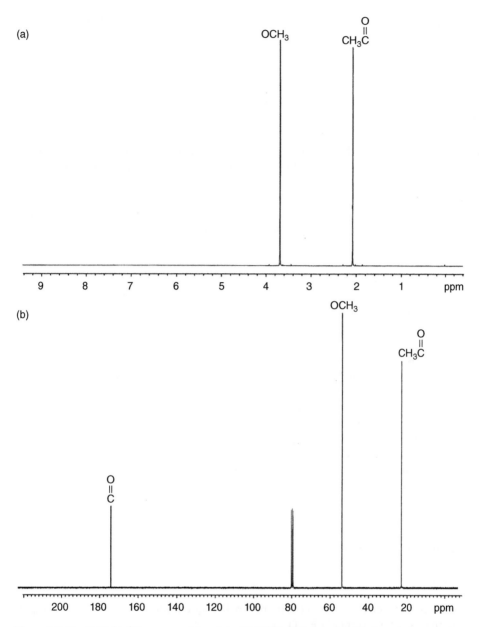

Figure 1.7 The 300 MHz ^{1}H spectrum (a) and the 75.45 MHz ^{13}C spectrum (b) of methyl acetate in CDCl$_3$. The resonance at δ 77 is from the solvent. The ^{13}C spectrum has been decoupled from the protons.

contains separate resonances for the three types of carbons (O—CH$_3$, C—CH$_3$, and carbonyl) (Figure 1.8).

The proton resonances may be assigned on the basis of the electron-withdrawing abilities, or electronegativities, of the neighboring atoms. The ester oxygen is more electron withdrawing than the carbonyl group, so the O—CH$_3$ resonance occurs at a higher

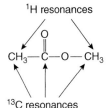

Figure 1.8 The resonances that are expected from methyl acetate.

frequency than (and to the left of) the C—CH$_3$ resonance. By convention, frequency in the spectrum increases from right to left, for consistency with other forms of spectroscopy. Therefore, shielding increases from left to right, because of the negative sign before σ in Eq. (1.3).

The system of units depicted in Figure 1.7 and used throughout this book has been developed to overcome the fact that chemical information often is found in small differences between large numbers. An intuitive system might be absolute frequency—for example, in hertz (Hz, which corresponds to cycles per second or cps). At the common field of 7.05 T, for instance, all protons resonate in the vicinity of 300 MHz. A scale involving numbers like 300.000 764, however, is cumbersome. Moreover, frequencies would vary from one B_0 field to another (Eq. (1.3)). Thus, for every element or isotope, a reference material has been chosen and assigned a relative frequency of zero. For both protons and carbons, the substance is tetramethylsilane [$(CH_3)_4Si$], usually called TMS, which is soluble in most organic solvents, is unreactive, has a strong signal, and is volatile. In addition, the low electronegativity of silicon means that the protons and carbons are surrounded by a relatively high density of electrons. Hence, they are highly shielded and resonate at very low frequency, to the right in the spectrum. Shielding by silicon is so strong, in fact, that the proton and carbon resonances of TMS are placed at the right extreme of the spectrum, providing a convenient spectral zero. In Figures 1.5 and 1.7, the position marked "0 ppm" is the hypothetical position of TMS.

The chemical shift may be expressed as the distance from a chemical reference standard by writing Eq. (1.3) twice—once for an arbitrary nucleus i as in Eq. (1.4a),

$$\nu_i = \frac{\gamma B_0 (1 - \sigma_i)}{2\pi} \tag{1.4a}$$

and again for the reference, that is, TMS, as Eq. (1.4b).

$$\nu_r = \frac{\gamma B_0 (1 - \sigma_r)}{2\pi} \tag{1.4b}$$

The distance between the resonances in the NMR frequency unit (Hz, which equals cps) then is given by the formula in Eq. (1.5).

$$\Delta \nu = \nu_i - \nu_r = \frac{\gamma B_0 (\sigma_r - \sigma_i)}{2\pi} = \frac{\gamma B_0 \Delta \sigma}{2\pi} \tag{1.5}$$

This expression for the frequency differences still depends on the magnetic field B_0. In order to have a common unit at all B_0 fields, the chemical shift of nucleus i is defined by Eq. (1.6),

$$\delta = \frac{\Delta \nu}{\nu_r} = \frac{\sigma_r - \sigma_i}{1 - \sigma_r} \sim \sigma_r - \sigma_i, \tag{1.6}$$

in which the frequency difference in Hz (Eq. (1.5)) is divided by the reference frequency in MHz (Eq. (1.4b)). In this fashion, the constants including the field B_0 cancel out. The δ scale is thus in units of Hz/MHz or parts per million (ppm) of the field. Because the reference shielding is chosen to be much less than 1.0, that is, $(1 − σ_r) \sim 1$, δ corresponds to the differences in shielding of the reference and the nucleus. An increase in $σ_i$, therefore, results in a decrease in $δ_i$, in accordance with Eq. (1.6).

As seen in the ^{1}H spectrum of methyl acetate (Figure 1.7), the δ value for the C—CH$_3$ protons is 2.07 ppm (always written as "δ 2.07" without "ppm," which is understood) and that for the O—CH$_3$ protons is 3.67 ppm (δ 3.67). These values remain the same in spectra taken at any B_0 field, such as either 1.41 T (60 MHz) or 24.0 T (1020 MHz), which represent the extremes of spectrometers currently in use. Chemical shifts expressed in Hz, however, vary from field to field. Thus, a resonance that is 90 Hz from TMS at 60 MHz is 450 Hz from TMS at 300 MHz, but always has a δ value of 1.50 ppm (δ = 90/60 = 450/300 = 1.50). Note that a resonance to the right of TMS has a negative value of δ. Also, since TMS is insoluble in water, other internal standards (δ = 0) are used for this solvent, including the sodium salts of 3-(trimethylsilyl)-1-propanesulfonic acid (also called 4,4-dimethyl-4-silapentane-1-sulfonic acid or DSS) [(CH$_3$)$_3$Si(CH$_2$)$_3$SO$_3$Na] and 3-(trimethylsilyl)propionic acid [(CH$_3$)$_3$SiCH$_2$CH$_2$CO$_2$Na].

In the first generation of commercial spectrometers, the range of chemical shifts, such as those in the scale at the bottom of Figures 1.5 and 1.7, was generated by varying the B_0 field while holding the B_1 field, and hence the resonance frequency $ν_0$, constant. As Eq. (1.3) indicates, an increase in shielding (σ) requires B_0 to be raised in order to keep $ν_0$ constant. Since nuclei with higher shielding resonate at the right side of the spectrum, the B_0 field in this experiment increases from left to right. Consequently, the right end came to be known as the high field, or upfield, end of the spectrum, and the left end as the low field, or downfield, end. This method was termed *continuous-wave* (CW) *field sweep*. Although the method is rarely used today, its vestigial terms such as "upfield" and "downfield" remain inappropriately in the NMR vocabulary.

Modern spectrometers vary the B_1 frequency, while the B_0 field is kept constant. An increase in shielding (σ) lowers the right side of Eq. (1.3) so that $ν_0$ must decrease in order to maintain a constant B_0. Thus, the right end of the spectrum, as noted before, corresponds to lower frequencies for more shielded nuclei. The general result is that *frequency increases from right to left and field increases from left to right*. Figure 1.9 summarizes the terminology. The right end of the spectrum still is often referred to as the

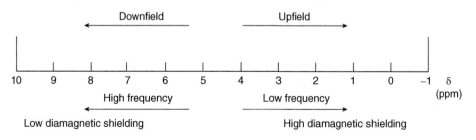

Figure 1.9 Spectral conventions.

high field or upfield end, in deference to the old field-sweep experiment, although it is more appropriate to call it the low-frequency or more shielded end.

The chemical shift is considered further in Chapter 3.

1.3 Excitation and Relaxation

To understand the NMR experiment more fully, it is useful to consider Figure 1.3 again—this time in terms of a collection of nuclei (Figure 1.10). At equilibrium, the $I_z = +\frac{1}{2}$ nuclei precess around the $+z$-axis, and the $-\frac{1}{2}$ nuclei precess around the $-z$-axis. Only 20 spins are shown on the surface of the double cone in the figure, and the excess of $+\frac{1}{2}$ over $-\frac{1}{2}$ nuclei is exaggerated (12 to 8). The actual ratio of populations of the two states is given by the Boltzmann equation (Eq. (1.1)). Inserting the numbers for $B_0 = 7.04$ T yields the result that, for every million spins, there are only about 50 more with $+\frac{1}{2}$ that $-\frac{1}{2}$ spin. If the magnetic moments are added vectorially, there is a net vector in the $+z$-direction because of the excess of $+\frac{1}{2}$ over $-\frac{1}{2}$ spins. The sum of all the individual spins is called the *magnetization* (M). The boldface arrow pointing along the $+z$-direction in Figure 1.10 represents the resultant M. Because the spins are distributed randomly (or incoherently) around the z-axis, there is no net x or y magnetization, that is, $M_x = M_y = 0$, and hence, $M = M_z$.

Figure 1.10 also shows the vector that represents the B_1 field placed along the x-axis. When the B_1 frequency matches the Larmor frequency of the nuclei, some $+\frac{1}{2}$ spins turn over and become $-\frac{1}{2}$ spins so that M_z decreases slightly. The component of the vector magnetic field (B) in the x-direction (B_1) exerts a force on M, the result of which is perpendicular to both vectors (inset at lower right of the figure); the force arises from the cross product $F = M \times B$. If B_1 is turned on just briefly, the magnetization vector M

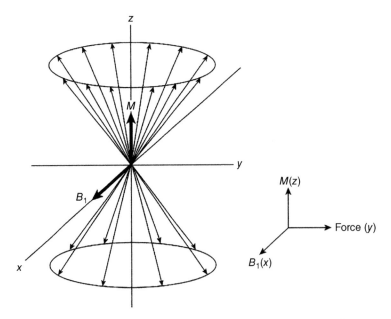

Figure 1.10 Spin-½ nuclei at equilibrium, prior to application of the B_1 field.

Figure 1.11 Spin-½ nuclei immediately after application of the B_1 field.

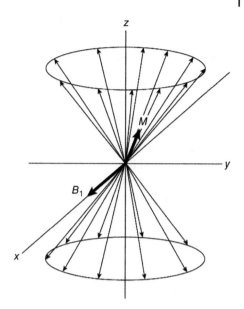

tips only slightly off the z-axis, moving toward the y-axis, which represents the mutually perpendicular direction. Figure 1.11 illustrates the result.

The 20 spins of Figure 1.10 include 12 nuclei of spin $+½$ and 8 of spin $-½$. This same set of nuclei, as shown in Figure 1.11, comprise 11 of spin $+½$ and 9 of spin $-½$, after application of the B_1 field. Thus, only one nucleus has changed its spin in this model. The decrease in M_z is exaggerated in the figure, but the tipping of the magnetization vector off the axis is clearly apparent. The positions on the circles, or the *phases*, of the 20 nuclei no longer are random, because the tipping requires bunching of the spins. The phases of the spins now have some *coherence*, and there are x and y components of the magnetization. The xy component of the magnetization is the signal detected electronically as the resonance. It is important to appreciate that the so-called absorption of energy as $+½$ nuclei become $-½$ nuclei is not measured directly.

The B_1 field in Figures 1.10 and 1.11 oscillates back and forth along the x-axis. As Figure 1.12 illustrates from a view looking down the z-axis, B_1 may be considered either (i) to oscillate linearly along the x-axis at so many times per second (with frequency v) or (ii) to move circularly in the xy plane with angular frequency ω ($2\pi v$) in radians per second. The two representations are vectorially equivalent. See Appendix B, and Figure B.1 in particular, for an expansion of these concepts. Resonance occurs when the frequency and phase of B_1 match that of the nuclei precessing at the Larmor frequency.

Figure 1.12 Analogy between linearly and circularly oscillating fields.

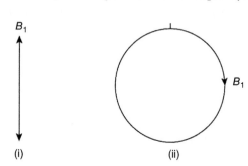

Figure 1.11 represents a snapshot in time, with the motion of both the B_1 vector and the precessing nuclei frozen. In real time, each nuclear vector is precessing around the z-axis so that the magnetization M also is precessing around that axis. Another way to look at the freeze frame is to consider that the x- and y-axes are rotating at the frequency of the B_1 field. In terms of Figure 1.12, the axes are following the circular motion. Consequently, B_1 appears to be frozen in position along the x-axis, instead of oscillating in the fashions shown in that figure. This *rotating coordinate system* is used throughout the book to simplify magnetization diagrams. In the rotating frame, the individual nuclei and the magnetization M no longer precess around the z-axis but are frozen for as long as they all have the same Larmor frequency as the frequency of the rotating x- and y-axes, corresponding to that of the B_1 field.

Application of B_1 at the resonance frequency results in both energy absorption ($+\frac{1}{2}$ nuclei become $-\frac{1}{2}$) and emission ($-\frac{1}{2}$ nuclei become $+\frac{1}{2}$). Because, initially, there are more $+\frac{1}{2}$ than $-\frac{1}{2}$ nuclei, the net effect is absorption. As B_1 irradiation continues, however, the excess of $+\frac{1}{2}$ nuclei decreases and eventually disappears so that the rates of absorption and emission eventually become equal. Under these conditions, the sample is said to be approaching *saturation*. The situation is ameliorated, however, by natural mechanisms whereby nuclear spins are returned from saturation to equilibrium. Any process that returns the z-magnetization to its equilibrium condition with the excess of $+\frac{1}{2}$ spins is called *spin–lattice*, or *longitudinal*, *relaxation* and is usually a first-order process with time constant T_1 (the time constant is the reciprocal of the rate constant but is the traditional quantity used in this context). For a return to equilibrium, relaxation also is necessary to destroy magnetization created in the xy plane. Any process that returns the x- and y-magnetizations to their equilibrium condition of zero is called *spin–spin*, or *transverse*, *relaxation* and is usually a first-order process with time constant T_2.

Spin–lattice relaxation (T_1) derives primarily from the existence of local oscillating magnetic fields in the sample that correspond to the resonance frequency. The primary source of these random fields is other magnetic nuclei that are in motion. As a molecule tumbles in solution in the B_0 field, each nuclear magnet generates a field caused by its motion. If this field is at the Larmor frequency, excess spin energy of neighboring spins can pass to this motional energy as $-\frac{1}{2}$ nuclei become $+\frac{1}{2}$ nuclei. The resonating spins are relaxed back to their initial state, and the absorption experiment can continue, thereby increasing signal intensity.

For effective spin–lattice relaxation, the tumbling magnetic nuclei must be spatially close to the resonating nucleus. When ^{13}C is under observation, attached protons ($^{13}C-^{1}H$) provide effective spin–lattice relaxation. A carbonyl carbon or a carbon attached to four other carbons thus relaxes very slowly and is more easily saturated because the attached atoms are nonmagnetic (motion of the nonmagnetic nuclei such as ^{12}C and ^{16}O provides no relaxation). Protons are relaxed by their nearest neighbor protons. Thus, protons within CH_2 or CH_3 groups are relaxed by geminal protons (HCH) within the group, but the CH entity must rely on vicinal (CH—CH) or more distant protons.

Spin–lattice relaxation also is responsible for generating the initial excess of $+\frac{1}{2}$ nuclei when the sample is first placed in the probe. In the absence of the B_0 field, all spins have the same magnetic energy. When the sample is immersed in the B_0 field, magnetization begins to build up as spins flip from the effects of interactions with surrounding magnetic nuclei in motion, eventually creating the equilibrium ratio with an excess of $+½$ over $-½$ spins.

For x- and y-magnetization to decay toward zero (spin–spin, or T_2, relaxation), the phases of the nuclear spins must become randomized (Figures 1.10 and 1.11). The mechanism that gives the phenomenon its name involves the interaction of two nuclei with opposite spin. The process whereby one spin goes from $+½$ to $-½$ while the other goes from $-½$ to $+½$ involves no net change in z-magnetization and hence no spin–lattice relaxation. The switch in spins, however, results in dephasing, because the new spin state has a different phase from the old one. In terms of Figure 1.11, a spin vector disappears from the surface of the upper cone and reappears on the surface of the lower cone (and vice versa) at a new phase position. As this process continues, the phases become randomized around the z-axis, and xy magnetization disappears. This process of two nuclei simultaneously exchanging spins is sometimes called the flip-flop mechanism.

Spin–spin relaxation also arises when the B_0 field is not perfectly homogeneous. Again, in terms of Figure 1.11, if the spin vectors are not located in exactly identical B_0 fields, they differ slightly in Larmor frequencies and hence precess around the z-axis at different rates. As the spins move faster or more slowly relative to each other, eventually their relative phases become randomized. When various nuclei resonate over a range of Larmor frequencies, the line width of the signal naturally increases. The spectral line width at half height and the spin–spin relaxation are related by the expression $w_{½} = 1/(\pi T_2)$. Both mechanisms (flip-flop and field inhomogeneity) can contribute to T_2 in the same sample.

The subject of relaxation is discussed further in Section 5.1 and Appendix E.

1.4 Pulsed Experiments

In the pulsed NMR experiment, the sample is irradiated close to the resonance frequency with an intense B_1 field for a very short time. For the duration of the pulse, the **B** vector on the x-axis (B_1) in the rotating coordinate system exerts a force (see inset in Figure 1.10) on the **M** vector, which is on the z-axis, pushing the magnetization toward the y-axis. Figures 1.13a,b, respectively, simplify Figures 1.10 and 1.11 by eliminating the individual spins. Only the net magnetization vector **M** is shown in Figure 1.13.

As long as the strong B_1 field is on, the magnetization vector **M** continues to rotate, or precess, around B_1 on the x-axis. The strength of the B_1 field is such that, when it is on, it forces precession to occur preferentially around its direction (x), rather than around the natural direction (z) of the weaker B_0 field. Consequently, the primary field present at the nuclei is B_1, so the expression for the precession frequency becomes $\omega = \gamma B_1$. More precisely, this equation holds at the resonance frequency $\omega_0 = \gamma B_0$. Farther and farther from the resonance frequency, the effect of B_1 wanes, and precession around B_0 returns. A full mathematical treatment requires the inclusion of terms in both B_0 and B_1, but, qualitatively, our interest focuses on the events at the resonance frequency.

The angle θ of rotation of the magnetization **M** increases as long as B_1 is present (Figure 1.13). A short pulse might leave the magnetization at a 30° angle relative to the z-axis (Figure 1.13b). A pulse three times as long (90°) aligns the magnetization along the y-axis (Figure 1.13c). A pulse of double this duration (180°) brings the magnetization along the $-z$-direction (Figure 1.13d), meaning that there is an excess of $-½$ spins, or

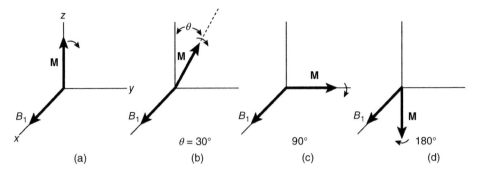

Figure 1.13 (a) Net magnetization M before application of the B_1 field. (b) After a 30° pulse. (c) After a 90° pulse. (d) After a 180° pulse.

a population inversion, a truly unnatural situation. The exact angle θ caused by a pulse thus is determined by its duration t_p. The angle θ is therefore ωt_p, in which ω is the precession frequency in the B_1 field. Since $\omega = \gamma B_1$, it follows that $\theta = \gamma B_1 t_p$.

If the B_1 irradiation is halted when the magnetization reaches the y-axis (a 90° pulse), and if the magnetization along the y-direction is detected over time at the resonance frequency, then the magnetization would be seen to decay (Figure 1.14). Alignment of the magnetization along the y-axis is a nonequilibrium situation. After the pulse, the x- and y-magnetization (collectively, xy) decays by spin–spin relaxation (T_2). At the same time, z magnetization reappears by spin–lattice relaxation (T_1). The reduction in y magnetization with time shown in the figure is called the *Free Induction Decay* (FID) and is a first-order process with time constant T_2.

The illustration in Figure 1.14 is artificial, because it involves only a single nucleus for which the resonance frequency γB_0 corresponds to the frequency of rotation of the x- and y-axes. Most samples have quite a few different types of protons or carbons so that several resonance frequencies are involved ($\gamma B_0(1 - \sigma_i)$), but the rotating frame can have only a single, or reference, frequency. What happens when there are nuclei with resonance frequencies different from the reference frequency? First, imagine again the case of a single resonance, whose Larmor frequency corresponds to the reference frequency, such as the protons of benzene in Figure 1.5. At the time the 90° B_1 pulse is turned off, the spins are lined up along the y-axis (Figure 1.15a). The nuclei then return to precessing about the z-axis at the Larmor frequency $\omega_0 = \gamma B_0$. In the rotating coordinate system, the x- and y-axes are rotating at γB_0 around the z-axis. Nuclei with the same resonance frequency γB_0 appear not to precess about the z-axis because their frequency of rotation matches that of the rotating frame so that they remain lined up along the z-axis (Figure 1.15a).

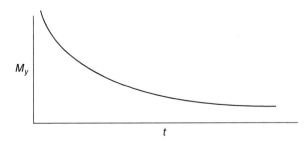

Figure 1.14 Time dependence of the magnetization M_y following a 90° pulse (the free induction decay).

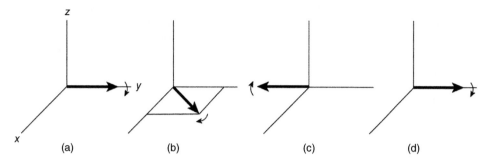

Figure 1.15 Induced magnetization along the y-axis as a function of time after a 90° pulse.

In the case of a second nucleus with a different Larmor frequency ($\omega \neq \omega_0 = \gamma B_0$), the nuclear magnets move off the y-axis with the xy plane (Figure 1.15b). Only nuclei precessing at the reference frequency γB_0 appear to be stationary in the rotating coordinate system. As time progresses, the magnetization continues to rotate within the xy plane, reaching the −y-axis (Figure 1.15c) and eventually returning to the +y-axis (Figure 1.15d). The detected y-magnetization during this cycle first decreases and then falls to zero as it passes the $y = 0$ point. Next it moves to a negative value in the −y-region (Figure 1.15c), and finally returns to positive values (Figure 1.15d). The magnitude of the magnetization thus varies periodically like a cosine function. When it is again along the +y-axis (Figure 1.15d), the magnetization is slightly smaller than at the beginning of the cycle, because of spin–spin relaxation (T_2). Moreover, it has moved out of the xy plane (not shown in the figure) as z-magnetization returns through spin–lattice relaxation (T_1). The magnetization varies as a cosine function with time, continually passing through a sequence of events illustrated by Figure 1.15.

Figure 1.16a shows what the FID looks like for the protons of acetone, when the reference frequency ω_0 is not the same as the Larmor frequency ω of acetone. The horizontal distance between adjacent maxima is the reciprocal of the difference between the Larmor frequency ω of acetone and the B_1 frequency ω_0 ($(\Delta\omega)^{-1} = (\omega - \omega_0)^{-1}$). The intensities of the maxima decrease as y magnetization is lost through spin–spin relaxation. Because the line width of the spectrum is determined by T_2, the FID contains all the necessary information to display a spectrum: frequency, line width, and overall intensity.

Now consider the case of two nuclei with different resonance frequencies, each different from the reference frequency. Their decay patterns are superimposed, reinforcing, and interfering to create a complex FID, as in Figure 1.16b for the protons of methyl acetate [$CH_3(C=O)OCH_3$]. By the time there are four frequencies, as in the carbons of 3-hydroxybutyric acid [$CH_3CH(OH)CH_2CO_2H$] shown in Figure 1.16c, it is impossible to unravel the frequencies visually. The mathematical process called Fourier analysis matches the FID with a series of sinusoidal curves and exponential functions to obtain from them the frequencies, line widths, and intensities of each component. The FID is a plot in time (see Figures 1.14 and 1.16), so the experiment is said to occur in the *time domain*. The experimentalist, however, wants a plot of frequencies, so the spectrum must be transformed to a *frequency domain*, as shown in Figures 1.5 and 1.7. The Fourier transformation (FT) from time to frequency domain is carried out rapidly and automatically by computer, and the experimentalist does not need to examine the FID.

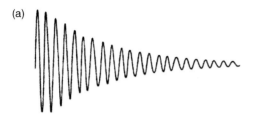

Figure 1.16 The free induction decay for the ^{1}H spectra of (a) acetone and (b) methyl acetate. (c) The free induction decay for the ^{13}C spectrum of 3-hydroxybutyric acid. All samples are without solvent.

1.5 The Coupling Constant

The form of a resonance can be altered by the presence of a distinct, neighboring magnetic nucleus. In 1-chloro-4-nitrobenzene (**1.1**),

1.1

for example, there are two types of protons, labeled A and X, which are, respectively, ortho to nitro and ortho to chloro. For the time being, we will ignore any effects from the identical A and X protons across the ring. Each proton has a spin of ½ and therefore can exist in two I_z spin states, $+½$ and $-½$, which differ in population only in parts per million. Almost exactly half the A protons have $+½$ X neighbors, and half have $-½$ X neighbors. The magnetic environments provided by these two types of X protons are not identical, so the A resonance is split into two peaks (Figures 1.17a and 1.18). By the same token, the X nucleus exists in two distinct magnetic environments, because the A proton has two unequal spin states. The X resonance also is split into two peaks for the same reasons (Figures 1.17b and 1.18). Quadrupolar nuclei, such as those of the chlorine and nitrogen atoms in molecule **1.1**, often act as if they are nonmagnetic and

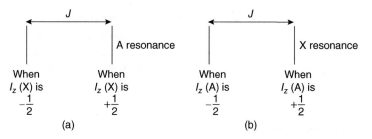

Figure 1.17 The four peaks of a first-order, two-spin system (AX): (a) the A portion and (b) the X portion.

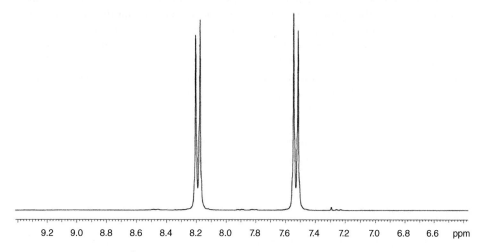

Figure 1.18 The 300 MHz ^{1}H spectrum of 1-chloro-4-nitrobenzene in CDCl$_3$.

may be ignored in this context. Thus, the proton resonance of chloroform (CHCl$_3$) is a singlet; the proton resonance is not split by chlorine. This phenomenon is considered in detail in Section 5.1.

The influence of neighboring spins on the multiplicity of peaks is called *spin–spin splitting, indirect coupling,* or *J-coupling*. The distance between the two peaks for the resonance of one nucleus split by another is a measure of how strongly the nuclear spins influence each other and is called the *coupling constant J*, measured in the energy unit Hz. In 1-chloro-4-nitrobenzene (**1.1**), the coupling between A and X is 10.0 Hz, a relatively large value of *J* for two protons. In general, when there are only two nuclei in the coupled system, the resulting spectrum is referred to as AX. Notice that the splitting in both the A and the X portions of the spectrum is the same (Figure 1.18), since *J* is a measure of the interaction between the nuclei and must be identical for both nuclei. Moreover, *J* is independent of B_0, because the magnitude of the interaction depends only on nuclear properties and not on external quantities such as the field. Thus, in 1-chloro-4-nitrobenzene (**1.1**), *J* is 10.0 Hz when measured either at 7.05 T (300 MHz, Figure 1.18) or at 14.1 T (600 MHz).

For two nuclei to couple, there must be a mechanism whereby information about spin is transferred between them. The most common such mechanism involves the interaction of electrons along the bonding path between the nuclei (see Figure 1.19 for an

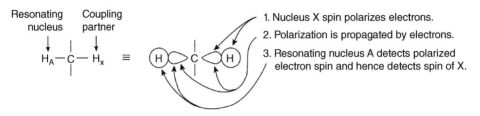

Figure 1.19 The mechanism for indirect spin–spin coupling.

abbreviated coupling pathway over two bonds). In the same fashion as protons, electrons act like spinning particles and have a magnetic moment. The X proton (H_X) influences, or polarizes, the spins of its surrounding electrons, making the electron spins favor one I_z state very slightly. Thus, a proton of spin $+½$ polarizes the electron to $-½$. The electron in turn polarizes the other electrons of the C—H bond, and so on, finally reaching the resonating A proton (H_A). This mechanism is discussed further in Section 4.3. Because the nuclei do not interact directly but via the electron pathway, the interaction is called *indirect coupling*. As an interaction through bonds, it is a useful parameter in drawing conclusions about molecular bonding, such as bond order and stereochemistry.

Additional splitting occurs when a resonating nucleus is close to more than one nucleus. For example, 1,1,2-trichloroethane (**1.2**)

$$ClCH_2{}^A\text{—}CH^XCl_2$$

1.2

has two types of protons, which we label H_A (CH_2) and H_X (CH). The A protons are subject to two different environments, from the $+½$ and $-½$ spin states of H_X, and therefore are split into a 1 : 1 doublet, analogous to the situations shown in Figures 1.17 and 1.18. The X proton, however, is subject to *three* different magnetic environments, because the spins of H_A must be considered collectively: both may be $+½$ (++), both may be $-½$ (−−), and one may be $+½$, while the other is $-½$, for which there are two equivalent possibilities, (+−) and (−+). The three different A environments—(++), ((+−)/(−+)), and (−−)—therefore result in three X peaks in the ratio 1 : 2 : 1 (Figure 1.20). Thus, the spectrum of **1.2** contains a 1 : 1 doublet and a 1 : 2 : 1 triplet and is referred to as A_2X (or AX_2 if the labels are switched) (Figure 1.21). The value of J is found in three different spacings in the spectrum between any two adjacent peaks of either multiplet (Figures 1.20 and 1.21).

As the number of neighboring spins increases, so does the complexity of the spectrum. The two identical ethyl groups in diethyl ether form an A_2X_3 spectrum (Figure 1.22). The methyl protons are split into a 1 : 2 : 1 triplet by the neighboring methylene protons, as in the X resonance of Figure 1.20. Because the methylene protons are split by three methyl protons, there are four peaks in the methylene resonance. The neighboring methyl protons can have all positive spins (+++), two positive spins and one negative (in three ways: ++−, +−+, and −++), one positive spin and two negative (also in three ways: +−−, −+−, and −−+), or all spins negative (−−−). The result is a 1 : 3 : 3 : 1 quartet (Figure 1.23). The triplet–quartet pattern seen in Figure 1.22 is a reliable and general indicator for the presence of an ethyl group.

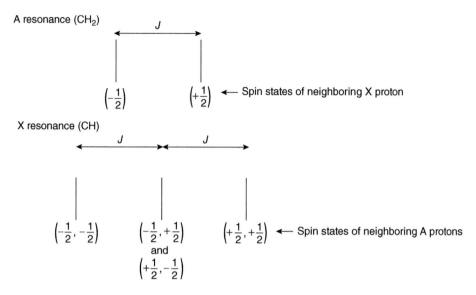

Figure 1.20 The five peaks of a first-order, three-spin system (A₂X).

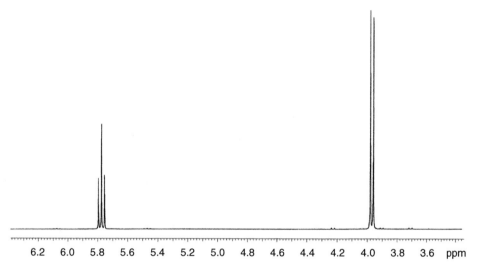

Figure 1.21 The 300 MHz ^{1}H spectrum of 1,1,2-trichloroethane in CDCl$_3$.

The splitting patterns of larger spin systems may be deduced in a similar fashion. If a nucleus is coupled to n equivalent nuclei with $I = \frac{1}{2}$, there are $n + 1$ peaks, unless second-order effects, discussed in Chapter 4, are present. The intensity ratios, to a first-order approximation, correspond to the coefficients in the binomial expansion and may be obtained from Pascal's triangle (Figure 1.24), since arrangements of the two I_z states are statistically independent events. Pascal's triangle is constructed by summing two horizontally adjacent integers and placing the result one row lower and between the two integers. Zeros are imagined outside the triangle. The first row (1) gives the resonance multiplicity (an unsplit singlet) when there is no neighboring spin,

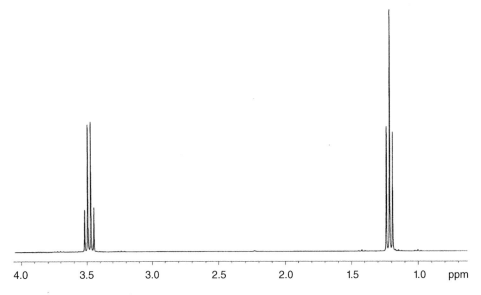

Figure 1.22 The 300 MHz ^{1}H spectrum of diethyl ether in CDCl$_3$.

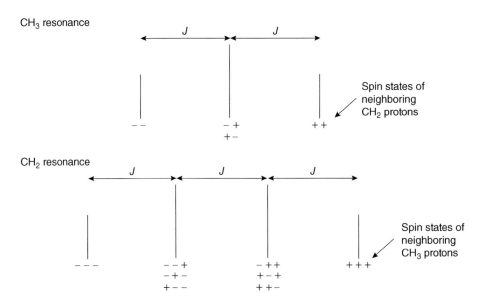

Figure 1.23 The seven peaks of the first-order A$_2$X$_3$ spectrum.

the second row (1 : 1) gives the multiplicity when there is one neighboring spin, and so on. We already have seen that two neighboring spins give a 1 : 2 : 1 triplet and three give a 1 : 3 : 3 : 1 quartet, as shown, respectively, in the third and fourth rows of Pascal's triangle (Figure 1.24). Four neighboring spins are present for the CH proton in the arrangement —CH$_2$—CHX—CH$_2$— (X is nonmagnetic), and the CH resonance is a 1 : 4 : 6 : 4 : 1 (the fifth row) quintet (AX$_4$). The CH resonance from an isopropyl

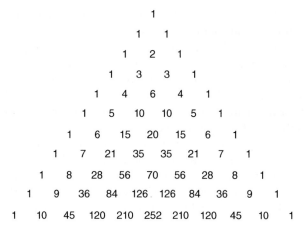

Figure 1.24 Pascal's triangle.

Table 1.1 Common first-order spin–spin splitting patterns.

Spin system	Molecular substructure	A multiplicity	X multiplicity
AX	—CH^A—CH^X—	Doublet (1:1)	Doublet (1:1)
AX$_2$	—CH^A—CH$_2^X$—	Triplet (1:2:1)	Doublet (1:1)
AX$_3$	—CH^A—CH$_3^X$	Quartet (1:3:3:1)	Doublet (1:1)
AX$_4$	—CH$_2^X$—CH^A—CH$_2^X$—	Quintet (1:4:6:4:1)	Doublet (1:1)
AX$_6$	CH$_3^X$—CH^A—CH$_3^X$	Septet (1:6:15:20:15:6:1)	Doublet (1:1)
A$_2$X$_2$	—CH$_2^A$—CH$_2^X$—	Triplet (1:2:1)	Triplet (1:2:1)
A$_2$X$_3$	—CH$_2^A$—CH$_3^X$	Quartet (1:3:3:1)	Triplet (1:2:1)
A$_2$4	—CH$_2^X$—CH$_2^A$—CH$_2^X$—	Quintet (1:4:6:4:1)	Triplet (1:2:1)

group, —CH(CH$_3$)$_2$, is a 1:6:15:20:15:6:1 (the seventh row) septet (AX$_6$). Several common spin systems are given in Table 1.1.

Except in cases of second-order spectra (Sections 4.1 and 4.7), coupling between protons that have the same chemical shift does not lead to splitting in the spectrum. It is for this reason that the spectrum of benzene in Figure 1.1 is a singlet, even though each proton is coupled to its two ortho neighbors. For the same reason, protons within a methyl group normally do not cause splitting of the methyl resonance. Other examples of unsplit spectra (singlets) include those of acetone, cyclopropane, and dichloromethane. The absence of splitting between coupled nuclei with identical resonance frequencies is quantum mechanical in nature and is explained in Appendix C.

Almost all the coupling examples given so far have been between vicinal protons, over three bonds (H—C—C—H). Coupling over four or more bonds usually is small or unobservable. It is possible for geminal protons (—CH$_2$—) to split each other, provided that each proton of the methylene group has a different chemical shift. Such geminal splittings are observed when the carbon is part of a ring with unsymmetrical substitution on the upper and lower faces (—CH$_2$—CXY—), when there is a single chiral center in the molecule (—CH_2—CXYZ), or when an alkene lacks an axis of symmetry (CH_2=CXY). Such couplings are discussed further in Sections 4.2 and 4.4.

Coupling can occur between ^1H and ^{13}C, as well as between two protons. Because ^{13}C is in such low natural abundance (about 1.1%), these couplings are not important in analyzing ^1H spectra. In 99 cases out of 100, protons are attached to nonmagnetic ^{12}C atoms. Small satellite peaks from the 1.1% of ^{13}C sometimes can be seen in ^1H spectra. In the ^{13}C spectrum, the carbon nuclei are coupled to nearby protons. The largest couplings occur with protons that are directly attached to the carbon. Thus, the ^{13}C resonance of a methyl carbon is split into a quartet, that of a methylene carbon into a triplet, and that of a methine carbon (CH) into a doublet. A quaternary carbon lacks an attached proton and hence does not exhibit a one-bond coupling. Figure 1.25a shows the ^{13}C spectrum of 3-hydroxybutyric acid ($CH_3CH(OH)CH_2CO_2H$), which contains a carbon resonance with each type of multiplicity. From right to left are seen a quartet (CH_3), a triplet (CH_2),

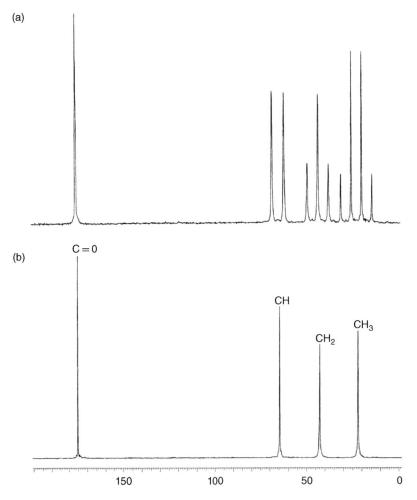

Figure 1.25 (a) The 22.6 MHz ^{13}C spectrum of 3-hydroxybutyric acid, $CH_3CH(OH)CH_2CO_2H$, without solvent. (b) The ^{13}C spectrum of the same compound with proton decoupling.

a doublet (CH), and a singlet (CO$_2$H). Thus, the splitting pattern in the ^{13}C spectrum is an excellent indicator of each of these types of groupings within a molecule.

Instrumental procedures, called *decoupling*, are available by which spin–spin splittings may be removed. These methods, discussed in Section 5.3, involve irradiating one nucleus with an additional field (B_2) while observing another nucleus resonating in the B_1 field. Because a second field is being applied to the sample, the experiment is called *double resonance*. This procedure was used to obtain the ^{13}C spectrum of methyl acetate in Figure 1.7b and the spectrum of 3-hydroxybutyric acid at the bottom of Figure 1.25. It is commonly employed to obtain a very simple ^{13}C spectrum, in which each carbon gives a singlet. Measurement of both decoupled and coupled ^{13}C spectra then produces in the first case, a simple picture of the number and types of carbons and in the second case, the number of protons to which they are attached (Figure 1.25). Coupling is treated in more detail in Chapter 4.

1.6 Quantitation and Complex Splitting

The signal that is detected when nuclei resonate is directly proportional to the number of spins present. Thus, the protons of a methyl group (CH$_3$) produce three times the signal of a methine proton (CH). This difference in intensity can be measured through electronic integration and exploited to elucidate molecular structure. Figure 1.26 illustrates electronic integration for the ^{1}H spectrum of ethyl *trans*-crotonate (CH$_3$CH=CH(CO)OCH$_2$CH$_3$). The vertical displacement of the continuous line above or through each resonance provides a relative measure of the area under the peaks. The vertical displacements show that the doublet at δ 5.84, the quartet at δ 4.19, and

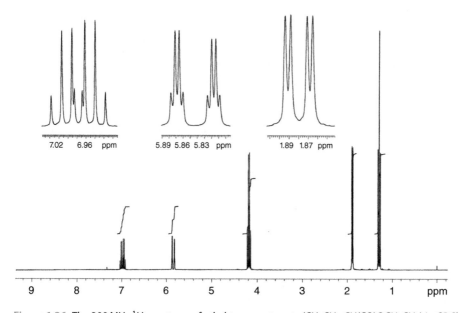

Figure 1.26 The 300 MHz ^{1}H spectrum of ethyl *trans*-crotonate (CH$_3$CH=CH(CO)OCH$_2$CH$_3$) in CDCl$_3$. The upper multiplets are expansions of three resonances. The continuous lines above or through a resonance is its integral.

the triplet at δ 1.28 are, respectively, in the ratio 1 : 2 : 3. Quantitative integrals are provided digitally by the spectrometer. The integration provides only relative intensity data, however, so that the experimentalist must select a resonance with a known or suspected number of protons and normalize the other integrals to it.

Each of the peaks in the ^{1}H spectrum of ethyl crotonate illustrated in Figure 1.26 may be assigned by examining the integrals and splitting patterns. The triplet at the lowest frequency (the highest field, δ 1.28) has a relative integral of 3 and must come from the methyl part of the ethyl group. Its J value corresponds to that of the quartet in the middle of the spectrum at δ 4.19, whose integral is 2. This latter multiplet then must come from the methylene group. The mutually coupled methyl triplet and methylene quartet form the resonances for the ethyl group attached to oxygen (—OCH$_2$CH$_3$). The methylene resonance is at a higher frequency than the methyl resonance because CH$_2$ is closer than CH$_3$ to the electron-withdrawing oxygen atom.

The remaining resonances in the spectrum come from protons coupled to more than one type of proton. Coupling patterns in such cases are more complex, as seen in the three spectral expansions in Figure 1.26. The highest-frequency (lowest-field) resonance (δ 6.98) has an intensity of unity and comes from one of the two alkenic (—CH=) protons. This resonance is split into a doublet ($J = 16$ Hz) by the other alkenic proton, and then each member of the doublet is further split into a quartet ($J = 7$ Hz) by coupling to the methyl group on carbon with a crossover of the inner two peaks. Stick diagrams (often called a tree) are useful in analyzing complex multiplets, as in Figure 1.27, for the resonance at δ 6.98.

The resonance of unit integral at δ 5.84 is from the other alkenic proton and is split into a doublet ($J = 16$ Hz) by the proton at δ 6.98. There is a small coupling (1 Hz) over four bonds to the methyl group, giving rise to a quartet (Figure 1.28). The significance of these differences in the magnitude of couplings is discussed in Chapter 4. The resonance at δ 6.98 can be recognized as originating from the proton closer to the methyl group. Thus J is larger because it is over three, rather than four, bonds.

The resonance at δ 1.88 has an integral of 3 and hence comes from the remaining methyl group, attached to the double bond. Because it is split by both of the alkenic protons, but with unequal couplings (7 and 1 Hz), four peaks result (Figure 1.29). This grouping is called a *doublet of doublets*; the term *quartet* normally is reserved for 1 : 3 : 3 : 1 multiplets. The two unequal couplings in the resonance at δ 1.88 correspond precisely to the quartet splittings found, respectively, in the two alkenic resonances.

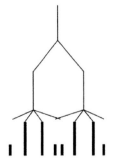

Resonance at δ 6.98 if there were no coupling

Splitting by the other CH=(δ 5.84) into a 1 : 1 doublet (J = 16 Hz)

Splitting by each member of the doublet into a 1 : 3 : 3 : 1 quartet (J = 7 Hz) by coupling to CH$_3$ (δ 1.88) (note the crossover of the two middle peaks)

Figure 1.27 Overlapping peaks of the resonance δ 6.98, which arise when nuclei are unequally coupled to more than one other set of spins.

Figure 1.28 Overlapping peaks of the resonance at δ 5.84 (not to scale).

Resonance at δ 5.84 if there were no coupling

Splitting by the other CH═ (δ 6.98) into a 1:1 doublet (J = 16 Hz) (note same splitting as for δ 6.98 resonance)

Splitting by each member of the doublet (J = 1 Hz) into a 1:3:3:1 quartet by coupling to CH₃ (δ 1.88)

Figure 1.29 Overlapping peaks of the resonance at δ 1.88 (not to scale).

Resonance at δ 1.88 if there were no coupling

Splitting by CH═ with J = 7 Hz (δ 6.98)

Splitting by CH═ with J = 1 Hz (δ 5.84)

The final assignments are as follows:

δ 1.9 7.0   5.8   4.2 1.3

CH_3CH_2═$CHCO_2CH_2CH_3$

Integration also may be used as a measure of the relative amounts of the components of a mixture. In this case, after normalizing for the number of protons in a grouping, the proportions of the components may be calculated from the relative integrals of protons in different molecules. An internal standard with a known concentration may be included. Comparisons of other resonances with those of the standard then can provide a measure of the absolute concentration.

1.7 Commonly Studied Nuclides

Which nuclei (in this context, "nuclides") are useful in chemical problems? The answer depends on one's area of specialty. Certainly, for the organic chemist, the most important elements are carbon, hydrogen, oxygen, and nitrogen. The biochemist would add phosphorus to the list. The organometallic or inorganic chemist would focus on whichever elements are of potential use in a particular subfield, possibly boron, silicon, tin, mercury, platinum, or some of the low-intensity nuclei, such as iron and potassium.

Table 1.2 NMR properties of common nuclei.

Nuclide	Spin	Natural abundance (N_a)(%)	Natural sensitivity (N_s) (for equal numbers of nuclei) (vs ^{1}H)	Receptivity (vs ^{13}C)	NMR frequency (at 7.05 T)	Reference substance
Proton	1/2	99.985	1.00	5680	300.00	$(CH_3)_4Si$
Deuterium	1	0.015	0.00965	0.0082	46.05	$(CD_3)_4Si$
Lithium-7	3/2	92.58	0.293	1540	38.86	LiCl
Boron-10	3	19.58	0.0199	22.1	32.23	$Et_2O \cdot BF_3$
Boron-11	3/2	80.42	0.165	754	96.21	$Et_2O \cdot BF_3$
Carbon-13	1/2	1.108	0.0159	1.00	75.45	$(CH_3)_4Si$
Nitrogen-14	1	99.63	0.00101	5.69	21.69	$NH_3(l)$
Nitrogen-15	1/2	0.37	0.00104	0.0219	30.42	$NH_3(l)$
Oxygen-17	5/2	0.037	0.0291	0.0611	40.68	H_2O
Fluorine-19	1/2	100	0.833	4730	282.27	CCl_3F
Sodium-23	3/2	100	0.0925	525	79.36	NaCl (aq)
Aluminum-27	5/2	100	0.0206	117	78.17	$Al(H_2O)_6^{3+}$
Silicon-29	1/2	4.70	0.00784	2.09	59.61	$(CH_3)_4Si$
Phosphorus-31	1/2	100	0.0663	377	121.44	85% H_3PO_4
Sulfur-33	3/2	0.76	0.00226	0.0973	23.04	CS_2
Chlorine-35	3/2	75.53	0.0047	20.2	29.40	NaCl (aq)
Chlorine-37	3/2	24.47	0.00274	3.8	24.47	NaCl (aq)
Potassium-39	3/2	93.1	0.000509	2.69	14.00	K^+
Calcium-43	7/2	0.145	0.00640	0.0527	20.19	$CaCl_2$ (aq)
Iron-57	1/2	2.19	0.0000337	0.0042	9.71	$Fe(CO)_5$
Cobalt-59	7/2	100	0.277	1570	71.19	$K_3Co(CN)_6$
Copper-63	3/2	69.09	0.0931	365	79.58	$Cu(C_3CN)_4^+ BF_4^-$
Selenium-77	1/2	7.58	0.00693	2.98	57.22	$Se(CH_3)_2$
Rhodium-103	1/2	100	0.0000312	0.177	9.56	Rh metal
Tin-119	1/2	8.58	0.0517	25.2	37.29	$(CH_3)_4Sn$
Tellurium-125	1/2	7.0	0.0315	12.5	78.51	$Te(CH_3)_2$
Platinum-195	1/2	33.8	0.00994	19.1	64.38	Na_2PtCl_6
Mercury-199	1/2	16.84	0.00567	5.42	53.73	$(CH_3)_2Hg$
Lead-207	1/2	22.6	0.00920	11.8	62.57	$Pb(CH_3)_4$

Source: For a more complete list, see Ref. [1].

The success of the experiment depends on several factors, which are listed in Table 1.2 for a variety of nuclides and are described in the following sections:

Spin. The overall spin of a nucleus (second column in Table 1.2) is determined by the spin properties of the protons and neutrons, as discussed in Section 1.1. By and large, spin-½ nuclei exhibit more favorable NMR properties than quadrupolar nuclei

($I > \frac{1}{2}$). Nuclei with odd mass numbers have half-integral spins ($\frac{1}{2}$, 3/2, etc.), whereas those with even mass and odd charge have integral spins (1, 2, etc.). Quadrupolar nuclei have a unique mechanism for relaxation that can result in extremely short relaxation times, as is discussed in Section 5.1. The relationship between lifetime (Δt) and energy (ΔE) is given by the Heisenberg uncertainty principle: $\Delta E \Delta t \geq \hbar$ (the normal statement of this principle in terms of position and momentum can be converted to this expression in terms of energy and time through conversion of units). When the lifetime of the spin state, as measured by the relaxation time, is very short, the larger uncertainty in energies (ΔE) implies a larger band of frequencies (Δt), or a broadened signal, in the NMR spectrum. The relaxation time, and hence the broadening of the spectral lines, also depends on the distribution of charge within the nucleus, as determined by the quadrupole moment. For example, quadrupolar broadening makes ^{14}N ($I = 1$) a generally less useful nucleus than ^{15}N ($I = \frac{1}{2}$), even though ^{14}N is far more abundant.

Natural abundance. Nature provides us with nuclides in varying amounts (third column in Table 1.2). Whereas ^{19}F and ^{31}P are 100% abundant and ^{1}H nearly so, ^{13}C is present only to the extent of 1.1%. The most useful nitrogen (^{15}N) and oxygen (^{17}O) nuclides occur to the extent of much less than 1%. The NMR experiment naturally is easier with nuclides with higher natural abundance. Because so little ^{13}C is present, there is a very small probability of having two ^{13}C atoms at adjacent positions in the same molecule ($0.011 \times 0.011 = 0.00012$, or about 1 in 10 000). Thus, *J*-couplings are not easily observed between two ^{13}C nuclei in ^{13}C spectra, although procedures to measure them have been developed.

Natural sensitivity. Nuclides have differing sensitivities to the NMR experiment (fourth column in Table 1.2), as determined by the gyromagnetic ratio γ and the energy difference ΔE ($= \gamma \hbar B_0$) between the spin states (Figure 1.6). The larger the energy difference, the more nuclei are present in the lower spin state (see Eq. (1.1)), and hence the more are available to absorb energy. With its large γ, the proton is one of the most sensitive nuclei, whereas ^{13}C and ^{15}N, unfortunately, are rather weak (Figure 1.6). Tritium (^{3}H) is useful to the biochemist as a radioactive label. It has $I = \frac{1}{2}$ and is highly sensitive. Since it has zero natural abundance, it must be introduced synthetically. As a hydrogen label, deuterium also is useful, but it has very low natural sensitivity and a spin $I = 1$ and also must be introduced synthetically. Nuclei that are of interest to the inorganic chemist vary from poorly sensitive, such as iron and potassium, to highly sensitive, such as lithium and cobalt. Thus, it is important to be familiar with the natural sensitivity of a nucleus before designing an NMR experiment.

Receptivity. The intensity of the signal for a spin-$\frac{1}{2}$ nucleus is determined by both the natural abundance (in the absence of synthetic labeling) and the natural sensitivity of the nuclide. The mathematical product of these two factors is a good measure of how amenable a specific nucleus is to the NMR experiment. Because chemists are quite familiar with the ^{13}C experiment, the product of natural abundance and natural sensitivity for a nucleus is divided by the product for ^{13}C to give the factor known as the *receptivity* (fifth column in Table 1.2). Thus, the receptivity of ^{13}C is, by definition, 1.00. The ^{15}N experiment then is seen to be about 50 times less sensitive than that for ^{13}C, since the receptivity of ^{15}N is 0.0219. In addition to these factors, Table 1.2 also contains the NMR resonance frequency at 7.05 T (the sixth column). The last column contains the reference substance for each nuclide, for which in most cases $\delta = 0$.

1.8 Dynamic Effects

According to the principles outlined in the previous sections, the ^{1}H spectrum of methanol (CH_3OH) should contain a doublet of integral 3 for the CH_3 group (coupled to OH) and a quartet of integral 1 for the OH group (coupled to CH_3). Under conditions of high purity or low temperature, such a spectrum is observed (Figure 1.30b). The presence of a small amount of an acidic or basic impurity, however, can catalyze the intermolecular exchange of the hydroxyl proton. When this proton becomes detached from the molecule by any mechanism, information about its spin states is no longer available to the rest of the molecule. For coupling to be observed, the rate of exchange must be considerably slower than the magnitude of the coupling, in Hz (s^{-1}). Thus, a proton could exchange a few times per second and still maintain coupling. If the rate of exchange is faster than J, no coupling is observed between the hydroxyl proton and the methyl protons. Hence, at high temperatures (Figure 1.30a), the ^{1}H spectrum of methanol contains only two singlets. If the temperature is lowered or the amount of acidic or basic catalyst is decreased, the exchange rate slows down. The coupling constant continues to be absent until the exchange rate reaches a critical value at which the proton resides sufficiently long on oxygen to permit the methyl group to detect the spin states. As can be seen from the figure, the transition from *fast exchange* (upper) to *slow exchange* (lower) can be accomplished for methanol over a temperature range of 80 °C. Under most spectral conditions, minor amounts of acid, or base impurities are present, so hydroxyl protons do not usually exhibit couplings to other nuclei. The integral is still unity for the OH group, because the amount of catalyst is small. Sometimes the exchange rate is intermediate, between fast and slow exchange, and broadened peaks are observed. Amino (NH or NH_2), ammonium (H_3N^+), and thiol (SH) protons exhibit similar behavior.

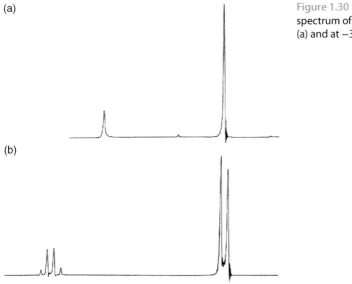

Figure 1.30 The 60 MHz ^{1}H spectrum of CH_3OH at +50 °C (a) and at −30 °C (b).

A process that averages coupling constants also can average chemical shifts. A mixture of acetic acid and benzoic acid can contain only one ^{1}H resonance for the CO_2H groups from both molecules. The carboxyl protons exchange between molecules so rapidly that the spectrum exhibits only the average of the two. Moreover, when the solvent is water, exchangeable protons such as OH in carboxylic acids or alcohols do not give separate resonances. Thus, the ^{1}H spectrum of acetic acid (CH_3CO_2H) in water contains two, not three, peaks: the water and carboxyl protons appear as a single resonance whose chemical shift falls at the weighted average of those of the pure materials. If the rate of exchange between $-CO_2H$ and water could be slowed sufficiently, separate resonances would be observed. Carboxyl protons give separate resonances in organic solvents such as $CDCl_3$. In dry dimethyl sulfoxide (DMSO, usually deuterated, as in $(CD_3)_2SO$) exchange often is slow enough to give separate resonances when there are multiple hydroxyl groups in the molecule and even exhibit H—O—C—H vicinal couplings).

Intramolecular (unimolecular) reactions also can influence the appearance of the NMR spectrum if the rate is comparable to that of chemical shift differences. The molecule cyclohexane [$(CH_2)_6$], for example, contains distinct axial and equatorial protons, yet the spectrum exhibits only one sharp singlet at room temperature. There is no splitting, because all protons have the same average chemical shift. Flipping of the ring interconverts the axial and equatorial positions. When the rate of this process is greater (in s^{-1}) than the chemical-shift difference between the axial and equatorial protons (in Hz, which, of course, is s^{-1}), the NMR experiment does not distinguish the two types of protons, and only one peak is observed. Again, this situation is called fast exchange. At lower temperatures, however, the process of ring flipping is much slower. At −100 °C, the NMR experiment can distinguish the two types of protons, so two resonances are observed (slow exchange). At intermediate temperatures, broadened peaks are observed that reflect the transition from fast to slow exchange. Figure 1.31 illustrates the spectral changes as a function of temperature for cyclohexane, in which all protons but one have been replaced by deuterium to remove vicinal proton–proton couplings to simplify the spectrum (**1.3**, in which the lone hydrogen on the left is axial and the lone hydrogen on the right is equatorial).

1.3

Processes that bring about averaging of spectral features occur reversibly, whether by acid-catalyzed intermolecular exchange as in methanol or by unimolecular reorganization as in cyclohexane. NMR is one of the few methods able to examine effects of reaction rates when a system is at equilibrium. Most other kinetic methods require that one substance be transformed irreversibly into another. The dynamic effects of the averaging of chemical shifts or coupling constants provide a nearly unique window into processes that occur on the order of a few times per second. The subject is examined further in Section 5.2.

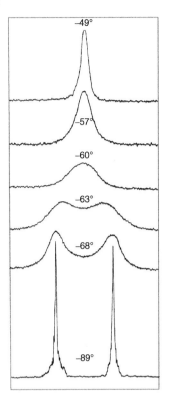

Figure 1.31 The 60 MHz ^{1}H spectrum of cyclohexane-d_{11} as a function of temperature. Source: Bovey et al. 1964 [2]. Reproduced with permission from American Institute of Physics.

1.9 Spectra of Solids

All of the examples and spectra illustrated thus far have been for liquid samples. Certainly, it would be useful to be able to take NMR spectra of solids, so why have we avoided discussing such samples? Under conditions normally used for liquids, the spectra of solids are broad and unresolved, providing only modest amounts of information. There are two primary reasons for the poor resolution of solid samples. In addition to the indirect spin–spin (*J*) interaction that occurs between nuclei through bonds, nuclear magnets also can couple through the direct interaction of their nuclear dipoles. This *dipole–dipole, dipolar, direct,* or *D-coupling* occurs through space, rather than through bonds. The resulting coupling is designated with the letter *D* and is much larger than *J*-coupling.

In solution, dipoles are continuously reorienting themselves through molecular tumbling. Just as two bar magnets have no net interaction when averaged over all mutual orientations, two nuclear magnets have no net dipolar interaction because of the randomizing effect of tumbling. Thus, the *D*-coupling normally averages to zero in solution. The indirect *J*-coupling does not average to zero, because tumbling cannot average out an interaction that takes place through bonds. In the solid phase, however, nuclear dipoles are held rigidly in position so that the *D*-coupling does not average to zero. The dominant interaction between nuclei in solids is, in fact, the *D*-coupling, which is on the order of several hundred to a few thousand hertz. Dipolar interactions also can become apparent in the spectra of very large biological molecules, which move very slowly in solution. The magnitude of the interaction depends on the angle between the nuclear dipoles. Each dipole can assume any relative angle in the solid, so the actual value of the dipolar coupling varies from zero to the maximum value in the optimal arrangement for a solid sample. Since such interactions assume a range of values, are much larger than *J*-coupling and most chemical shifts, and do not average out, very broad signals are produced.

As with the *J*-coupling, the *D*-coupling may be eliminated by the application of a strong B_2 field. Power levels required for the removal of *D* must be much higher than those for *J* decoupling, since *D* is 2–3 orders of magnitude larger than *J*. High-powered decoupling is used routinely to reduce the line width of the spectra of solids. In practice, dipolar decoupling is most easily brought about between different spin-$\frac{1}{2}$ nuclei, as when ^{13}C is observed with decoupling of ^{1}H. The analogous all-proton experiment is more difficult, since both observed and irradiated nuclei have the same frequency range. Thus, the acquisition of solid-state ^{13}C spectra is much simpler than that of ^{1}H spectra. Quadrupolar nuclei such as ^{27}Al also are more difficult to observe in the solid.

The second factor that contributes to line broadening for solids is *chemical shielding anisotropy*. The term "chemical shift anisotropy" should be avoided in this context, since, strictly speaking, the chemical shift is a scalar quantity and cannot be anisotropic. In solution, the observed chemical shift is the average of the shielding of a nucleus over all orientations in space, as the result of molecular tumbling. In a solid, shielding of a specific nucleus in a molecule depends on the orientation of the molecule with respect to the B_0 field. Consider the carbonyl carbon of acetone. When the B_0 field is parallel to the C=O bond, the nucleus experiences a different shielding from when the B_0 field is perpendicular to the C=O bond (Figure 1.32). The ability of electrons to circulate and give rise to shielding varies according to the arrangement of bonds in space. Differences between the abilities of electrons to circulate in the arrangements shown in the figure, as well as in all other arrangements, generate a range of shieldings and hence a range of resonance frequencies. Such anisotropy is largest for unsaturated carbons (C=O, C=C). For saturated carbons, shielding is similar for most orientations in the field, so anisotropy is low.

Double irradiation does not average chemical shielding anisotropy, since the effect is entirely geometrical. The problem is largely removed by spinning the sample to mimic the process of tumbling. The effects of spinning are optimized when the axis of spin is set at an angle of 54°74′ to the direction of the B_0 field. This is the angle between the edge of a cube and the adjacent solid diagonal. Spinning of a cube along this diagonal averages each Cartesian direction by interconverting the *x*-, *y*-, and *z*-axes, just as tumbling in solution does. When the sample is spun at that angle to the field, the various arrangements of Figure 1.32 average, and the chemical shieldings are reduced to the isotropic chemical shift. The technique therefore has been called *Magic Angle Spinning* (MAS). Because shielding anisotropies are generally a few hundred to several thousand hertz, the rate of spinning must exceed this range in order to average all orientations. Typical

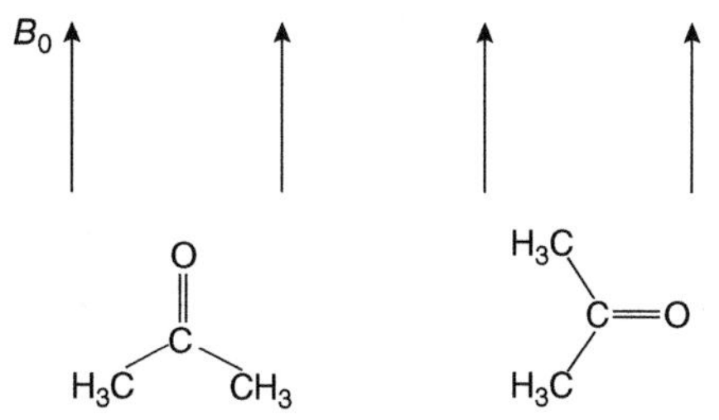

Figure 1.32 Anisotropy of shielding in the solid state.

minimum spinning rates are 2–5 kHz, but rates up to 50 kHz are possible. Spinning at the magic also reduces dipolar and quadrupolar interactions.

The combination of strong irradiation to reduce dipolar couplings and MAS to eliminate shielding anisotropy results in ^{13}C spectra of solids that are almost as high in resolution as those of liquids. Spectra of protons or of quadrupolar nuclei in a solid can be obtained but require more complex experiments. Figure 1.33 shows the ^{13}C spectrum of polycrystalline β-quinol methanol clathrate. The broad, almost featureless spectrum at the top (Figure 1.33a) is typical of solids. Strong double irradiation (Figure 1.33b) reduces dipolar couplings and brings out some features. MAS in addition to decoupling (Figure 1.33c) produces a truly high-resolution spectrum.

Relaxation times are extremely long for nuclei in the solid state because the motion necessary for spin–lattice relaxation is slow or absent. Carbon-13 spectra could take a very long time to record because the nuclei must relax for several minutes between pulses. The problem is solved by taking advantage of the more favorable properties of the protons that are coupled to the carbons. The same double irradiation process that eliminates J- and D-couplings is used to transfer some of the proton's higher magnetization and faster relaxation to the carbon atoms. The process is called *cross polarization* (CP) and is standard for most solid spectra of ^{13}C. After the protons are moved onto the y-axis by a 90° pulse, a continuous y field is applied to keep the magnetization precessing about that axis, a process called *spin locking*. The frequency of this field ($\gamma_H B_H$) is controlled by the spectrocopist. When the ^{13}C channel is turned on, its frequency ($\gamma_C B_C$) can be set to equal the ^{1}H frequency (the *Hartmann–Hahn condition*, $\gamma_H B_H = \gamma_C B_C$). Both protons and carbons then are precessing at the same frequency and hence have the same net magnetization, which, for carbon, is increased over that used in the normal pulse experiment. Carbon resonances thus have enhanced intensity and faster (proton-like) relaxation. When carbon achieves maximum intensity, B_C is turned off (ending the *contact time*) and carbon magnetization is acquired, while B_H is retained for dipolar decoupling and other beneficial effects.

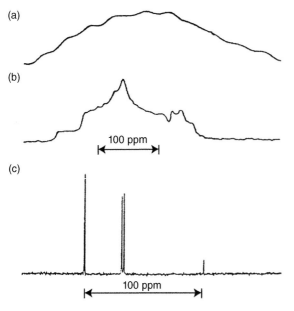

Figure 1.33 The ^{13}C spectrum of polycrystalline β-quinol methanol clathrate (a) without dipolar decoupling, (b) with decoupling, and (c) with both decoupling and magic angle spinning. Source: Terao 1983 [3]. Reproduced with permission from JEOL News.

The higher resolution and sensitivity of the experiment with cross polarization and magic angle spinning (CP/MAS) opened vast new areas to NMR investigations. Inorganic and organic materials that do not dissolve could be subjected to NMR analysis. Synthetic polymers and coal were two of the first materials to be examined. Biological and geological materials, such as wood, humic acids, and biomembranes, became general subjects for NMR study. Problems unique to the solid state—for example, structural and conformational differences between solids and liquids—also may be examined.

Problems

In the following problems, assume fast rotation around all single bonds.

1.1 Determine the number of chemically different hydrogen atoms and their relative proportions in the following molecules. Do the same for carbon atoms.

(a) methylcyclohexene with CH_3 substituent
(b) methylenecyclohexane with $=CH_2$
(c) $C_6H_5CH_2OCCH_3$ (with C=O)
(d) norbornane-type structure with H and OH

1.2 What is the expected multiplicity for each proton resonance in the following molecules?

(a) $ClCH_2CH_2CH_2Cl$
(b) $BrCH(CH_3)_2$
(c) $C_6H_5OCCH_2CH_3$ (with C=O)
(d) structure with H O Cl / H H

1.3 Predict the multiplicities for the ^1H and the ^{13}C resonances in the absence of decoupling for each of the following compounds. For the ^{13}C spectra, give only the multiplicities caused by coupling to attached protons. For the ^1H spectra, give only the multiplicities caused by couplings to vicinal protons (HCCH).

(a) $CH_3CH_2CH_2OCCH_3$ (with C=O)
(b) phenyl-CH_2CH_2Br
(c) tetrahydropyran-type ring with O
(d) $N(CH_2CH_3)_3$
(e) H_3C and H on one carbon, H and CO_2CH_3 on other carbon of C=C

1.4 For each of the following 300 MHz ^1H spectra, carry out the following operations. (i) From the elemental formula, calculate the unsaturation number

$U (= C + 1 - \frac{1}{2}(X - N)$, in which U is the number of unsaturations (1 for a double bond, 2 for a triple bond, and 1 for each ring), C is the number of tetravalent elements (C, Si, etc.), X is the number of monovalent elements (H and the halogens), and N is the number of trivalent elements (N, P, etc.)). (ii) Calculate the relative integrals for each group of protons. Then convert the integrals to absolute numbers by selecting one group to be of a known integral. (iii) Assign a structure to the compound. Be sure that your structure agrees with the spectrum in all aspects: number of different proton groups, integrals, and splitting patterns.

(a) C_4H_9Br

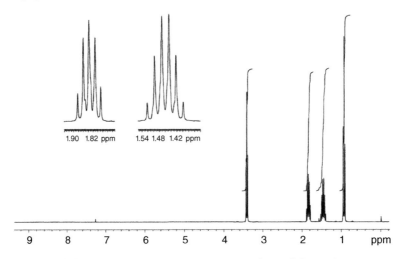

(b) $C_7H_{16}O_3$ (The resonance at δ 1.2 is a 1:2:1 triplet and that at δ 3.6 is a 1:3:3:1 quartet.)

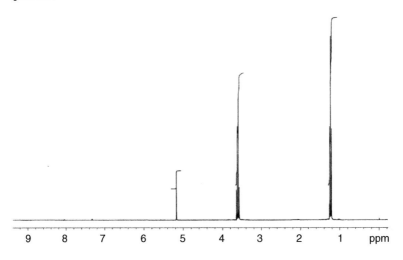

(c) $C_5H_8O_2$ (Ignore stereochemistry at this stage.)

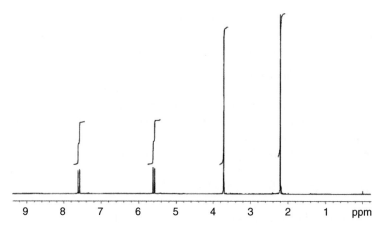

(d) $C_9H_{11}O_2N$ (*Hint:* The highest-frequency resonances (δ 6.6–7.8) come from a *para*-disubstituted phenyl ring. They are doublets. The resonance at δ 4.3 is a quartet, and that at δ 1.4 is a triplet.)

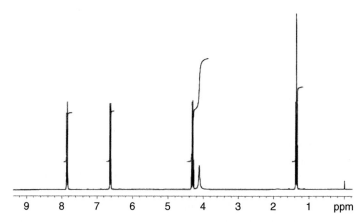

(e) C_5H_9ON (The resonances at δ 1.2 and 2.6 are triplets.)

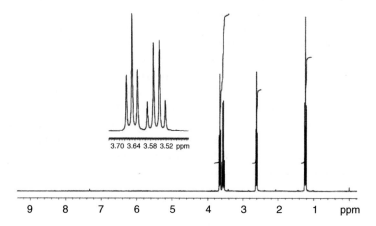

Tips on Solving NMR Problems

Spoiler alert: Some of these tips may be used directly in solving the foregoing problems and therefore should be studied only after you have made every effort to complete the problems. Many of these tips anticipate material that is covered in Chapters 3 and 4.

1) The elemental formula of a compound is obtained only rarely from elemental analysis. Usually, it is obtained from high-resolution mass spectrometric experiments. Often, the information is not available and must be inferred from the NMR spectra. The ^{13}C spectrum, for example, provides a reasonable count of all carbon atoms, and the relative amounts of equivalent carbons. As carbon functionalities are deduced from the NMR spectrum, they should be written down for later assembly of the entire molecule out of its constituents. Evidence for heteroatoms is obtained from mass spectrometry and from an analysis of chemical shifts.
2) If the elemental formula is provided in a problem, the first step is to calculate the unsaturation number as given in Problem 1.4. As unsaturations are deduced during analysis of the NMR spectra, they should be enumerated and compared with the unsaturation number until all are accounted for.
3) Upon completion of a problem, double check that every chemical shift, coupling constant, and integral is in agreement with the concluded structure.
4) The first overview of the ^1H spectrum should determine whether there are aromatics, alkenes, and saturated functionalities (with and without electron-withdrawing groups). A similar overview of the ^{13}C spectrum should indicate whether there are carbonyl groups. The spectral ranges for these classes of functionalities are developed in Chapter 3.
5) For a given type of substituent X, methyl groups (CH_3—X) are found at the lowest frequency (highest field), followed by methylene groups (—CH_2—X) and methinyl groups (>CH—X), both in ^1H and ^{13}C spectra.
6) Ethyl groups are indicated by a 1 : 2 : 1 triplet and a 1 : 3 : 3 : 1 quartet with a common coupling constant, when isolated from coupling with other groups. Isopropyl groups are indicated by a 1 : 1 doublet and a 1 : 6 : 15 : 20 : 15 : 6 : 1 septet. Sometimes the smallest peaks at either end of the septet are too small to be observed, so the resonance at first glance resembles a quintet.
7) Electron-withdrawing groups such as NO_2, Cl, Br, OH, CN, NH_2, (C=O)R, CHO, CO_2H, and C=C shift saturated ^1H and ^{13}C resonances according to their respective electronegativities. Throughout this text, the letter R is considered to be an alkyl substituent, whereas Ar is used for an aromatic substituent. The effect is attenuated rapidly with distance so that in 1-bromopropane ($BrCH_2CH_2CH_3$), for example, the effect is largest for the nearest hydrogen or carbon and decreases thereafter.
8) Unsaturated (NO_2, CN, C=O) or lone-pair-bearing (R_2N:, RO:, Cl:) substituents attached to double bonds and aromatic rings exert resonance effects that can shift alkenic and aromatic resonances to either higher or lower frequency, depending on whether the effect is electron withdrawal (unsaturated substituents) or election donation (lone-pair-bearing substituents) by resonance. Aromatic groups as substituents may be either electron withdrawing or electron donating. The effect of alkyl substituents R is described in Tip 5.

9) Primary ethers (CH$_2$—OR) and alcohols (CH$_2$—OH) are found typically at around δ 3.7, with secondary cases (CH—O) at higher frequency and methyl cases (CH$_3$—O) at lower frequency, in line with Tip 5. Primary amines (CH$_2$N) are found typically at around δ 2.7 with the appropriate variations for secondary and methyl cases. Methyls next to carbonyl groups are found at about δ 2.1, and those next to double bonds with only hydrocarbon substituents are at about δ 1.7.

10) Cyclopropane methylene protons (C—CH$_2$—C) are found at the lowest frequency (highest field) of any hydrocarbon, usually at about δ 0.2. Cyclopropane methine protons (C—CHR—C) are not normally in this region.

11) Ester methylene groups [R(C=O)—O—CH$_2$] are found at higher frequency (δ c. 4.2) than corresponding ether or alcohol methylene groups (δ c. 3.7) because the ester oxygen is more electron withdrawing. Resonance withdrawal by the carbonyl group places a formal positive charge on the ester oxygen [(RC—O$^-$)=O$^+$—CH$_2$].

12) Aldehyde protons [H(C=O)] resonate at a nearly unique position near δ 9.8.

13) Carboxyl protons [HO(C=O)] resonate at the very high frequency range of δ 12–14. Consequently, they may be out of the normally observed range of δ 0–10. If a carboxyl group is suspected, the spectral width should be adjusted accordingly. A wide range of δ −1 to +15 should include all resonances, in the absence of paramagnetic impurities. For ^{13}C, a window of δ −5 to +225 similarly should be inclusive.

14) Protons on nitrogen or oxygen in amines and alcohols (OH, NH, NH$_2$) are usually observed in chloroform solution as broad, unsplit peaks in the range δ 1–4, because of exchange phenomena. In organic solvents such as CDCl$_3$, the range is δ 1–2 for dilute alcohols (ROH), but at higher concentrations or without solvent, the peak moves to δ 3–5. Phenols (ArOH) resonate at δ 5–7 in CDCl$_3$ and δ 9–11 in DMSO.

15) Exchangeable protons may be identified by adding a drop of D$_2$O to a chloroform solution. Two layers are formed, with the small aqueous layer on top. The tube is shaken, and the OH and NH protons exchange with D$_2$O and become OD and ND, which are not observed in the spectrum. Thus the spectrum is recorded before and after addition of the drop of D$_2$O, and resonances from exchangeable protons disappear. The aqueous layer at the top should be out of the receiver coils and not detected.

16) The $n+1$ rule for identifying the number of neighboring protons has limited utility. Currently, proton connectivities are more likely to be determined by the two-dimensional correlation spectroscopy (COSY) experiment (Chapter 6) and carbon substitution patterns (CH$_3$ vs CH$_2$ vs CH vs C) by the distortionless enhancement by polarization transfer (DEPT) or related experiments (Chapter 5).

References

1.1 Lambert, J.B. and Riddell, F.G. (1983). *The Multinuclear Approach to NMR Spectroscopy*. Dordrecht: D. Reidel.

1.2 Bovey, F.A., Hood, F.P. III, Anderson, E.W., and Kornegay, R.L. (1964). *J. Chem. Phys.* 41: 2042.

1.3 Terao, T. (1983). *JEOL News* 19: 12.

Further Reading

Abraham, R.J., Fisher, J., and Loftus, P. (1992). *Introduction to NMR Spectroscopy*. New York: Wiley.

Becker, E.D. (2000). *High Resolution NMR*, 3e. New York: Academic Press.

Bovey, F.A., Jelinski, L.W., and Mirau, P.A. (1988). *Nuclear Magnetic Resonance Spectroscopy*, 2e. San Diego: Academic Press.

Breitmaier, E. (2002). *Structure Elucidation by NMR in Organic Chemistry: A Practical Guide*, 3e. New York: Wiley.

Brevard, C. and Granger, P. (1981). *Handbook of High Resolution Multinuclear NMR*. New York: Wiley.

Brey, W.S. (ed.) (1988). *Pulse Methods in 1D and 2D Liquid-Phase NMR*. New York: Academic Press.

Claridge, T.D.W. (2016). *High-Resolution NMR Techniques in Organic Chemistry*, 3e. New York: Elsevier.

Derome, A.E. (1987). *Modern NMR Techniques*. Oxford, UK: Pergamon Press.

Duddeck, H. and Dietrich, W. (1992). *Structure Elucidation by Modern NMR*, 2e. New York: Springer Verlag.

Farrar, T.C. (1989). *Pulse Nuclear Magnetic Resonance Spectroscopy*, 2e. Chicago: Farragut Press.

Field, L.D. and Sternhell, S. (ed.) (1989). *Analytical NMR*. Chichester, UK: Wiley.

Findeisen, M. and Berger, S. (2014). *50 and More Essential NMR Experiments: A Detailed Guide*. New York: Wiley-VCH.

Freeman, R. (1988). *A Handbook of Nuclear Magnetic Resonance*. New York: Longman Scientific & Technical.

Friebolin, H. (2011). *Basic One- and Two-Dimensional NMR Spectroscopy*, 5e. New York: Wiley-VCH.

Fyfe, C.A. (1983). *Solid State NMR for Chemists*. Guelph, ON: C.F.C. Press.

Günther, H. (2013). *NMR Spectroscopy*, 3e. New York: Wiley.

Harris, R.K. (1983). *Nuclear Magnetic Resonance Spectroscopy*. London: Pitman Publishing Ltd.

Harris, R.K. and Wasylishen, R.E. (ed.) (2012). *Encyclopedia of NMR* (10 volume set). New York: Wiley.

Hore, P.J. (2015). *Nuclear Magnetic Resonance*, 2e. New York: Oxford University Press.

Jackman, L.M. and Sternhell, S. (1969). *Applications of Nuclear Magnetic Resonance Spectroscopy in Organic Chemistry*, 2e. Oxford, UK: Pergamon Press.

Laws, D.D., Bitter, H.-M.L., and Jerschow, A. (ed.) (2002). *Angew. Chem. Int. Ed.* 41: 3096–3129.

Ning, Y.-C. and Ernst, R.R. (2005). *Structural Identification of Organic Compounds with Spectroscopic Techniques*. Weinheim, Germany: Wiley-VCH.

Sanders, J.K.M. and Hunter, B.K. (1993). *Modem NMR Spectroscopy*, 2e. Oxford: Oxford University Press.

2

Introductory Experimental Methods

Now that the principles of nuclear magnetic resonance (NMR) spectroscopy have been introduced, we will see how NMR spectra of the two most common nuclei, protons, and carbon-13, are obtained. The principles described for carbon-13 are applicable to many other spin-$\frac{1}{2}$ nuclei such as nitrogen-15, fluorine-19, silicon-29, and phosphorus-31. Topics to be discussed include the components of a typical NMR spectrometer, sample preparation, signal optimization techniques, spectral acquisition, processing parameter selection, spectral presentation, and spectrometer calibration.

2.1 The Spectrometer

Although there is a wide variety of NMR instrumentation available, certain components are common to all, including (i) a magnet to supply the B_0 field, (ii) devices to generate the B_1 pulse and receive the resulting NMR signal, (iii) a probe for positioning the sample in the magnet, (iv) hardware for stabilizing the B_0 field and optimizing the signal, and (v) computers for controlling much of the operation and for processing the NMR signals.

The first NMR spectrometers relied on electromagnets and operated in the *continuous wave* (CW) mode much like many of today's infrared and ultraviolet–visible spectrophotometers. Although a generation of chemists used them, the magnets had low sensitivity and poor stability. More modern, tabletop, permanent magnets are simpler to maintain but still have low sensitivity. Most research-grade instruments today use a superconducting magnet and operate in the *pulse Fourier transform* (commonly referred to as FT-NMR, Section 1.4) mode. These magnets have field strengths of 7.0–23.5 T (300–1000 MHz for protons) and provide high sensitivity and stability. Possibly most important, the very high fields produced by superconducting magnets result in better separation of resonances because chemical shifts and hence chemical shift differences (both expressed in Hz) increase with field strength. The superconducting magnet has a double Dewar-jacket arrangement, which resembles a solid cylinder with a central axial hole, with the outer Dewar filled with liquid nitrogen and the inner Dewar filled with liquid helium. The solenoid coils are immersed in liquid helium and kept at c. 4.2 K. The central bore tube on current solenoids has a diameter of 53 or 89 mm, the latter being considerably more expensive. The bore tube is maintained at room temperature, and the direction of the B_0 field (z) is aligned with the axis of the cylinder. While most samples are examined as liquids in cylindrical NMR tubes, the study of solid-state samples has become increasingly important for those materials that either cannot be dissolved in suitable NMR solvents or are more appropriately examined in the solid state. For liquid samples, the axis of the tube is placed parallel to the z-axis of the superconducting cylinder while magic angle spinning is used for solid samples (Section 1.9).

Separate from the magnet is a console that contains, among other components, two or more transmitter channels. Typically, a proton channel is paired with either a dedicated carbon channel or a broadband (BB) channel, each with a frequency synthesizer and power amplifier. A BB channel can be tuned to one of several lower frequency nuclei such as ^{15}N, ^{31}P, ^{29}Si in addition to ^{13}C. Most spectrometers are designed to record the resonances of several nuclides (multinuclear spectrometers). The console also contains a BB receiver and preamplifiers necessary to magnify the inherently weak NMR signals (relative to other spectroscopic methods such as infrared and ultraviolet) from the probe. Other standard features include a variable-temperature controller, a shim coil power supply, and an analog-to-digital converter. Additional components included in almost all modern instruments include pulsed field generators to implement pulsed gradients and waveform generators on one or more channels to produce shaped pulses and decoupler pulse sequences needed to irradiate specific regions of the spectrum. Modern spectrometers typically have a dedicated acquisition controller/processor in addition to a main computer used for interaction with the spectrometer and recording data. Ideally, one or more workstations are linked to the host computer to permit data processing and spectral plotting away from the spectrometer.

The sample is placed in the most homogeneous region of the magnetic field by means of an adjustable probe. The probe contains (i) a holder for the sample, (ii) mechanical means for adjusting its position in the field, (iii) two (or more) transmitter coils for supplying the B_1 and B_2 (double resonance) fields and for receiving the NMR signals, (iv) coils for the field/frequency lock circuitry, and (v) devices for improving magnetic field homogeneity. The arrangement of receiver coils depends on the primary purpose of the probe. A probe used mainly for proton observation has an inner coil for ^{1}H detection, located closer to the sample to maximize sensitivity, and an outer X-nucleus coil. For nuclei, including ^{13}C, other than protons, the X-nucleus coil is positioned on the inside while the ^{1}H-coil is on the outside. Probes are available for sample tubes ranging in diameter from 30 mm down to 1.7 mm. The most common-size sample tube, however, is 5 mm, which requires a volume of 500–650 µl of solvent. Sample tubes larger than 10 mm are generally used only for biological samples.

If sample is in short supply and solubility is not a problem, microtubes (120–150 µl) and submicrotubes (25–30 µl), which require much smaller volumes, can be used. With their receiver coils placed very close to the minute samples, these microprobes are excellent at scavenging the signals of their small, but concentrated, samples. Conversely, use of wider diameter tubes, such as 10 or 15 mm, is appropriate for the following situations: (i) a relatively large amount of sample exists and can be readily put in solution, (ii) a relatively small quantity of sample is present but cannot be adequately dissolved, and (iii) the experimental examination of low-sensitivity nuclei for which microprobes have not been developed. For commonly studied nuclei, microtubes should be considered when one is *sample* limited while large-diameter tubes should be employed when one is *solubility* limited.

There are two additional approaches for increasing the NMR signal of either small or relatively dilute (due to high molecular weight) samples. One is very old and consists of increasing the magnetic field strength because sensitivity increases with the $3/2$ power of B_0. More recently, cryoprobes have been developed in which the receiver coils and preamplifiers are kept at very low temperatures, using liquid helium or liquid nitrogen, to minimize noise and, thereby, increase the signal-to-noise ratio.

2.2 Sample Preparation

The first important step in sample preparation is the selection of good-quality NMR tubes. Tube quality is especially important for higher field spectrometers (400 MHz and up), for which only very high quality, such as, New Era Enterprises MP5, HP5, and UP5; Wilmad 528, 535, and 541; and Aldrich Z569348, Z569364, and Z569380, tubes should be used. Previously used tubes should be carefully cleaned prior to reuse. A suitable procedure is as follows: (i) washing with a glassware detergent solution, (ii) thoroughly rinsing (10 times or so to remove all detergent) with water, (iii) rinsing with acetone and then diethyl ether (although methanol and then $CDCl_3$ also are used), and (iv) air drying while inverted or drying on a vacuum pump. Alternatively, NMR sample tube cleaners, which spray a jet of solvent up into the tube, are available commercially from sources like New Era Enterprises, MilliporeSigma and Wilmad-LabGlass. They are connected easily to a laboratory aspirator and permit large quantities of NMR tubes to be cleaned in a very short time. Tubes that cannot be completely cleaned should be discarded. Most problems with tube cleaning arise from solvent evaporation from samples whose NMR spectra were determined several weeks or months previously. An obvious solution, which requires some discipline, is to make it a policy of transferring samples made up in volatile NMR solvents to inexpensive vials soon after their spectra have been recorded and then cleaning the NMR tubes immediately after transfer. Chromic acid–sulfuric acid ("cleaning solution") should never be used for NMR tubes, since paramagnetic chromium ions that remain on the glass will broaden the signals (especially proton) of the next samples to be placed in the tubes. In addition, oven-drying of NMR tubes should be avoided since this practice may warp the tubes.

The sample must have good solubility in a solvent, which must have no resonances in the regions of interest. NMR solvents generally are deuterated to provide a ^2H lock signal (Section 2.3.3). The most commonly used organic NMR solvent is $CDCl_3$. For polar compounds that are sparingly soluble in $CDCl_3$, CD_3OD and acetone-d_6 are good choices. DMSO-d_6 (dimethyl sulfoxide) is an excellent solvent for polar compounds and compounds containing hydroxyl groups. DMSO-d_6 solutions, however, rapidly absorb water, and good vacuum pumps are required to recover samples from them. D_2O is a good solvent for highly polar and ionic compounds. If the solution to be observed is dilute, however, a water-suppression technique may have to be used because of the intense monodeuterated water (HOD) solvent signal that arises from D_2O exchange with H_2O in the atmosphere. Lastly, if the spectrum is to be recorded above or below room temperature, the solvent chosen must not boil or freeze during the experiment.

Because magnetic field homogeneity is very critical to the NMR experiment (Section 2.3.4), the depth to which sample tubes are filled is important. Tubes must be neither underfilled nor overfilled. The probe or spectrometer manual should indicate the optimum sample solution depth in the NMR tube. Underfilling the sample tube adversely affects the field homogeneity in the region of the sample while overfilling results in poorer signal to noise since some of the sample will be outside of the receiver coils. If the NMR tube is accidentally overfilled, there are several options. If the sample in the tube represents just part of the total solution, i.e. with more still remaining in a vial or test tube, the excess can simply be withdrawn. If, however, the material in the tube (i) represents the entire sample, (ii) is a relatively small quantity, or (iii) there is not sufficient time to evaporate off the excess solvent, then the first option may not be attractive. In this case, consideration should be given as to how the sample tube will be placed in the magnet. This subject is addressed in Section 2.3.1.

Solid particles remaining in the sample solution after preparation, such as dust, dirt, or undissolved sample, should be filtered off before the solution is transferred to the NMR tube. If the sample needs to be filtered, it can be taken up into a syringe and passed through a Millipore™, or equivalent, filter assembly. Contaminants such as phthalates, however, can be a problem with this method. Suspected particles in the NMR tube can be detected by inverting the sample tube, turning it right-side-up, and then holding it at a c. 45° angle. Any suspended particles can be observed easily in the swirling solvent at the bottom of the sample tube as the solvent runs back down the NMR tube.

If ultra-high resolution NMR, spectra are required, particularly for small molecules, degassing the sample to remove dissolved oxygen may be helpful. To avoid the glass-blowing difficulties associated with sealing NMR tubes, screw-cap NMR tubes are commercially available. Degassing must be done when absolute (not effective) relaxation times (Section 7.1) are being determined.

In practice, ^{1}H NMR spectra can be obtained on less than 1 μg of material, although the result depends on the molecular weight of the compound in question, magnet field strength, probe design, and sample preparation, among other factors. ^{13}C NMR spectra typically require samples in the milligram range. Micro- and submicroprobes have recently pushed this limit down to the nanogram and microgram ranges for proton and carbon spectra, respectively.

2.3 Optimizing the Signal

2.3.1 Sample Tube Placement

Positioning of the NMR sample tube in the probe, which itself is in the magnet, is another factor to be considered. Correct NMR tube placement in the *spinner turbine* is usually accomplished in one of two ways (a discussion of why sample tubes are spun during spectral acquisition is presented in Section 2.3.4). Most instrument and probe manufacturers provide depth charts, which show a drawing of the spinner turbine containing an NMR tube. A much better means of ensuring proper tube placement is through the use of a *depth gauge*. These devices can be purchased from spectrometer manufacturers or made by a machine shop. In either case, the NMR tube is inserted into the spinner turbine, and the turbine either is placed against the depth chart or itself is inserted into the depth gauge. The sample tube is then pushed down until the bottom either matches the drawing on the depth chart or touches the bottom of the depth gauge.

Proper placement of the NMR tube in the spinner turbine is important. If the tube is not inserted far enough, some of the sample will be outside of the receiver coils (similar to overfilling the sample tube). If the tube is inserted too deeply, the consequences can be far worse. Temperature sensors, glass inserts, etc., are located in nearly all probes at depths slightly below where the bottom of a sample tube should be. An NMR tube that has been inserted too far could easily damage the very expensive probe and, again, place some of the sample outside of the receiver coils. In the previous section, the question was raised concerning accidental overfill of an NMR tube when removal of the excess sample is not a viable option. An answer may present itself in the nature of the depth chart or gauge. Some of these guides indicate the location of the center of the receiver coils. If this information is given, the overfilled sample tube may be able to be positioned so that the receiver coils are centered with respect to the solution height. In this manner, some of the sample will be above the radio frequency (RF) coils and an equal amount below. It is important, however, to remember that this maneuver can be considered only if further insertion of the NMR tube will not take it below the maximum tube depth.

2.3.2 Probe Tuning

The NMR experiment is plagued by the dual problems of sensitivity and resolution. In order to achieve maximum sensitivity, modern probes are designed to give optimum performance over a narrow frequency range. Therefore, they must be carefully adjusted to match the specific frequency and solution dielectric constant. Probes usually have a *tuning* capacitor and a *matching* capacitor for each coil. The first part of the process (tuning) sets the coil to the radiofrequency of the nucleus being studied (Table 1.2). This operation is analogous to rotating a radio knob to the desired station. Proper tuning is necessary for optimum probe sensitivity. The second part of the overall tuning process (matching) makes the total effective resistance to an alternating current (impedance) of the coil, solvent with sample, and NMR tube equal to that of the transmitter and receiver. Matching is required for passage of the maximum possible radiofrequency energy from the transmitter to the sample and then on to the receiver. Adjustment of the two capacitors is interactive, and at least two cycles of probe tuning are typically required. The tuning procedure is carried out by adjusting capacitors to give either (i) a minimum level on a probe-tuning meter or an oscilloscope display or (ii) a V-shaped signal that is horizontally (tuning error) or vertically (matching error) displaced from its correct position. Probe-tuning circuits are delicate, and tuning should be done very carefully.

As was discussed earlier in the chapter, probes typically include two coils: ^{1}H and X nucleus, e.g. ^{13}C or ^{15}N. The inner (or observe) coil is more sensitive and requires more careful adjustment. Normal ^{1}H NMR spectra usually have such good signal-to-noise ratios that probe tuning is not critical for relatively concentrated samples. Tuning, however, is very important for both 1D and 2D, X-nucleus detected experiments and many ^{1}H-detected, two-dimensional techniques. A surprising number of these experiments have failed simply because the X-coil was not tuned to the correct nucleus. In addition, if X-nucleus detection is to be conducted with proton BB decoupling, as is usually the case (Section 1.5), then it is important that the ^{1}H decoupling coil also be tuned optimally.

2.3.3 Field/Frequency Locking

All magnets are subject to field drift, whose effects can be compensated by electronically locking the field to the resonance of a substance contained in the sample. In pulsed experiments, the nucleus to whose signal the magnetic field is to be locked cannot be of the same type that is being either observed or decoupled. Thus ^{1}H cannot be used for locking. Deuterium, however, is an excellent candidate for the role of lock nucleus. Almost all common organic solvents have protons that can be replaced by deuterons. A large variety of deuterated solvents are commonly used for this purpose and are available commercially from sources like Cambridge Isotope Laboratories and MilliporeSigma. The type of lock just described is called an *internal* lock. In some instruments, the field is locked to a sample contained in a separate tube located permanently elsewhere in the probe. This type of *external* lock is usually found only in spectrometers designed for a highly specific use, such as taking only ^{1}H spectra in the CW mode or studying solid-state samples.

In the former case, an internal lock is established at the deuterium frequency of the solvent by adjusting the frequency of the lock transmitter until it matches this frequency. The operator typically observes a decreasing number of interference-pattern sine waves as the lock transmitter frequency approaches that of the solvent deuterium nuclei. A null appears when the two frequencies are identical, and the operator then turns the lock control to "On." Autolocking procedures also are available on most modern spectrometers that search for the deuterium resonance and automatically lock the spectrometer when the signal is found.

The deuterium lock system can be regarded as another NMR experiment being conducted simultaneously, and largely invisibly, with the desired one. Just as there is an observe transmitter and receiver for the nucleus of interest, so there is a lock transmitter and lock receiver for the deuterium nucleus. The stability of the lock signal depends on several factors: (i) the magnetic field homogeneity, (ii) the power of the lock transmitter, and, to a much lesser degree, and (iii) the phase of the lock signal. Certain lock signal behavior is indicative of specific problems. If the lock signal bounces up and down *erratically*, there may be suspended particles in the sample that have to be filtered off. If the lock signal bounces in a *rhythmic* fashion, the lock power is set too high, and *saturation* of the lock signal is occurring. A simple test for saturation is to lower the lock power and observe the lock level. If saturation is taking place, the lock level will initially fall but then climb back to a higher level than before.

It is important to avoid lock instability due to either saturation or suspended particles because it interferes with magnet field regulation by the lock channel and, in severe cases, can result in loss of the lock signal altogether. When a stable lock level is achieved somewhere around midrange, the *lock phase* (discussed in Section 2.3.4) should be maximized. Only an approximately maximum lock phase, however, is desired at this point because the lock phase is dependent on the magnetic field homogeneity, which is addressed in the next section.

2.3.4 Spectrometer Shimming

As stated previously, the NMR experiment suffers inherently from both sensitivity and resolution problems. Peak separations of <0.5 Hz may need to be resolved, so the B_0 field homogeneity must be uniform to a very high degree (for a separation of 0.4 Hz at 400 MHz or 0.6 Hz at 600 MHz, field homogeneity must be better than 1 part in 10^9). The stringency of this homogeneity requirement can, perhaps, be appreciated by considering it in optical terms. The late James Shoolery, formerly of Varian NMR Systems, suggested that such resolution corresponds to training a telescope on the moon and being able to distinguish two cats that are separated by one cat length. Corrections to field homogeneity are made for small gradients in B_0 by the use of *shim coils*, and this process of homogeneity adjustment is called *shimming*. The name derives from the early days of NMR when small pieces of nonmagnetic metal (shims) were used to improve the homogeneity of electromagnets by changing the distance between the halves of the magnet. For today's magnets, the field along the z-direction usually is slightly greater at one point than another. Such a gradient may be compensated for by applying a small current through a shim coil built into the probe. Shim coils are available for correcting gradients in all three Cartesian coordinates (primary gradients, x, y, and z) as well as higher order (z^2, z^3, z^4, etc.) and combination gradients (xz, yz, x^2-y^2, etc.).

The amount of shimming required depends on a number of factors: (i) how well the sample was prepared (Section 2.2), (ii) the quality of the initial shim set, and (iii) whether very high resolution is required for the sample in question. Therefore, after placing a sample in the magnet, tuning the probe, and engaging the field/frequency lock, the operator adjusts the various homogeneity controls to achieve a highly homogeneous magnetic field in the region of the sample. One approach is to optimize the primary and secondary axial shim gradients (z and z^2) and take a quick proton spectrum to assess the magnet homogeneity. If the proton spectrum is reasonably good in terms of symmetrical line shape and overall appearance, the operator can proceed to shim the magnet as described below. If the ^{1}H NMR spectrum indicates that the homogeneity is quite poor, however, the operator might recall a stored shim set (for the *same* probe from either (i) an earlier ^{1}H spectrum taken with, ideally, the same solvent or (ii) the NMR laboratory shim library, if such a library has been created). In either case, this set can provide a starting shim set.

Shimming may be carried out by either maximizing the lock level or by shimming on a strong signal in the sample. The former procedure is simpler to carry out while the latter is better if very high resolution is critical. Shimming in the latter case is done by maximizing the area of the free induction decay (FID) signal and not by actually observing a resonance in the spectrum. In either case, it is important to remember that shimming is an interactive process. As the field homogeneity is improved, and the lock level thus increased, the phase of the lock signal (analogous to that of NMR signals, Section 2.6.1) also must be readjusted for a maximum lock signal level at the start of each cycle of the shimming procedure.

The sample can be spun around its z-axis for 1D experiments at a rate of 20–25 Hz by an air flow to improve magnetic field homogeneity further. Spinning improves resolution because a nucleus at a particular location in the tube experiences a field that is averaged over a circular path. In a superconducting magnet, the axis of the NMR tube is in the z-direction. Spinning does not average gradients along the axis of the NMR tube, so shimming is required primarily for z gradients. If there are significant spinning sidebands (signals on either side of an intense signal at a distance from the central signal that is equal to the spinning frequency), however, then adjustment of the nonspinning gradients (xy or xz) is recommended. Automatic shimming of nonaxial shim coils has made this practice unnecessary in most modern spectrometers.

Shimming is an art, and sometimes there seem to be almost as many shimming routines as there are NMR spectroscopists. Before magnet shimming is undertaken, however, the lock level should be set to midrange. Just as saturation of the lock signal due to excess lock power (Section 2.3.3) must be avoided, lock levels that are too high because of the lock gain setting also are undesirable. It is difficult to see if the shimming process is improving the lock level if that level is very close to the maximum value. An exhaustive shimming procedure is given by Braun et al. [1]. An abbreviated method for superconducting magnets, which suffices in most cases, is presented below. (i) With the sample spinning, adjust Z1 and Z2 (*coarse* and then *fine*) interactively for a maximum. (ii) With the spinner turned off, interactively maximize Z1 and the lower-order, nonspinning controls (X, Y, XZ, and YZ). (iii) Turn the spinner on again, and, when the spinning speed has returned to its original value, interactively adjust the higher-order *z* controls (Z3, Z4, and Z5, if present). (iv) With the spinner again turned off, interactively maximize the higher-order, nonspinning controls (X2–Y2, X3, Y3, etc.). If good proton spectra can be recorded in eight scans or fewer, it is helpful to take such spectra between the steps listed above to see how, or if, the spectra are improving with shimming. If major changes in lock level or FID area are noted for the steps beyond the Z1, Z2 optimization, the first step and possibly others, depending on the appearance of test spectra, should be repeated. However, for two-dimensional spectra, which are performed *without* sample spinning, the z controls should, likewise, be adjusted with the sample not spinning.

In practice, with a well-prepared sample and a good starting shim set, it is frequently not necessary to go beyond the first step. If additional shimming is required, then the computerized shimming routines included in the spectrometer software can be of very great help. These automatic procedures iteratively optimize sets of interacting gradient shim coils. They are particularly useful when dealing with lock signals that are relatively insensitive to small changes since the operator often has little idea as to whether a specific shim control is being moved in the correct direction during the shimming operation.

The results of shimming are best judged by observing spectral (most often proton) line shapes. Examples of NMR signals in which one or more shims are misadjusted are shown in Figure 2.1.

If one of the odd-order gradients is misset, the resulting line is *symmetrically* broadened, and considerably so for Z1. If Z3 or Z5 is misadjusted, as in (a) and (b), respectively, the amount of broadening not only is less but occurs closer to the baseline for increasingly higher-order gradients. Thus Z5 line broadening in (b) can be seen to appear lower on the signal than Z3 broadening in (a).

If Z2 or Z4 is misset, as in (c) and (d), respectively, the resulting signal is *asymmetrically* broadened. The same inverse relationship between the height of the line broadening effect and gradient order is observed for these even-order gradients. Likewise, Z2 broadening in (c) appears higher on the signal than that of Z4 in (d). In addition, the *direction* of the asymmetry (to higher or lower frequency) is related to both the sign of the gradient misadjustment and the manufacturer of the magnet. In the examples shown in Figure 2.1c,d, asymmetric broadening occurs on the low frequency (upfield) side of the signal when Z2 or Z4 is misset in a positive sense (too large). If Z2 or Z4 is misadjusted in a negative sense (too small), then the asymmetry appears on the high-frequency (downfield) side of the resonance line. For magnets having the opposite polarity, the Z2 and Z4 asymmetries described above are reversed.

When the sample is poorly adjusted without spinning, x and y gradients produce unusually large spinning sidebands. These sidebands occur on either side of a signal (commonly a sharp single line) and are separated from it by multiples of the spinning speed, e.g. 20 Hz or 20 and 40 Hz. If a first-order shim (X, Y, XZ, or YZ) is misset (e), then disproportionally large first-order (inner) spinning sidebands are observed. Conversely,

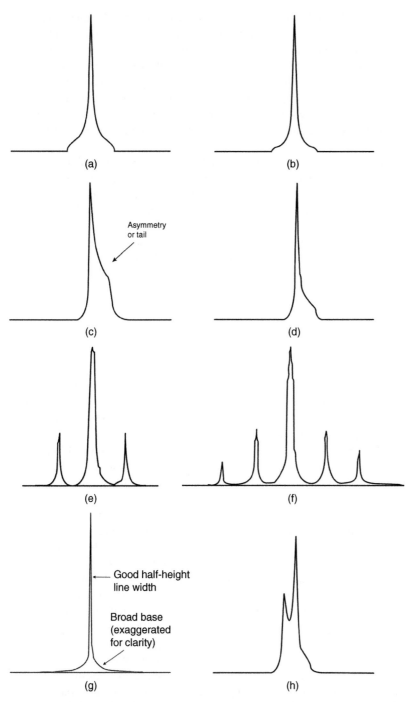

Figure 2.1 Effects of misadjusted shim settings. (a) Z3 misset, (b) Z5 misadjusted, (c) Z2 misset, and (d) Z4 misadjusted. (e) First-order spinning sidebands: X, Y, XZ, and YZ misset. (f) Second-order spinning sidebands: XY and X2–Y2 misadjusted on top of first-order sidebands. (g) High-order nonspin shims: X3, Y3, Z3X, and Z3Y misset. (h) Z1, Z2, and Z4 misadjusted. These spectral effects have been exaggerated for the purpose of illustration. Source: Courtesy of Varian Inc. Technical Publications.

if XY or X2–Y2 is misadjusted, then atypically large second-order (outer) sidebands appear. First- and second-order spinning sidebands are seen (f) when members of both of the above shim groups are misset. Lastly, misadjustment of a higher order, nonspinning shim (X3, Y3, Z3X, Z3Y) results in broad signal bases (g). In most cases, of course, several shims are misset, giving rise to complex lineshapes and complicating the above visual analysis. Such an example is (h), in which Z1 can be misadjusted in either direction, Z2 is too small, and Z4 is too large. Note that for well-maintained NMR spectrometers, usually nothing more than Z1 and Z2 adjustment is required for each sample.

An alternative, rapid, and very effective approach to shimming is available for spectrometers with pulse field gradient accessories and probes with gradient coils (Section 6.6). The procedure is known as *gradient shimming*. The pulse gradient accessory can be used to generate a field "map" of inhomogeneity along a particular axis and then determine the optimum settings for the shim coils relevant to that axis. In most cases, probes contain only a z gradient coil and, therefore, only the Z1–Z5 controls can be optimized in this manner. These controls, however, are the most critical, and this method is highly recommended if the appropriate gradient accessories are available.

2.4 Determination of NMR Spectral-Acquisition Parameters

In the FT-NMR experiment, a number of acquisition parameters must be considered before NMR data are collected. The first of these is *spectral resolution* (SR), which is controlled directly by the amount of time taken to acquire the signal. To distinguish two signals separated by Δv (in Hz), acquisition of data must continue for at least $1/\Delta v$ seconds. For example, a desired resolution of 0.5 Hz in a ^{13}C spectrum requires an acquisition time of 1/0.5 Hz or 2.0 seconds. Sampling for a longer time would improve resolution; for example, acquisition for 4.0 seconds would yield a resolution of 0.25 Hz. Thus, longer acquisition times are necessary to produce narrower lines until the natural linewidth is reached.

A second parameter concerns the range of frequencies to be detected (the *spectral width*). It is determined by how often the detector samples the value of the FID: the *sampling rate*. The FID is made up of a collection of sinusoidal signals (Figure 1.16). A single specific signal must be sampled at least twice within one sinusoidal cycle to determine its frequency (the Nyquist condition). For a collection of signals up to a frequency of N Hz, the FID thus must be sampled at a rate of $2N$ Hz. For example, for a ^{13}C spectral width of 20 000 Hz (200 ppm at 100 MHz), the signal must be sampled 40 000 times per second. Figure 2.2 illustrates this situation. The top signal (a) is sampled exactly twice per cycle (each dot is a sampling). The higher frequency signal in (b) is sampled at the same rate, but not often enough to determine its frequency. In fact, the lower frequency signal in (c) gives exactly the same collection of points as in (b). A real signal from the points in (b) is indistinguishable from an *aliased* or *foldover signal* (discussed in Section 2.4.2) with the frequency in (c). For the above ^{13}C example, if the spectral width is reduced to 10 000 Hz (100 ppm at 100 MHz) and sampled only 20 000 times per second, a signal with a frequency of 150 ppm (15 000 Hz) appears as a

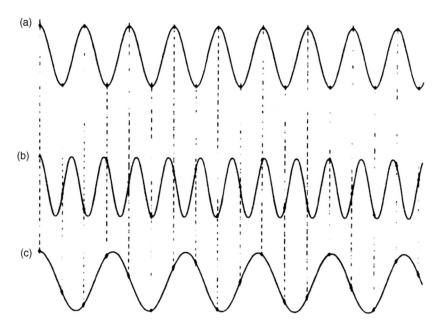

Figure 2.2 (a) Sampling a sine wave exactly two times per cycle. (b) Sampling a sine wave less than twice per cycle. (c) The lower frequency sine wave that contains the same points as in (b) and is sampled more than twice per cycle. The frequency of (b) is not detected, but appears as an aliased peak at the frequency of (c). Source: Cooper and Johnson 1986 [3]. Reprinted with permission from IBM Instruments.

distorted peak in the 0–100 ppm region. In this case, the highest frequency that can be accurately determined is half of the sampling rate or 10 000 Hz. Since the test frequency is greater than the maximum allowed frequency, the degree of aliasing of this signal is then calculated: 15 000 Hz$_{actual}$ − 10 000 Hz$_{limit}$ = 5000 Hz = 50 ppm at 100 MHz.

If a signal is sampled 20 000 times per second, the detector spends 50 μs on each point. The reciprocal of the sampling rate is called the *dwell time*, which signifies the amount of time between sampling. Reducing the dwell time means that more data points are collected in the same period of time, so that a larger computer memory is required. If the acquisition time is 4.0 seconds (for a resolution of 0.25 Hz) and the sampling rate is 20 000 times per second (for a spectral width of 10 000 Hz) the computer must store 80 000 data points. Making do with fewer points because of computer limitations would require either lowering the resolution or decreasing the spectral width.

In the early days of FT-NMR, the dedicated spectrometer computers were quite slow by today's standards and had very limited memory. Consequently, NMR spectroscopists had to consider the tradeoff of spectral width and resolution mentioned above. Today, we are fortunate to have dedicated computers that are very fast and have abundant memory. For the most part, NMR spectroscopists now can let science, rather than computer limitations, dictate how they set the parameters of their experiments.

Like the shim libraries of many NMR laboratories discussed in Section 2.3.4, spectrometers usually have standard sets of proton and ^{13}C acquisition parameters that suffice in most routine instances. It is important, however, to have an understanding of these spectral parameters and their interaction.

2.4.1 Number of Data Points

The computer algorithms that carry out the discrete Fourier transform calculation work most efficiently if the number of data points (np) is an integral power of 2. Generally, for basic spectra, at least 16 384 (for ^{1}H, referred to as "16 K") data points, and 32 768 points (for ^{13}C, "32 K") should be collected for full spectral windows. With today's higher field instruments and large-memory computers, larger data sets are now commonly used.

2.4.2 Spectral Width

The spectral width (sw) is the range of frequencies over which NMR signals are to be detected and, of course, represents the range of frequencies over which these signals are expected. For ^{1}H NMR this range is usually less than 10 ppm but, on rare occasions, can be as large as 20 ppm (Chapter 3). For ^{13}C NMR the normal range of frequencies can exceed 220 ppm (also Chapter 3), and ^{13}C spectral widths are typically set at 220 ppm. Which spectral width is used for ^{1}H NMR depends on whether the sample is a known or an unknown compound. If the former is with no unusually deshielded protons (resulting in signals at very high frequency, or downfield), a 10-ppm spectral width almost always suffices. If the identity of the sample is unknown to the operator, however, an initial 15-ppm spectral width might be appropriate just in case there are any unexpected high frequency signals, for example, from carboxylic acids. If there is none, succeeding ^{1}H NMR spectra can be determined using a standard 10-ppm spectral width.

In the beginning of this section, the concept of signal aliasing or foldover was discussed in terms of the sampling rate. A signal whose frequency lies outside of the spectral window was described as appearing as a distorted peak at a frequency inside the spectral window. Just how such signal aliasing occurs depends on the manner in which the detection of the signals is done. Quadrature phase detection (Section 5.8) is now the standard practice on all modern spectrometers and is accomplished in one of two ways: (i) using two phase-sensitive detectors whose reference phases differ by 90° or (ii) using only one detector and incrementing the phase of the receiver by 90° after each measurement. It is important for the operator to know which type of quadrature detection is being used because these two methods produce very different results with respect to aliased signals.

NMR signals that are aliased appear in the spectral window in one of the two following ways as illustrated in Figure 2.3.

If two phase detectors are employed, signals wrap into the spectral window on the *opposite* side from where they are located. For example, resonances that lie outside of the spectral window to higher frequencies wrap into the window at the low-frequency end (Figure 2.3b), while those at lower frequencies wrap in at the high-frequency end. Conversely, if a single-phase detector is used, signals fold into the spectral window on the *same* side on which they are situated. Higher frequency resonances fold in at the high-frequency end and vice versa (Figure 2.3c).

When the operator suspects the presence of an aliased signal, the testing procedure for such aliasing also depends on the type of quadrature detection being used. For either type of quadrature phase detection system, the sw parameter can be significantly increased. The position of the putative aliased resonance then changes in relation to the rest of the signals.

In addition, for single-phase detection systems, the transmitter offset (Section 2.4.5) can be changed by several hundred hertz. The position of a supposed folded signal now

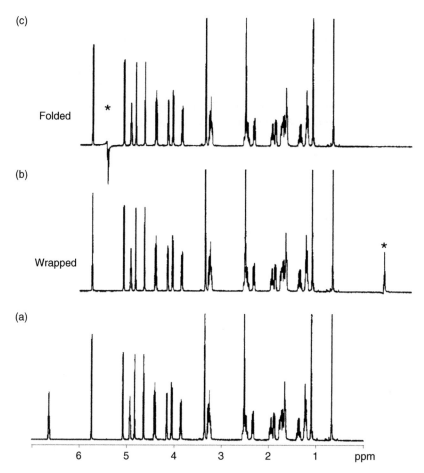

Figure 2.3 Aliasing of signals. (a) Normal spectrum, with all of the signals at their correct chemical shifts. (b) The highest-frequency signal has wrapped around into the spectrum at the low-frequency end. (c) The same signal has folded back into the spectrum at the high-frequency end. The appearance of aliased signals (denoted by an asterisk) is distorted in both cases. Source: Claridge 1999 [4]. Reprinted with permission from Elsevier.

moves in a direction opposite to that of the rest of the resonance lines. The transmitter offset-changing test procedure assumes that folding has occurred an *odd* number of times. Multiple folding is virtually unheard of for ^{1}H and ^{13}C NMR. It is something, however, for which the operator should be watchful in the investigation of less common nuclei where signals tend to be few in number and chemical shift ranges often are very large. Detection of folding when it takes place an *even* number of times is more difficult because the change of transmitter-offset test may not work (all of the resonances may move in the *same* direction). Increasing the sw may also not be a definitive test, for either type of quadrature detection system, if the multiply (even-numbered) aliasing signal is the only one in the spectrum. Here, the sw has to be reduced, perhaps several times, and the position of the resonance noted until the signal is made to fold, or wrap, an odd number of times.

2.4.3 Filter Bandwidth

In addition to signals, which lie accidentally outside of the spectral window, noise also folds, or wraps, back into the spectrum. Such noise can be greatly reduced by placing filters near the edges of the desired spectral window. Care must be taken, however, in positioning these filters since they do not cut off sharply and will reduce the intensities of signals that are near the ends of the spectrum. Positioning of the filters is controlled by a parameter known as the *filter bandwidth*, which typically has values 1.1–1.25 sw. The filter bandwidth is commonly automatically set by the spectrometer software once the spectral width has been selected.

2.4.4 Acquisition Time

The acquisition time (t_a) is related to the two previous parameters by Eq. (2.1).

$$np = 2(sw)t_a \qquad (2.1)$$

For ^{1}H spectra determined with np = 32 K and sw = 4000 Hz (10 ppm at 400 MHz), t_a is c. 4.1 seconds. For ^{13}C spectra recorded with np = 32 K and sw = 22 000 Hz (220 ppm at 100 MHz), t_a is c. 0.75 seconds. The acquisition time is normally set by the spectrometer after the number of data points and spectral width have been selected.

2.4.5 Transmitter Offset

The transmitter offset describes the location of the observation frequency and is closely related to the spectral width. With quadrature phase detection of sample signals (Section 5.8), the frequency of the transmitter is positioned in the middle of the spectral width. By so doing, the operator has the best chance of irradiating, with equal intensity, those nuclei whose resonances are both close to and far from the transmitter frequency. With their small chemical shift range, this is not a problem for protons, but it can be for nuclei with large chemical shift ranges (Chapter 3).

Standard sets of acquisition parameters include typical transmitter offset values. If wider spectral widths are required, simple trial and error with concentrated samples or standards permits the operator to widen the spectral window in a selective manner. For example, we may suspect the presence of highly deshielded ^{1}H signals and wish to open the ^{1}H spectral width from the usual 10 ppm (4000 Hz at 400 MHz) to 15 ppm (6000 Hz). In order to add all of the additional 5 ppm (2000 Hz) worth of sw capacity to the *high frequency* (downfield) end of the spectral range, i.e. from 10 to 15 ppm, the transmitter offset value is increased by c. 2.5 ppm (1000 Hz at 400 MHz). This procedure also keeps the transmitter offset positioned in the middle of the widened spectral window.

2.4.6 Flip Angle

The flip angle (α) parameter has generated considerable debate. The root of the problem can be seen in Figure 1.13. If we have a *given* amount of time in which to record spectral data, signal acquisition can be approached in basically two ways: (i) we can use a series of 90° pulses (as in Figure 1.13c) and wait a certain length of time (the acquisition time plus a relaxation delay time (DT)) between pulses for the equilibrium magnetization

(M_{z0}) to be reestablished or (ii) we can employ a series of shorter pulses (as in Figure 1.13b where α might be 30°) without any waiting time between them. With the latter approach, M_{z0} has to be entirely reestablished during t_a.

The flip angle problem can be resolved by the following analysis. Let us subject a generalized nuclear vector to a series of pulses of various duration as shown in Figure 1.13 and determine how much of the original magnetization is deposited on the y-axis for detection by the receiver. We have already seen that a 90° pulse aligns *all* of the starting magnetization along the y-axis (Figure 1.13c). Conversely, a 30° pulse places the magnetization at a 30° angle to the z-axis (similar to Figure 1.13b). Trigonometry shows that the quantity of magnetization deposited on the y-axis bears a sine relationship to the angle α. Therefore, a 30° pulse places *half* of the original magnetization (sin 30°) on the y-axis but also leaves c. 87% (cos 30°) remaining aligned along the z-axis. Consequently, recovery of the z magnetization is considerably faster if 30°, rather than 90°, pulses are used. The 30° pulses can be administered more rapidly than their 90° counterparts. Of course, less magnetization is detected per pulse with the former method.

Richard Ernst (Nobel Prize in Chemistry, 1991) conclusively demonstrated that the second method of using <90° pulses with no DTs is superior [2]. He calculated a T/T_1(max) ratio and related it to the optimum angle of rotation α^0 (the Ernst angle). In this expression, T is the total time between pulses, which includes (i) the acquisition time and (ii) any relaxation DT. T_1(max) is the longest longitudinal relaxation time *of interest*. Figure 2.4 contains a plot of various T/T_1(max) ratios against α^0. To use this figure, divide the estimated, or known, T_1(max) into the t_a and then read the optimum flip angle that corresponds to this ratio. If a series of 90° pulses is being used, $T = 1.27\, T_1$(max).

While the Ernst angle relationship has worked very well for many years, problems can arise if the T/T_1(max) ratio moves, for a variety of reasons, off scale (to the left) or if the resulting α^0 necessitates a pulse width that is too short for the spectrometer. The following situations require the use of relaxation delays. First, the T_1(max) might be unusually long. Second, t_as can run the risk of becoming too short at the higher magnetic field strengths (14.1 T; 600 MHz, ^{1}H), which are increasingly prevalent. When B_0 increases

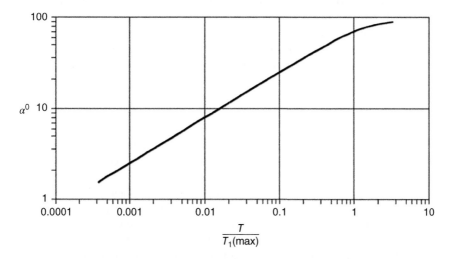

Figure 2.4 Plot used to identify optimum flip angle (Ernst angle).

so does sw, expressed in Hz, and from Eq. (2.1), t_a is seen to decrease as sw increases. In such instances, introduction of a relaxation delay effectively moves the $T/T_1(\max)$ ratio back on scale.

Even if the t_a does not appear to be too short for the relaxation times involved, we saw in the beginning Section 2.4 that spectral resolution (SR) and t_a are inversely related. Desirable ^{1}H and ^{13}C SR values are c. 0.25 and 1 Hz, respectively, and such numbers, in turn, require t_a's of c. 4 and 1 seconds, respectively. One remedy for the problem of short t_a's is the use of a larger number of data points. If np doubles from 32 768 (32 K) to 65 536 (64 K), t_a also is doubled (assuming that sw remains constant). This approach, however, increases both the memory and speed requirements of the spectrometer computer.

The consequences of this relationship are considerably different for ^{1}H and ^{13}C NMR. We have seen in Section 2.4.4 that proton t_a's are c. 4 seconds for standard 10 ppm spectral widths at 400 MHz, while those for ^{13}C are c. 0.75 seconds for common 220 ppm spectral windows at 100 MHz. The numerators of the $T/T_1(\max)$ ratio for ^{1}H and ^{13}C, therefore, are considerably different if T is essentially equal to t_a (t_p is in µs, and its contribution to T can be ignored). The denominators of this ratio are less different for ^{1}H and ^{13}C than one might imagine, when quaternary carbons are ignored. The T_1 values of protons and many *protonated* carbons are similar in magnitude, <2–3 seconds if very small molecules are discounted (Section 5.1). Nonprotonated carbons are another matter and have significantly longer T_1 values. The T_1s of these carbons, however, are usually considered only indirectly when NMR spectroscopists determine $T/T_1(\max)$ ratios.

Because ^{1}H and ^{13}C T_1s are not generally known, the selection of α often is done in the following manner. For ^{1}H NMR experiments with $T = t_a \sim$ 4 seconds and $T_1(\max) = $ 2–3 seconds, the α^0's are 76–83°. Since these T_1 values are approximations, many operators just take $\alpha = 90°$. In this approach, the equilibrium magnetization of slower relaxing protons will not be quite complete and their signal intensities somewhat diminished, but this seldom causes problems.

For ^{13}C NMR experiments, the situation is less well defined. With $T = t_a \sim 0.75$ seconds and using a protonated carbon T_1 range of 2–3 seconds, the α^0s are 40–45°. If there are nonprotonated carbons in the molecule, however, and assuming that we wish to observe them, a smaller α should be used so that they will appear in the spectrum, albeit with reduced intensities. Many operators compromise and set $\alpha = 45°$ for situations in which the longest T_1s of the nonprotonated carbons are expected to be about 10 seconds. When longer T_1s are anticipated, the procedure is to set $\alpha = 30°$.

2.4.7 Receiver Gain

The degree to which the receiver is turned on to detect the faint, transient signals that the previously excited, excess nuclei produce as they lose phase coherence and reestablish equilibrium z magnetization is determined empirically. A proper receiver gain setting is important for the following reasons: (i) too high a setting causes baseline distortions of the signals after Fourier transformation and (ii) too low a value causes precious sample signal to be lost. Modern spectrometers have *auto-gain setting* routines in which a pulse, at the selected flip angle, is first delivered with full receiver gain. If this pulse causes a receiver overload, the gain is decreased by 10% and the spectrometer pulsed again. This process is repeated until a suitable gain is found that does not cause receiver overload.

2.4.8 Number of Scans

The number of scans (ns), or transients, that is collected depends not only on the desired quality of the spectrum but also, to some degree, on the amount of spectrometer time available to the operator and its cost. Moreover, consideration must be given to the manner in which spectral signal to noise (S/N) is improved in multiple-scan, or signal-averaging, NMR experiments. In this procedure, a digitized FID is stored in the computer memory. Additional FIDs are then recorded a certain number of times and added to the same memory locations. Any signals present are reinforced while noise tends to be canceled out. If n scans are carried out and added digitally, the theory of random processes states that the signal amplitude is proportional to n and the noise is proportional to $n^{1/2}$. The S/N, therefore, increases by $n/n^{1/2}$ or $n^{1/2}$. For example, for ^{1}H NMR, if ns is initially set to a very large number like 1024, four scans might be taken for a preliminary spectrum. If this ns is found to be too few, an additional 60 scans could be added and another FT performed. The S/N in the two experiments would be increased as follows: $S/N = (64/4)^{1/2} = 4$. Multiple scanning is routine part of FT-NMR, and nuclei such as ^{13}C and ^{15}N cannot be observed at natural abundance without it.

For ^{1}H NMR spectra, the number of scans should be a multiple of 4 since this is the length of the CYCLOPS phase cycle used to minimize imperfections associated with quadrature signal detection (Section 5.8). Anywhere from 4 to 128 scans are usually sufficient to obtain a good spectrum with a relatively flat baseline. This number, however, depends heavily on sample concentration. Nevertheless, accumulation times of over one hour are relatively rare.

In this regard, a useful feature of many spectrometers is the *block size* parameter. For ^{1}H NMR and in general, block sizes are set to 4 or a multiple of 4, again depending on sample concentration. The number of scans is set to an accumulation time that might correspond to the total time, which the operator has reserved (modern spectrometers also have programs that calculate the total experimental time from the number of scans, t_a, and any DTs). At the end of each block, the summed FIDs (see above) are written into the computer memory where they can be Fourier transformed. When the spectrum displays a sufficient S/N, the acquisition is halted. Remember, from the square-root relationship discussed above, in order to double the S/N we must quadruple ns.

For ^{13}C and other low-sensitivity nuclei, the number of scans required to obtain a good spectrum is much larger than for ^{1}H. A procedure similar to that of ^{1}H NMR can be employed, by which the number of scans is again set to the equivalent of the maximum spectrometer time (now several hours) which the operator has reserved. Because of the larger number of scans, the block size now is typically set to 32 or 64. For small-sized samples, whose spectra cannot be determined in two hours, it is better to think in terms of an overnight experiment (again, doubling a two hoursS/N requires eight hours, and even this accumulation time might not be enough).

2.4.9 Steady-State Scans

Steady-state, or dummy, scans are used to allow a sample to come to equilibrium before data collection begins. A number of scans are taken as in a regular experiment, but data are not collected during what would be the normal acquisition time. Steady-state scans are usually performed before the start of an experiment but, for certain experiments on older instruments, may be acquired before the start of each incremented time value. This technique is not necessary in typical 1D NMR experiments but is employed in 1D methods that involve spectral subtraction, e.g. distortionless enhancement by polarization transfer (DEPT) (Section 7.1.2.2), and virtually all 2D experiments.

Steady-state scan numbers given in the experimental sections of Chapters 2 and 7 are performed at the beginning of the subject experiment unless stated otherwise.

2.4.10 Oversampling and Digital Filtration

We saw in Section 2.4.7 that the receiver gain must be set neither too high nor too low. A potential problem arises especially in the ^{1}H NMR spectroscopy of mixtures when one wishes to observe the small signals of minor components in the presence of much larger ones due to major components. In such a situation, the receiver gain is, of course, set to a low value to avoid overload; the signals of the minor components thus tend to be weak.

This difficulty has recently been addressed by *oversampling* the FID and involves digitizing spectral data at a rate that is much faster than that required to satisfy the Nyquist condition (Section 2.4). Since we usually think of appropriate signal digitization in terms of adequate spectral widths rather than sufficient sampling rates, oversampling can be thought of as the acquisition of NMR data with much larger spectral width than would ordinarily be necessary. The decrease in *digitization noise* (not the same as *thermal noise* encountered in Section 2.4.8) follows a similar pattern to that of S/N enhancement observed in Section 2.4.8: oversampling by a factor of N reduces digitization noise by a factor of $N^{1/2}$. Common oversampling factors for ^{1}H NMR are 16–32.

One potential problem that appears to arise from this solution is related to data processing. As we saw in Eq. (2.1) of Section 2.4.4, if the spectral width is increased to several hundred kilohertz, the number of data points required to describe the FID is commensurately augmented. This, in turn, necessitates much greater computer data storage capacity and results in considerably slower data processing. Modern spectrometers avoid these limitations by a combination of oversampling and *digital signal filtration* techniques. The FID is then reduced to the usual number of data points (16 or 32 K) prior to storage by averaging the oversampled data. Data points are taken at intervals that satisfy the Nyquist condition for the desired spectral width. The final FID thus has the same number of data points as would have been taken if it had been sampled in the usual manner.

A second potential problem concerns aliased noise due to the increased spectral width. Digital filters, when combined with analog ones (Section 2.4.3), do an excellent job of keeping unwanted noise (and signals too!) from aliasing into the selected spectral window. This very efficient filtration can become a liability if spectral widths are not carefully chosen because the presence of signals that have been accidentally left outside of the spectral window is not revealed by the appearance of aliased resonances.

2.4.11 Decoupling for X Nuclei

As we saw in Section 1.5, spectra of X nuclei (e.g. ^{13}C or ^{15}N) are usually recorded and presented under conditions of ^{1}H decoupling for several reasons, the main one being sensitivity. X-nucleus spectra are, generally speaking, inherently weak due to both the low natural abundance of most X nuclei (Section 1.7) and the presence of both direct and longer range X–H couplings (Section 1.5). The effect of these two factors is to divide intrinsically small, singlet signals into even smaller multiplets, and this problem is especially serious for ^{13}C. ^{1}H decoupling not only removes this difficulty for X-nuclear spectra but also considerably enhances the resulting singlet signals by means of a phenomenon known as the nuclear Overhauser effect (NOE, Section 5.4). ^{1}H decoupling also is carried out for ^{13}C spectra to eliminate spectral congestion. For instance, if we consider the methylene region of a fatty acid, it often is difficult to determine the exact number of CH_2 carbons in ^{1}H-decoupled ^{13}C spectra. This task would be impossible in the presence of severely overlapping multiplets.

In some respects, the proton-decoupling operation can be viewed as a simultaneous ^{1}H NMR experiment. Just as there is a transmitter offset that positions the X-nucleus observation frequency, so there is a corresponding decoupler offset for the proton-decoupling frequency. In many spectrometers, there are both transmitter and decoupler power levels. There also are three parameters, however, that are specific to decoupling and that have no counterparts among the spectral observation parameters that have been discussed.

There are several types of decoupling (Section 5.3), but the one considered here is nonselective, or broad-banded, in nature. What is meant by these terms is that the entire ^{1}H NMR spectrum is irradiated in a continuous fashion, so that the X nuclei do not experience spin coupling effects from their ^{1}H neighbors. An example of such ^{1}H decoupling is shown in Figure 1.25. For NMR spectrometers whose magnetic field strengths are 7.05 T or below (or whose ^{1}H frequencies are 300 MHz or less), the removal of proton effects can be accomplished by standard *BB decoupling*.

For field strengths of 9.4 T or above, corresponding to ^{1}H frequencies of 400 MHz or more, BB decoupling cannot be employed because too much heat is generated in the decoupling process. Several proton-decoupling techniques, however, have been developed that are appropriate for higher field instruments. The most popular of these is the wideband, alternating-phase, low-power technique for zero residual splitting (WALTZ) decoupling scheme of Freeman, which is discussed in Section 5.8. At this point, let us just state that a *decoupler field strength* has to be determined for WALTZ decoupling. The procedures by which this WALTZ parameter and various transmitter pulse times (used for generating flip angles of varying magnitude) are determined are discussed in Section 2.7.2.

2.4.12 Typical NMR Experiments

Now that the various spectral parameters have been introduced, let us consider an ordinary ^{1}H FT-NMR experiment. The pulse sequence used in this experiment is shown in Figure 2.5, in which the duration of the observation t_p pulse (called the *pulse width* (Section 2.7.1) in microseconds), is greatly exaggerated. The sequence begins with an optional relaxation DT (in seconds), in which equilibrium z magnetization can be restored. This period is followed by the observation pulse (t_p), during which the transmitter is turned on for a certain length of time, and the z magnetization is rotated through the desired flip angle (Section 2.4.6). Finally, the receiver detects the signal during the acquisition time (t_a, in seconds, Section 2.4.4). This sequence is repeated as

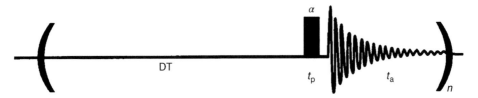

Figure 2.5 A typical ^{1}H one-dimensional NMR pulse sequence.

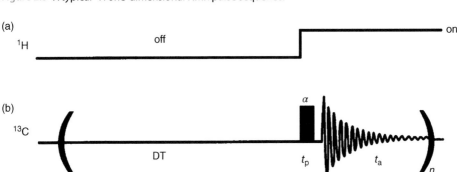

Figure 2.6 A typical one-dimensional pulse sequence for the observation of ^{13}C (b) while irradiating ^{1}H (a).

necessary to provide, upon Fourier transformation, a spectrum of desired S/N (Section 2.4.8).

In a typical ^{13}C FT-NMR experiment, performed with proton decoupling, we must consider both nuclei in the pulse sequence shown in Figure 2.6.

The ^{13}C part of the sequence (bottom) is the same as that pictured in Figure 2.5. The ^{1}H part (top) indicates the status of the decoupler during the experiment. There are four types of ^{13}C NMR experiments in which the decoupler can be turned on or off during both t_p and t_a and similarly during DT: (i) standard decoupling where the decoupler is ON continuously producing ^{1}H-decoupled ^{13}C spectra (Figure 1.25b) with NOE enhancement of individual signal intensities (Section 5.4), (ii) decoupler OFF continuously yielding fully ^{1}H-coupled ^{13}C spectra without NOE enhancements (Figure 1.25a), (iii) *gated* decoupling where the decoupler is ON during DT and OFF during t_p and t_a giving ^{1}H-coupled ^{13}C spectra with NOE enhancements, and (iv) *inverse-gated* decoupling where the decoupler is OFF during DT and ON during t_p and t_a producing ^{1}H-decoupled ^{13}C spectra without NOE enhancements. Experiments (iii) and (iv) are employed either to take advantage of, or to suppress, respectively, the NOE. The former yields ^{13}C spectra that have greater signal intensities than those of experiment (ii) while the latter is useful for counting the number of carbons that give rise to each signal in a ^{13}C spectrum.

2.5 Determination of NMR Spectral-Processing Parameters

After data acquisition has been completed, it is a good habit to save the NMR data in the operator's disk space. In this way, the original data are safe (especially after long accumulations such as overnight or over-weekend runs) from a power failure, accidental deletion, or simply forgetting to save the data after work-up (it happens).

2.5.1 Exponential Weighting

The FID, which is displayed on the monitor after data acquisition is complete (Section 1.4 and Figure 1.16), has two distinct regions. The front part, on the left, contains most of the intensity of the signals to be observed in the spectrum after Fourier transformation. Conversely, the tail of the FID, on the right, contains mostly noise mixed with those parts of the signals that give rise to narrow lines and is, therefore, the resolution part of the FID. There are two basic weighting functions that can be applied to the FID before Fourier transformation.

If sensitivity is a primary concern, as is usually the case for ^{13}C NMR, an *exponential weighting function*, Figure 2.7a, can be applied to the later portions of the FID. This process improves S/N at the expense of resolution. The use of such weighting necessarily results in some line broadening, and concomitant loss of resolution, because it attenuates the resolution portion of the FID. For this reason, weighting functions are usually listed in spectrometer manuals as *line broadening functions*. ^{13}C NMR spectra are typically presented with a 0.5–3 Hz line broadening.

When resolution is a primary concern, as is almost always the case for ^{1}H NMR, exponential weighting is typically not done. After viewing the spectrum, however, the operator may possibly wish to use a *resolution enhancement function*. Such a function improves resolution at the expense of S/N by attenuating the beginning part of the FID and emphasizing the latter part. An example of a resolution function is shown in Figure 2.7b. It is the resultant (iii) of a negative line broadening function (i), the opposite of Figure 2.7a, and a Gaussian function (ii). The effects of line broadening and resolution enhancement on ^{1}H spectra are seen in Figures 2.8 and 2.9, respectively. Ordinarily such line broadening is not introduced into ^{1}H NMR spectra, but it can be helpful in quantitative work (Section 2.6.3).

2.5.2 Zero Filling

After the application of any weighting functions (primarily in ^{13}C NMR), the next step in data processing is to *zero fill* the data to at least a factor of 2 (called one level of zero filling). The reason for this step is that the complex Fourier transform of np data points consists of a real part (from the cosine part of the FT) and an imaginary part (from the sine part of the FT), with each containing np/2 points in the frequency domain. Therefore, the actual displayed spectrum is described by only *half* of the original number of points. The technique of zero filling effectively allows all of the experimental data points to be used to generate the real spectrum. It accomplishes this goal by retrieving information that would otherwise be lost in discarding the imaginary spectrum and thus halves (improves) the *digital resolution* (DR). In this method, np is doubled by adding an equal number of zeros after the collected data points. If, as is usually the case, np = 32 768, then, for one level of zero filling, the Fourier transform number (Fn) = 2np = 65 536. In this manner, the number of points that describe the spectrum is 32 768 rather than just 16 384. The only experimental requirement for zero filling is that the FID must have decayed to zero by the end of the acquisition time.

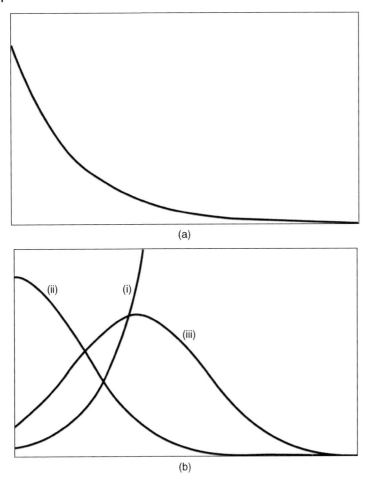

Figure 2.7 Weighting functions. (a) Sensitivity enhancement. (b) Resolution enhancement.

Zero filling beyond a factor of 2 also can be done but does not produce further improvement in line width. It is useful, however, in quantitative work (Section 2.6.3) because it provides more data points to define line shapes and signal positions.

2.5.3 FID Truncation and Spectral Artifacts

Acquisition times (Section 2.4.4, Eq. (2.1)) depend inversely on the field strength of the magnet, as (i) the number of points used to determine a spectrum are largely the same irrespective of nucleus and (ii) the spectral widths for ^{1}H and ^{13}C NMR are directly proportional to magnetic field strength. Distorted NMR signals can arise if the FID does not decay to zero by the end of the acquisition time. In this case, the FID is said to be *truncated*, and symmetrical oscillations appear at the base of strong signals, as illustrated in Figure 2.10.

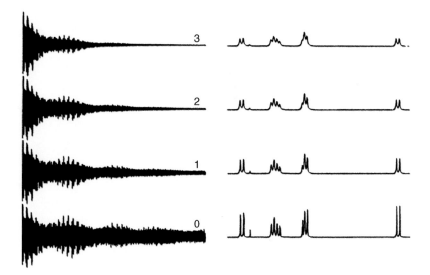

Figure 2.8 The effect of multiplying a free-induction decay by a line-broadening factor of 3, 2, 1, and 0 Hz. Source: Cooper and Johnson 1986 [3]. Reprinted with permission from IBM Instruments, Inc.

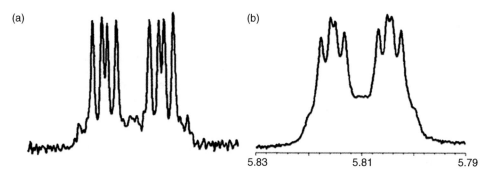

Figure 2.9 The effect of multiplying a free-induction decay by a resolution enhancement function (a) and the normal spectrum (b).

If such truncation oscillations, or wiggles are observed, the situation is remedied by applying an exponential weighting function to the FID in order to drive the tail of the FID to zero by the end of t_a. This process is called *apodization* (from Greek roots that mean "removing the feet").

Truncation artifacts are rarely a problem in ^{1}H NMR because acquisition times usually are long enough (2–4 seconds) to permit nearly complete decay of the FID. The situation is quite different for ^{13}C NMR with much larger spectral widths and t_a's typically in the range of 0.5–1 seconds. FIDs often are truncated, especially those of carbons with relatively long T_2's in small molecules. The types of truncation effects shown in Figure 2.10, however, usually are not observed even here because apodization is performed by the sensitivity-enhancing weighting functions, e.g. Figure 2.7a, which almost

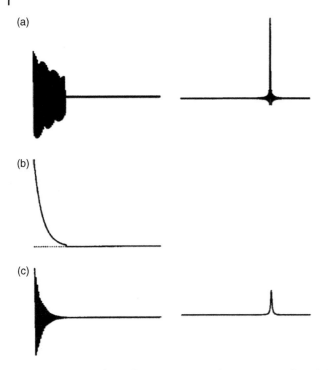

Figure 2.10 The effects of truncating a signal. (a) A truncated FID (left) showing truncation wiggles at the base of the Fourier-transformed signal (right). (b) Exponential weighting function for apodization. (c) The apodized FID (left) and Fourier-transformed signal (right). While the truncation wiggles have been removed, the resulting signal has been broadened. Source: Hoch and Stern 1996 [5]. Reprinted with permission from John Wiley & Sons.

always are applied to ^{13}C FIDs. As described in Chapter 7, apodization is very important in 2D NMR experiments.

2.5.4 Resolution

Several terms pertaining to resolution are used widely in NMR spectroscopy. It is appropriate to define them further at this point to avoid confusion. SR concerns the acquisition time employed to acquire the spectral data. We saw in Section 2.4 that this number is the reciprocal of t_a and is expressed in Hz. If np = 32 768 and sw = 4000 Hz, then t_a = 4.1 seconds, and SR is given by Eq. (2.2).

$$SR = 1/t_a = 1/4.1 \text{ s} = 0.24 \text{ Hz} \tag{2.2}$$

DR is related to both sw (in Hz) and the number of points actually used to describe the final Fourier-transformed spectrum. It is an important concept because the distinction of two signals whose separation is Δv (these can be two single lines or a coupling constant) necessitates a DR of c. ½(Δv), e.g. for $\Delta v \sim$ 1 Hz, a DR of ½ ($\sim$1 Hz) or c. 0.5 Hz is required. For the above SR example, DR is calculated from Eq. (2.3).

$$DR = sw/(np/2) = 4000 \text{ Hz}/(32\ 768/2 \text{ points}) = 0.24 \text{ Hz/point} \tag{2.3}$$

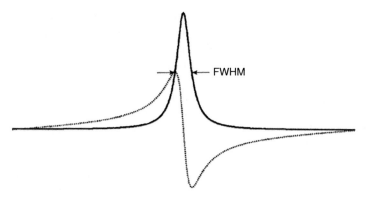

Figure 2.11 Comparison of shapes of absorption (solid line) and dispersion (dotted line) signals. Spectra are usually phase corrected to give pure absorption-mode peaks. The arrows indicate the full width at half maximum. Source: Hoch and Stern 1996 [5]. Reprinted with permission from John Wiley & Sons.

With the very large data sets currently used in 1D experiments (32 or 64 K), DR is seldom a problem. Equation (2.3) also demonstrates that SR and DR are identical. Both are derived from Eq. (2.1) in Section 2.4.4. DR, however, provides the entry to zero filling and the recovery of lost data points (Section 2.5.2). Again, for the above example, with one level of zero filling to 65 536 total points, DR is calculated from Eq. (2.4).

$$\text{DR} = \text{sw}/(\text{np}/2) = 4000\,\text{Hz}/(65\ 536/2\,\text{points}) = 0.12\,\text{Hz/point} \quad (2.4)$$

Resolution, of course, also refers to *magnetic field homogeneity*. It is a measure of how well the magnet has been shimmed (Section 2.3.4) and is unrelated to the above terms. This concept of resolution is assessed in terms of the narrowness of a line at half-height and is shown as the FWHM (full width at half maximum) in Figure 2.11.

Spectroscopists describe shimming results with expressions such as half-Hz (acceptable) or quarter-Hz (good) resolution. These numbers assume that a suitable t_a (for SR) and np (for DR) have been selected and, perhaps, even zero filling has been performed to allow the operator actually to observe the level of resolution sought.

2.6 Determination of NMR Spectra: Spectral Presentation

2.6.1 Signal Phasing and Baseline Correction

When the accumulated FIDs have been zero filled, perhaps exponentially weighted, and Fourier transformed, the next step is to phase correct the resulting resonances to give a spectrum consisting of pure absorption mode signals (see below and Appendix B). In an NMR experiment, there usually are many vectors such as those pictured in Figure 1.13. They correspond to nuclei with varying chemical shifts and, as such, have varying frequencies. These vector components are initially placed partially (Figure 1.13b) or completely (Figure 1.13c) along the *y*-axis by the observation pulse. They begin to evolve (rotate away from the *y*-axis in the *xy* plane), however, during both the finite pulse and the delay period between the pulse and when the receiver is gated on.

Signal phasing is usually necessary for the following reasons. For simplicity, let us consider the magnetization vectors referred to above as rotating (Figure 1.13c) in the *xy* plane. When the receiver is turned on, their initial phases generally do not exactly match that of the receiver. Since these phase differences are approximately the same for all of the vectors, i.e. independent of their resonance frequencies, the correction applied is referred to as a *zero-order* phase correction. A further complication arises, however, as the vectors become dispersed in the *xy* plane. Those vectors with the greatest frequencies move the farthest, and, if they do not lap the slowest-moving vectors, require the largest phase corrections. Since this correction is dependent on the resonance frequencies of the vectors, it is known as a *first-order* phase correction.

The two basic types of signals that can be detected in an NMR experiment are an *absorption* signal (due to detection of a signal perpendicular to B_1) and a *dispersion* signal (due to a signal parallel to B_1), as shown in Figure 2.11 and discussed in Appendix B. For analysis purposes, NMR spectra are displayed with completely absorptive signals. Phase errors are manifested by the dispersive character that they introduce into these signals prior to phase correction. Zero- and first-order phase errors are illustrated in Figure 2.12.

Resonances with zero-order phase errors exhibit the *same* degree of dispersive character across the spectrum (Figure 2.12a) and are adjusted to give absorption signals with the *zero-frequency* phase control (also called the *zero-order* or *right-phase* control). Those resonances with first-order phase errors display *varying* degrees of dispersive character across the spectrum (Figure 2.12b) and are phase corrected by applying a *first-order* or *left-phase*, correction that varies linearly with frequency.

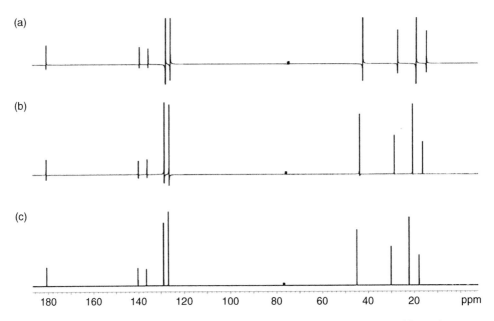

Figure 2.12 ¹H spectra showing zero- and first-order phase errors. (a) The spectrum of Ibuprofen (2-(4-isobutylphenyl)propanoic acid) with frequency-independent (zero-order) phase errors. (b) The spectrum with frequency-dependent (first-order) phase errors. (c) The correctly phased spectrum.

Typically both types of phase correction have to be performed, and the usual phasing procedure is first to correct the phase of the very low-frequency signals with the zero-order control. The operator then moves across the spectrum, to increasingly higher frequencies, phasing groups of resonances with the first-order control. The process is interactive so that by the time that the highest-frequency signals have been phase corrected, the lowest-frequency resonances may require further correction. Two to three passes through the spectrum usually suffice to correct the phase of all of the signals, and the correctly phased spectrum is displayed in Figure 2.12c. In addition, most modern spectrometers have automatic-phasing routines, which generally do a relatively good job of phasing spectra instantaneously.

A situation that appears to be a phasing problem but actually is a case of FID truncation (Section 2.5.4) is illustrated in Figure 2.13a.

Note that most of the signals can be phased properly, but slight distortions (some with opposite phase) are observed for the resonances at δ c. 17, 30, 44, 136, and 140. Attempted phasing will convince the operator that these distortions are not phasing errors. Comisarow has demonstrated that they are, in fact, due to the truncation of certain FIDs owing to acquisition times that are too short. These effects are most often found in the ^{13}C spectra of small molecules for the signals of those carbons that have relatively long T_2's. FID truncation effects most easily are eliminated by either (i) application of an even greater line-broadening weighting function (Section 2.5.1) or (ii) introduction of a relaxation DT (Section 2.4.6), as was done in Figure 2.13b.

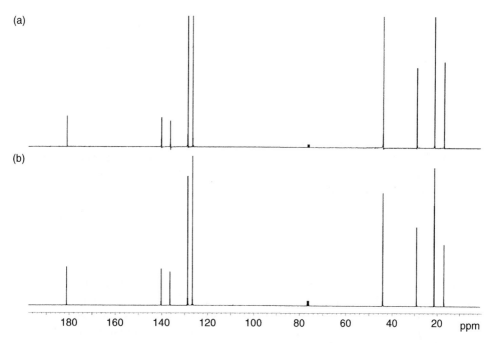

Figure 2.13 Apparent phase errors due to signal-truncation effects. (a) The distorted spectrum of Ibuprofen caused by an acquisition time that is too short (DT = 0 second). (b) The distortion-free spectrum after the introduction of a relaxation delay time (DT = 1 second).

Lastly, if the spectrum displayed on the monitor is not level, a *baseline correction* can be applied. This procedure is accomplished with a simple spectrometer command, and the correction, like auto-phasing, is done immediately.

2.6.2 Zero Referencing

The last matter to be dealt with before the spectrum is plotted is selecting an appropriate zero reference for the signals in the spectrum. This subject was mentioned briefly in Section 1.2 where we saw that reference materials have been agreed upon, for the most part, for each nuclide (Table 1.2) and assigned a relative frequency of zero. We also learned that the compound tetramethylsilane (TMS) serves as an internal zero reference for protons, carbon and silicon.

TMS is an almost ideal reference material in that it appears at such low frequency (high field) that very few proton and carbon signals appear to its right. Nevertheless, its great volatility presents a real problem in that it is very difficult to add a small amount of TMS (even right out of the freezer) to a sample tube, although common solvents such as chloroform may be purchased with TMS already present. An attractive alternative to the use of TMS as a primary reference is its use as a secondary reference. In this method, a ^{1}H spectrum is referenced to the signal of the residual proton(s) in a deuterated solvent, e.g. the small amount of $CHCl_3$ at δ 7.24 in $CDCl_3$, while a ^{13}C spectrum is referenced to the carbon(s) of the solvent, e.g. the middle line of the 3-line resonance of $CDCl_3$ at δ 77.23. The chemical shifts of the carbons and residual protons of the more common solvents are presented in Table 2.1.

The two water-soluble, internal, proton and carbon standards mentioned in Section 1.2 (the sodium salts of 3-(trimethylsilyl)propionic acid and 3-(trimethylsilyl)-1-propanesulfonic acid) also present some problems. Again, with today's very sensitive spectrometers, only small amounts of these powdery compounds need be added to NMR sample tubes. Handling these materials is thus somewhat inconvenient. Even worse is the fact that the chemical shifts of the trimethylsilyl groups of both compounds are appreciably temperature-dependent. As we saw previously with TMS, an alternative, secondary reference material exists for both protons and carbons: 1,4-dioxane. It is water soluble, and one drop in a 5-mm tube is sufficient to produce a signal that is easily visible but not too intense. The eight protons of 1,4-dioxane appear as a singlet at δ 3.68 and its four carbons also as a singlet at δ 67.06. These chemical shift values are also somewhat temperature-dependent but less so than those of the solid standards mentioned above.

2.6.3 Determination of Certain NMR Parameters

2.6.3.1 Chemical Shifts and Coupling Constants

After NMR spectra have been zero referenced, the determination of ^{13}C chemical shifts is straightforward, since the spectra are almost always collected under proton-decoupling conditions and, therefore, appear as single lines. Modern spectrometers have *peak-picking* programs in which a threshold value is first set and then the line positions of all signals that exceed this threshold are recorded.

^{1}H chemical shifts are another matter, since the signals of most protons are multiplets. Strictly speaking, spectral analysis is required to determine the chemical shifts

Table 2.1 Chemical shift data of common deuterated NMR solvents.[a]

Solvent	^{1}H chemical shift(s)[b]	^{13}C chemical shift(s)
Acetic acid-d_4	2.04 (5)	20.0 (7)
	11.65 (1)	178.99 (1)
Acetone-d_6	2.05 (5)	29.92 (7)
	—	206.68 (1)
Acetonitrile-d_3	1.94 (5)	1.39 (7)
	—	118.69 (1)
Benzene-d_6	7.16 (1)	128.39 (3)
Chloroform-d	7.24 (1)	77.23 (3)
Deuterium oxide	4.81[c]	NA
Dichloromethane-d_2	5.32 (3)	54.00 (5)
N,N-dimethyl-formamide-d_7	2.75 (5)	29.76 (7)
	2.92 (5)	34.89 (7)
	8.03 (1)	163.15 (3)
Dimethyl sulfoxide-d_6	2.50 (5)	39.51 (7)
1,4-Dioxane-d_8	3.53 (*m*)	66.66 (5)
Methanol-d_4	3.31 (5)	49.15 (7)
	4.87 (1)	—
Pyridine-d_5	7.22 (1)	123.87 (3)
	7.58 (1)	135.91 (3)
	8.74 (1)	150.35 (3)
Tetrahydrofuran-d_8	1.73 (1)	25.37 (5)
	3.58 (1)	67.57 (5)

a) Multiplicities in parentheses.
b) Signals of residual solvent protons.
c) Chemical shift highly dependent on pH and temperature.
Source: Courtesy of Cambridge Isotope Laboratories, Andover, MA.

and coupling constants of those protons whose signals display second-order behavior (Chapter 4 and Appendix D). For those protons whose spectra are first order or even pseudo-first order, chemical shifts and couplings can be determined from the spectra by inspection, but some care should be exercised in this endeavor.

^{1}H spectra first should be spread out so that only one multiplet or one region containing several multiplets is displayed on the monitor. Vertical cursors, or arrows, should be set so that they are either (i) between the middle signals of a multiplet containing an even number of lines or (ii) on top of the middle resonance of a multiplet containing an odd number of signals. Cursor/arrow positions then can be recorded in the laboratory notebook.

Peak-picking programs often are misused in the determination of coupling constants. The reason is that coupling constants have to be just that – *constant*. The spin coupling that is determined for the methyl triplet (c. δ 1.2) of diethyl ether in Figure 1.22 has to be the same as that for the methylene quartet (c. δ 3.5), both being c. 7 Hz. If we request a line listing for the seven lines of the diethyl ether spectrum and do the subtraction, however, we quickly discover that the line separations, and thus the apparent coupling constants, are almost always *not* exactly the same. The reason for this inconsistency is that the peak-picking routine reads either the highest data point that describes each spectral line or an interpolated maximum as the position of that line. This procedure is adequate for the determination of ^{13}C chemical shifts, because the horizontal scale is in thousands of hertz, and moving the shift-determining data point(s) a small amount to either side does not make a significant difference. ^{1}H coupling constants, however, generally are less than 20 Hz, and corresponding data point movement is now sufficient to produce unequal spacings. The spreading out and averaging procedure described above for the determination of ^{1}H chemical shifts avoids this problem but can be tedious.

An attractive alternative in which spectrometer line listing routines are used is as follows. After the spectral data have been Fourier transformed and saved, the proton FID is zero filled three to four times. If np = 32 K, then *Fn* should be set to 256 or 512 K. The resulting spectral lines should then have sufficient definition for peak picking. This expectation can easily be tested by examining a known multiplet such as a triplet or quartet. As stated above, the two- or three-line spacings are, of course, identical and this must be reflected in the line listings displayed on the monitor. No matter what technique is used, *J* values between spin-coupled nuclei should always be reported as identical and averaged if necessary.

2.6.3.2 ^{1}H Integration

As we saw in Section 1.6, ^{1}H NMR spectra are almost always recorded in such a manner that the detected signals are directly proportional to the number of protons giving rise to those signals. For most ^{1}H-decoupled ^{13}C spectra, however, a spread of T_1s and the NOE (Section 5.4) combines to produce signals whose intensities are *not* proportional to the number of carbons present.

The process of measuring the intensities of NMR signals is called *integration* and is normally performed only for ^{1}H spectra. The integral of the total spectrum is a line that starts at the left side of the spectrum and increases in a stepwise fashion at each ^{1}H signal. The areas of the sections into which this continuous line is cut correspond to the number of protons giving rise to the signals below them and are know as *integrals*. This name derives from the fact that for the elementary compounds that were first studied by NMR, the ratios of the signal areas were simple integers. An example of a typical, integrated ^{1}H spectrum can be seen in Figure 8.1 for the fungal metabolite T-2 toxin.

Proton spectra, for which integration is to be performed, commonly are determined in one of two ways, depending upon whether the sample in question is predominantly a single compound or a mixture of two or more compounds. In the former case, integration is much more easily carried out because the signal areas must be integral ratios of one another. These spectra can be acquired in the manner described above, e.g. with acquisition times of c. 4 seconds, because proton relaxation is essentially complete at the end of t_a. For mixtures, however, integration must be performed with much greater care, because the integral values are seldom simple multiples of one another. For good quantitative results, it is critical that proton relaxation be essentially complete between scans. Since proton T_1s tend to be short (millisecond to a few seconds), this condition can best be achieved by using small flip angles and relaxation DTs of several seconds.

The initial approach to integration is to determine where the spectrum can be cut between groups of signals and to identify the largest continuous run of signals for which the integral of that group cannot be cut to the baseline. This region likely contains the largest number of proton signals. It is now spread out on the screen and its integral amplitude essentially maximized, i.e. set to a round number like, say, 2000 which can easily be doubled, tripled, etc. The idea is next to cut the rest of the signals into regions (which are generally less spread out than in the previous section for the determination of ^{1}H chemical shifts and coupling constants) that can be integrated accurately.

If the operator so wishes, the integral amplitudes of these smaller regions also can be maximized so they are multiples of the first amplitude. In this approach, integrals that would be less accurate because they are small, compared with the largest integral, can be increased in size and better compared. By keeping track of the multiplication factors, all of the integrals can be recorded and then scaled to the same integral amplitude. This technique is especially helpful if mixtures of compounds are being analyzed, since the integrals infrequently are simple 1 : 2 : 3, etc. ratios of one another.

Like shimming, there are many ways to perform integration. The following parameters of Gard, Pagel, and Yang represent a general approach to obtaining precise ^{1}H NMR quantitative measurements.

Acquisition Parameters:

1) Set $\alpha = $ c. 30° with DT = 1 seconds.
2) Increase the filter bandwidth by 500–1000 Hz if there are signals near the edges of the spectral window.
3) Take enough scans to ensure a good S/N and thus smoother integrals.
4) Adjust receiver dead time (the time between the end of the pulse and the beginning of signal acquisition) to minimize pulse breakthrough, which is manifested by a baseline roll. Certain spectrometers carry out this operation in the following way. First, a spectrum is recorded and phased as usual. A software command then calculates the dead time such that the first-order (left) phase control equals zero in a repeated spectrum.

Processing Parameters:

1) Set line broadening = 0.2–0.3 Hz.
2) Use three levels of zero filling (8 np), e.g. if np is 32 K, then *Fn* is 256 K.

Spectral Display:

1) Careful phasing is very important.
2) Make integral regions 5–7 linewidths, if possible, to get the full integral. It is even more critical, however, that all integral regions be the *same width* and extend the same distance on either side of each signal.
3) Use baseline corrections on each integral cut if spectral tilt is noticeable. Extensive zero filling (Step (2) of *Processing Parameters*) is useful because it provides more data points to define line shapes better (Section 2.5.2). Slight line broadening produces smoother spectral and integral lines with less baseline noise.
4) Consistently include, or exclude, ^{13}C satellite signals (Section 1.5).

2.7 Calibrations

In Section 2.4, we encountered a number of spectral acquisition parameters. The selection of these parameters is relatively direct and straightforward, with the exception of the flip angle α and the decoupler modulation frequency (for WALTZ decoupling). Neither α nor the decoupler modulation frequency is entered directly, and both parameters require spectrometer calibration. Automated procedures are implemented for many or all of these on newer instruments but an understanding of the process is important.

2.7.1 Pulse Width (Flip Angle)

In Section 1.4, it was demonstrated that magnetization vectors could be rotated along the x-axis through various angles θ. As we have seen, a 90° pulse places all of the original z magnetization along the y-axis and leads to detection of a full signal (Figure 1.13c). Conversely, a 180° pulse places the entire magnetization vector along the $-z$-axis, and, therefore, results in no signal at all (Figure 1.13d). A 270° pulse aligns the z magnetization along the $-y$-axis, producing a full, negative signal, and a 360° pulse returns the magnetization vector to its original position along the z-axis. As θ increases from 0° to 360°, the detected magnetization exhibits the behavior shown in Figure 2.14. The maximum signal is, therefore, obtained when $\theta = 90°$.

The length of time that the transmitter has to be turned on, i.e. the *pulse width* (t_p), to achieve a certain vector rotation is determined empirically in the following manner. In principle, any particular sample can be used for the ^{1}H 90° pulse width determination. In practice, it should have at least one well-defined signal, e.g. ideally an intense singlet, which can be observed with one pulse. More important is the requirement that the T_1 of the test signal not be excessively long, because the z magnetization must be reestablished between pulses. This important criterion is further considered later in this section. A sample of 10% acetone in CDCl$_3$ is a possible ^{1}H 90° pulse-width standard. It is better if

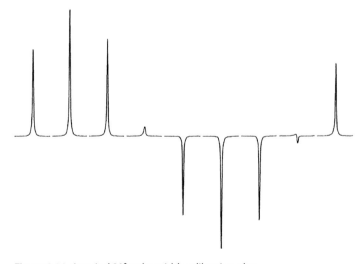

Figure 2.14 A typical 90° pulse width calibration plot.

this sample is *not* a sealed and degassed reference standard because the T_1 value of the methyl protons in these latter standards is c. 30 seconds.

With the test sample in the magnet, the probe is tuned to ^{1}H and the magnet homogeneity optimized. A test spectrum is next determined, in which the t_p used is unimportant. This spectrum then serves as a starting point for another spectrum that has a reduced spectral width. The original sw is now reduced to c. 500 Hz and the transmitter offset adjusted so that it is in the middle of the reduced sw. Many spectrometers have programs that do both operations with one command. Since sw has been considerably reduced, the number of data points should also be decreased to c. 4000 to maintain an acquisition time of c. 4 seconds.

A reduced, one-scan spectrum is then taken and displayed in the *absolute intensity* mode, with t_p set to c. one half of the value given for the 90° t_p in the instrument specifications, e.g. set $t_p = 5\,\mu s$ for a 90° t_p of 10 μs. The resulting one-line spectrum should be processed with 1-Hz line broadening, phased properly, and presented in such a way that the baseline appears at about midscreen and the signal height adjusted so that it occupies about one half of the vertical size of the screen.

A series of experiments is then queued in which t_p is increased from this first value to a final one that is c. five times the specified 90° t_p. For the example given in Figure 2.14, the suggested 90° t_p is c. 6 μs. The queued t_p's then are the following sequence: 3, 6, 9, 12, 15, 18, 21, 24, and 27 μs. In addition, a relaxation DT estimated to be $3T_1$ of the test proton(s) (c. 30–45 seconds) was inserted before each pulse in the series. Modern spectrometers have programs for displaying the accumulated signals, which appear like those shown in Figure 2.14.

A plot like Figure 2.14 gives us only an approximate idea of the 90° t_p, but that is all right because we are really interested in the 360° t_p. Determining a 360° t_p, rather than a 90° t_p, is a better procedure for several reasons. First, it is much easier to see where a series of signals crosses over from negative to positive, in the case of a series of t_p's arrayed around 360°, e.g. 350° to 10°, than to decide on a maximum for a corresponding series arrayed around 90°, e.g. 80–100°. Second, if the T_1 of the test proton(s) is not known, the relaxation DT might be insufficient to permit complete reestablishment of z magnetization between pulses. If complete relaxation does not occur, the resulting signal amplitudes are decreased from their true intensities and displaced toward smaller pulse widths as incomplete z-magnetization recovery increases and reaches a steady-state condition. The problem of incomplete relaxation is much more severe for a series of t_p's arrayed around 90°, where nearly complete z magnetization has to be reestablished after each pulse, than for a corresponding series arrayed around 360°, where the z magnetization is little disturbed from its equilibrium value with each pulse.

A second series of experiments now is carried out, in which the t_p's are arrayed around 360° and the relaxation DT is set to c. 30 seconds. Additional series of t_p's can be used as one closes in on the 360° t_p. Values of t_p do not change appreciably between common organic solvents, and, therefore, do not have to be determined for each sample. Since ionic samples require retuning of a probe with respect to nonionic ones, the t_p's of the former have to be redetermined relative to those of the latter. It must be pointed out, however, that the 90° pulse width thus determined is likely to be slightly longer than 360°/4 since the rise time (the time required for the transmitter to reach full strength after it is turned on) of a 90° pulse is slightly less than that of a 360° pulse. Nevertheless, the potential errors cited above, due both to incomplete relaxation and the difficulty of determining signal maxima that are encountered for the determination of 90° pulse widths and avoided for 360° pulses, generally make the latter procedure the method of choice.

^{13}C 90° t_p's are determined in much the same manner. Since carbon signals are much weaker than proton resonances, however, only ^{13}C-enriched or very concentrated samples, which give good one-scan spectra, are suitable as ^{13}C 90° t_p reference materials. One such candidate is the ^{13}C sensitivity reference standard, 60% C_6D_6/40% dioxane, in which the dioxane resonance is observed. Since reference standards are, as a rule, degassed, special care must be taken to ensure that suitable relaxation DTs (c. 45–60 seconds) are used between initial spectral acquisitions. Shorter DTs can be employed for t_p's arrayed around 360°. With the test sample in the magnet, the probe is tuned to both ^{13}C (transmitter) and ^{1}H (decoupler), and the magnet homogeneity is maximized. ^{13}C spectra are recorded with full ^{1}H decoupling and are displayed in the same manner (with, perhaps, a line broadening of 3–4 Hz) as described above for ^{1}H 90° t_p calibration. Values of t_p for both nuclei should be recalibrated at least once a month.

As a general rule the transmitter and decoupler power levels are set so that both the ^{1}H and ^{13}C 90° t_p's are in the range of 5–10 μs. Pulse widths less than 5 μs are inaccurate because pulse rise and fall times are in the microsecond range. Probes having t_p's appreciably greater than 10 μs may be unable to execute those pulse sequences that require rapid switching on and off of several transmitters.

2.7.2 Decoupler Field Strength

We saw in Section 2.4.11 that (i) ^{13}C NMR spectra are usually recorded with WALTZ proton decoupling for spectrometers whose field strengths are 9.4 T or above (or whose ^{1}H frequencies are 400 MHz and above) and (ii) one of the components of WALTZ decoupling, the *decoupler field strength* (γB_2), has to be determined for each individual spectrometer. The quantity γB_2 is a function of the decoupler power level and characteristics of the probe. In addition, γB_2 even can depend on the sample itself if highly ionic solutions are being examined, which, as we saw above, require different probe tuning relative to common organic solvents.

The standard method for calibrating γB_2 is a technique known as *off-resonance decoupling* (Section 5.3 and Figure 5.8). In Figure 1.25, we saw that methyl carbons appear as quartets, methylene carbons as triplets, and methine carbons as doublets in fully proton-coupled ^{13}C NMR spectra. In off-resonance, proton-coupled ^{13}C spectra, the same patterns are observed (doublets, triplets, etc.), but the proton-carbon coupling constants are *reduced*. The ^{13}C sensitivity reference standard, 60% C_6D_6/40% dioxane (to which we were introduced in the previous section) is a good test sample for this calibration.

In the off-resonance decoupling method, the reduced, carbon–hydrogen coupling constant (J_R) of the methylene group of dioxane is measured. Two sequential, off (proton)-resonance experiments are performed, one with the decoupler situated at a higher frequency than the chemical shift of the methylene protons and a second with the decoupler located at a lower frequency. This technique results in two carbon spectra, which are both triplets, and in which the values of the reduced carbon–hydrogen couplings are a function of both γB_2 and the two distances that the decoupler is positioned off resonance (Δv).

With the test sample in the magnet, the probe is tuned to both ^{13}C (observe) and ^{1}H (decoupling) and the magnet homogeneity maximized. A test spectrum is then determined, and, as in the 90° t_p calibrations above, sw is reduced to c. 500 Hz and the transmitter offset adjusted so that it is in the middle of this reduced sw. Since sw has been considerably reduced, np is also decreased to c. 4000 (for a t_a of c. 4 seconds). The resulting spectrum should be processed with a 3–4 Hz line broadening, phased properly, and presented so that the height of the middle signal of the triplet occupies the full vertical size of the screen and the three lines of the spectrum occupy c. 3/4 of the horizontal size of the screen.

The first partially decoupled spectrum is acquired with the decoupler frequency situated c. +1000 Hz off resonance. The reduced coupling is best determined by measuring the distance between the outer lines of the reduced triplet and dividing this result by two (the coupling constant appears twice in a triplet, Section 1.5). This high-frequency, reduced coupling is then recorded in the lab notebook. The second spectrum is acquired with the decoupler frequency located c. −1000 Hz off resonance and the low-frequency, reduced coupling similarly measured and noted. Most modern spectrometers have programs that calculate γB_2 from these two reduced couplings and the value of the full carbon-hydrogen coupling constant of dioxane (J_0), which is 142 Hz. γB_2 also can be calculated from Eq. (2.5),

$$\Delta v = \gamma B_2 J_R / 2\pi (J_0^2 - J_R^2)^{1/2} \quad (2.5)$$

in which J_R is the reduced, C—H coupling constant of dioxane described above, J_0 the natural, one-bond, C—H coupling of dioxane (c. 142 Hz), and Δv the difference between the resonance frequency of the equivalent dioxane carbons and the decoupler frequency. The *decoupler modulation frequency* employed in WALTZ decoupling is $4(\gamma B_2)$.

Problems

2.1 (a) What is the spectral resolution of an experiment whose acquisition time is 0.66 seconds?
(b) What is the minimum sampling rate required to avoid aliasing for a spectral width of 25 kHz?

2.2 (a) How many data points are required for an experiment whose digital resolution is 1.22 Hz/point and whose spectral width is 20 kHz?
(b) What acquisition time is required for an experiment in which 32 768 data points are used and whose spectral width is 20 kHz?

2.3 What is the difference in the amounts of signal acquired for the following two experiments: (a) four 90° pulses with a three-second delay time between pulses and (b) twelve 30° pulses with no delay time between pulses? Assume in both cases that $t_p = 1$ second and that relaxation is complete by the time the next pulse is delivered.

2.4 (a) By how much would a signal increase if the number of scans went from 4 to 16?

(b) How much would the signal of a 15-hour experiment increase if the operator decided to continue acquisition for another hour? What conclusion can be drawn from these two observations?

References

2.1 Braun, S., Kalinowski, H.-O., and Berger, S. (1998). *150 and More Basic NMR Experiments*, 2e. New York: Wiley-VCH.
2.2 Ernst, R.R. and Anderson, W.A. (1966). *Rev. Sci. Instrum.* 37: 93.
2.3 Cooper, J.W. and Johnson, R.D. (1986). *FT NMR Techniques for Organic Chemists*. IBM Instruments, Inc.
2.4 Claridge, T.D.W. (2016). *High-Resolution NMR Techniques in Organic Chemistry*, 3rd edition. Elsevier Science.
2.5 Hoch, J.C. and Stern, A.S. (1996). *NMR Data Processing*. Wiley.

Further Reading

Fukushima, E. and Roeder, S.B.W. (1981). *Experimental Pulse NMR*. Reading, MA: Addison-Wesley.

Martin, M.L., Martin, G.J., and Delpeuch, J.-J. (1980). *Practical NMR Spectroscopy*. Philadelphia: Heyden.

3

The Chemical Shift

3.1 Factors That Influence Proton Shifts

Interpreting the location of a resonance in the ^{1}H spectrum in terms of molecular structure requires understanding several contributing factors. Chemical shifts vary according to structure because nuclei experience different degrees of shielding by magnetic fields produced by their surrounding electrons. In the absence of shielding, nuclei experience a field with the value of B_0, but when shielded by their surrounding electrons, the field experienced by a nucleus becomes $B_0(1-\sigma)$, in which the quantity σ is called the *shielding* (Figure 3.1a,b). Because the magnetic field ($-B_0 \cdot \sigma$) induced by the electrons opposes the static B_0 field, the effect is said to be *diamagnetic* (represented by the symbol σ^d).

3.1.1 Local Fields

Shielding by the electrons that surround the resonating nuclei is said to arise from *local fields*, which may be assessed by considering electron density. For the proton, the electronic effects of physical organic chemistry (electronegativity and conjugation) conveniently describe the role of structure vis-à-vis electron density. In this way, both the atom to which the proton is attached and more distant atoms can modulate the electron density at the proton and hence alter the shielding effect.

The effects of electronegativity, usually called polar or inductive effects, are manifested in the following fashion. An attached or nearby electron-withdrawing group such as —OH or —CN decreases the electron density and hence the diamagnetic shielding, with the result that the resonance of the attached proton moves toward the left of the chart (to a higher frequency, or downfield; Figure 1.9). By contrast, an electron-donating atom or group increases the diamagnetic shielding and moves the resonance toward the right of the chart (to a lower frequency, or upfield). Although the effects of shielding on chemical shift are more properly described in terms of frequency, we include field terminology parenthetically, since it still enjoys wide, though inappropriate, usage.

Progressive replacement of hydrogen with chlorine on a methane molecule moves the chemical shift to higher frequency (downfield) because of the ability of chlorine to remove electron density from the nearby protons: δ 0.23 for CH_4, 3.05 for CH_3Cl, 5.30 for CH_2Cl_2, and 7.27 for $CHCl_3$. The trend for a series of methyl resonances often can be explained in the same fashion by the polar effect. The chemical shifts for the series

Nuclear Magnetic Resonance Spectroscopy: An Introduction to Principles, Applications, and Experimental Methods, Second Edition. Joseph B. Lambert, Eugene P. Mazzola, and Clark D. Ridge.
© 2019 John Wiley & Sons Ltd. Published 2019 by John Wiley & Sons Ltd.

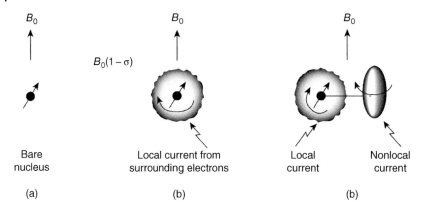

Figure 3.1 (a) An unshielded nucleus. Nuclei shielded by (b) local and (c) nonlocal currents.

CH_3X for which X may be F, HO, H_2N, H, Me_3Si, or Li, respectively, are δ 4.26, 3.38, 2.47, 0.23, 0.0 (tetramethylsilane (TMS), the standard), and −0.4 (this last value is considerably dependent on the solvent; the minus sign indicates a lower frequency than TMS). This trend follows the electronegativity of the atom attached to CH_3.

Electron density is influenced by conjugation (also called resonance in organic chemistry), as well as by polar effects, as seen particularly in unsaturated molecules such as alkenes and aromatics. Donation of electrons through resonance by a methoxy group increases the electron density at the β position of a vinyl ether (**3.1**,

$$CH_2\!=\!\underset{H}{\overset{OCH_3}{C}}\;\;\overset{4.1}{}\quad\longleftrightarrow\quad {}^-CH_2\!-\!\underset{H}{\overset{{}^+OCH_3}{C}}$$

3.1

in which the chemical shift in δ units is included next to the appropriate protons) and at the para position of anisole ($C_6H_5OCH_3$). Thus, the chemical shift of the β protons in **3.1** is at about δ 4.1, in comparison with δ 5.28 in ethene. The resonance frequency decreases, as expected with the increased shielding from electron donation. The electron-withdrawing polar effect of CH_3O is overpowered by the resonance effect. Groups such as nitro, cyano, and acyl withdraw electrons by both resonance and induction, so they can bring about significant shifts to high frequency (downfield). Ethyl *trans*-crotonate (**3.2**,

$$\underset{CH_3}{\overset{7.0\;H}{}}\!\!\!\!\!C\!=\!C\!\!\!\!\!\underset{\underset{5.8}{H}}{\overset{EtO}{}}\!\!\!\!\!C\!=\!O \quad\longleftrightarrow\quad \underset{CH_3}{\overset{H}{}}\!\!\!\!\!C\!\!\!\overset{+}{-}\!\!\!C\!\!\!\!\!\underset{H}{\overset{EtO}{}}\!\!\!\!\!C\!-\!O^-$$

3.2

Section 1.6) illustrates this effect. The electron-withdrawing group shifts the β proton strongly to high frequency (δ 7.0). Although the α proton is not subjected to this strong resonance effect, it is close enough to the electron-withdrawing carboethoxy group to be shifted slightly to higher frequency (δ 5.8) by the polar effect.

Hybridization of the carbon to which a proton is attached also influences electron density. As the proportion of s character increases from sp^3 to sp^2 to sp orbitals, bonding electrons move closer to carbon and away from the protons, which then become deshielded. For this reason, methane and ethane resonate at δ 0.23 and 0.86, respectively, but ethene resonates at δ 5.28. Ethyne (acetylene) is an exception in this regard, as we shall see presently. Hybridization contributes to shifts in strained molecules, such as cyclobutane (δ 1.98) and cubane (δ 4.00), for which hybridization is intermediate between sp^3 and sp^2.

3.1.2 Nonlocal Fields

Induction, conjugation, and hybridization modulate electron density immediately surrounding a proton, as the result of local electron currents around the nucleus (Figure 3.1b). In the absence of changes in electron density, purely magnetic effects of substituents also can have major effects on proton shielding, but only when the groups have a nonspherical shape. The drawing in Figure 3.1c represents the combined effects of local fields and nonlocal fields. The group giving rise to the nonlocal field could be, for example, methyl, phenyl, or carbonyl, and the resonating nucleus need not be attached directly to the group. To see why a spherical or isotropic ("same in all directions") group contributes no nonlocal effect, consider a proton attached to such a group, for example, chlorine. The local effect arises from the electrons that surround the resonating proton. The electrons in the substituent, which are not around the proton, also precess in the applied field (Figure 3.1c). They induce a magnetic field that opposes B_0 and that can have a nonzero value at the position of the proton.

The nonlocal induced field may be represented by magnetic lines of force as shown in Figure 3.2. If the bond from the spherical substituent to the resonating proton is parallel to the direction of B_0, as shown in Figure 3.2a, the lines of force from the induced field oppose B_0 at the proton, thereby shielding it. If the bond from the substituent to the

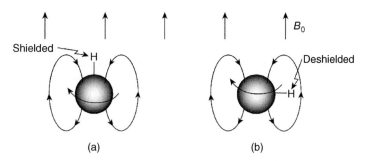

Figure 3.2 Shielding by a spherical (isotropic) group. (a) When the C—H bond is parallel to the direction of the magnetic field. (b) When the C—H bond is perpendicular to the direction of the magnetic field.

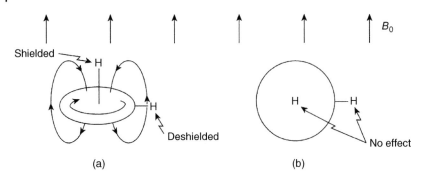

Figure 3.3 Shielding by an oblate ellipsoid, the model for aromatic rings. (a) When the plane of the oblate ellipsoid is perpendicular to the direction of the magnetic field. (b) When the plane of the oblate ellipsoid is parallel to the direction of the magnetic field.

proton is perpendicular to B_0, as in Figure 3.2b, the induced lines of force reinforce those of B_0, deshielding the proton. Because the group is isotropic, the two arrangements are equally probable. As the molecule tumbles in solution, the effects of the induced field cancel out. Other orientations cancel each other in a similar fashion. Thus, an isotropic substituent has no effect over and above what it provides to local currents from induction or resonance.

Most substituents, however, are not spherical. The flat shape of an aromatic ring, for example, resembles an oblate ellipsoid (flat and shaped like a dish), and the elongated shape of single or triple bonds resembles a prolate ellipsoid (shaped like a rod). For a proton situated at the edge of an oblate ellipsoid such as a benzene ring, there again are two extremes (Figure 3.3). When the flat portion is perpendicular to the static field (Figure 3.3a), a proton at the edge is deshielded, since the induced lines of force reinforce the B_0 field. For the same geometry (Figure 3.3a), a proton situated over the middle of the ellipsoid is shielded, as the induced lines of force oppose B_0. For this geometry, the induced field is large because aromatic electrons circulate easily above and below the ring. When the ring is parallel to B_0 (Figure 3.3b), however, induced currents would have to move from one ring face to the other. As a result, little current or field is induced from this geometry. The cancelation seen for the sphere as the molecule tumbles in solution does not occur for aromatic rings. A group that has appreciably different currents induced by B_0 from different orientations in space is said to have *diamagnetic anisotropy* (anisotropy—"not same in all directions"). Because an oblate ellipsoid has the larger effect for the geometry shown in Figure 3.3a, a proton at the edge of an aromatic ring is deshielded and one at or over the center is shielded. It is for this reason that benzene resonates at an unusually high frequency (low field, δ 7.27), compared with the frequency of alkenes such as ethene at δ 5.28.

Just as the local effect can result in either shielding (from electron donation) or deshielding (from electron withdrawal), the nonlocal effect also can have either sign, depending on whether the nonlocal field enhances or diminishes the static magnetic

Figure 3.4 Shielding geometry for a benzene ring.

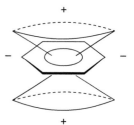

field B_0. Figure 3.4 illustrates how the diamagnetic effect of benzene is shielding (+) above and below the ring, but deshielding (−) around the edge. This effect was modeled quantitatively by McConnell as the influence of a magnetic dipole on the point in space at which a proton resides. He derived the formula of Eq. (3.1)

$$\sigma_A(r, \theta) = \frac{(\chi_L - \chi_T)\,(3\cos^2\theta - 1)}{3r^3} \tag{3.1}$$

for the shielding σ_A by an anisotropic group (represented by the dipole X–Y) of a hydrogen atom at an arbitrary point in space with polar coordinates (r, θ). In this equation, r is the distance from the midpoint of X–Y to that point, θ is the angle between the line connecting X and Y and the line from the midpoint of X–Y to the point (r, θ), and χ_L and χ_T are the diamagnetic susceptibilities of the group along its longitude and its transverse, as illustrated by Figure 3.3a,b, respectively. The effect changes sign at a null point at which the angle from the ring is 54°74′, the so-called magic angle at which the expression $(3\cos^2\theta - 1)$ goes to zero.

Although the protons of benzene reside in the deshielded portion of the cone, molecules have been constructed to explore the full range of the effect. The methylene protons of methano[10]annulene (**3.3**)

3.3

are constrained to positions above the 10π electron system of the aromatic system and consequently are shielded to a position (δ −0.5) at even lower frequency (higher field) than TMS. [18]Annulene (**3.4**) has one set of protons around the edge of the aromatic ring that resonates at a deshielded position of δ 9.3 and a second set located toward the center of the ring that resonates at a shielded position of δ −3.0.

The presence of $(4n + 2)$ π electrons is a requirement for the existence of a diamagnetic circulation of electrons, the so-called Hückel rule for aromaticity. Pople showed that an external magnetic field can induce an opposite, or paramagnetic, circulation in a $4n$ π electron (antiaromatic) system. Under such circumstances, the conclusions drawn from the configuration in Figure 3.3a and structure **3.4** are reversed, i.e. outer protons are shielded and inner protons are deshielded. The spectrum of [16]annulene is consistent with this interpretation (inner protons at δ 10.3, outer at 5.2). The most dramatic of such an example is the [12]annulene (**3.5**).

The bromine atom was included to prevent a conformational interconversion of the inner and outer protons. The indicated inner proton of **3.5** with $4n$ π electrons resonates at δ 16.4, compared with δ −3.0 for the inner protons of [18]annulene (**3.4**) with $(4n + 2)$ π electrons.

The two arrangements of a prolate ellipsoid, used as a model for a single or a triple bond, may be considered in a similar fashion (Figure 3.5). In this case, it is not always clear which arrangement produces a stronger induced current (or has higher diamagnetic susceptibility χ). The π electrons of acetylene (ethyne) provide one clear-cut example. When the axis of the molecule is parallel to the B_0 field (as in Figure 3.5a), the π electrons are particularly susceptible to circulation around the cylinder. The alternative arrangement in Figure 3.5b is ineffective for acetylene and therefore does not provide a canceling effect. The acetylenic proton is attached to the end of this array of electrons and hence is shielded. For this reason, the acetylene resonance (δ 2.88) falls

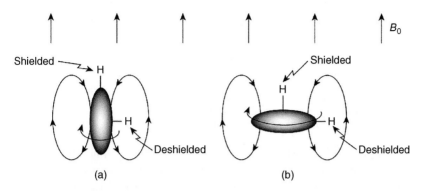

Figure 3.5 Shielding by a prolate ellipsoid, the model for a chemical bond. (a) When the triple bond is parallel to the direction of the magnetic field. (b) When the triple bond is perpendicular to the direction of the magnetic field.

between those of ethane (δ 0.86) and ethene (δ 5.28). The effects of hybridization thus are superseded by those of C≡C diamagnetic anisotropy. In terms of the McConnell equation (Eq. (3.1)), the longitudinal susceptibility is much greater than the transverse susceptibility, i.e. $\chi_L > \chi_T$. Thus, shielding is positive at the end of the bond ($\theta = 0°$), resulting in a shift to lower frequency (higher field).

Circulation of charge within a carbon–carbon single bond is less strong than that of a triple bond, and the larger effect occurs when the axis of the bond is perpendicular to B_0 (as in Figure 3.5b, i.e. $\chi_T > \chi_L$ for single bonds). Thus, a proton at the side of a single bond is more shielded than one along its end (Figure 3.6a). The axial and equatorial protons of a rigid cyclohexane ring exemplify these arrangements (Figure 3.6b). The two protons illustrated are equivalently positioned with respect to the 1,2 and 6,1 bonds, which thus produce no differential effect. The 1-axial proton, however, is in the shielding region of the farther removed 2,3 and 5,6 bonds (darkened), whereas the 1-equatorial proton is in their deshielding region. In general, axial protons are more shielded and resonate at a lower frequency (higher field) than do equatorial protons, typically by about 0.5 ppm.

The high-frequency position for methine (CH) compared with methylene (CH_2) protons and for methylene compared with methyl (CH_3) protons (as in $(CH_3)_2CHX$ to CH_3CH_2X to CH_3X) for a single X group may be attributed to the anisotropy of the additional C—C bonds (Figure 3.7), although changes in hybridization also may contribute.

The highly shielded position of cyclopropane resonances (δ 0.22 vs 1.43 in cyclohexane) may be attributed either to an aromatic-like ring current or to the anisotropy of the

Figure 3.6 (a) Shielding region around a carbon–carbon single bond. (b) Geometry of a cyclohexane ring.

Figure 3.7 Shielding properties of methyl, methylene, and methine groups.

Methyl deshielded by one bond

Methylene deshielded by two bonds

Methine deshielded by three bonds

C—C bond that is opposite to a CH$_2$ group in the three-membered ring. The effect is much larger than the indicated 1.2 ppm (1.4–0.2) because the sp^2 cyclopropane carbon orbital to hydrogen (compared with the sp^3 orbital in cyclohexane) deshields the proton. A cyclopropane ring also can shield more distant hydrogens. In spiro[2.5]octane (**3.6**),

3.6

the indicated equatorial proton resonates at 1.2 ppm lower frequency (higher field) than does the axial proton. Since H$_{ax}$ normally is about 0.5 ppm lower frequency than H$_{eq}$, the differential effect is 1.7 ppm. In **3.6**, H$_{eq}$ is perched over the shielding region of the cyclopropane ring, so it undergoes a very strong shift to a lower frequency (upfield).

Most common single bonds containing a heteroatom (C—O, C—N) have shielding properties that parallel those of the C—C bond. There appears to be a sign reversal, however, for the C—S bond for the value of ($\chi_L - \chi_T$). In all these heteroatomic cases, however, the geometry is more complex than that for the C—C bond. In some instances, the lone electron pairs can have a special effect. In N-methylpiperidine (**3.7**),

3.7

the axial lone pair shields the vicinal H$_{ax}$ by an n–σ* interaction without any effect on H$_{eq}$. As a result, $\Delta\delta_{ae}$ increases to about 1.0 ppm or more from the normal value of 0.5.

The anisotropy of double bonds is more difficult to assess, because they have three nonequivalent axes. Thus the McConnell equation, with only two axes, does not apply in its simple form. Protons situated over double bonds in general are predicted by the McConnell model to be more shielded than those in the plane (Figure 3.8), both for alkenes and for carbonyl groups. The proton numbered 9 that is pointing toward the double bond (syn) in structure **3.8**,

The Chemical Shift | 83

```
        Shielded                               Shielded
                                               R⁗
Deshielded  \C=C/  Deshielded      Deshielded    \C=O
            /    \                               /
        Shielded                          H
                                               Shielded
           (a)                                (b)
```

Figure 3.8 Diamagnetic anisotropic properties of (a) the C=C double bond and (b) the C=O double bond. These properties are altered by perturbations of the electronic structure by nearby atoms.

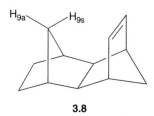

3.8

from the work of Marchand and Rose, is at the optimal position for shielding and is shifted to δ 0.48, compared with δ 1.2 for the 7 protons of norbornane. When the position of the proton vis-à-vis the double bond is altered, by moving it closer to, farther from, or off-center to the side of the double bond, the effect may be diminished or even reversed. Such effects may be caused not only by modulation of anisotropic shielding of the double bond within the cone but also by introduction of a new effect from van der Waals shielding (σ_W), to be discussed presently.

The anisotropic effects of aromatic rings and ethynes are stronger and maintain the models of Figures 3.3–3.5. The highly deshielded position of aldehydes (δ 9.8) is attributed to a combination of a strong inductive effect, with possible contributions from the diamagnetic anisotropy of the carbonyl group.

The nonspherical array of lone pairs of electrons may exhibit diamagnetic anisotropy, although, alternatively, the effect may be considered a perturbation of local currents. A proton that is hydrogen bonded to a lone pair invariably is deshielded. Thus, the hydroxyl proton in ethanol as a dilute solute in a nonhydrogen-bonding solvent such as CCl_4 resonates at δ 0.7, but in pure ethanol with extensive hydrogen bonding, it resonates at δ 5.3. Carboxylic protons (CO_2H) resonate at extremely high frequency (low field, δ 11–14) because every proton is hydrogen bonded within a dimer or higher aggregate. Lone-pair anisotropy also has been invoked to explain trends in heteroatom-substituted ethyl groups (CH_3CH_2X). The resonance positions of the CH_2 group attached to X are well explained by the polar effect (for X = F (δ 4.36), Cl (3.47), Br (3.37), I (3.16)), but the trend for the more distant methyl group is opposite (in the same order, δ 1.24, 1.33, 1.65, and 1.86). As the size of X increases, the lone pair moves closer to the methyl group and deshields it more strongly.

In summary, functional group effects on proton chemical shifts may be explained largely by the two general effects just described. (i) Electron withdrawal or donation by induction (including hybridization) or by conjugation alters the electron density and hence the local field around the resonating proton. Higher electron density shields the proton and moves its resonance position to lower frequency (downfield). (ii) Diamagnetic anisotropy of nonspherical substituents is largely responsible for the proton resonance positions of aromatics, acetylenes, cyclopropanes, cyclohexanes, and, possibly, hydrogen-bonded species.

Two other, less general diamagnetic effects also influence chemical shifts. As discussed, a substituent may deshield a neighboring proton, i.e. it may decrease σ^d, by polar electron withdrawal. If the substituent is distant by a sufficient number of bonds, the effect becomes negligible. When a substituent atom is held rigidly at a distance from the resonating nucleus that is less than the sum of the van der Waals radii, the atom repels electrons from the vicinity of the resonating atom. The net effect, therefore, is a decrease in σ^d. The nucleus is deshielded, and its resonance is shifted to higher frequency (downfield). The phenomenon arises from the mutual repulsion of induced dipoles (by van der Waals, London, or dispersion forces). The magnitude of the effect (σ_W) falls off very rapidly with increasing internuclear distance and depends critically on the size and polarizability of the nuclei. A proton is not deshielded by another proton until they are within about 2.5 Å of each other. A bromine atom can deshield a proton from a much greater distance, and a fluorine atom is intermediate.

The 0.2-ppm high-frequency shift of the *tert*-butyl protons in *ortho*-di-*tert*-butylbenzene with respect to the *tert*-butyl resonances in the meta and para isomers has been attributed to a van der Waals effect. A very dramatic example is in the partial cage compound **3.9**,

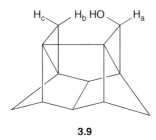

3.9

in which the chemical shifts of H_b and H_c are δ 3.55 and 0.88, respectively. By comparison, the methylene protons of cyclohexane resonate at about δ 1.4. The oxygen atom therefore deshields H_b by over 2 ppm. Interestingly, the electron density displaced from H_b is in part shifted to H_c, which is shielded by about 0.5 ppm.

The second additional diamagnetic effect involves polar bonds that can generate an electric field with an appreciable value at the position of a nearby resonating nucleus. This electric field distorts the electronic structure around the nucleus and causes, usually, deshielding by diminishing σ^d. Unlike the through-bond polar effect, the electric-field effect can be derived from a polar group that is many bonds removed from the resonating nucleus, given a favorable stereochemical arrangement. For a significant value of σ_E (Eq. (3.5)), the polar bond must be reasonably close to the nucleus, but need not be in van der Waals contact.

Many examples of electric-field shielding have come from ^{19}F spectroscopy, since the larger shifts and high electronegativity of fluorine nuclei magnify the effect. The ^{19}F resonance of 1-chloro-2-fluorobenzene (**3.10**)

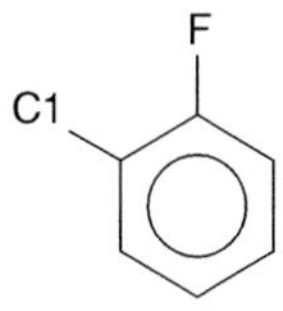

3.10

is more than 20 ppm to a higher frequency than a simple polar effect could explain. The interpretation that this large additional shift is due to the electric field of the C—Cl bond has been substantiated by calculations. The 18-ppm chemical shift between the axial and equatorial fluorines in perfluorocyclohexane (**3.11**)

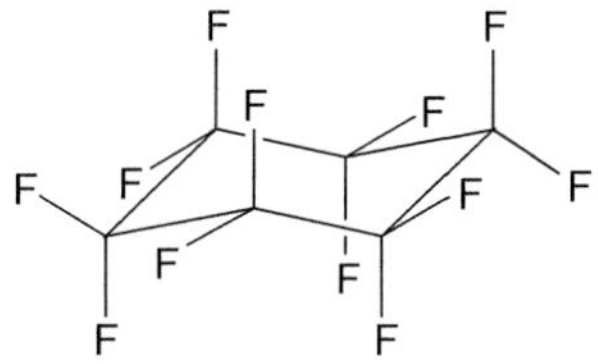

3.11

compares with a 0.5-ppm difference for the corresponding protons in cyclohexane. The effect of diamagnetic anisotropy (σ_A) is not sufficient to explain the large ^{19}F separation. The electric fields at axial and equatorial fluorine atoms in **3.11** are different from the electric field at the other polar bonds in the molecule. For both of these atoms, there may be a van der Waals contribution as well, since the nuclei are relatively close. An analysis of ^{19}F chemical shifts is not complete unless both effects are taken into consideration. The importance of these effects has not been fully explored in many systems or for many nuclei, so much work remains to be done.

3.2 Proton Chemical Shifts and Structure

Assignment of a structure on the basis of nuclear magnetic resonance (NMR) spectra requires knowledge of the relationship between chemical shifts and functional groups, as developed from the foregoing principles. Normally, both proton and carbon spectra are recorded and analyzed. This section considers the relationship between proton resonances and structure. Figure 3.9 summarizes the resonance ranges for common proton functionalities.

3.2.1 Saturated Aliphatics

3.2.1.1 Alkanes

Cyclopropane has the lowest-frequency position (δ 0.22) of any simple hydrocarbon because of a ring current or the anisotropy of the carbon–carbon bonds. Unsubstituted methane has essentially the same chemical shift (δ 0.23). Progressive addition of saturated carbon–carbon bonds to methane results in a shift to higher

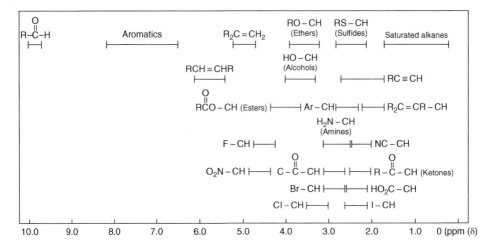

Figure 3.9 Proton chemical shift ranges for common structural units. The symbol CH represents methyl, methylene, or methine, and R represents a saturated alkyl group. The range for —CO$_2$H and other strongly hydrogen-bonded protons is off scale to the left. The indicated ranges are for common examples; actual ranges can be larger.

frequency (downfield), as in the series consisting of ethane (CH$_3$CH$_3$, δ 0.86), propane (CH$_3$CH$_2$CH$_3$, δ 1.33), and isobutane ((CH$_3$)$_3$CH, δ 1.56) (Figure 3.7). The resonance position of cyclobutane is at an unusually high frequency (δ 1.98) because of the lower s character of the carbon orbitals. Cyclic structures other than cyclopropane and cyclobutane have resonance positions similar to those of open-chain systems, for example, δ 1.43 for cyclohexane. In complex natural products such as steroids or alkaloids, a large number of structurally similar alkane protons leads to overlapping resonances in the region δ 0.8–2.0, the analysis of which requires the highest possible field.

3.2.1.2 Functionalized Alkanes

The presence of a functional group alters the resonance position of neighboring protons according to the polar effect of the group and its diamagnetic anisotropy. Ethane (δ 0.86) is a useful point of reference for methyl groups. Replacement of one methyl group in ethane with hydroxyl yields methanol (CH$_3$OH), with a resonance position of δ 3.38. The electron-withdrawing effect of the oxygen atom is the primary cause of the large shift to a higher frequency (lower field). Just as in unfunctionalized alkanes, methylene groups (CH$_3$CH$_2$OH, δ 3.56) and methine groups ((CH$_3$)$_2$CHOH, δ 3.85) are found at progressively higher frequencies. In general, methylene and methine protons resonate, respectively, about 0.3 and 0.7 ppm to higher frequency than do analogous methyl groups (CH$_3$X vs —CH$_2$X vs >CHX). There is considerable variation from one case to another, depending on the remainder of the structure, so that resonances for a given functionality can range over 1 ppm. Ether resonances are similar to those for alcohols (CH$_3$OCH$_3$, δ 3.24). Ester alkoxy groups, however, usually resonate at even higher frequency (CH$_3$O(CO)CH$_3$, δ 3.67), because the attached oxygen is more electron withdrawing as the result of ester resonance (**3.12**).

$$\underset{CH_3O}{\overset{O}{\underset{\|}{C}}}\underset{CH_3}{} \longleftrightarrow \underset{CH_3\overset{+}{O}}{\overset{O^-}{\underset{\|}{C}}}\underset{CH_3}{}$$

3.12

Since nitrogen is not so electron withdrawing as oxygen, amines resonate at somewhat lower frequency (higher field) than ethers: δ 2.42 for methylamine (CH_3NH_2 in aqueous solution). Introducing a positive charge through quaternization induces increased electron withdrawal and causes a shift to high frequency, as with $(CH_3)_4N^+$ (δ 3.33), compared with $(CH_3)_3N$ (δ 2.22). An intermediate charge, as is produced in amides through resonance, results in an intermediate shift, as with N,N-dimethylformamide (**3.13**, δ 2.88).

$$\underset{(CH_3)_2N}{\overset{O}{\underset{\|}{C}}}\underset{H}{} \longleftrightarrow \underset{(CH_3)_2\overset{+}{N}}{\overset{O^-}{\underset{\|}{C}}}\underset{H}{}$$

3.13

The lower electronegativity of sulfur means that sulfides are at lower frequency (higher field), δ 2.12 for dimethyl sulfide (CH_3SCH_3). Halogens move resonances to higher frequency according to the electronegativity of the atom: δ 2.15 for CH_3I, 2.69 for CH_3Br, 3.06 for CH_3Cl, and 4.27 for CH_3F. In all these cases, the shifts probably are affected by the anisotropy of the C—X bond, but this factor is hard to assess and is somewhat diminished by free rotation in open-chain systems. Other electron-withdrawing substituents also cause shifts to higher frequencies, as for example cyano in acetonitrile (CH_3CN, δ 2.00) and nitro in nitromethane (CH_3NO_2, δ 4.33). Electron-donating atoms such as silicon in TMS (δ 0.00) cause shifts to lower frequencies (higher field).

Methyl groups attached to unsaturated carbons are usually found in the region δ 1.7–2.5, as for the allylic protons in isobutylene (($CH_3)_2C=CH_2$), δ 1.70), the propargylic protons in methylacetylene (($CH_3C\equiv CH$), δ 1.80), and the benzylic protons in toluene ($C_6H_5CH_3$, δ 2.31). Methyl groups on carbon–oxygen double bonds are found in the region δ 2.0–2.7, as in acetone ($CH_3(CO)CH_3$, δ 2.07), acetic acid (CH_3CO_2H, δ 2.10), acetaldehyde (CH_3CHO, δ 2.20), and acetyl chloride ($CH_3(CO)Cl$, δ 2.67). These functionalities exhibit an appreciable range determined by further substitution.

3.2.2 Unsaturated Aliphatics

3.2.2.1 Alkynes

The anisotropy of the triple bond results in a relatively low-frequency (upfield) position for protons on sp-hybridized carbons. For acetylene (ethyne) itself, the chemical shift is δ 2.88, and the range is about δ 1.8–2.9.

3.2.2.2 Alkenes

The increased electronegativity of the sp² carbon and the modest anisotropy of the carbon–carbon double bond result in a high-frequency (low-field) position for protons on alkene carbons. The range is quite large (δ 4.5–7.0), as the exact resonance position depends on the nature of the substituents on the double bond. The value for ethene is δ 5.28. 1,1-Disubstituted hydrocarbon alkenes (vinylidenes), including exomethylene groups on rings (C=CH₂), resonate at a somewhat lower frequency, as in isobutylene ((CH₃)₂C=CH₂, δ 4.73). The CH₂ part of a vinyl group, —CH=CH₂, also is usually at a lower frequency than δ 5.0. 1,2-Disubstituted alkenes, as found, for example, in endocyclic ring double bonds (—CH=CH—) and in trisubstituted double bonds, generally resonate at even higher frequency than δ 5.0 (*trans*-2-butene (CH₃CH=CHCH₃), δ 5.46). Angle strain on the double bond moves the resonance position to a higher frequency, as with norbornene (**3.14**,

5.94   5.78   6.42
3.14   **3.15**   **3.16**

δ 5.94). Conjugation also usually moves the resonance position to a higher frequency, as in 1,3-cyclohexadiene (**3.15**, δ 5.78). The double bonds in 1,3-cyclopentadiene (**3.16**) are both strained and conjugated, so the chemical shift is at an even higher frequency, δ 6.42. The phenyl ring of styrene (C₆H₅CH=CH₂) withdraws electrons from the double bond by the polar effect, so the position of the nearer (α) CH proton is moved to a high frequency, δ 6.66. The more distant (β) CH₂ protons are nonequivalent and resonate at δ 5.15 and 5.63. The anisotropy of the aromatic ring is largely responsible for the difference between the β protons. The closer cis proton is shifted to a higher frequency. The nonaromatic portion of styrene, —CH=CH₂, is a *vinyl* group, and the term should be restricted to that structure. The term *alkenic*, rather than vinylic, should be used generally for protons on double bonds.

Carbonyl groups are strongly electron withdrawing by both induction and resonance. Thus, the β protons on double bonds that are conjugated with a carbonyl group resonate at very high frequencies (downfield), for example, δ 6.83 in the α,β-unsaturated ester *trans*-CH₃CH₂O₂C—CH=CH—CO₂CH₂CH₃ (note that both protons are beta in this symmetrical structure). Compounds **3.17** and **3.18**

6.37, 4.65   ⇌   5.93, 6.88   ⇌
3.17   **3.18**

illustrate the effects of conjugation on alkene chemical shifts. Whereas the alkenic protons of cyclohexene resonate at a normal δ 5.59, the oxygen atom in the unsaturated ether **3.17** donates electrons to the β position by resonance and moves the β proton to a lower frequency (δ 4.65). The oxygen atom withdraws electrons inductively from the α position, whose proton resonance moves to a higher frequency (δ 6.37). In contrast, the carbonyl group in the unsaturated ketone **3.18** withdraws electrons from the β position by resonance and thus moves the β proton to higher frequency (δ 6.88). In this case, the polar effect of the carbonyl group causes a small shift to higher frequency for the α proton (δ 5.93).

These effects were quantified in the empirical approach of Tobey and of Pascual, Meier, and Simon, who used the formula of Eq. (3.2)

$$\delta = 5.28 + Z_{gem} + Z_{cis} + Z_{trans} \qquad (3.2)$$

to calculate the chemical shift of a proton on a double bond. Substituent constants Z_i for groups geminal, cis, or trans to the proton under consideration are added to the chemical shift of ethene. Empirical calculations of this type have been subsumed into computer programs, which are discussed in Section 3.2.5.

3.2.2.3 Aldehydes
The aldehydic proton is shifted to very high frequency (low field) by induction and possibly diamagnetic anisotropy of the carbonyl group. For acetaldehyde (CH_3CHO), the value is δ 9.80, and the range is relatively small, generally δ 10 ± 0.3.

3.2.3 Aromatics

Diamagnetic anisotropy of the benzene ring augments the already deshielding influence of the sp^2 carbon atoms to yield a very high frequency (low field) position for benzene, δ 7.27. Polar and resonance effects of substituents are similar to those in alkenes. For toluene ($C_6H_5CH_3$), the electronic effect of the methyl group is small, and all five aromatic protons resonate at about δ 7.2. A narrow range is typical for saturated hydrocarbon substituents (arenes). Conjugating substituents, however, result in a large spread in the aromatic resonances and in spectral multiplicity from spin–spin splitting. For nitrobenzene (Figure 3.10), the polar effect of the nitro group (**3.19**)

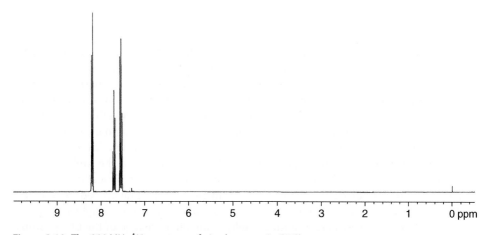

Figure 3.10 The 300 MHz ^1H spectrum of nitrobenzene in $CDCl_3$.

moves all resonances to a higher frequency with respect to benzene or toluene, but the ortho and para protons are further shifted through electron withdrawal by conjugation. In contrast, the methoxy group in anisole (**3.20**) donates electrons by conjugation, so the ortho and para positions are at a lower frequency than that for benzene. The α protons in heterocycles generally are shifted to high frequency, as in pyridine (**3.21**) and pyrrole (**3.22**), largely because of the polar effect of the heteroatom.

Aromatic proton resonances also may be treated empirically, provided that no two substituents are ortho to each other (producing steric effects). The shift of a particular proton is obtained by adding substituent parameters to the chemical shift of benzene, as shown in the relationship of Eq. (3.3).

$$\delta = 7.27 + \Sigma S_i \tag{3.3}$$

Jackman and Sternhell compiled substituent parameters from numerous sources to provide this useful approach to estimate aromatic proton chemical shifts (see Jackman and Sternhell, 1969, Table 3.6.1, with an opposite sign convention from that of Eq. (3.3)).

3.2.4 Protons on Oxygen and Nitrogen

The NMR properties of protons attached to highly electronegative atoms such as oxygen or nitrogen are strongly influenced by the acidity, basicity, and hydrogen-bonding properties of the medium. For hydroxyl protons, minute amounts of acidic or basic impurities can bring about rapid exchange, as illustrated in Figure 1.30. The chemical shifts of such protons then are averaged with those of other exchangeable protons, either in the same molecule or in other molecules such as the solvent. Only a single resonance is observed for all the exchangeable protons at a weighted-average position. In addition, no coupling is observed of the exchanging proton with other protons in the molecule. The resonance may vary from being quite sharp to having a characteristically broadened shape, depending on the exchange rate. A convenient experimental procedure to identify hydroxyl resonances in organic solvents such as $CDCl_3$ is to add a couple of drops of D_2O to the NMR tube. Shaking the tube briefly results in exchange of the OH protons with deuterium, which is in molar excess. The aqueous layer separates out, usually to the top of the NMR tube in halogenated solvents, and is located above the receiver coil. Consequently, hydroxyl resonances in the original spectrum may be identified by their absence in the two-phase case (they are converted largely to OD). In highly purified, basic solvents such as dimethyl sulfoxide exchange is slow and coupling between OH and adjacent protons can be observed.

At infinite dilution in CCl_4 (no hydrogen bonding), the OH resonance of alcohols may be found at about δ 0.5. Under more normal conditions of 5–20% solutions, hydrogen bonding results in resonances in the range δ 2–4. More acidic phenols (ArOH, in which Ar represents a general aromatic group) have resonances at higher frequency (lower field), δ 4–8. If the phenolic hydroxyl can hydrogen bond fully with an ortho group, the position moves to δ 10 or higher. Most carboxylic acids (RCO_2H) exist as hydrogen-bonded dimers or oligomers, even in dilute solution. Because essentially every OH proton is hydrogen-bonded, the acid protons resonate in the very high frequency range of δ 11–14 (δ 11.37 for acetic acid, CH_3CO_2H). Other highly hydrogen-bonded protons also may be found in this range, such as sulfonic acids (RSO_3H) or the OH proton of enolic acetylacetone. Because of the variable position and variable appearance of hydroxy resonances, including that of water, one must be very careful of spectral assignments for the group.

Protons on nitrogen have similar properties, but the slightly lower electronegativity of nitrogen, compared with oxygen, results in lower frequency (higher field) shifts than those of analogous OH protons: δ 0.5–3.5 for aliphatic amines, δ 3–5 for aromatic amines (anilines), δ 4–8 for amides, pyrroles, and indoles, and δ 6–8.5 for ammonium salts. The most common nuclide of nitrogen is ^{14}N, which is quadrupolar and possesses unity spin (Sections 1.7 and 5.1). The three resulting spin states conceivably could split the resonance of attached protons into a 1 : 1 : 1 triplet, but such a pattern is seen only in highly symmetrical cases, such as NH_4^+ or NMe_4^+. Otherwise, rapid relaxation of quadrupolar nuclei averages the spin states. The resonance of a proton on nitrogen thus can vary from a triplet to a sharp singlet, depending on the relaxation rate, but the most common result is a broadened resonance representing incomplete averaging. In some cases, the broadening can render NH resonances almost invisible. In addition, amino protons can exchange rapidly with solvent or other exchangeable protons to achieve an averaged position.

3.2.5 Programs for Empirical Calculations

Huge quantities of data are available on substituted alkane, alkene, and aromatic structures, from which programs have been developed for the empirical calculation of proton chemical shifts. Such calculations, however, are never simple. Corrections must be applied in order to avoid nonadditivity caused primarily by steric effects. Thus, three groups on a single carbon atom, two large groups cis to each other on a double bond, or two groups ortho to each other on an aromatic ring can cause deviations from the standard parameters. If sufficient model compounds are available, corrections can be applied. Empirical calculations are possible for any structural entity so that the eclipsing strain in cyclobutanes, the variety of steric interactions in cyclopentanones, or the variations in angle strain in norbornanes may be taken into account.

Commercial software for carrying out these calculations, based on hundreds of thousands of chemical shifts in various databases, is widely available (see Refs. [1, 2]). The procedure is begun by drawing the structure of the compound under study. The program then searches the database for molecules with protons whose structural environment resembles that of the compound under study. From the available data, the program calculates and displays the expected proton spectrum. Such information is extremely valuable, because the amount of empirical data available from the program vastly exceeds the amount resident in the minds of most experimentalists or even in all published compilations.

The approach, however, is subject to several limitations. (i) The specific skeleton or functional groups may not exist in the database. (ii) The database may not include sufficient information to assess steric effects that can lead to nonadditivity within a particular series. (iii) Solvent effects (Section 3.3) are not fully taken into consideration. (iv) Coupling constants are calculated from simple relationships, such as the Karplus equation (Section 4.6). Because the calculations are not quantitatively reliable, couplings generally are represented more poorly than chemical shifts by these commercial programs. Usually, the program provides a list of the compounds used to calculate chemical shifts so that the experimentalist can judge their relevancy. Sometimes, the compound under study in fact proves to be in the database so that the real spectrum is reproduced. If not, the experimentalist always should review the structures of the compounds used for the calculations and decide whether they are sufficiently similar to trust the calculations. Despite considerable progress in this field, these calculations still do not accurately reproduce most proton spectra, say producing resonances that are within <0.1 ppm of the observed positions. Semiempirical and ab initio calculations of spectra do not suffer from most of these drawbacks but have yet to be applied comprehensively to all structural possibilities. Density functional theory (DFT) in particular has been used with success by Bagno (Ref. [3]).

3.3 Medium and Isotope Effects

3.3.1 Medium Effects

The observed shielding of a particular nucleus consists of intramolecular components σ_{intra} (discussed in regard to protons in Section 3.2) and intermolecular components σ_{inter}. The total shielding then is given by Eq. (3.4).

$$\sigma = \sigma_{intra} + \sigma_{inter} \tag{3.4}$$

Buckingham, Schaefer, and Schneider enumerated five sources of intermolecular shielding, given in Eq. (3.5).

$$\sigma_{inter} = \sigma_B + \sigma_W + \sigma_E + \sigma_A + \sigma_S \tag{3.5}$$

Several of these terms correspond to already-discussed intramolecular contributors, including the van der Waals effect (σ_W), the electric field effect (σ_E), and the effect of diamagnetic anisotropy σ_A. We consider each intermolecular contribution in turn and then give several illustrations of these effects.

The solvent has a bulk diamagnetic susceptibility that is dependent on the shape of the sample container. The solvent in a spherical container shields the solute to an extent that is slightly different from the shielding afforded by solvent in a cylindrical container. The shielding is given by Eq. (3.6),

$$\sigma_B = \left(\frac{4}{3}\pi - \alpha\right)\chi_V \tag{3.6}$$

in which α is a geometric parameter and χ_V is the volume susceptibility of the solvent. For a sphere, $\alpha = 4\pi/3$, so there is no effect. The effect does not disappear for a cylinder, $\alpha = 2\pi/3$, for which σ_B is $2\pi\chi_V/3$. Normally, the solute and the standard (TMS) are present in the same solution (the case of an internal standard). Under these circumstances, they experience parallel bulk effects, and no correction for σ_B on the relative shift is necessary. Since internal standards are common, the effect of bulk susceptibility is largely ignored. A correction would be necessary only if chemical shifts had to be compared in examining data obtained without an internal standard from containers of different shapes, e.g. the normal cylinder and a spherical microtube, or from solvents with different volume susceptibilities.

Close approach of the solute and the solvent can distort the shape of the electron cloud around a proton and deshield it, even when both components are nonpolar. Such a phenomenon (σ_W in Eq. (3.5)) is analogous to the van der Waals effect on the chemical shift. The magnitude is rarely more than 0.3 ppm. If chemical shifts are measured from the resonance of an internal standard, the contribution from σ_W should affect the solute and the standard similarly. Chemical shifts so measured should be largely independent of σ_W.

A polar solute, or even a nonpolar molecule with polar groups, induces an electric field in the surrounding dielectric medium. This reaction field, proportional to $(\varepsilon - 1)/(\varepsilon + 1)$ (ε is the dielectric constant of the medium), can influence the shielding of protons elsewhere in the molecule. Generally, the effect (σ_E in Eq. (3.5)) is largest for protons closest to the polar group. The sign can be either positive or negative because of an angular dependence, but more often it is negative (indicating deshielding). This effect is not compensated for by the use of an internal standard. Even within the solute molecule, the effect can be quite variable for different protons. For polar molecules in solvents of high dielectric constant, σ_E can range up to 1 ppm. The effect may be minimized by the use of solvents with small dielectric constants.

An anisotropic solvent will not orient itself completely randomly with respect to the solute. Thus, even a nonpolar solute such as methane may be exposed preferentially to the shielding face of benzene or to the deshielding side of acetonitrile. In general, aromatic (dishlike) solvents induce shifts (σ_A in Eq. (3.5)) to lower frequency (upfield) and rodlike solvents (acetylene, nitriles, and CS_2) induce shifts to higher frequency. In the absence of any special solute–solvent interaction σ_S (charge transfer, dipole–dipole, or hydrogen bond), the internal standard and the solute exhibit similar anisotropic shifts so that σ_A is compensated for in the δ value. As often as not, however, there is a special interaction between the solvent and a polar group in the solute molecule. As a result, the anisotropic solvent has a different effect on different protons, and solute resonances can undergo real shifts up to 0.5 ppm with respect to the internal standard. In some cases, only certain protons near a functional group are affected. The solvent effect then can be used to bring about differential shifts within the spectrum of a molecule. This technique is useful in spectral analysis to alter, for example, a case of accidental overlap. Because dishlike and rodlike solvents cause shifts in opposite directions, the investigator has some control over the movement of the resonances.

The chemical shift alterations caused by aromatics have been termed *Aromatic Solvent-Induced Shifts*, with the acronym ASIS. The chemical shift in the aromatic solvent is compared with the resonance position in $CDCl_3$, as in Eq. (3.7).

$$\Delta_{C_6H_6}^{CDCl_3} = \delta_{CDCl_3} - \delta_{C_6H_6} \qquad (3.7)$$

Because the shift is usually to lower frequency (upfield), Δ normally is positive. Although anisotropic shifts frequently are strongest close to a polar group, they may be differentiated from electric-field effects by the dependence of the latter on the dielectric constant of the solvent as opposed to the shape of the solvent.

Specific interactions between solvent and solute, such as hydrogen bonding, can cause quite large effects (σ_S). It is not known whether the ASIS is caused by a time-averaged cluster of solvent molecules about a polar functional group or by a 1 : 1 solute–solvent charge transfer complex. In the latter case, the ASIS is more legitimately classified under σ_S than under σ_A.

In an early study of solvent shifts, Buckingham, Schaefer, and Schneider examined the solute methane. For this molecule, σ_E and σ_S as are zero and σ_S may be calculated or allowed to cancel by use of an internal standard. Thus, only σ_W and σ_A should affect the solute chemical shift. The authors plotted the difference between the chemical shift of methane in a given solvent and that in the gas phase vs the heat of vaporization of the solvent at the boiling point. The latter quantity was taken as a measure of the van der Waals interaction. More than a dozen solvents, including neopentane, cyclopentane, hexane, cyclohexane, the 2-butenes, ethyl ether, acetone, $SiCl_4$, and $SnCl_4$, fell on a straight line with a small negative slope. The shifts to higher frequency (downfield) from gaseous methane ranged from 0.13 for neopentane to 0.32 ppm for $SnCl_4$, largely as a function of atomic polarizability. The linear relationship with ΔH_v indicates that these shifts are due solely to σ_W. Well above this line are the dish-shaped aromatic molecules: benzene, toluene, and chlorobenzene, but also nitromethane and nitroethane, whose nitro group serves as an anisotropic oblate ellipsoid. Below the line are the rodlike molecules: acetonitrile, methylacetylene, dimethylacetylene, butadiyne, and carbon disulfide. These deviations from the Δv–vs–ΔH_v line are due to true anisotropic shifts (σ_A), since the nonpolar, isotropic methane molecule has no direct (σ_S) interactions with the solvent. The largest positive displacement from the line was nitrobenzene (0.72 ppm), and the largest negative displacement was N≡C—C≡C—N≡C (0.53 ppm).

Molecule **3.23**

3.23

provides an interesting example of the electric-field effect. Table 3.1 shows the resonance positions of the methyl groups on C-8, C-10, and C-13 in cyclohexane ($\varepsilon = 2.02$) and in methylene chloride ($\varepsilon = 9.1$). These solvents were chosen for a low σ_A effect; σ_W and

Table 3.1 Methyl chemical shifts (δ) as a function of solvent (Laszlo).

Solvent	ε	v_{10}	v_8	v_{12}
Cyclohexene	2.02	0.855	1.01	1.12
CH_2Cl_2	9.1	0.852	1.07	1.20

σ_B are discounted by the use of an internal standard. The methyl groups close to the ether linkage (C-8, C-13) are shifted 0.067 ppm to higher frequency by an electric-field effect. There is little or no shift of the C-10 methyl resonance, since the reaction field diminishes rapidly with distance.

A large ASIS appears on the relative chemical shifts of the nonequivalent methyl groups in N,N-dimethylformamide (**3.24**).

3.24

The distance between the two methyl peaks increases by up to 1.7 ppm upon replacement of $CHCl_3$ by benzene. Interestingly, the lower frequency peak is responsible for almost all of the shift. This observation was explained in terms of a short-lived 1 : 1 solvent–solute complex in which one of the methyl groups is situated over the benzene ring and the other is directed away from the ring. The model of a 1 : 1 complex between cyclohexanone and benzene has been used to justify the shift of the 2,6-axial protons (located above the benzene ring in the complex) to a lower frequency and the negligible shift of the 2,6-equatorial protons (near the region of no effect at $\theta = 55°74'$). These shifts also have been explained in terms of a time-averaged solvent cluster, which produces the same effect without recourse to a short-lived 1 : 1 complex.

3.3.2 Isotope Effects

Isotopic changes within a molecule can alter the chemical shifts of neighboring nuclei. The methyl group in toluene is 0.015 ± 0.002 ppm higher frequency than the corresponding protons in toluene-α-d_1. The protons in cyclohexane are 0.057 ppm higher frequency than the single proton in cyclohexane-d_{11}. The effect is larger for other nuclei, such as ^{19}F, but falls off very rapidly with distance. The isotope shift is caused by differences in zero-point vibrational energy differences between the H and D systems. Differential isotope effects have been exploited in a study of the ring reversal of 1,4-dioxane (Eq. (3.8)).

(3.8)

In undeuterated dioxane, H_{eq} and H_{ax} coincidentally have the same chemical shift (at the field studied), so they could not be differentiated at low temperatures (Sections 1.8 and 5.2). In 1,4-dioxane-d_7 (an impurity in commercial 1,4-dioxane-d_8), both H_{ax} and H_{eq} exhibit isotope shifts to lower frequency, but H_{ax} is shifted somewhat more. As a result, the axial and equatorial protons give separate resonances at low temperatures, in contrast to the undeuterated material. Because of a chlorine isotope effect, chloroform is a poor substance for an internal lock or a resolution standard at fields above about 9.4 T. At high resolution, the chloroform proton resonance shows up as several closely spaced peaks due to $CH(^{35}Cl)(^{37}Cl)_2$, $CH(^{35}Cl)_2(^{37}Cl)$, $CH(^{35}Cl)_3$, and $CH(^{37}Cl)_3$.

3.4 Factors That Influence Carbon Shifts

Carbon is the defining element in organic compounds, but its major nuclide (^{12}C) has a spin of zero. The advent of pulsed Fourier transform methods in the late 1960s made examination of the low-abundance nuclide ^{13}C (1.11%) a practical spectroscopic technique. The low probability of having two adjacent ^{13}C nuclei in a single molecule ($(0.0111)^2 = 0.0001$, or 0.01%) removes complications from carbon–carbon couplings. When carbon–hydrogen couplings are removed by decoupling techniques (Section 5.3), the spectrum is essentially free of spin–spin couplings, and one singlet arises for each distinct type of carbon. Spectral analysis therefore is simpler for ^{13}C than for ^1H spectra. Integration, however, is less reliable because carbons have a much larger range of relaxation times than do protons and because the decoupling process perturbs intensities (Chapter 5). Interpretation of ^{13}C chemical shifts nonetheless is straightforward and often more useful than that of proton shifts.

Diamagnetic shielding (σ^d), which is responsible for proton chemical shifts, is caused by circulation of the electron cloud about the nucleus, as depicted in Figure 3.1b. Hindrance to free electron circulation creates an additional contribution called *paramagnetic shielding* (σ^p) (the overall $\sigma = \sigma^d + \sigma^p$). Although s electrons circulate freely, 2p electrons can have angular momentum that hinders free circulation. Because σ^p serves to reduce σ^d, the two terms have opposite signs. Higher diamagnetic shielding results in a shift to lower frequency (higher field), whereas higher paramagnetic shielding results in a shift to higher frequency (lower field). Protons are surrounded solely by s electrons (lacking angular momentum) and consequently exhibit only diamagnetic shielding. Carbon nuclei (and almost all other nuclides as well) additionally are surrounded by p electrons and experience both forms of shielding. The term *paramagnetic* is appropriate because the effect is opposite in sign to diamagnetic contributions. In this context, the word should not be confused with its common usage to describe molecules with unpaired electrons. Some authors have referred to shielding in closed shell molecules for nuclides with p electrons as the *second-order paramagnetic effect*, yielding priority to phenomena associated with unpaired electrons, which become the *first-order paramagnetic effect*.

The paramagnetic component can be quite large. Whereas the chemical shift range for protons is only a few parts per million (Figure 3.9), paramagnetic shifts can extend over a range of hundreds or even thousands of parts per million for other nuclei. Qualitatively, angular momentum can arise from excited electronic states and from π bonding.

The effects are larger when electron density about the nucleus increases. These three considerations were gathered by Ramsey, Karplus, and Pople into the simple empirical relationship of Eq. (3.9).

$$\sigma^P \propto -\frac{1}{\Delta E}\langle r^{-3}\rangle \sum Q_{ij} \quad (3.9)$$

The quantity ΔE is the average energy of excitation of certain electronic transitions such as the $n \to \pi^*$ transition for many ^{13}C and ^{15}N nuclei. The radial term $<r^{-3}>$ is the average distance r from the nucleus of the 2p electrons (for second row elements like carbon). This term serves as a measure of electron density. Finally, ΣQ_{ij} is a measure of π bonding to carbon. The negative sign in Eq. (3.9) indicates that paramagnetic shielding is in the opposite direction from σ^d.

Structural changes can affect all three components of the equation. The quantity ΔE in Eq. (3.9) represents the weighted-average energy difference between the ground and certain excited states. Because of symmetry considerations, the $\pi \to \pi^*$ transition often is excluded. Low-lying excited states (with small ΔE) make the largest contribution, since ΔE appears in the denominator. Saturated molecules, such as alkanes, typically have no low-lying excited states (and hence possess a large ΔE) so that σ^P is small and alkane carbon resonances are found at very low frequency (high field). Remember that paramagnetic shielding causes shifts to high frequency, whereas diamagnetic shielding causes shifts to a low frequency. Similarly, the nitrogen atoms in aliphatic amines and the oxygen atoms in aliphatic ethers or alcohols have no low-lying excited states, so their ^{15}N and ^{17}O resonances also are found at low frequency. Carbonyl carbons, C=O, have a low-lying excited state involving the movement of electrons from the oxygen lone pair to the antibonding π orbital that generates a paramagnetic current. This $n \to \pi^*$ transition causes the large shift to high frequency that characterizes carbonyl groups, up to 220 ppm from the zero of TMS. Even larger shifts to high frequency, to δ 335, have been observed for carbocations, R_3C^+.

The radial term in Eq. (3.9) is responsible for effects related to electron density that parallel polar effects on proton chemical shifts. Paramagnetic shielding is larger when the p electrons are closer to the nucleus. Thus, substituents that donate or withdraw electrons influence the paramagnetic shift. Electron donation increases repulsion between electrons, which can be relieved by an increase in r. The paramagnetic shielding then decreases, causing a shift to lower frequency (upfield). Similarly, electron withdrawal permits electrons to move closer to the nucleus, increasing the paramagnetic shielding and causing a shift to higher frequency. Hence, placement of a series of electron-withdrawing atoms on carbon results in progressively higher frequency shifts, as in the series CH_3Cl (δ 25), CH_2Cl_2 (δ 54), $CHCl_3$, (δ 78), and CCl_4 (δ 97). The situation is qualitatively similar to that for protons, but the numbers are much larger because the shift is from the paramagnetic term. Substituent effects in both cases generally follow the electronegativity of groups attached to carbon.

Electronegativity is a measure of the ability of a nucleus to attract electrons. A highly electronegative element such as oxygen attracts p electrons more than does carbon and reduces the value of r. Thus, ^{17}O chemical shifts are correspondingly larger than ^{13}C shifts. A plot of the ^{17}O shifts of aliphatic ethers vs the ^{13}C shifts of the analogous alkanes is linear with a slope of about three. The linearity shows that oxygen and carbon chemical shifts are sensitive to the same structural factors, and the slope indicates that oxygen is more sensitive to these factors, because its 2p electrons are closer to the nucleus.

The third factor in Eq. (3.9), ΣQ_{ij}, is related to charge densities and bond orders and can be considered a measure of multiple bonding. The greater the degree of multiple bonding, the greater is the shift to high frequency (low field). This term provides a rationale for the series ethane (δ 6), ethene (δ 123), and the central sp-hybridized carbon of allene ($CH_2=C=CH_2$, δ 214). Arene shifts are similar to those of alkenes (benzene, δ 129), in contrast to the situation with the hydrogen nucleus. The effects of diamagnetic anisotropy on carbon chemical shifts are similar in magnitude to those on protons but are small in relation to the range of carbon shifts induced by paramagnetic phenomena. The chemical shifts of alkynes do not follow this pattern, but are at an intermediate position (δ 72 for acetylene), because their linear structure has zero angular momentum about the C≡C axis.

Interpretation of the chemical shifts of most elements other than hydrogen is accomplished by analyzing the three factors in Eq. (3.9): accessibility of certain excited states, distance of the p electrons, and multiple bonding. For carbon, the shifts of alkanes, alkenes, arenes, and carbonyl groups and the effects of electron-donating or electron-withdrawing groups may be interpreted in this fashion. There are exceptions in addition to acetylene, the most prominent being the effect of heavy atoms. The series CH_3Br (δ 10), CH_2Br_2 (δ 22), $CHBr_3$ (δ 12), and CBr_4 (δ −29) defies any explanation based on electronegativity, unlike the analogous series given previously for chlorine. The same series with iodine is monotonic to lower frequency (δ −290 for CI_4), that is, opposite to the chlorine series. This so-called *heavy atom effect* has been attributed to a new source of angular momentum from *spin-orbit coupling*. These anomalous shifts to a low frequency (high field) can be expected when nuclei other than hydrogen have a heavy atom substituent.

3.5 Carbon Chemical Shifts and Structure

Figure 3.11 illustrates general ranges for ^{13}C chemical shifts.

3.5.1 Saturated Aliphatics

3.5.1.1 Acyclic Alkanes

The absence of low-lying excited states and of π bonding minimizes paramagnetic shielding and places alkane chemical shifts at very low frequencies (high field). Methane itself resonates at δ −2.5. The series ethane (CH_3CH_3), propane ($CH_3CH_2CH_3$), and isobutane [$(CH_3)_3CH$] follows a steady trend to higher frequency (δ 5.7, 16.1, 25.2), similar to the trend for proton shifts in the series methyl, methylene, methine series. Replacement of H by CH_3 adds about 9 ppm to the chemical shift of the attached carbon. The effect is similar for replacement by saturated CH_2, CH, or C. Because the added methyl group is attached directly to the resonating carbon, the shift has been termed the α *effect* (**3.25**).

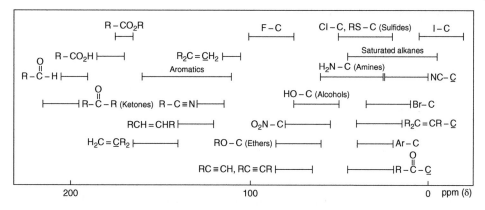

Figure 3.11 Carbon chemical-shift ranges for common structural units. The symbol C represents methyl, methylene, methine, or quaternary carbon; R represents a saturated alkyl group. The indicated ranges are for common examples; actual ranges can be larger.

The effect is not restricted to replacement of H by carbon. Any group X that replaces hydrogen on a resonating carbon atom causes a relatively constant shift that depends primarily on the electronegativity of X.

Replacement of hydrogen by CH_3 (or by CH_2, CH, or C) at a β position (**3.26**) also causes a constant shift of about +9 ppm. Thus, the central carbon in pentane ($CH_3CH_2CH_2CH_2CH_3$) is shifted by the α effects of the two methylene groups and by the *β effects* of the two methyl groups to δ 34.7. Replacement of a γ hydrogen (**3.27**) by CH_3 (or by CH_2, CH, or C) causes a shift of about −2.5 ppm (to lower frequency, or upfield). Unlike the α and β effects, this *γ effect* has an important stereochemical component that produces the negative sign. Because of α, β, and γ effects, the alkane chemical shift range is relatively large. Methyl resonances in alkanes are typically found at δ 5–15, depending on the number of β substituents; methylene resonances are at δ 15–30, and methine resonances are at δ 25–45.

$$\overset{\alpha}{CH_3}\overset{\beta}{CH_2}\overset{\gamma}{CH_2}\overset{\delta}{CH_2}CH_3 \qquad \delta = -2.5 + 9.1 + 9.4 - 2.5 + 0.3 = 13.8 \qquad (\text{obs. } 13.9)$$

$$\overset{\alpha}{CH_3}\overset{}{\mathbf{CH_2}}\overset{\alpha}{CH_2}\overset{\beta}{CH_2}\overset{\gamma}{CH_3} \qquad \delta = -2.5 + (9.1 \times 2) + 9.4 - 2.5 = 22.6 \qquad (\text{obs. } 22.8)$$

$$\overset{\beta}{CH_3}\overset{\alpha}{CH_2}\overset{}{\mathbf{CH_2}}\overset{\alpha}{CH_2}\overset{\beta}{CH_3} \qquad \delta = -2.5 + (9.1 \times 2) + (9.4 \times 2) = 34.5 \qquad (\text{obs. } 34.7)$$

Figure 3.12 Calculation of the ^{13}C chemical shifts of pentane.

Carbon-13 chemical shifts lend themselves conveniently to empirical analysis, because these shifts are easily measured and tend to have well-defined substituent effects. For saturated, acyclic hydrocarbons, Grant developed the formula of Eq. (3.10)

$$\delta = -2.5 + \Sigma A_i n_i \qquad (3.10)$$

as an empirical measure of chemical shifts. For any resonating carbon, a substituent parameter A_i for each other carbon atom in the molecule, up to a distance of five bonds, is added to the chemical shift of methane (δ −2.5). There are different substituent parameters for carbons (whether CH_3, CH_2, CH, or C) that are α (9.1), β (9.4), γ (−2.5), δ (0.3), or ε (0.1) to the resonating carbon. We already have alluded to the first three figures. If more than one α carbon is present, the substituent parameter is multiplied by the appropriate number n_i, and similar factors are applied for multiple substitution at other positions. Figure 3.12 illustrates the calculation for each carbon in pentane. The methyl chemical shift is calculated by adding contributions from single α, β, γ, and δ carbons to the shift (−2.5) of methane. The shift of carbon 2 is calculated by adding contributions for two α carbons, one β carbon, and one γ carbon. Usually, the observed shifts are calculated to within 0.3 ppm, providing a reliable means for spectral assignment.

There are complications, however. Grant found that corrections must be applied if there is branching, because Eq. (3.10) applies rigorously only to straight chains. The resonance position of a methyl group is corrected for the presence of an adjacent tertiary (CH) carbon by adding −1.1 and for an adjacent quaternary carbon by adding −3.4. Methylene carbons have corrections of −2.5 and −7.2, respectively, for adjacent tertiary and quaternary carbons. Methine carbons have respective corrections of −3.7, −9.5, and −1.5 for adjacent secondary, tertiary, and quaternary carbons, respectively. Finally, quaternary carbons have corrections of −1.5 and −8.4 for adjacent primary and secondary carbons. Corrections for adjacent tertiary and quaternary carbons undoubtedly are significant, but are not known accurately. For example, the methyl group in isobutane (first calculation in Figure 3.13) is adjacent to a tertiary carbon. The methyl chemical shift is calculated by adding the contributions for one α carbon, two β carbons, and the correction of −1.1, since the methyl group is adjacent to a tertiary (3°) center. In the second calculation in Figure 3.13 (for neopentane), the methyl group is adjacent to a quaternary (4°) center.

It is noteworthy that the γ effect of a carbon substituent is negative (−2.5). A γ carbon can be either gauche or anti (Figure 3.14) to the resonating carbon, and the proportion of conformers can vary from molecule to molecule. The γ effect reported by Grant of −2.5 is a weighted average for open-chain conformers and does not serve accurately for

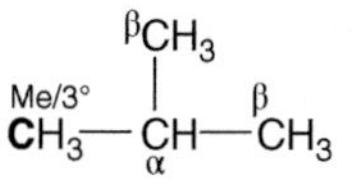

$$\underset{\text{Me/3°}}{CH_3}-\underset{\alpha}{CH}-\overset{\beta CH_3}{\underset{|}{CH}}-CH_3 \qquad \delta = -2.5 + 9.1 + (9.4 \times 2) - 1.1 = 24.3 \quad \text{(obs. 24.3)}$$

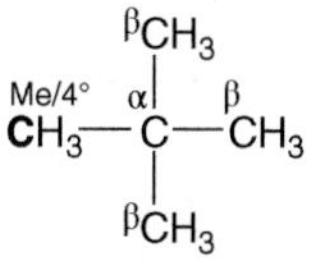

$$\underset{\text{Me/4°}}{CH_3}-\overset{\overset{\beta CH_3}{|}}{\underset{\underset{\beta CH_3}{|}}{C}}-\overset{\beta}{CH_3} \qquad \delta = -2.5 + 9.1 + (9.4 \times 3) - 3.4 = 31.4 \quad \text{(obs. 31.7)}$$

Figure 3.13 Calculation of the ^{13}C chemical shifts of the indicated carbon in 2-methylpropane (isobutane) and in 2,2-dimethylpropane (neopentane).

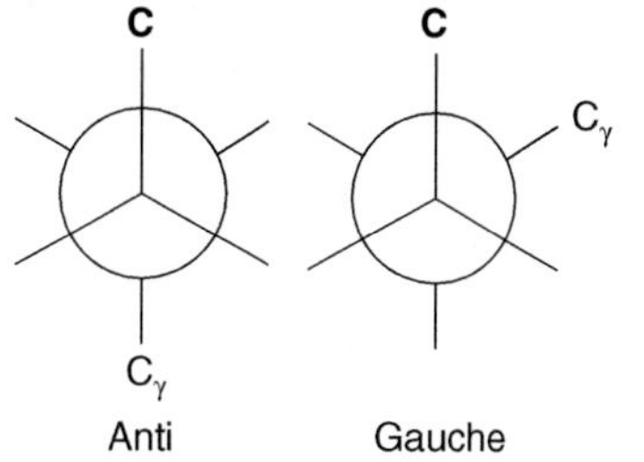

Figure 3.14 The anti and gauche geometries in a butane fragment.

all situations. For a pure γ-*anti effect*, the shift is about +1, and for a pure γ-*gauche effect*, it is about −6. The average value of −2.5 measured by Grant clearly indicates a mix of the two conformations. Hydrocarbons with unusually large deviations from the average mix may give poor results with Eq. (3.10). The α and β effects are determined by fixed geometries and have no stereochemistry component.

3.5.1.2 Cyclic Alkanes

With a resonance position of δ −2.6, cyclopropane has the lowest frequency resonance of hydrocarbons. Cyclobutane resonates at δ 23.3, and the remaining cycloalkanes generally resonate within 2 ppm of cyclohexane, which resonates at δ 27.7. The fixed stereochemistry represented by cyclohexane requires an entirely new set of empirical parameters that depend on the axial or equatorial nature of the substituent, as well as on the distance from the resonating carbon, which have been published by Grant.

3.5.1.3 Functionalized Alkanes

Replacement of a hydrogen on carbon with a heteroatom or an unsaturated group usually results in shifts to higher frequency (downfield) because of polar effects of the radial term. The effect parallels the same structural change on the ^1H chemical shifts but arises from a different mechanism. Strongly electron-withdrawing groups thus have large, positive α effects. In the halogen series CH_3X, the methyl chemical shifts are δ 75.4 for X = fluorine, 25.1 for chlorine, 10.2 for bromine, and −20.6 for iodine. Multiple substitution results in larger effects, such as δ 77.7 for $CHCl_3$. Recall that the α effect of heavy atoms such as iodine or bromine is influenced by a spin-orbit mechanism and hence does not follow the simple order of electronegativity. The general range for the α halogen effect in hydrocarbons extends from the values given for the simple CH_3X systems to about 25-ppm higher frequency (downfield) for CH_2X and CHX systems, since α and β effects of the unspecified hydrocarbon pieces contribute to the shift to higher frequency.

Methanol (CH_3OH) resonates at δ 49.2, and the range for hydroxy-substituted carbons is δ 49–75. Dimethyl ether [$(CH_3)_2O$] resonates at δ 59.5, and the range for alkoxy-substituted carbons is δ 59–80. The ether range is translated a few parts per million to higher frequency (downfield) from alcohols because each ether must have one additional β effect in comparison with the analogous alcohol.

The lower electronegativity of nitrogen, compared with oxygen, moves the amine range somewhat to lower frequency (upfield). Methylamine in aqueous solution resonates at δ 28.3, the range for amines extending some 30 ppm to higher frequency. The amine range is larger than the alcohol range because nitrogen can carry up to three substituents, with the possibility of more α and β effects. Dimethyl sulfide resonates at δ 19.5, acetonitrile at δ 1.8, and nitromethane at δ 62.6, with the respective ranges for thioalkoxy, cyano, and nitro substitution extending some 25 ppm to higher frequency. The anomalous low-frequency position for cyano substitution is related to the cylindrical shape of the group with its reduced angular momentum and lower paramagnetic term.

A double bond has only a small effect on an attached methyl group. The position for the methyls of trans-2-butene (*trans*-$CH_3CH=CHCH_3$) is δ 17.3, and that for the methyl of toluene ($C_6H_5CH_3$) is δ 21.3. The range for carbons on double bonds is about δ 15–40. Methyls on carbonyl groups are at slightly higher frequency (lower field): δ 30.2 for acetone and 31.2 for acetaldehyde, with a range of about δ 30–45.

Introducing heteroatoms or unsaturation into alkane chains requires completely new sets of empirical parameters that depend on the substituent, on its distance from the resonating carbon (α, β, or γ), and on whether the substituent is terminal (**3.28**) or internal (**3.29**)

Terminal
3.28

Internal
3.29

(Table 3.2). These numbers represent the effect on a resonating carbon of replacing a hydrogen atom at the respective position with a group X. With the usual exception of cyano, acetyleno, and the heavy atom iodine, the α effects are determined largely by the electronegativity of the substituent. It is interesting that the β effects are all positive and generally of similar magnitude (6–11) and that the γ effects are all negative and generally of similar magnitude (−2 to −5). Although the details are not entirely understood, it is clear that simple polar considerations do not dominate the β and γ effects.

To use the substituent parameters given in Table 3.2, one adds the appropriate values to the chemical shift of the carbon in the unsubstituted hydrocarbon analogue, rounding off to the nearest parts per million. As seen in Figure 3.15, the chemical shift of carbon 1 of 1,3-dichloropropane may be calculated from the value (δ 16) for the methyl carbon of propane and from the figures in Table 3.2. The chemical shift of the β carbon of cyclopentanol similarly may be calculated from the value (δ 27) for cyclopentane.

Table 3.2 Carbon substituent parameters for functional groups.

X	Terminal X (3-28)			Internal X (3-29)		
	α	β	γ	α	β	γ
F	68	9	−4	63	6	−4
Cl	31	11	−4	32	10	−4
Br	20	11	−3	25	10	−3
I	−6	11	−1	4	12	−1
OH	48	10	−5	41	8	−5
OR	58	8	−4	51	5	−4
OAc	51	6	−3	45	5	−3
NH_2	29	11	−5	24	10	−5
NR_2	42	6	−3			−3
CN	4	3	−3	1	3	−3
NO_2	63	4		57	4	
$CH=CH_2$	20	6	−0.5			−0.5
C_6H_5	23	9	−2	17	7	−2
C≡CH	4.5	5.5	−3.5			−3.5
(C=O)R	30	1	−2	24	1	−2
(C=O)OH	21	3	−2	16	2	−2
(C=O)OR	20	3	−2	17	2	−2
$(C=O)NH_2$	22		−0.5	2.5		−0.5

Source: From Ref. [4].

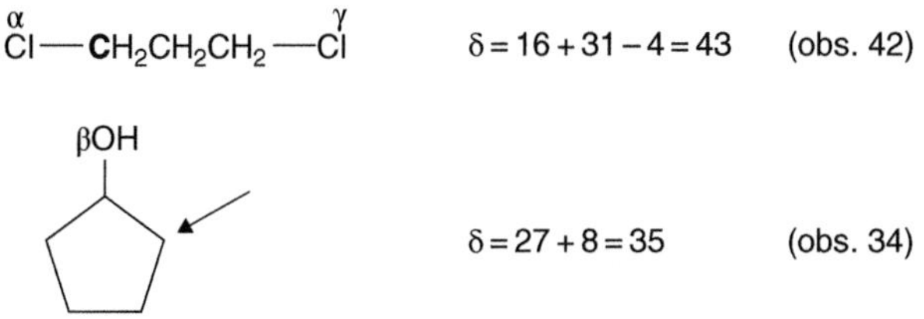

$\overset{\alpha}{\text{Cl}}-\text{CH}_2\text{CH}_2\text{CH}_2-\overset{\gamma}{\text{Cl}}$ δ = 16 + 31 − 4 = 43   (obs. 42)

[cyclopentanol structure with βOH label]   δ = 27 + 8 = 35   (obs. 34)

Figure 3.15 Calculation of the ^{13}C chemical shifts of the indicated carbons in 1,3-dichloropropane and in cyclopentanol.

3.5.2 Unsaturated Compounds

The effects of diamagnetic anisotropy on a carbon and a proton have similar magnitudes, but the much larger paramagnetic shielding renders anisotropy relatively unimportant for carbon. Thus, benzene (δ 128.4) and the alkenic carbons of cyclohexene (δ 127.3) have almost identical carbon resonance positions, in contrast to the situation with their protons. The full range of alkene and aromatic carbon resonances is about δ 100–170.

3.5.2.1 Alkenes
Alkenic carbons that bear no substituents ($=CH_2$) resonate at low frequency (high field), such as isobutylene [$(CH_3)_2C=CH_2$] at δ 107.7, and have a range of about δ 104–115

for hydrocarbons. Alkenic carbons that have one substituent (=CHR), like those in *trans*-butene (δ 123.3), resonate in the range δ 120–140. Finally, disubstituted alkenic carbons (=CRR′), like that in isobutylene (δ 146.4), resonate at the highest frequency (δ 140–165). Polar substituents on double bonds, particularly those in conjugation with the bond, can alter the resonance position appreciably. α,β-Unsaturated ketones such as **3.30** and **3.31**,

cyclohexenone: 128.4, 149.8 — **3.30**
cyclopentenone: 132.9, 164.2 — **3.31**

have lower frequency α resonances and higher frequency β resonances. The effect is reduced in acyclic molecules. Electron donation, as in enol ethers, reverses the effect: $CH_2^\beta=CH^\alpha OCH_3$ (δ(α) 153.2, δ(β) 84.2). Electron donation or withdrawal alters the radial term through delocalization.

3.5.2.2 Alkynes and Nitriles

An alkyne carbon that carries a hydrogen (≡CH) generally resonates in the narrow range δ 67–70. An alkyne carbon that carries a carbon substituent (≡CR) resonates at slightly higher frequency (δ 74–85) because of α and β effects from the R group. Effects of conjugating, polar substituents expand the total range to δ 20–90. Nitriles resonate in the range δ 117–130 (acetonitrile, $CH_3C\equiv N$, δ 117.2). The $n \rightarrow \pi^*$ transition pushes the range to high frequency.

3.5.2.3 Aromatics

Alkyl substitution, as in toluene (**3.32**),

toluene: 137.8, 129.3, 128.5, 125.6 — **3.32**
nitrobenzene: 148.3, 123.4, 129.5, 134.7 — **3.33**

has its major (α) effect on the ipso carbon. Because this carbon has no attached proton, its relaxation time is much longer than those of the other carbons, and its intensity is usually lower. Conjugating substituents like nitro (**3.33**) have strong perturbations on the aromatic resonance positions, as the result of a combination of traditional α, β, and γ effects, plus changes in electron density through delocalization (**3.19**, **3.20**). A similar interplay of effects is seen in the resonance positions of pyridine (**3.34**) and pyrrole (**3.35**).

3.34 (pyridine): 135.9, 123.9, 150.2

3.35 (pyrrole): 108.0, 118.4

3.5.3 Carbonyl Groups

Carbonyl groups have no direct representation in proton NMR spectra, so carbon NMR spectroscopy provides unique information for their analysis. The entire carbonyl chemical shift range, δ 160–220, is well removed to high frequency from those of almost all other functional groups. Like nitriles and aromatic ipso carbons, carbonyl carbons other than those in aldehydes carry no attached protons and hence relax more slowly and tend to have low intensities.

Aldehydes resonate toward the middle of the carbonyl range, at about δ 190–205, with acetaldehyde (CH_3CHO) at δ 199.6. Unsaturated aldehydes, in which the carbonyl group is conjugated with a double bond or phenyl ring, are shifted to lower frequency (upfield): benzaldehyde (C_6H_5CHO) at δ 192.4 and acrolein (propenal, $CH_2=CHCHO$ at δ 192.2. The α, β, and γ effects of substituents on ketones add to the carbonyl chemical shift and hence are found at the high-frequency end of the carbonyl range. Their overall range is δ 195–220, with acetone at δ 205.1 and cyclohexanone at δ 208.8. Again, unsaturation shifts the resonances to lower frequency.

Carboxylic derivatives occur in the range δ 155–185. The resonances for the series carboxylate (CO_2^-), carboxyl (CO_2H), and ester (CO_2R) often are well defined, as for sodium acetate (δ 181.5), acetic acid (δ 177.3), and methyl acetate (δ 170.7). The range for esters is about δ 165–175 and that for acids is δ 170–185. Acid chlorides are at slightly lower frequency (higher field): δ 160–170, with δ 168.6 for acetyl chloride [$CH_3(CO)Cl$]. Anhydrides have a similar range: δ 165–175, with δ 167.7 for acetic anhydride [$CH_3(CO)O(CO)CH_3$]. Lactones overlap the ester range, with the six-membered lactone at δ 176.5. Amides also have a similar range: δ 160–175, with δ 172.7 for acetamide [$CH_3(CO)NH_2$]. Oximes have a larger range, extending from δ 145 to 165. The central carbon of allenes ($R_2C=C=CR_2$) falls into the ketonic range, δ 200–215, although the outer carbons have a much lower frequency range, δ 75–95.

3.5.4 Programs for Empirical Calculations

The facility and accuracy of empirical calculations for carbon have been exploited through commercial computer programs for general predictions of ^{13}C chemical shifts. As with ^1H calculations (Section 3.2.5), the results are only as good as the data set used in their creation. The programs assume that the effects of multiple substitution are additive, unless specific connections have been incorporated. Unconsidered or nonadditive phenomena, such as conformational and other steric effects, can cause unexpected deviations between observed and calculated chemical shifts.

3.6 Tables of Chemical Shifts

Structural analysis of an unknown organic material normally begins with an examination of the ^{1}H and ^{13}C spectra. Resonance positions are analyzed, if possible, with the benefit of knowledge of the molecular formula and structural information based on synthetic precursors. Representative chemical shifts are given in Tables 3.3–3.7, drawn from references at the end of this chapter.

Table 3.3 Methyl and methylene groups.

	δ (^{1}H)		δ (^{13}C)	
	CH$_2$	CH$_3$	CH$_2$	CH$_3$
CH$_3$Li		−0.4		−13.2
CH$_3$CH$_3$		0.86		5.7
(CH$_3$)$_3$CH		0.89		25.2
(CH$_3$)$_4$C		0.94		31.7
(CH$_3$)$_3$COH		1.22		29.4
CH$_3$CH=CH$_2$		1.72		18.7
CH$_3$C≡CH		1.80		−1.9
(CH$_3$)$_3$P=O		1.93		18.6
CH$_3$CN		2.00		0.3
CH$_3$CO$_2$CH$_3$		2.01		18.7
CH$_3$(CO)CH$_3$		2.07		30.2
CH$_3$CO$_2$H		2.10		18.6
(CH$_3$)$_2$S		2.12		19.5
CH$_3$I		2.15		−20.6
CH$_3$CHO		2.20		31.2
(CH$_3$)$_3$N		2.22		47.3
CH$_3$C$_6$H$_5$		2.31		21.3
CH$_3$NH$_2$		2.42		30.4
CH$_3$(SO)CH$_3$		2.50		40.1
CH$_3$(CO)Cl		2.67		32.7
CH$_3$Br		2.69		10.2
(CH$_3$)$_4$P$^+$		2.74		11.3
CH$_3$(SO$_2$)CH$_3$		2.84		42.6
(CH$_3$)$_2$NCHO		2.88		36.0
		2.97		30.9
CH$_3$Cl		3.06		25.1
(CH$_3$)$_2$O		3.24		59.5
(CH$_3$)$_4$N$^+$		3.33		55.6
CH$_3$OH		3.38		49.2
CH$_3$CO$_2$**CH$_3$**		3.67		51.0

Table 3.3 (Continued)

	δ (¹H)		δ (¹³C)	
	CH$_2$	CH$_3$	CH$_2$	CH$_3$
CH$_3$OC$_6$H$_5$		3.73		54.8
CH$_3$F		4.27		75.4
CH$_3$NO$_2$		4.33		57.3
(CH$_3$CH$_2$)$_2$S	2.49	1.25	26.5	15.8
CH$_3$CH$_2$NH$_2$	2.74	1.10	36.9	19.0
CH$_3$CH$_2$C$_6$H$_5$	2.92	1.18	29.3	16.8
CH$_3$CH$_2$I	3.16	1.86	0.2	23.1
CH$_3$CH$_2$Br	3.37	1.65	28.3	20.3
CH$_3$CH$_2$Cl	3.47	1.33	39.9	18.7
(CH$_3$CH$_2$)$_2$O	3.48	1.20	67.4	17.1
CH$_3$CH$_2$OH	3.56	1.24	57.3	15.9
CH$_3$CH$_2$F	4.36	1.24	79.3	14.6
CH$_3$CH$_2$NO$_2$	4.37	1.58	70.4	10.6
BrCH$_2$CH$_2$Br	3.63		32.4	
HOCH$_2$CH$_2$OH	3.72		63.4	
ClCH$_2$CH$_2$Cl	3.73		51.7	

Table 3.4 Saturated ring systems.

		¹H	¹³C
Cyclopropane		0.22	−2.6
Cyclobutane		1.98	23.3
Cyclopentane		1.51	26.5
Cyclohexane		1.43	27.7
Cycloheptane		1.53	29.4
Cyclopentanone	(α)	2.06	37.0
	(β)	2.02	22.3
Cyclohexanone	(α)	2.22	40.7
	(β)	1.8	26.8
	(γ)	1.8	24.1
Oxirane		2.54	40.5
Tetrahydrofuran	(α)	3.75	69.1
	(β)	1.85	26.2
Oxane (tetrahydropyran)	(α)	3.52	68.0
	(β)	1.51	26.6
	(γ)		23.6

(Continued)

Table 3.4 (Continued)

		^1H	^{13}C
Pyrrolidine	(α)	2.75	47.4
	(β)	1.59	25.8
Piperidine	(α)	2.74	47.5
	(β)	1.50	27.2
	(γ)	1.50	25.5
Thiirane		2.27	18.9
Tetrahydrothiophene	(α)	2.82	31.7
	(β)	1.93	31.2
Sulfolane	(α)	3.00	51.1
	(β)	2.23	22.7
1,4-Dioxane		3.70	66.5

Table 3.5 Alkenes.

		^1H	^{13}C
$CH_2=CHCN$	(α)	{5.5 – 6.4}	107.7
	(β)		137.8
$CH_2=CHC_6H_5$	(α)	6.66	112.3
	(β)	5.15, 5.63	135.8
$CH_2=CHBr$	(α)	6.4	115.6
	(β)	5.7–6.1	122.1
$CH_2=CHCO_2H$	(α)	6.5	128.0
	(β)	5.9–6.5	131.9
$CH_2=CH(CO)CH_3$	(α)	{5.8 – 6.4}	138.5
	(β)		129.3
$CH_2=CHO(CO)CH_3$	(α)	7.28	141.7
	(β)	4.56, 4.88	96.4
$CH_2=CHOCH_2CH_3$	(α)	6.45	152.9
	(β)	3.6–4.3	84.6
$\overset{4}{C}H_3\overset{3}{C}H=\overset{2}{C}CH_3=\overset{1}{C}H=CH_2$	(1)	5.02	
	(2)	6.40	
	(4)	5.70	
$(CH_3)_2C=CHCO_2CH_3$	(α)	—	114.8
	(β)	5.62	155.9
Cyclopentene		5.60	130.6
Cyclohexene		5.59	127.2
1,3-Cyclopentadiene		6.42	132.2, 132.8

Table 3.5 (Continued)

		¹H	¹³C
1,3-Cyclohexadiene		5.78	124.6, 126.1
2-Cyclopentenone	(α)	6.10	132.9
	(β)	7.71	164.2
2-Cyclohexenone	(α)	5.93	128.4
	(β)	6.88	149.8
exo-Methylenecyclohexane	(=CH$_2$)	4.55	106.5
	(C=)	—	149.7
Allene	(=CH$_2$)	4.67	74.0
	(=C=)	—	213.0

Table 3.6 Aromatics.

	¹H			¹³C			
	o	m	p	i	o	m	p
C$_6$H$_5$CH$_3$	7.16	7.16	7.16	137.8	129.3	128.5	125.6
C$_6$H$_5$CH=CH$_2$	7.24	7.24	7.24	138.2	126.7	128.9	128.2
C$_6$H$_5$SCH$_3$	7.23	7.23	7.23	138.7	126.7	128.9	124.9
C$_6$H$_5$F	6.97	7.25	7.05	163.8	114.6	130.3	124.3
C$_6$H$_5$Cl	7.29	7.21	7.23	135.1	128.9	129.7	126.7
C$_6$H$_5$Br	7.49	7.14	7.24	123.3	132.0	130.9	127.7
C$_6$H$_5$OH	6.77	7.13	6.87	155.6	116.1	130.5	120.8
C$_6$H$_5$OCH$_3$	6.84	7.18	6.90	158.9	113.2	128.7	119.8
C$_6$H$_5$O(CO)CH$_3$	7.06	7.25	7.25	151.7	122.3	130.0	126.4
C$_6$H$_5$(CO)CH$_3$	7.91	7.45	7.45	136.6	128.4	128.4	131.6
C$_6$H$_5$CO$_2$H	8.07	7.41	7.47	130.6	130.0	128.5	133.6
C$_6$H$_5$(CO)Cl	8.10	7.43	7.57	134.5	131.3	129.9	136.1
C$_6$H$_5$CN	7.54	7.38	7.57	109.7	130.1	127.2	130.1
C$_6$H$_5$NH$_2$	6.52	7.03	6.63	147.9	116.3	130.0	119.2
C$_6$H$_5$NO$_2$	8.22	7.48	7.61	148.3	123.4	129.5	134.7

	¹H			¹³C		
	α	β	Other	α	β	Other
Naphthalene	7.81	7.46	—	128.3	126.1	—
Anthracene	7.91	7.39	8.31	130.3	125.7	132.8
Furan	7.40	6.30	—	142.8	109.8	—
Thiophene	7.19	7.04	—	125.6	127.4	—
Pyrrole	6.68	6.05	—	118.4	108.0	—
Pyridine	8.50	7.06	7.46	150.2	123.9	135.9

Table 3.7 Carbonyl compounds.

	^{1}H(CH$_3$)	^{1}H(other)	^{13}C(C=O)
H(CO)OCH$_3$	3.79	8.05 (HCO)	160.9
CH$_3$(CO)Cl	2.67	—	168.6
CH$_3$(CO)OCH$_2$CH$_3$	2.02 (CH$_3$CO)	4.11 (CH$_2$), 1.24(CH$_3$C)	169.5
CH$_3$(CO)N(CH$_3$)$_2$	2.10 (CH$_3$CO)	2.98 (CH$_3$N)	169.6
CH$_3$CO$_2$H	2.10	1.37 (HO)	177.3
CH$_3$CO$_2^-$Na$^+$	—	—	181.5
CH$_3$(CO)C$_6$H$_5$	2.62	—	196.0
CH$_3$(CO)CH=CH$_2$	2.32	5.8–6.4 (CH=CH$_2$)	197.2
H(CO)CH$_3$	2.20	9.80 (HCO)	199.6
CH$_3$(CO)CH$_3$	2.07	—	205.1
2-Cyclohexenone	—	5.93, 6.88 (CH$_\alpha$=CH$_\beta$)	197.1
2-Cyclopentanone	—	6.10, 7.71 (CH$_\alpha$=CH$_\beta$)	208.1
Cyclohexanone	—	1.7–2.5	208.8
Cyclopentanone	—	1.9–2.3	218.1

Problems

3.1 A trisubstituted benzene possessing one bromine and two methoxy substituents exhibits three aromatic resonances, at δ 6.40, 6.46, and 7.41. What is the substitution pattern?

3.2 Octahedral cobalt complexes, CoL$_6$, have three filled t_{2g} and two empty e_g molecular orbitals. The ^{59}Co chemical shifts of several such complexes exhibit a linear relationship with the wavelength of the first absorption (longest wavelength) in the UV/visible spectrum. Explain the linearity in terms of Ramsey's Eq. (3.9) for shielding.

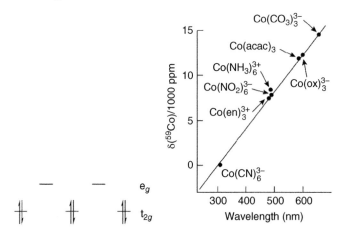

3.3 Calculate the expected ^{13}C resonance positions for all the carbon atoms in the following molecules. Ignore the δ and ε effects.
(a) $CH_3CH_2CH(CH_3)CH(CH_3)_2$
(b) $ICH_2CH_2CH_2Br$
(c) $CH_3CH_2CH(NO_2)CH_3$
(d) $(CH_3)_3CCN$

3.4 Of the α and β protons of naphthalene (see structure), which should resonate at a higher frequency? Why? Compare both resonance positions with that of benzene.

3.5 The —OH proton resonance is found at δ 5.80 for phenol in dilute CDCl$_3$ and at δ 10.67 for 2-nitrophenol in dilute CDCl$_3$. Explain.

3.6 Derive the structures of the compounds that have the ^{1}H (300 MHz) and ^{13}C (75 MHz) spectra shown in parts (a)–(g). The 1 : 1 : 1 triplet at δ 78 in the ^{13}C spectra is from the solvent CDCl$_3$ (used in all cases except (f)).
(a) $C_4H_6O_2$

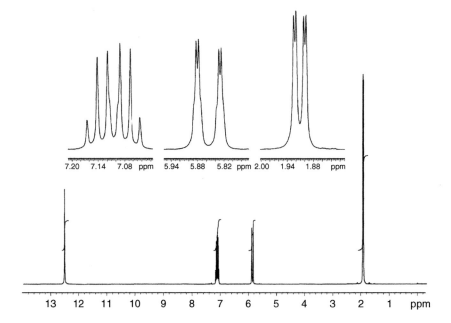

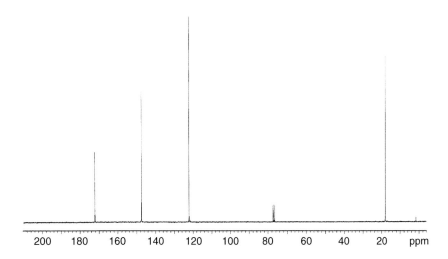

(b) $C_4H_8Cl_2$

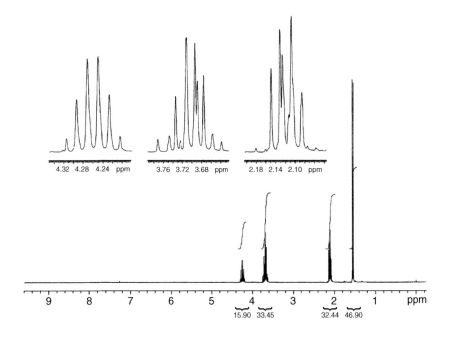

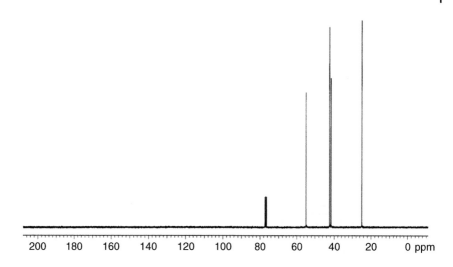

(c) C$_5$H$_9$OCl

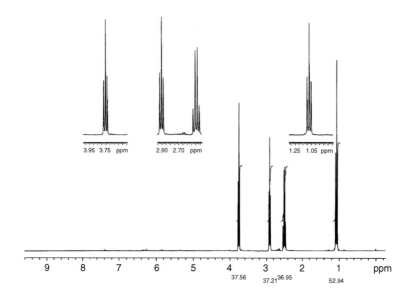

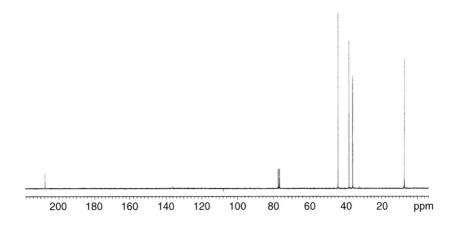

(d) $C_4H_8O_2$

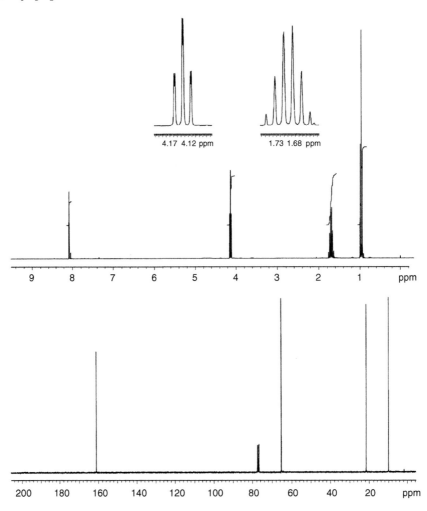

(e) $C_{10}H_{12}O_3$ (The peak at δ 6.87 disappears after addition of D_2O and shaking.)

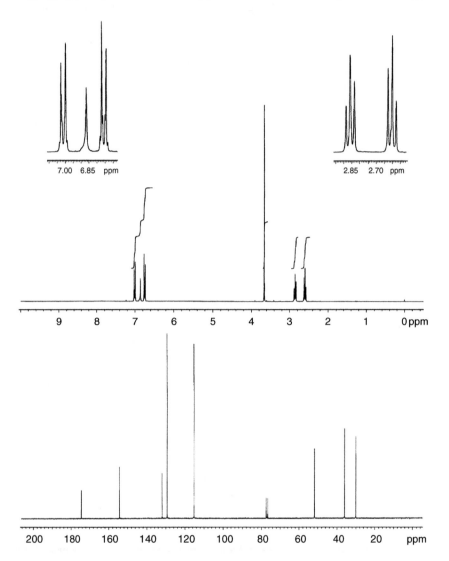

(f) $C_8H_7BrO_3$ (A ^1H resonance of unit integral at δ 12 is not shown; the ^1H signals at δ 2.05 and the ^{13}C signals at δ 30 are from the solvent acetone-d_6.)

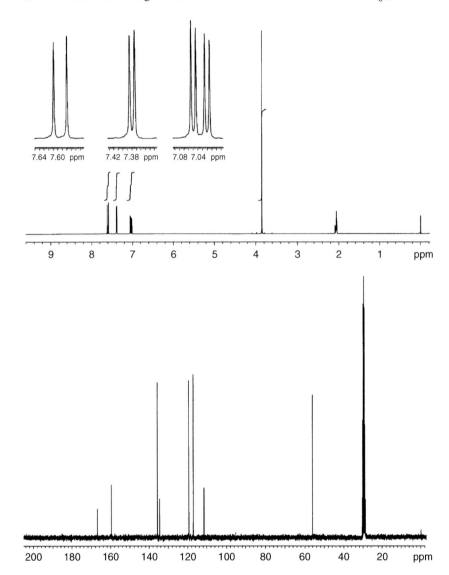

(g) C_6H_7NO

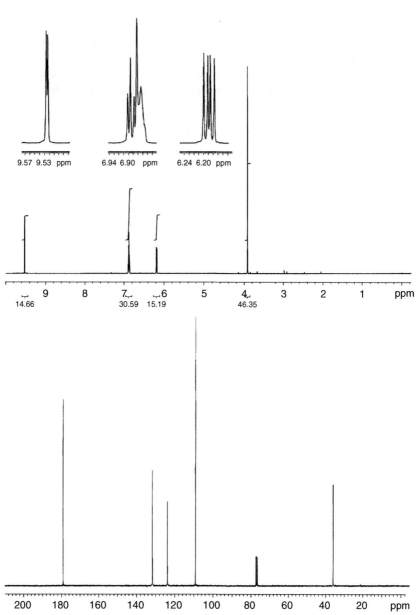

3.7 The aromatic and alkenic portions of the 500 MHz ^{1}H spectrum of the potassium channel activator bimakalin are given below, along with the structure of the molecule.
(a) Assign these eight resonances to the appropriate protons strictly by the coupling patterns. Discuss your assignments.
(b) Explain the portion of the three resonances clustered around δ 7.6.
(c) Why is the proton with chemical shift δ 6.12 found at so low a frequency?

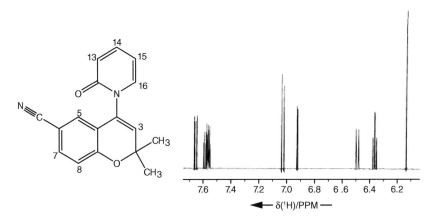

Reproduced with permission from K. G. R. Pachler, *Magn. Reson. Chem.*, **36**, 437 (1998).

3.8 The bark resin of *Commiphora mukul*, called the guggul tree, is thought to lower cholesterol and fat intake. The primary active components were found to be the five molecules below, belonging to a class of compounds called cembrenes.

The Chemical Shift

From the following table of the already assigned ^{1}H chemical shifts for these compounds, assign structures **A–E** to NMR spectra. In the table, for each proton the chemical shift is given in units of δ, then the multiplicity, and finally the values of any coupling constants in Hz. Also try this exercise without any prior knowledge of the peak assignments.

Proton	Cpd1	Cpd2	Cpd3	Cpd4	Cpd5
1	2.06 m	1.63 m	n/a	1.28 m	1.45 m
2	2.02 m 1.98 m	5.19 dd 15.5, 9.8	2.22 m 2.18 m	4.58 d 8.7	2.72 m
3	5.22 t 7.6	6.09 d 15.5	5.27 t 7.4	5.33 d 8.7	5.69 d 12.5
4	n/a	n/a	n/a	n/a	n/a
5	2.14 m 2.19 m	5.55 t 7.6	2.22 m	2.17 m 2.20 m	2.09 m 2.23 m
6	2.19 m 2.29 m	2.44 dd 3.4, 1.5 3.07 br s	2.12 m 2.31 m	2.12 m 2.20 m	2.05 m 2.42 m
7	5.00 t 6.2	5.13 d 10.6	4.91 t 6.8	5.04 t 5.8	4.91 d 10.5
8	n/a	n/a	n/a	n/a	n/a
9	2.08 m	2.06 m 2.23 m	1.95 m 2.14 m	1.94 m 2.13 m	2.11 m 2.26 m
10	2.14 m	2.03 m 2.32 m	2.10 m	2.07 m 2.18 m	1.33 m 1.49 m
11	5.08 t 6.2	4.90 d 6.8	5.01 t 7.4	4.94 dd 7.9, 7.2	2.27 m
12	n/a	n/a	n/a	n/a	n/a
13	1.96 m	2.02 m	1.94 m	2.05 m	1.86 m
14	1.69 s	1.69 m	1.65 m	1.69 m	2.00 m
15	n/a	1.51 m	1.72 m	1.79 m	n/a
16	1.68 s	0.88 d 6.8	0.96 d 6.8	0.99 d 6.8	0.81 s
17	4.68 s 4.74 s	0.84 d 6.8	0.94 d 6.8	0.96 d 6.8	0.74 s
18	1.59 s	1.82 s	1.56 s	1.60 s	1.56 s
19	1.62 s	1.62 s	1.59 s	1.59 s	1.53 s
20	1.57 s	1.54 s	1.61 s	1.56 s	1.28 s

Further Tips on Solving NMR Problems

See the Spoiler Alert in Chapter 2, Tips on Solving NMR Problems.

1) Empirical correlations provide a useful guideline for peak assignments but should not be construed as completely reliable. Problems of nonadditivity can be resolved only through the introduction of interaction factors, which generally have been ignored in calculations of proton chemical shifts. Most reliable for protons probably are the calculations of aromatic chemical shifts, provided that no two substituents are ortho to each other (see Problem 3.1). Empirical correlations for ^{13}C chemical shifts have been more successful (Problem 3.3), and many examples of interaction factors have been introduced. With commercial programs for the empirical calculation of spectra, the results can never be more accurate than the structural models on which they were based. Consequently, any such calculation should always include direct assessment of the model molecules on which the calculation was based.

2) A double of quartets (dq) might be expected when a CH_n group is coupled to both a methyl and a methine group, as in $CH_3CH—CH$ or $CH_3CH=CH$. The resulting pattern, however, can vary from four to eight peaks, depending on the relative values of $J(CH_3CH)$ and $J(CH—CH)$. If the latter coupling is zero, the resulting pattern is a quartet. Such an event could occur if the CH—CH dihedral angle is close to a 90° angle, according to the Karplus relationship (Section 4.6). If the CH—CH coupling is considerably larger than the $CH—CH_3$ coupling, as could occur for $CH_3CH=CH$ with a trans arrangement between the alkenic protons (again, Section 4.6), the maximum of eight peaks (dq) would be observed. When the two couplings are comparable, anything from five to eight peaks can occur, depending on the precise relative values. For the example in Problem 3.6a, four of the peaks overlap in the middle to reduce the pattern to six peaks. This sort of situation is nicely analyzed by the use of stick diagrams or trees as described in Section 1.6 and is treated thoroughly in Hoye et al. (1994) in Chapter 4.

3) The proton integrals and the carbon peak intensities in Problem 3.6a illustrate the reliability of the former and the unreliability of the latter. Although each carbon peak represents a single carbon, the intensities vary by more than a factor of 2. The carbonyl carbon of the CO_2H group has no attached proton, relaxes more slowly, and has the smallest intensity. The intensities of the carbons bearing protons vary with the number of attached and neighboring protons and with spectral parameters such as the delay time (Section 2.4.6).

4) Carbon-13 spectra measured in $CDCl_3$ usually include a peak from the solvent at about δ 78. This resonance is a 1 : 1 : 1 triplet, because the carbon is coupled to the deuterium atom, which exists in three nearly equally populated spin states (+1, 0, and −1).

5) A pair of 1 : 2 : 1 triplets often is a diagnostic for the isolated bismethylene group, $—CH_2CH_2—$. Deviations from first-order conditions can distort the intensities of the inner and outer peaks of both triplets.

6) The H(C=O) group (formyl) usually is associated with aldehydes, which exhibit a sharp singlet at about δ 9.8. Formates, however, contain the same grouping: $H(C=O)OCH_3$ (methyl formate). A formate formyl proton is found at lower frequency (higher field) than the aldehyde proton, typically around δ 8.0 (Problem 3.6d).

7) The relative locations of alkyl groups in oxygen-containing molecules often are not clear from the splitting patterns (Problem 3.6e). The distinctions to be drawn are between ether/ester ($CH_3O—$) functionalities and carbonyl [$CH_3(C=O)—$] functionalities. These may be distinguished by the observation of resonances at δ c. 2.1 for carbonyl functionalities, at δ c. 3.7 for ethers, and at δ c. 4.1 for esters. Higher frequencies are observed with more substitution on carbon, e.g. $RCH_2O—$ or $RCH_2(C=O)—$.

8) To distinguish ketones and aldehydes from esters, acids, amides, and their ilk, it is useful to remember that δ 190 provides something of a borderline. Ketones and aldehydes usually are at higher frequency than 190 and the carboxylic acid family usually is at lower frequency than 190. Problem 3.6a,c–f provide examples.

9) Disubstituted aromatics (Problem 3.6e) can occur with ortho, meta, or para patterns:

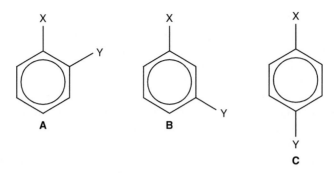

These may be distinguished by the knowledge that the H—H couplings J(ortho) > J(meta) > J(para). This order follows the number of bonds (three, four, and five, respectively), but the more complicated actual situation is discussed in Sections 4.4–4.6. The para coupling, over five bonds, is almost always negligible. In the para-substituted case (**C**), meta couplings (HCCCH) are small and between equivalent protons, so the spectrum is dominated by the large ortho couplings (HCCH). The spectrum usually reduces to a simple two-spin AX spectrum of four peaks (Section 1.5 and Figure 1.18), sometimes with complications from magnetic nonequivalence (Section 4.2). If the para X and Y substituents are identical, the spectrum of course becomes a singlet. The meta-substituted case (**B**) often exhibits a singlet or a closely spaced triplet for the proton between the substituents (the smaller splitting is from coupling to two meta protons). The spectrum for the remaining three protons is either an AMX spectrum (two doublets and a triplet) or an A₂X spectrum (a doublet and a triplet), depending on whether the X and Y substituents are identical. The doublets from the outer protons can be further split by meta couplings. The ortho-substituted case is the most complicated, with an AGRX spectrum when X and Y are different and an AA'XX' spectrum when X and Y and the same. The AA'XX' spectrum is an example of second-order effects discussed in Section 4.2 and exemplified by 1,2-dichlorobenzene in Figure 4.3.

10) Trisubstituted aromatics (Problem 3.6f) can occur with 1,2,3, 1,2,4, or 1,3,5 patterns:

There is great variability in the appearances of these spectra, and we will consider only the cases when all three substituents are different. The cases when all three substituents are the same or when only two are the same can be worked out easily. For 1,3,5 substitution (**C**), the spectrum contains three singlets or closely spaced doublets (from meta couplings). For 1,2,3 substitution (**A**), the AMX spectrum contains two widely spaced doublets (or doublets split into closely spaced doublets from the meta coupling) for the protons adjacent to X and Z. The middle proton (para to Y) is a 1:2:1 triplet from two equal ortho couplings. The appearance of the 1,2,4 spectrum (**B**) is highly diagnostic for this substitution pattern. The peak between Y and Z is a singlet or a closely spaced doublet from the meta coupling. The proton next to Z is a widely spaced doublet from the ortho coupling, usually split by the meta coupling into closely spaced doublets (hence, dd). The proton next to X is a simple doublet from the ortho coupling, without any further coupling. The spectrum in Problem 3.6f shows this characteristic of widely spaced doublet, narrowly spaced doublet, and doublet of doublets. These three substitution patterns all offer several structural isomers. The relative locations of the X, Y, and Z groups may not be assigned with coupling constant analysis, but empirical chemical shift correlations sometimes can accomplish this task (Problem 3.1).

11) Problem 3.6e illustrates the difficulties of getting a handle on OH and NH protons, which can give broad resonances at variable positions. The ^{14}N atom (Problem 3.6g) also broadens the resonance of the α protons on the ring. Thus a broadened resonance should not automatically be assumed to be from OH or NH functionalities, which must be assigned by the D_2O test in $CDCl_3$.

12) Although N—CH_3 protons usually resonate at δ c. 2.4, this is the case only for amines. Changing the charge on or the hybridization of the nitrogen alters the chemical shift. Thus methyl groups in ammonium salts [(R_3N—CH_3)$^+$] resonate at δ c. 3.3, methyl groups in amides resonate at δ c. 2.9, and methyl groups on pyrroles at δ c. 3.7, as in Problem 3.6g. The ^{13}C chemical shifts of N—CH_3 groups exhibit similar variations.

References

3.1 ACD NMR Predictor. http://www.acdlabs.com/ (accessed 21 March 2018).
3.2 gNMR (2003). http://www.adeptscience.co.uk/tag/gnmr (accessed 21 March 2018).
3.3 Bagno, A. (2001). *Chem. Eur. J.* 7: 1652–1661.
3.4 Wehrli, F.W., Marchand, A.P., and Wehrli, S. (1988). *Interpretation of Carbon-13 NMR Spectra*, 2e. Chichester: Wiley.

Further Reading

See also References Harris et al. (2008), Haigh and Mallion (1979), Pascual et al. (1966), Friedrich and Runkle (1984), Beauchamp and Marquez (1997), Tilley et al. (2002), Lodewyk et al. (2012), Homer (1975), and Berger (1990).

Chemical Shift Conventions

Harris, R.K., Becker, E.D., Cabral de Menezes, S.M. et al. (2008). *Magn. Reson. Chem.* 46: 582–596.

Diamagnetic Anisotropy

Chen, Z., Wannere, C.S., Corminboeuf, C. et al. (2005). *Chem. Rev.* 105: 3842.
Haddon, R.C. (1971). *Fortschr. Chem. Forsch.* 16: 105.
Haigh, C.W. and Mallion, R.B. (1979). *Prog. Nucl. Magn. Reson. Spectrosc.* 13: 303.
Lazzeretti, P. (2000). *Prog. Nucl. Magn. Reson. Spectrosc.* 36: 1.

Van der Waals Effects

Rummens, F.H.A. (1975). *NMR Basic Princ. Progr.* 10: 1.

Empirical Correlations

Beauchamp, P.S. and Marquez, R. (1997). *J. Chem. Educ.* 74: 1483–1485.
Brown, D.W. (1985). *J. Chem. Educ.* 62: 209–212.
Craik, D.J. (1983). *Annu. Rep. NMR Spectrosc.* 15: 1.
Friedrich, E.C. and Runkle, K.G. (1984). *J. Chem. Educ.* 61: 830.
Lodewyk, M.W., Siebert, M.R., and Tantillo, D.J. (2012). *Chem. Rev.* 112: 1839.
Martin, G.J. and Martin, M.L. (1972). *Prog. Nucl. Magn. Reson. Spectrosc.* 8: 163.
Pascual, C., Meier, J., and Simon, W. (1966). *Helv. Chim. Acta* 49: 164.
Shine, H.J. and Rangappa, P. (2007). *Magn. Reson. Chem.* 45: 971–979.
Tilley, L.J., Prevoir, S.J., and Forsyth, D.A. (2002). *J. Chem. Educ.* 79: 593–600.
Tobey, S.W. (1969). *J. Org. Chem.* 34: 1281.

Chemical Shifts of Solvents

Gottlieb, H.E., Kotlyar, V., and Nudelman, A. (1997). *J. Org. Chem.* 62: 7512–7515.

Solvent Effects

Homer, J. (1975). *Appl. Spectrosc. Rev.* 9: 1.
Laszlo, P. (1967). *Prog. Nucl. Magn. Reson. Spectrosc.* 3: 231.
Ronayne, J. and Williams, D.H. (1969). *Annu. Rev. NMR Spectrosc.* 2: 83.

Isotope Effects

Batiz-Hernandez, H. and Bernheim, R.A. *Prog. Nucl. Magn. Reson. Spectrosc.* 3: 63, 919670.
Berger, S. (1990). *NMR Basic Princ. Progr.* 22: 1.

Carbon-13 Chemical Shifts

Breitmaier, E. and Völter, W. (1987). *Carbon-13 NMR Spectroscopy*, 3e. Weinheim: Wiley-VCH.
Bremser, W., Franke, B., and Wagner, H. (1982). *Chemical Shift Ranges in ^{13}C NMR*. Weinheim: Wiley-VCH.
Kalinowski, H.-D., Berger, S., and Braun, S. (1988). *Carbon-13 Spectroscopy*. Chichester: Wiley.
Levy, G.C., Lichter, R.L., and Nelson, G.L. (1980). *Carbon-13 Nuclear Magnetic Resonance Spectroscopy*, 2e. New York: Wiley.
Pihlaja, K. and Kleinpeter, E. (1994). *Carbon-13 NMR Chemical Shifts in Structural and Stereochemical Analysis*. New York: Wiley.
Stothers, J.B. (1973). *Carbon-13 NMR Spectroscopy*. New York: Academic Press.

Other Nuclei (General)

Berger, S., Braun, S., Kalinowski, H.-O., and Becconsall, J.K. (1997). *NMR Spectroscopy of the Nonmetallic Elements*. New York: Wiley.
Brevard, C. and Granger, P. (1981). *Handbook of High Resolution Multinuclear NMR*. New York: Wiley.
Chandrakumar, N. (ed.) (1996). *Spin-1 NMR*, NMR Basic Principles and Progress, vol. 34.
Harris, R.K. and Mann, B.E. (1978). *NMR and the Periodic Table*. London: Academic Press.
Lambert, J.B. and Riddell, F.G. (ed.) (1983). *The Multinuclear Approach to NMR Spectroscopy*. Dordrecht: D. Reidel.
Laszlo, P. (ed.) (1983). *NMR of Newly Accessible Nuclei*. New York: Academic Press.
Mann, B.E. (1991). *Annu. Rev. NMR Spectrosc.* 23: 141–207.
Mason, J. (ed.) (1987). *Multinuclear NMR*. New York: Plenum Press.

Other Nuclei (Specific)

C. D. Schaeffer, Jr,. NMR Bibliography. http://www.wiredchemist.com/nmr/bibliography, Part V (accessed 21 March 2018).

Programs for Empirical Calculations

ChemNMR. http://www.cambridgesoft.com/software/details/?ds=7 (accessed 21 March 2018).
ChemWindow. http://www.bio-rad.com/en-us/product/chemical-structure-drawing-software (accessed 21 March 2018).

4

The Coupling Constant

4.1 First- and Second-order Spectra

Most spectra illustrated up to this point are said to be *first order*. For a spectrum to be first order, the frequency difference ($\Delta\nu$) between the chemical shifts of any given pair of nuclei must be much larger than the value of the coupling constant J between them, $\Delta\nu/J >$ c. 10. In addition, an important symmetry condition discussed in the next section must hold. First-order spectra exhibit a number of useful and simple characteristics:

- Spin–spin multiplets are centered on their resonance frequency.
- Spacings between adjacent components of a spin–spin multiplet are equal to the coupling constant J.
- Multiplicities that result from coupling exactly reflect the $n + 1$ rule for $I = \frac{1}{2}$ nuclei ($2nI + 1$, in general). Thus, two equivalent, neighboring protons split the signal from the resonating nucleus into three peaks.
- The intensities of spin–spin multiplets correspond to the coefficients of the binomial expansion given by Pascal's triangle for spin-$\frac{1}{2}$ nuclei (Figure 1.24).
- Nuclei with the same chemical shift do not split each other, even when the coupling constant between them is nonzero.

When the chemical-shift difference between two nuclei is less than about 10 times J ($\Delta\nu/J \leq 10$), the nuclei are said to be *closely coupled* and *second-order* effects appear in the spectrum. These include deviations in intensities from the binomial pattern and other exceptions from the preceding characteristics (Section 4.8). By the Pople notation, nuclei that have a first-order relationship are represented by letters that are far apart in the alphabet (AX), and those that have a second-order relationship are represented by adjacent letters (AB). Figure 4.1 illustrates the progression for two spins from AB almost to AX. When $\Delta\nu/J$ is 0.4, the spectrum is practically a singlet. Intensity distortions increase peak heights toward the center of the multiplet. A second-order multiplet typically leans toward the resonances of its coupling partner. The peak intensities within a multiplet are not equal even when $\Delta\nu/J = 15$. Thus, the commonly quoted first-order criterion of $\Delta\nu/J = 10$ is not always adequate to eliminate second-order effects. With the wide availability of proton frequencies of 300 MHz and higher, first-order spectra have become common, but by no means exclusive.

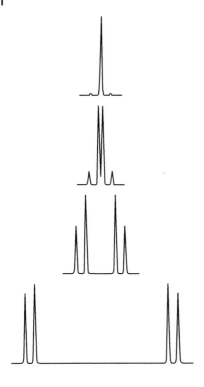

Figure 4.1 The two-spin spectrum with Δν/J values of 0.4 (top), 1.0, 4.0, and 15.0.

4.2 Chemical and Magnetic Equivalence

In addition to meeting the requirement that compares chemical-shift differences with coupling constants (Δν/J), first-order spectra must pass a symmetry test. *Any two chemically equivalent nuclei must have the same coupling to any other specific nucleus.* Nuclear pairs that fail this test are said to be *magnetically nonequivalent*, and their spectra are second order. To apply the test, it is useful first to understand the role of symmetry in the NMR spectrum.

Nuclei are *chemically equivalent* if they can be interchanged by any symmetry operation of the molecule. Thus, the two protons in 1,1-difluoroethene (**4.1**)

4.1   **4.2**   **4.3**

or in difluoromethane (**4.2**) may be interchanged by a 180° rotation. Nuclei that are interchangeable by rotational symmetry are said to be *homotopic*. Rotation about carbon–carbon single bonds is so rapid that the chemist rarely considers the fact that the three methyl protons in CH_3CH_2Br are not in fact equivalent by symmetry (compare nuclei A and X in **4–3**). Rapid C—C rotation, however, results in an average environment in which they are equivalent. Dynamic effects are considered more thoroughly in Section 5.2. Do not confuse the symmetry operation of rotation and the physical operation of rotation. The former involves no actual motion of the molecular constitutions but rather an entirely mental operation, whereas the latter involves physical movement of atoms or groups, as in methyl rotation.

Nuclei related by a plane of symmetry are said to be *enantiotopic*, provided that there is no rotational axis of symmetry. For example, the protons in bromochloromethane (**4.4a**)

4.4a   **4.4b**

are chemically equivalent and enantiotopic because they are related by the plane of symmetry containing C, Br, and Cl. If the molecule is placed in a chiral environment, this statement no longer holds true. Such an environment may be created by using a solvent composed of an optically active material or by placing the molecule in the active site of an enzyme. This kind of environment may be represented by **4.4b**, in which bromochloromethane has a small hand placed to one side. The protons are no longer equivalent because the hand is a chiral object. Since the plane of symmetry is lost in a chiral environment, the nuclei are not enantiotopic. They have become chemically nonequivalent (no symmetry operation can interchange them). Enantiotopic nuclei may be expected to become chemically nonequivalent and give distinct resonances in an optically active solvent. In a biological context, enantiotopic protons may be rendered nonequivalent by an enzyme and may exhibit distinct chemical properties, such as acidity or reaction rates.

The term *enantiotopic* was coined because the replacement of one proton of the pair by another atom or group, such as deuterium, produces the enantiomer (the nonsuperimposable mirror image, **4.4c**)

4.4c   **4.4d**

of the molecule that results when the other proton is replaced by the same group (**4.4d**). A pair of homotopic nuclei treated in this fashion produces identical molecules with superimposable mirror images. Enantiotopic or homotopic protons need not be on the same carbon atom. Thus, the alkenic protons in cyclopropene (**4.5**)

4.5   **4.6**

are homotopic, but those in 3-methylcyclopropene (**4.6**) are enantiotopic. Chemically equivalent nuclei (either homotopic or enantiotopic) are represented by the same letter in the spectral shorthand of Pople. Cyclopropene (**4.5**) is A_2X_2, as is difluoromethane (**4.2**), since the two fluorine atoms have spins of ½. The ring protons of 3-methylcyclopropene (**4.6**) constitute an AX_2 group.

To be *magnetically equivalent*, nuclei must be chemically equivalent and have the same coupling constant to any other nucleus in the molecule. This test is more stringent than that for chemical equivalence, because it is necessary to go beyond considering just the overall symmetry of the molecule. The first two molecules discussed in this chapter provide contrasting results. In difluoromethane (**4.2**), each of the two hydrogen atoms has the same coupling to any specific fluorine atom, because both hydrogens have the same spatial relationship to that fluorine. Consequently, the protons are magnetically equivalent. By the same token, the two fluorine atoms also are magnetically equivalent, by reference to coupling to either proton, and the spin system is labeled A_2X_2.

In 1,1-difluoroethene (**4.1**), however, the two protons do not have the same spatial relationship with respect to a given fluorine. Therefore, they have different couplings [J(HCCF)], one a J_{cis} and the other a J_{trans} (**4.7**),

$$H_A \diagdown \hspace{1em} \diagup F_X$$
$$C=C$$
$$H_{A'} \diagup \hspace{1em} \diagdown F_{X'}$$

with J_{cis} between H_A and F_X, and J_{trans} between $H_{A'}$ and F_X.

4.7

and are magnetically nonequivalent. The group of spins is represented by the notation AA′XX′ so that the two couplings may be denoted by J_{AX} (J_{cis}) and $J_{AX'}$ (J_{trans}). In contrast, an A_2X_2 system such as difluoromethane (**4.2**) or cyclopropene (**4.5**) has only one coupling, J_{AX}. In an AA′XX′ system, J_{AX} and $J_{A'X'}$ are the same, as are $J_{AX'}$ and $J_{A'X}$. Any spin system that contains nuclei that are chemically equivalent but magnetically nonequivalent is, by definition, second order. Moreover, raising the magnetic field does not alter basic structural relationships between nuclei so that the spectrum remains second order at the highest accessible fields. Nuclei that do not have the same chemical shift (*anisochronous*) also are magnetically nonequivalent because they resonate at different resonance frequencies (*magnetic nonequivalence by the chemical shift criterion*). *Isochronous* nuclei (nuclei with the same chemical shift) that are magnetically nonequivalent by having unequal couplings to another nucleus fail the coupling constant criterion (*magnetic nonequivalence by the coupling constant criterion*).

The AA′XX′ notation may be parsed as follows: The chemical shifts of the A and X nuclei are very far from each other (at opposite ends of the alphabet). The A and A′ nuclei are chemically equivalent (denoted by the same letter of the alphabet) but magnetically nonequivalent (indicated by the prime), as are the X and X′ nuclei. Figure 4.2 illustrates the proton AA′ part of the spectrum of 1,1-difluoroethene, in which 10 peaks are visible. This appearance is quite different from the simple 1 : 2 : 1 triplet expected in the first-order case, such as the methyl triplet in Figure 1.22, part of a first-order A_2X_3 spectrum. The multiplicity of peaks in Figure 4.2 even permits measurement of $J_{AA'}$, the coupling between the equivalent protons. Such a measurement is impossible in first-order systems such as that of Figure 1.22. The presence of splitting between equivalent nuclei in second-order spectra emphasizes that such couplings do exist but are not manifested in first-order spectra.

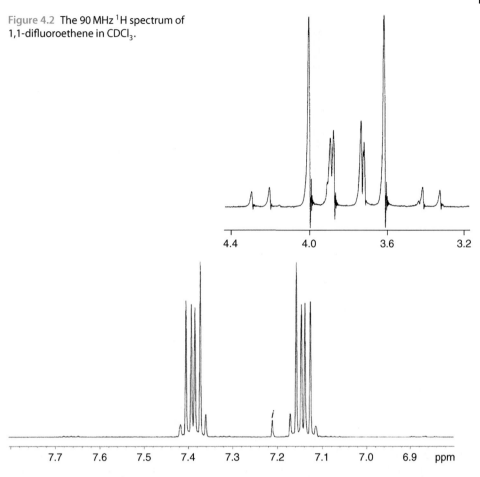

Figure 4.2 The 90 MHz ^{1}H spectrum of 1,1-difluoroethene in CDCl$_3$.

Figure 4.3 The 300 MHz ^{1}H spectrum of 1,2-dichlorobenzene in CDCl$_3$. An impurity is signified by the letter *i*.

Magnetic nonequivalence is not uncommon for isochronous nuclei. The spin systems for both para- and ortho-disubstituted benzene rings formally are AA'XX' (or AA'BB' if the chemical shifts are close). Figure 4.3 illustrates the proton spectrum of 1,2-dichlorobenzene (**4.8**),

in which H$_A$ and H$_{A'}$ are seen to have different couplings to H$_X$), which is AA'XX' and relatively complex. Constraints of a ring frequently convey magnetic nonequivalence, as, for example, in propiolactone (**4.9**, in which H$_A$ and H$_{A'}$ also have different couplings to H$_X$). Even open-chain systems such as 2-chloroethanol (ClCH$_2$CH$_2$OH, Figure 4.4)

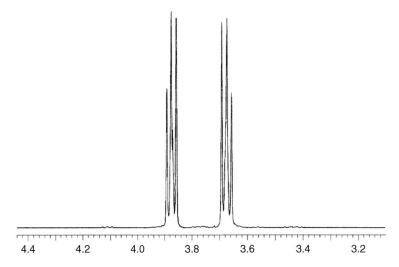

Figure 4.4 The 300 MHz ^{1}H spectrum of 2-chloroethanol (methylene resonances only) in CDCl$_3$.

contain magnetically nonequivalent spin systems, although they are understandable only by an examination of the contributing rotamers. (See two paragraphs down and the problems at the end of the chapter.) Propiolactone, chloroethanol, and both *o*- and *p*-dichlorobenzene thus all give AA'XX' (or AA'BB') spectra (if the hydroxyl proton is ignored in the alcohol).

In Figure 4.4, the second-order character of the spectrum is manifested in two ways. First, peaks do not have the binomial intensity relationship. Thus the inner peaks of each resonance are larger than the outer peaks. A first-order spectrum would have comprised two 1 : 2 : 1 triplets. More careful examination, however, shows that the $n + 1$ rule also fails. Instead of three peaks in each resonance, there are four. The fourth peak requires spectral expansion, but is easily seen on the right side of the central peak of the high-frequency (low-field) resonance and less obviously on the left side of the central peak of the low-frequency resonance.

When protons are on different carbons, it usually is straightforward to determine whether they are chemically equivalent on the basis of symmetry. Geminal protons (those on the same carbon, as in CH$_2$) can be more subtle. Consider the protons of ethylbenzene (C$_6$H$_5$CH$_2$CH$_3$) and those of its β-bromo-β-chloro derivative (C$_6$H$_5$CH$_2$CHClBr). Rotation about the saturated C—C bond generates three rotamers for each molecule, which may be represented by the Newman projections shown in 4.10 and 4.11.

For **4.10**, the three rotamers are identical. In the rotamer **4.10a**, H_A and $H_{A'}$ are chemically equivalent and enantiotopic by reason of the plane of symmetry. When methyl rotation is slow, H_A and $H_{A'}$ are magnetically nonequivalent, because each methylene proton would couple unequally with either H_X or H_Y. The plane of symmetry actually requires that H_Y should be labeled $H_{X'}$, but we retain the distinct lettering to illustrate the effect of methyl rotation. Thus the frozen structure **4.10a** would exhibit an AA'XX'Z spectrum. Rapid methyl rotation averages the X, Y (X'), and Z environments, so that the three methyl protons become chemically equivalent on average. The A and A' protons then have equal couplings to all three methyl protons on average, and hence become magnetically equivalent. On average, there is only one coupling constant, and the spectrum is A_2X_3, when the aromatic protons are ignored.

Molecule **4.11** contains a chiral or stereogenic center in place of the methyl group, so that the three rotamers now are distinct (**4.11a–c**). Moreover, no symmetry operation in any of them relates H_A to H_B. Consequently, even with rapid C—C rotation, H_A and H_B have different chemical shifts and exhibit a mutual coupling constant. They are magnetically nonequivalent by the chemical shift criterion. The spin system is ABX (AMX if the chemical-shift difference is large). The AB protons in **4.11** exemplify a particular class of chemically nonequivalent nuclei that are termed *diastereotopic*. Diastereoisomers are stereoisomers other than enantiomers. Replacement of H_A by deuterium gives **4.11d**,

a diastereoisomer of **4.11e**, which is formed when H_B is replaced by deuterium. The deuterated derivative has two stereogenic centers. In general, the protons of a saturated methylene group are diastereotopic when there is a stereogenic center elsewhere in the molecule because no symmetry operation relates the two protons. The protons in **4.4b** become diastereotopic because the hand provides the stereogenic center. Accidental degeneracy can occur when the chemical-shift difference is small, so that diastereotopic protons can appear to be equivalent in the spectrum.

Groups as well as atoms can be diastereotopic. The methyl groups in an isopropyl group can be diastereotopic when stereogenic centers are present in the molecule, as in α-thujene (**4.12**).

The proton resonance then appears as a pair of doublets (coupled to the methine proton), and the carbon resonance appears as two singlets (proton decoupled).

A stereogenic center is not necessary for methylene protons to be diastereotopic. The diethyl acetal of acetaldehyde (**4.13**) contains diastereotopic protons because the symmetry axis of the molecule is not a symmetry axis for the CH$_2$ protons. This situation may be understood by examining the rotamers or by replacing H$_A$ with deuterium. The latter operation simultaneously creates two stereogenic centers, —OCHD(CH$_3$) and —OCH(CH$_3$)O—, and the resulting molecule is a diastereoisomer of the molecule in which H$_B$ is replaced with deuterium. The methylene protons in *cis*-1,2-dichlorocyclopropane (**4.14a**)

4.14a   4.14b   4.14c

are diastereotopic because **4.14b** and **4.14c** are diastereoisomers. The axial and equatorial protons on a single carbon in ring-frozen cyclohexane are diastereotopic because cyclohexane-d_{axial} and cyclohexane-$d_{equatorial}$ are diastereoisomers. When ring flipping is fast on the NMR time scale, the geminal protons become equivalent on average. Thus, the diastereotopic nature of protons can depend on the rate of molecular interconversions.

4.3 Signs and Mechanisms of Coupling

Spin–spin coupling arises because information about nuclear spin is transferred from nucleus to nucleus via the electrons. Exactly how does this process occur? Several mechanisms have been considered, but the most important is the *Fermi contact mechanism*. Both nuclei and electrons are magnetic dipoles, whose mutual interactions normally are described by the point-dipole approximation, as used, for example, by McConnell in his analysis of diamagnetic anisotropy (Eq. (3.1)). Fermi found that this approximation breaks down when dipoles are very close (comparable to the radius of a proton). Under these circumstances, when the nucleus and the electron are essentially in contact, their interaction is described by a new mechanism, the Fermi contact term. The energy of the interaction is proportional to the gyromagnetic ratios of the nucleus and the electron, the scalar (dot) product of their spins (**I** for a nucleus, **S** for an electron), and the probability Ψ^2 that the electron is at the nucleus (the square of the electronic wave function evaluated with zero distance from the nucleus): $E_{FC} \propto -\gamma_n \gamma_e (\mathbf{I} \cdot \mathbf{S}) \Psi^2(0)$. Because the nuclear and electronic gyromagnetic ratios have opposite signs ($\gamma_{H,C} > 0$, $\gamma_e < 0$), the stabler arrangement is when the nucleus and the electron are antiparallel (spins paired).

According to this model, an electron in a bond X—Y, in which the nuclei of both X and Y are magnetic, spends a finite amount of time at the same point in space as nucleus X. If nucleus X has a spin $I_z = +\frac{1}{2}$, then by the Fermi contact mechanism, the opposite ($-\frac{1}{2}$) spin is favored for the electron. In this way, the nuclear spin polarizes the electron spin (gives one spin state a higher population). The electron in turn shares an orbital in the X—Y bond with another electron, which, by the Pauli Exclusion Principle, must have

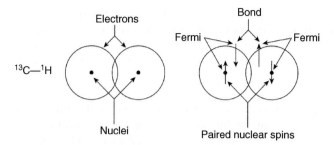

Figure 4.5 Diagram of the Fermi contact mechanism for the indirect coupling of two spins.

a spin of $+\frac{1}{2}$ when the spin of the first electron is $-\frac{1}{2}$. This second $(+\frac{1}{2})$ electron by the Fermi mechanism polarizes the Y nucleus to prefer a spin of $-\frac{1}{2}$ slightly. With two Fermi mechanisms, the energy of the coupling interaction then is proportional to $\gamma_A \gamma_X \gamma_e^2 (\mathbf{I}_A \cdot \mathbf{S}) \Psi_A^2 (0) (\mathbf{I}_X \cdot \mathbf{S}) \Psi_X^2 (0)$. All the constants move into the proportionality constant, designated as the coupling constant J, which then is proportional to the product of the two nuclear gyromagnetic ratios, $\gamma_A \gamma_X$. Thus, whenever nucleus X has a spin of $+\frac{1}{2}$, nucleus Y favors a spin of $-\frac{1}{2}$ slightly. The entire mechanism is illustrated in Figure 4.5 for a $^{13}\text{C}-^{1}\text{H}$ coupling, in which an upward-pointing arrow in the figure represents a $+\frac{1}{2}$ spin, and a downward-pointing arrow a $-\frac{1}{2}$ spin. Since the bonding electrons are used to pass the spin information, the contact term is not averaged to zero by molecular tumbling.

When one spin slightly polarizes another spin oppositely, as in the above model for coupling across X—Y, the coupling constant J between the spins is said by convention to have a positive sign. A negative coupling occurs when spins polarize each other in the same (parallel) direction. Qualitative models analogous to that shown in Figure 4.5 indicate that coupling over two bonds, as in H—C—H, is negative, while coupling over three bonds, as in H—C—C—H, is positive. There are numerous exceptions to this qualitative approach, but it is useful in understanding that J has sign as well as magnitude.

The magnetic dipoles of two magnetically active nuclei also may interact directly through space. This interaction, however, is averaged to zero for nuclei in solution by molecular tumbling, whereby all possible relative orientations are populated. Direct coupling (D, to distinguish it from the indirect J coupling) may be manifested in the spectrum when molecular motion is inhibited in some way: by an external electric field, in liquid crystal solvents, in the solid, or even in liquids under certain circumstances, such as high viscosity. The D coupling can be a nuisance and lead to line broadening, or it can be exploited to obtain information about internuclear distances or absolute signs of coupling. When the sample is subjected to an external electric field, the absolute signs of D can be obtained. The relative signs of D and J can be derived from spectra of a partially oriented molecule, as in a liquid crystal solvent. In this manner, the absolute sign of J can be inferred from the electric-field experiment. In electric-field experiments, Buckingham and McLauchlan showed that the ortho coupling in 4-nitrotoluene has an absolute positive sign. In this way, the signs of all J coupling constants have been related back to a few whose absolute signs actually were measured.

High-resolution NMR spectra normally are not dependent on the absolute sign of coupling constants. Reversal of the sign of every coupling constant in a spin system results in an identical spectrum. Many spectra, however, depend on the *relative signs* of component couplings. For example, the general ABX spectrum is determined in part by three couplings: J_{AB}, J_{AX}, and J_{BX} (Section 4.8). Different spectra can be obtained when J_{AX} and J_{BX} have the same sign (both positive or both negative) from when they have opposite signs (one positive and the other negative), even when the magnitudes are the same.

The usual convention for referring to a coupling constant is to denote the number of bonds between the coupled nuclei by a superscript to the left of the letter J and any other descriptive material by a subscript to the right or parenthetically. A two-bond (*geminal*) coupling between protons then is $^2J_{HCH}$ or $^2J(HCH)$, and a three-bond (*vicinal*) coupling between a proton and a carbon is $^3J_{HCCC}$ or $^3J(HCCC)$. Beyond three bonds, couplings between protons are said to be *long range*.

4.4 Couplings over One Bond

The one-bond coupling between carbon-13 and protons is readily measured from the ^{13}C spectrum when the decoupler is turned off. Although usually unobserved because of decoupling, this coupling provides useful information and illustrates several important principles. Because of the node of p orbitals, such electrons can never come in contact with the nucleus. Thus only electrons in s orbitals can contribute to the Fermi contact mechanism. Indeed, s orbitals have a maximum in electron density at the location of the nucleus. For protons, all electrons reside in the 1s orbital, but, for other nuclei, only that proportion of the orbital that has s character can contribute to coupling. When a proton is attached to an sp^3 carbon atom (25% s character), $^1J(^{13}C-^1H)$ is about half as large as that for a proton attached to an sp carbon atom (50% s character). The alkenic CH (sp^2, 33%) coupling is intermediate. The values of ^1J for methane (sp^3), ethene (sp^2), benzene (sp^2), and ethyne (sp) are 125, 157, 159, and 249 Hz, respectively. These numbers define the following linear relationship between the percentage of s character of the carbon orbital and the one-bond coupling (Eq. (4.1)).

$$\%s(C-H) = 0.2\, J(^{13}C-^1H) \tag{4.1}$$

The zero intercept of this equation indicates that there is no coupling when the s character is zero in agreement with the Fermi contact model.

One-bond CH couplings range from about 100 to 320 Hz and much of this variation may be interpreted in terms of the $J-s$ relationship of Eq. (4.1). The coupling constant in cyclopropane (162 Hz) demonstrates that the carbon orbital to hydrogen is approximately sp^2 hybridized. Intermediate values in hydrocarbons may be interpreted in terms of fractional hybridization. The $J = 144$ Hz for the indicated CH bond in tricyclopentane (**4.15**)

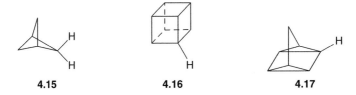

4.15   4.16   4.17

corresponds to 29% s character (sp$^{2.4}$), 160 Hz in cubane (**4.16**) to 32% s character (sp^2), and 179 Hz in quadricyclane (**4.17**) to 36% s character (sp$^{1.8}$). Although the J–s relationship works well for hydrocarbons, there is some question as to its applicability to polar molecules. Variations in the effective nuclei charge, in addition to hybridization effects, may alter the coupling constants.

Just as the resonance frequency of a nucleus is proportional to its gyromagnetic ratio γ (Eq. (1.3)), the coupling constant between two nuclei, as noted above, is proportional to the product of both gyromagnetic ratios, $J(X—Y) \propto \gamma_X \gamma_Y$. Nuclei with very small gyromagnetic ratios, such as ^{15}N, tend to have correspondingly small couplings. Furthermore, in this case, $\gamma(^{15}$N$)$ has a negative sign, whereas $\gamma(^{13}$C$)$ and $\gamma(^1$H$)$ are positive. As a result, one-bond couplings between nitrogen-15 and hydrogen have a negative sign. The sign does not represent an exception to the Fermi model described above (Figure 4.5) but reflects the negative sign of the gyromagnetic ratio.

One-bond couplings have been measured between protons and many other types of nuclei. The coupling between nitrogen and hydrogen ranges from 51 Hz in diphenylketimine [$(C_6H_5)_2C=NH$] to 130 Hz in protonated acetonitrile ($CH_3C\equiv NH^+$). Typical values for sp^2 nitrogen, as in acetamide [$CH_3(C=O)NH_2$] or protonated pyridine, are around 90 Hz, and typical values for sp^3 nitrogen, as in the ammonium ion, are 70–75 Hz. With the three points of hybridization provided by protonated acetonitrile, protonated pyridine, and ammonium, the relationship between coupling and hybridization can be established to be [s(N) = 0.43 $^1J(^{15}$N—H$)$ − 6], and the hybridization, for example, of nitrogen in ammonia (^1J = 61 Hz) can be calculated to be sp^4 (s = 20%). Because of the negative gyromagnetic ratio of nitrogen, all these couplings actually are negative, but the signs have been omitted.

One-bond coupling constants between ^{31}P and ^{1}H range from 186 Hz (PH_2—PH_2) to 707 Hz (H_3PO_3). Couplings of ^{11}B to ^{1}H range from 29 Hz (the bridging hydrogen in $H_2BHBH_2\cdot N(CH_3H)$ to 211 Hz in HBF_2. Some extremely large couplings have been observed lower in the periodic table, for example, $^1J(^{119}$Sn—^{1}H$)$ = 1931 Hz in SnH_4, $^1J(^{195}$Pt—^{1}H$)$ = 1307 Hz in [$P(C_2H_5)_3]_2PtHCl$, and $^1J(^{207}$Pb—^{1}H$)$ = 2379 Hz in $(CH_3)_3PbH$.

When neither nucleus is a proton, the coupling constant depends on the product of the s characters of the orbitals from both nuclei that form the bond. Thus, $^1J(^{13}$C—^{13}C$)$ depends on the hybridization of both carbons. With two unknown hybridizations and only one measured coupling, s character can be measured only when the symmetry of the molecule is such that the two carbons are identical. Measurements with molecules such as ethane, ethene, and ethyne led to the relationship $s^2(C) = 17.4\ ^1J(^{13}$C—^{13}C$) + 60$. The range of ^{13}C—^{13}C couplings is from 34 Hz in $C_6H_5CH_2CH_3$ to 176 Hz in $C_6H_5C\equiv CH$. The one-bond coupling between two carbon atoms is readily measured by the technique known as INADEQUATE (Section 5.7) and is extremely useful in mapping carbon connectivities in complex molecules. Coupling between two heavy nuclei with lone pairs, as in $^1J(^{31}$P—^{31}P$)$ or $^1J(^{13}$C—^{15}N$)$, appears to have complexities resulting from effects other than hybridization. Again, extremely large couplings have been observed between two nuclei when at least one is from lower in the periodic table (e.g. $^1J^{205}$Tl—^{19}F$)$ = 12 000 Hz in TlF and $^1J(^{195}$Pt—^{31}P$)$ = 700 Hz in $\{[(C_2H_5)_2O]_3P\}_2PtCl_2)$. Examples of one-bond couplings are summarized in Table 4.1 at the end of the chapter.

4.5 Geminal Couplings

The geminal coupling between two protons (H—C—H) may be measured directly from the spectrum when the coupled nuclei are chemically nonequivalent, thus constituting, for example, the AB or AM part of an ABX, AMX, or ABX_3, etc., spectrum. If the relationship is first order (AM), the coupling may be measured by inspection, but in second-order cases (AB), the spectrum must be simulated computationally, unless the two spins are isolated (a *two-spin system*, Section 4.7). When geminal nuclei are chemically equivalent but magnetically nonequivalent, as in the AA' part of an AA'XX' spectrum, the coupling constant often is accessible by computational methods.

Splittings are not observed between coupled nuclei when they are magnetically equivalent, but the coupling constant may be measured by replacing one of the nuclei with deuterium. For example, in dichloromethane-d ($CHDCl_2$), the geminal H—C—D coupling is seen as the spacing between the components of the 1 : 1 : 1 triplet (deuterium has a spin of 1). Since coupling constants are proportional to the product of the gyromagnetic ratios of the coupled nuclei, $J(HCH)$ may be calculated from $J(HCD)$ by Eq. (4.2).

$$J(HH) = \frac{\gamma_H}{\gamma_D} J(HD) = 6.51 J(HD) \tag{4.2}$$

Geminal couplings depend strongly on the angle formed by the three atoms, H—C—H, as seen in the saturated, cyclic hydrocarbon series [cyclohexane (−12.6 Hz), cyclopentane (−10.5 Hz), cyclobutane (−9 Hz), and cyclopropane (−4.3 Hz)], or by comparison of acyclic alkanes (methane, −12.4 Hz) with acyclic alkenes (ethene, +2.3 Hz). Note that the sign of the coupling constant is important. Although most geminal couplings are negative, many of those for sp^2 carbons are positive.

The typical range for alkanes is −5 to −20 Hz and for alkenes is +3 to −3 Hz (or 0–3 Hz if the sign is ignored). The c. 15 Hz difference between these ranges is due to significant changes in structure and hybridization. Rather than attempting to explain these differences, Pople and Bothner-By used the values of ethane and ethene as class standards and related deviations to two effects: polar or induction (σ effects) and hyperconjugation (π effects). A substituent that withdraws electrons from a CH_2 group by induction (σ withdrawal), they found, makes ^2J more positive, and a σ-donating substituent makes ^2J less positive (or more negative). In contrast, that found that a substituent that withdraws electrons by hyperconjugation (π withdrawal) makes ^2J less positive, but a π donor makes it more positive.

For alkanes, the negative coupling thus decreases in absolute value (becoming less negative) from −12.4 Hz for methane to −10.8 Hz for CH_3OH, −9.2 Hz for CH_3I, and −5.5 Hz for CH_2Br_2. Electron donation makes the coupling more negative (higher absolute value), as in −14.1 Hz in TMS (Me_4Si). Analogous substitution on sp^2 carbon changes the coupling profoundly, as in the effect of electron withdrawal in the structure $H_2C=X$, for example, +2.3 Hz when X is CH_2 (ethene), +17 Hz when X is N-(*tert*-butyl) (an imine), and +40 Hz when X is O (formaldehyde). These positive couplings are becoming more positive, paralleling the change to be less negative for negative couplings.

The σ effects (induction) can be augmented or decreased by the π effects of hyperconjugation. Lone pairs of electrons can donate electrons and make J more positive, whereas the π orbitals of double or triple bonds can withdraw electrons and make J less positive (or more negative). The above-mentioned large increase in the geminal coupling of imines or formaldehyde compared with that of ethene results from reinforcement of the effects of σ withdrawal and π donation, as illustrated in structure **4.18**.

The effect of π withdrawal also occurs for carbonyl, nitrile, and aromatic groups, as in the values for acetone (−14.9 Hz), acetonitrile (−16.9 Hz), and dicyanomethane (−20.4 Hz). The π effect is somewhat reduced by free rotation in open-chain systems, but particularly large effects are created by constraints of rings, as seen in **4.19** and **4.20**,

as well as in the α protons of cyclopentanones and cyclohexanones. Structure **4.21** illustrates an example of how π donation by lone pairs makes J more positive. This effect also explains the difference in the geminal couplings of three-membered rings: cyclopropane (−4.3 Hz) and oxirane [$(CH_2)_2O$, +5.5 Hz]. Although the difference in the absolute value of the couplings is only 1.2 Hz, when signs are taken into consideration, the difference is almost 9 Hz.

Geminal coupling between protons and other nuclei also has been studied. The H—C—^{13}C coupling responds to substituents in much the same way as does the H—C—H coupling. The HCC couplings are smaller because of the smaller gyromagnetic ratio of ^{13}C. Typical values in alkanes are −4.8 Hz in **H—CH$_2$—CH$_3$** and +1.2 Hz in **H—CCl$_2$—CHCl$_2$**. Couplings from hydrogen to sp^2 carbon (**H—CH$_2$—C=**) are typically −4 to −7 Hz, as in acetone [−5.9 Hz, **H—CH$_2$—(C=O)CH$_3$**]. When the intermediate carbon is sp^2 [**H—(C=X)—C**], the coupling becomes larger and positive, as in aldehydes (+26.7 Hz for acetaldehyde, **H—(C=O)—CH$_3$**, but typically 5–10 Hz in alkenes [**H—(C=CR$_2$)—C**].

Unlike the proton–proton case, the proton–carbon geminal-coupling pathway can include a double bond (**H—C=C**), and factors not considered by Pople and Bothner-By become important. For sp^2 carbons, these couplings often are small (−2.4 Hz for ethene, **H—CH=CH$_2$**). With proper substitution, however, stereochemical differences may be observed, as in *cis*-dichloroethene (**4.22**,

16.0 Hz) and *trans*-dichloroethene (**4.23**, 0.8 Hz). Such differences between alkene stereoisomers are common and may be exploited to prove stereochemistries. The geminal couplings in aromatics (**H—C=C**) are 4–8 Hz. For sp carbons, the coupling becomes quite large: 49.3 Hz in ethyne (**H—C≡CH**) and 61.0 Hz in **H—C≡C—O—Ph**.

The two-bond coupling between hydrogen and nitrogen-15 strongly depends on the presence and orientation of the nitrogen lone pair. The H—C—^{15}N coupling in imines is larger and negative when the proton is cis to the lone pair, but smaller and positive for a proton trans to the lone pair, as in **4.24**.

4.24   **4.25**

Thus, ^2J(HCN) is a useful structural diagnostic for syn–anti isomerism in imines, oximes, and related compounds. In saturated amines with rapid bond rotation, however, values typically are small and negative (−1.0 Hz for methylamine, CH$_3$NH$_2$). The cis relationship between the nitrogen lone pair and hydrogen also is found in heterocycles such as pyridine (**4.25**), in which the coupling constant is −10.8 Hz.

Two-bond couplings between ^{15}N and ^{13}C follow a similar pattern and also can be used for structural and stereochemical assignments. The carbon on the same side as the lone pair in imines again has a large negative coupling (−11.6 Hz in **4.26**).

4.26   **4.27**   **4.28**

The isomer shown in **4.26**, in which the methyl is syn to hydroxyl (anti to the lone pair), has a ^2J(CCN) of only 1.0 Hz. The two indicated carbons in quinoline (**4.27**) have couplings, respectively, of −9.3 and +2.7 Hz, as one is syn and the other anti to the nitrogen lone pair.

Couplings between ^{31}P and hydrogen also have been exploited stereochemically. The maximum positive value of ^2J(HCP) is observed when the H—C bond and the phosphorus lone pair are eclipsed (syn), and the maximum negative value when they are orthogonal or anti. The situation is similar to that for couplings between hydrogen and ^{15}N, but signs are reversed as a result of the opposite signs of the gyromagnetic ratios of ^{15}N and ^{31}P. The heterocycle **4.28** exhibits a coupling of +25 Hz between ^{31}P and H$_a$ (syn) and of −6 Hz between ^{32}P and H$_b$ (anti). The coupling also is structurally dependent, as it is larger for P(III) than for P(V): 27 Hz for (CH$_3$)$_3$P: and 13.4 Hz for (CH$_3$)$_3$P=O.

Geminal H—C—F couplings usually are close to +50 Hz for an sp^3 carbon (47.5 Hz for CH$_3$CH$_2$F) and +80 Hz for an sp^2 carbon (84.7 Hz for CH$_2$=CHF. Geminal F—C—F couplings are quite large (+150–250 Hz) for saturated carbon (240 Hz for 1,1-difluorocyclohexane), but less than 100 Hz for unsaturated carbon (35.6 Hz for CH$_2$=CF$_2$).

4.6 Vicinal Couplings

Coupling between protons over three bonds provided the most important early stereochemical applications of NMR spectroscopy. In 1961, Karplus derived a mathematical relationship between ^3J(HCCH) and the H—C—C—H dihedral angle ϕ. The simple formula of Eq. (4.3),

$$^3J = \begin{cases} A \cos^2\phi + C & (\phi = 0-90°) \\ A' \cos^2\phi + C' & (\phi = 90-180°) \end{cases} \quad (4.3)$$

as illustrated in Figure 4.6, offers chemists a general and easily applied qualitative tool. The cosine-squared relationship results from a strong coupling when orbitals are parallel. They can overlap at the synperiplanar and antiperiplanar geometries (respectively, $\phi = 0-30°$ and $150-180°$, with ϕ defined in Figure 4.6). When orbitals are staggered or orthogonal ($\phi = 60-120°$), coupling is weak.

The additive constants C and C' usually are neglected, as they are thought to be less than 0.3 Hz. When the constants A and A' can be evaluated, quantitation is possible. The inequality of A and A' ($A < A'$) means that J is different at the syn maximum on the left and the anti maximum on the right of Figure 4.6. Unfortunately, these multiplicative constants vary from system to system in the range of c. 8–14 Hz (larger for alkenes). Because of this variation, quantitative applications cannot be transferred easily from one structure to another.

The Karplus equation provides useful qualitative interpretations in a number of very fundamental systems. In chair cyclohexanes (Figure 4.7), the coupling constant between adjacent axial protons (J_{aa}) is large (8–13 Hz) because ϕ_{aa} is close to 180°, whereas the coupling between two equatorial protons ($J_{ee} = 0-5$ Hz) and between an axial and an equatorial proton ($J_{ae} = 1-6$ Hz) are small because ϕ_{ee} and ϕ_{ae} are close to 60°. With reference to the Karplus plot (Figure 4.6), a value of 180° is at its maximum, and 60° is close to its minimum. An axial proton that has an axial proton neighbor is identified

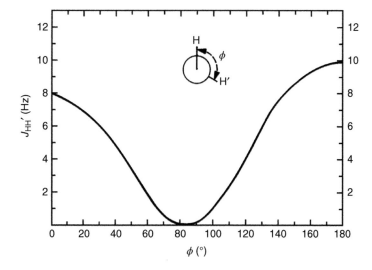

Figure 4.6 The vicinal H—C—C—H coupling constant as a function of the dihedral angle ϕ.

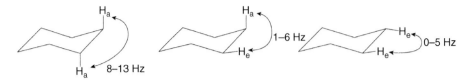

Figure 4.7 Coupling magnitudes between vicinal protons in chair six-membered rings (axial protons are designated by the letter a and equatorial protons by the letter e).

readily by its large J_{aa}. When cyclohexane rings are flipping between two chair forms, J_{aa} is averaged with J_{ee} to give an averaged J_{trans} in the range 4–9 Hz, and J_{ae} is averaged with J_{ea} to give an averaged, smaller J_{cis} in the range 1–6 Hz. Although the ranges overlap because of substituent effects, the average J_{trans} always is larger than the average J_{cis}. In complex spin systems, axial proton resonances sometimes can be recognized by their larger line widths or total spread than those of equatorial protons, because J_{aa} increases the width or spread more than does J_{ae} or J_{ee}. By way of example, the 2 proton in 2-bromocyclohexanone (**4.29**)

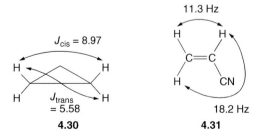

4.29a   4.29e

may be either axial (**4.29a**) or equatorial (**4.29e**). (The designations a and e refer to the location of the proton geminal to bromine, not to the bromine.) Because these two conformations are interconverting rapidly at room temperature, actual couplings of the protons at the 2 position with the two vicinal protons at the 3 position are therefore weighted averages, as given by Eq. 4.4a,

$$J_{trans} = aJ_{aa} + eJ_{ee} \qquad (4.4a)$$

$$J_{cis} = aJ_{ea} + eJ_{ae} \qquad (4.4b)$$

in which a is the fraction of **4.29a** and e is the fraction of **4.29e**. Similar considerations apply to acyclic conformations.

In three-membered rings (**4.30**),

$J_{cis} = 8.97$

$J_{trans} = 5.58$

4.30

11.3 Hz

18.2 Hz

4.31

J_{cis} ($\phi = 0°$) is always larger than J_{trans} ($\phi = 120°$), as reference to the Karplus plot in Figure 4.6 indicates. For the parent cyclopropane, $J_{cis} = 8.97$ Hz and $J_{trans} = 5.58$ Hz. In four-membered rings, the cis coupling usually is larger than the trans coupling, but the two quantities can be close enough to be ambiguous. In five-membered rings, either J_{cis} or J_{trans} can be the larger, because the dihedral angles are toward the center of the Karplus curve and vary with the complex conformational mix present in these rings.

In alkenes, J_{trans} ($\phi = 180°$) always is larger than J_{cis} ($\phi = 0°$), e.g. 18.2 and 11.3 Hz, respectively, in acrylonitrile (**4.31**). The spectrum of this compound is illustrated in Figure 4.8. The vinyl resonance is composed of three groupings, which constitute an AMX, ABX, or ABC spectrum, depending on the field strength. The three couplings (J_{AM}, J_{AX}, and J_{MX}) may be assigned by inspection for the AMX case with the knowledge that $^3J_{trans} > {}^3J_{cis} > {}^2J_{gem}$. The reader should carry out the measurements for the couplings in **4.31** from Figure 4.8. The 12 peaks are located at δ 5.647, 5.684, 5.708, 5.744, 6.090, 6.093, 6.128, 6.132, 6.211, 6.215, 6.273, and 6.278. Finally, ortho protons in aromatic rings have a dihedral angle of 0°, so $^3J_{ortho}$ generally is quite large (6–9 Hz) and can be distinguished from the smaller J_{meta} and J_{para}.

Despite the potentially general application of the Karplus equation to dihedral angle problems, there are quantitative limitations. The vicinal H—C—C—H coupling constant additionally depends on the C—C bond length or bond order, the H—C—C valence angles, the electronegativities of substituents on the carbon atoms, and the stereochemical orientation of these substituents, in addition to the H—C—C—H dihedral angles. All of these other factors contribute to the A and A' factors in Eq. (4.3). A properly controlled calibration series of molecules must be rigid (monoconformational) and have unvarying bond lengths and valence angles. Several approaches have been developed to take the only remaining factor, substituent electronegativity, into account. One approach is to derive the mathematical dependence of ^3J on electronegativity. Another is empirical allowance through the use of chemical shifts that depend on electronegativity in a similar fashion to ^3J. A third approach is to eliminate the problem through the use of the ratio (the R value) of two ^3J coupling constants that respond to the same or related

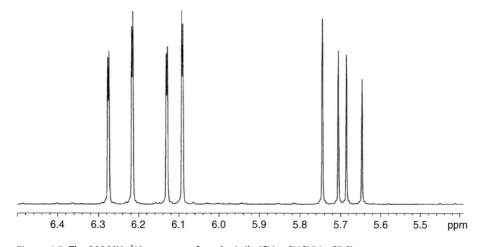

Figure 4.8 The 300 MHz ^{1}H spectrum of acrylonitrile (CH_2=CHCN) in $CDCl_3$.

dihedral angles and that have the same multiplicative dependence on substituent electronegativity. The ratio R then divides out the effect of electronegativity. These more sophisticated versions of the Karplus method have been used quite successfully to obtain reliable quantitative results.

The existence of factors other than the dihedral angle results in ranges of vicinal coupling constants at constant ϕ, even in structurally analogous systems. Saturated hydrocarbon chains exhibit vicinal couplings (H—C—C—H) in the range 3–9 Hz, depending on the substituent electronegativity and rotamer mixes (J = 3.06 Hz for $Cl_2CHCHCl_2$ and 8.90 for CH_3CH_2Li). Higher substituent electronegativity always lowers the vicinal coupling constant. In small rings, the variation is almost entirely the result of substituent electronegativity, with cis ranges of 7–13 Hz and trans ranges of 4–10 Hz in cyclopropanes. Coupling constants in oxiranes (epoxides) are smaller because of the effect of the electronegative oxygen atom. Couplings across a double bond, H—C=C—H, depend strongly on the valence angles, as well as on the electronegativities of the other two substituents. In cycloalkenes, the value varies from 1.3 Hz in cyclopropene to 8.8 Hz in cyclohexene, all with the dihedral angle $\phi = 0°$. In acyclic alkenes, J_{trans} has a range of 10–24 Hz, J_{cis} of 2–19 Hz. Because the ranges overlap, the distinction between cis and trans isomers is fully reliable only when the coupling constants are measured for both isomers. When bonds are intermediate between single and double bonds, ^3J is proportional to the overall bond order, as in $^3J_{12} = 8.6$ Hz and $^3J_{23} = 6.0$ Hz in naphthalene, although $\phi = 0°$ in both the cases.

The ortho coupling in benzene derivatives varies over the relatively small range of 6.7–8.5 Hz, depending on the resonance and inductive effects of the substituents. The presence of heteroatoms in the ring expands the range at the lower end down to 2 Hz, because of the effects of electronegativity (pyridines) and of smaller rings (furans and pyrroles).

When one carbon in the coupling pathway is sp^3 and one is sp^2 (**4.32**)

$$H-\underset{|}{\overset{|}{C}}-\overset{X}{\underset{}{\overset{\|}{C}}}-H \qquad H-\overset{X}{\underset{}{\overset{\|}{C}}}-\overset{Y}{\underset{}{\overset{\|}{C}}}-H$$

4.32   **4.33**

the range is 5–8 Hz for freely rotating acyclic hydrocarbons (X = CR$_2$) and 1–5 Hz for aldehydes (X = O). The value varies in hydrocarbon rings from −0.8 Hz in cyclobutene to +3.1 Hz in cyclohexene and +5.7 Hz in cycloheptene. For the central bond in dienes (**4.33**), the range is 10–12 Hz for transoid systems (X, Y = CR$_2$). When constrained to rings, the pathway is cisoid and the coupling is 1.9 Hz in cyclopentadiene and 5.1 Hz in 1,3-cyclohexadiene. In α,β-unsaturated aldehydes (**4.33**, X = O, Y = CR$_2$), the coupling is about 8 Hz when transoid and 3 Hz when cisoid.

The H—C—X—H (X = O, N, S, Si, etc.), H—C—C—C, H—C—C—F, H—C—N—F, H—C—X—P (X = C, O, S), and C—C—C—C couplings also follow Karplus-like relationships. The ^3J(H—C—O—P) couplings are useful in determining backbone conformations of nucleotides. The ^3J(C—C—C—C) couplings have a range of values (3–15 Hz) that is larger than that in the two-bond case. (The range for ^2J(C—C—C) is 1–10 Hz.) The F—C—C—F and H—C—C—P couplings appear not to follow the Karplus pattern.

4.7 Long-range Couplings

Coupling between protons over more than three bonds is said to be *long range*. Sometimes, coupling between ^{13}C and protons over two $^2J(CCH)$ or three $^3J(CCCH)$ bonds also is called long range, but the term is inappropriate. Long-range coupling constants between protons normally are less than 1 Hz and frequently are unobservably small. In at least the following three structural circumstances, however, such couplings commonly become significant.

4.7.1 σ–π Overlap

Interactions of C—H (σ) bonds with π electrons of double bonds, triple bonds, or aromatic rings along the coupling pathway often increase the magnitude of the coupling constant. One such case is the four-bond *allylic coupling*, HC—C=CH, with a range of about +1 to −3 Hz and typical values close to −1 Hz in freely rotating systems. Larger values are observed when the saturated C—H$_a$ bond (**4.34**)

is parallel to the π orbitals. σ–π Overlap in this arrangement enables coupling to be transmitted more effectively. When the C—H$_a$ bond is orthogonal to the π orbitals, there is no σ–π contribution and couplings are small (<1 Hz). In acyclic systems, the dihedral angle is averaged over both favorable and unfavorable arrangements, so an average ^4J is found, as in 2-methylacryloin (**4.35**, $^4J = |1.45|$ Hz). Ring constraints can freeze bonds into a favorable arrangement, as in indene (**4.36**, $^2J = -2.0$ Hz).

The five-bond doubly allylic coupling (also called *homoallylic*) HC—C=C—CH depends on the orientation of two C—H bonds with respect to the π orbitals. For acyclic systems such as the 2-butenes, ^5J typically is 2 Hz, with a range of 0–3 Hz. When both protons are well aligned, the coupling can be quite large, as in the planar 1,4-cyclohexadiene (**4.37**),

for which the cis coupling is 9.63 Hz and the trans coupling is 8.04 Hz. These couplings were measured by suitable deuterium labeling. It is not unusual for the doubly allylic coupling to be larger than the allylic, as in **4.38** [$^4J(CH_3—H_a) = 1.1$ Hz, $^5J(CH_3—H_b) = 1.8$ Hz).

Coupling constants are particularly large in alkynic and allenic systems, in which σ–π overlap can be very effective. In allene itself ($CH_2=C=CH_2$, **4.39**),

4.39   **4.40**

^4J is –7 Hz. In 1,1-dimethylallene (**4.40**), the five-bond ^5J is relatively large, 3 Hz. Allene stereochemistry is locked into a favorable arrangement for σ–π overlap, as is illustrated by the pair of arrows from the CH_2 group into the π orbitals of **4.39**. In both propyne (methylacetylene, $^4J = 2.9$ Hz) and 2-butyne (dimethylacetylene, $^5J = 2.7$ Hz), the long-range coupling is enhanced because the triple bond imposes no steric limitations on σ–π overlap. Appreciable long-range couplings have been observed over up to seven bonds in polyalkynes.

Protons on saturated carbon atoms attached to an aromatic ring ($CH_3-C_6H_5$) couple with all three types of protons on the ring (ortho, meta, and para). These *benzylic couplings* depend on the σ–π interaction between the substituent C—H σ bonds and the aromatic π electrons, much like the allylic coupling ($^4J_{ortho} = 0.6$–0.9 Hz, $^5J_{meta} = 0.3$–0.4 Hz, $^6J_{para} = 0.5$–0.6 Hz). A doubly benzylic coupling, analogous to the homoallylic coupling, can take place between protons on different saturated carbons, both of which are directly attached to the benzene ring, as in the xylene ($CH_3-C_6H_4-CH_3$, $^5J_{ortho} = 0.3$–0.5 Hz).

4.7.2 Zigzag Pathways

In the second major category of long-range coupling, enhanced values often are observed between protons that are related by a planar W or zigzag pathway. This geometry is seen, for example, in the 1,3-diequatorial arrangement between protons in chair six-membered rings (**4.41**,

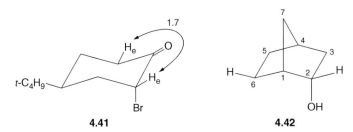

4.41   **4.42**

$^4J = 1.7$ Hz). The norbornane framework (**4.42**) contains several W arrangements, including that illustrated not only between the 2 and 6 exo protons but also between the bridgehead protons (1 and 4) and between the 3-endo and 7-anti protons.

In the planar, zigzag arrangement, there is a favorable overlap between parallel C—H and C—C bonds, analogous to the optimal vicinal coupling at $\phi = 180°$. The large magnitude also has been explained in terms of direct passing of spin information by interaction of the rear lobes of the C—H orbitals in the HC—C—CH pathway, called a *percaudal interaction* from the Latin for "through the tail." When the first and third carbons are particularly close because of ring strain or when there are multiple zigzag pathways, the coupling can be quite large, as in **4.43**

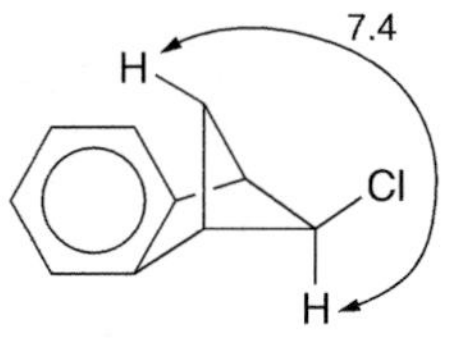

4.43 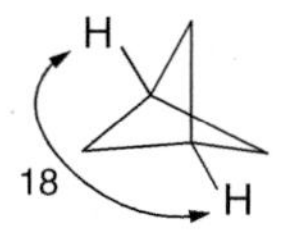 **4.44**

($^4J = 7.4$ Hz) and **4.44** ($^4J = 18$ Hz).

The zigzag pathway is entirely within the σ framework but can be important for many π systems, including aromatic meta couplings (hence the enhanced $^4J = 1.37$ Hz in benzene, **4.45**).

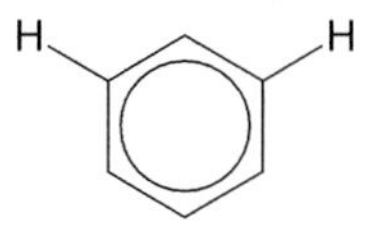

4.45 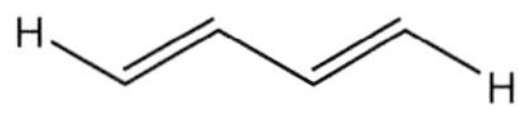 **4.46**

Despite the enhanced meta value, the ortho coupling still is larger (a Karplus maximum for $\phi = 0°$), so the order remains that $^3J_{ortho}$ (7.54 Hz) > $^4J_{meta}$ (1.37 Hz) > $^5J_{para}$ (0.69) in benzene itself. There is no enhanced mechanism for $^5J_{para}$, since there is neither a zigzag pathway nor an opportunity for σ–π overlap. In 1,3-butadiene, there are two four-bond (−0.86 and −0.83 Hz) and three five-bond (+0.60, +1.30, and +0.69 Hz) couplings. The enhanced value of one of the five-bond couplings derives from the zigzag pathway depicted in **4.46**. A similar pathway is present in quinoline (**4.47**)

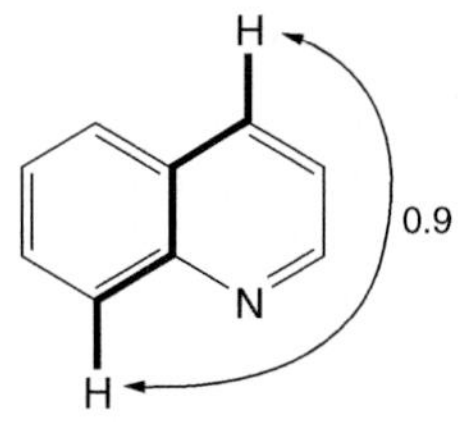

4.47

with $^5J = 0.9$ Hz coupling between the indicated protons. Zigzag pathways up to six bonds have been found to exhibit couplings. None of these couplings in butadiene and benzene involves σ–π overlap, as they all occur within a plane.

4.7.3 Through-Space Coupling

Although coupling information is always passed via electron-mediated pathways, in some cases part of the through-bond pathway may be skipped, as in allylic (**4.34**) and benzylic couplings with σ–π overlap. Two nuclei that are within van der Waals contact in space over any number of bonds can interchange spin information if at least one of the nuclei possesses lone-pair electrons. These *lone-pair-mediated, through-space couplings* are found most commonly, but not exclusively, in H—F and F—F pairs. The six-bond CH_3—F coupling is negligible in **4.48**

CH₃ F   CH₃ F
<0.5   8.3
4.48   **4.49**

(H—F distance 2.84 Å), but is 8.3 Hz in **4.49** (1.44 Å). (The sum of the H and F van der Waals radii is 2.55 Å.) In the latter case, coupling information is probably passed from the proton through the lone-pair electrons to the fluorine nucleus. Such a mechanism very likely is important in the geminal F—C—F coupling in saturated systems, which is unusually large. Values of ^2J(FCF) are larger for sp³ CF_2 (c. 200 Hz) than for sp² CF_2 (c. 50 Hz), as the smaller tetrahedral angle brings the fluorine atoms closer together.

4.8 Spectral Analysis

We have not said much about how coupling constants are extracted from spectra. Measurement is straightforward when the spectrum is first order, as chemical shifts correspond to the midpoint of a resonance multiplet. The midpoint falls between the components of a doublet from coupling to one other spin, is coincident with the middle peak of a triplet from coupling to two other spins, and so on. The coupling constant corresponds to the distance between adjacent peaks in the resonance multiplet. These ideal characteristics may fail in second-order spectra. Because most nuclei other than the proton have very large chemical shift ranges and because these nuclei often are in low natural abundance and hence do not couple with each other, second-order analysis is a consideration primarily for proton spectra alone. For protons, spectra measured above 500 MHz are usually first order, from the $\Delta \nu/J$ criterion. Magnetic nonequivalence (Section 4.2), however, is independent of the field and produces second-order spectra such as AA′XX′ even with the highest field superconducting magnet.

The AX spectrum consists of two doublets, all four components having equal intensities. The spectra in Figure 1.18 and at the bottom of Figure 4.1 are very close to first order. The doublet spacing is J_{AX}, and the midpoints of the doublets are ν_A and ν_X. The second-order, two-spin (AB) system also contains four lines, but the inner peaks are always more intense than the outer peaks (Figures 4.1 and 4.9). The coupling constant (J_{AB}) still is obtained directly and accurately from the spacings within the doublets, but no specific peak position or simple average correspond to the chemical shifts. The A chemical shift (ν_A) occurs at the weighted-average position of the two A peaks (and similarly for ν_B), weighted according to peak heights. The chemical shift difference, $\Delta \nu_{AB} = (\nu_B - \nu_A)$, is calculated easily from Eq. (4.5),

$$\Delta \nu_{AB} = (4C^2 - J^2)^{1/2} \tag{4.5}$$

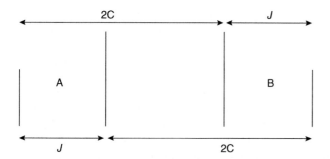

Figure 4.9 Notation for spacings in a second-order, two-spin system (AB).

in which $2C$ is the spacing between alternate peaks (Figure 4.9). The values of v_A and v_B then are readily found by adding $\frac{1}{2}\Delta v_{AB}$ to and subtracting it from the midpoint of the quartet. The ratio of intensities of the larger inner peaks to the smaller outer peaks is given by the expression $(1 + J/2C)/(1 - J/2C)$. All of these relationships are developed in Appendix C, which describes the quantum mechanical formulation of the two-spin system.

Three-spin systems can be analyzed readily by inspection only in the first-order cases AX_2 and AMX. The second-order AB_2 spectrum can contain up to nine peaks: four from spin flips of the A proton alone, four from spin flips of the B protons alone, and one from simultaneous spin flips of both the A and the B protons. The ninth peak is called a *combination line* and ordinarily is forbidden and of low intensity. Although sometimes these patterns may be analyzed by inspection, recourse normally is made to computer programs. The other second-order three-spin systems (ABB′, AXX′, ABX, and ABC) and almost all second-order systems of four spins (AA′BB′, AA′XX′, ABXY, etc.) or larger are seldom able to be analyzed by inspection, so that computer methods must be employed. Appendix D describes methods for analyzing some three- and four-spin systems.

Most spectrometers today provide software for calculation of spectra with up to seven spins. The first step is a trial-and-error procedure of approximating the chemical shifts and coupling constants in order to match the observed spectrum through computer simulation. Chemical shifts are varied until the widths and locations of the observed and calculated multiplets approximately agree. Then the coupling constants or their sums and differences are varied systematically until a reasonable match is obtained. This method is relatively successful for three and four spins but is difficult to employ with larger systems.

Refinements of direct calculations or of this trial-and-error procedure utilize iterative computer programs. The program of Castellano and Bothner-By (LAOCN-5 in its later version) iterates on peak positions, but it requires assignments of peaks to specific spin flips. The program of Stephenson and Binsch (DAVINS) operates directly on unassigned peak positions.

4.9 Second-order Spectra

4.9.1 Deceptive Simplicity

Second-order spectra are characterized by peak spacings that do not correspond to coupling constants, by nonbinomial intensities, by chemical shifts that are not at the midpoints of resonance multiplets, and by resonance multiplicities that do not follow the $n + 1$ rules (see Figures 4.1–4.3). Even when the spectrum has the appearance of being first order, it may not be. Lines can coincide in such a way that the spectrum assumes a simpler appearance than seems consistent with the actual spectral parameters (a situation called *deceptive simplicity*). For example, in the ABX spectrum, the X nucleus is coupled to two nuclei (A and B) that are themselves closely coupled ($\Delta v/J$ <c. 10). Under these circumstances, the A and B *spin states* are fully mixed, and X responds as if the nuclei were equivalent. Spin states are the quantum mechanical representation of the spins, as described in the appendices. Under conditions of extreme close coupling, the ABX spectrum can have the appearance of an A_2X spectrum, as if $J_{AX} = J_{BX}$. Figure 4.10 illustrates this situation. When $\Delta v_{AB} = 3.0$ Hz (Figure 4.10a), the calculated example looks like a first-order A_2X spectrum with one coupling constant, even though $J_{AX} \neq J_{BX}$. When $\Delta v_{AB} = 8.0$ Hz (Figure 4.10b), a typical ABX spectrum is obtained. Deceptive simplicity sometimes, but not always, can be removed by use of a higher field. When the spectrum is deceptively simple, only sums or averages of coupling constants may be measured. Actual coupling constants are impossible to obtain.

The AA'XX' spectrum often is observed as a deceptively simple pair of triplets, resembling A_2X_2. In this case, it is the A and A' nuclei that are closely coupling ($\Delta v_{AA'} = 0$ Hz and $J_{AA'}$ is large). Such deceptive simplicity is not eliminated by raising the field because A and A' are chemically equivalent. The chemist should beware of the pair of triplets that falsely suggests magnetic equivalence (A_2X_2) and equal couplings ($J_{AX} = J_{AX'}$), when the molecular structure suggests AA'XX'. Sometimes the couplings between A and X may be observed by lowering the field to turn the AA'XX' spectrum into AA'BB', with a larger number of peaks that may permit complete analysis.

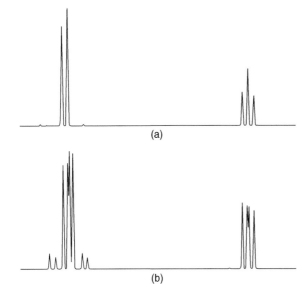

Figure 4.10 (a) A deceptively simple ABX spectrum: $v_A = 0.0$ Hz, $v_B = 3.0$ Hz, $v_X = 130.0$ Hz, $J_{AB} = 15.0$ Hz, $J_{AX} = 5.0$ Hz, and $J_{BX} = 3.0$ Hz. (b) The same parameters, except that $v_B = 8.0$ Hz. The larger value of Δv_{AB} removes the deceptive simplicity and produces a typical ABX spectrum.

4.9.2 Virtual Coupling

A particularly subtle example of second-order complexity occurs in the ABX spectrum (or, more generally, $A_xB_yX_z$) in those circumstances when A and B are very closely coupled, J_{AX} is large, and J_{BX} is zero. With no coupling to B, the X spectrum should be a simple doublet from coupling to A. Since A and B are closely coupled, however, the spin states of A and B are mixed, and the X spectrum is perturbed by the B spins. The phenomenon has been termed *virtual coupling*, which is somewhat of a misnomer, since B is not coupled to X. As an example in a slightly larger but analogous spin system, the CH and CH_2 protons of β-methylglutaric acid (**4.50**)

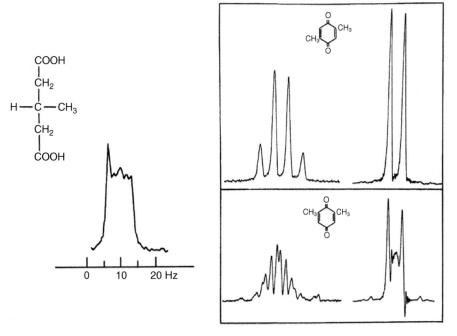

are closely coupled. Although the CH_3 group is coupled only to the CH proton, its resonance is much more complicated than a simple doublet (Figure 4.11). The CH and CH_2

Figure 4.11 (left) The 60 MHz methyl ^{1}H resonance of β-methylglutaric acid. (right) The 60 MHz spectra of the 2,5- and 2.6 dimethylquinones. Sources: (left) Anet 1961 [1]. Reproduced with permission from NRC Press. (right) Becker [2]. Reproduced with permission of Academic Press.

protons are closely coupled, so their spin states are mixed. The CH_3 group interacts with a mixture of CH and CH_2 spin states, even though $J = 0$ between CH_3 and CH_2. This problem is eliminated by a higher field, at which the CH and CH_2 resonances are well separated. Methyl, unmixed with the CH_2 spin states, then couples cleanly with CH.

The dimethylbenzoquinones provide a further example of so-called virtual coupling. The proton spectrum of the 2,5-dimethyl isomer (**4.51**) contains a first-order methyl doublet and an alkene quartet. The spectrum of the 2,6 isomer (**4.52**) is very complicated (Figure 4.11). The alkenic protons in both molecules are equivalent (AA'). In **4.51**, they are coupled only to the methyl groups ($J_{AA'} = 0$ Hz), but in **4.52**, they are closely coupled to each other because of the zigzag pathway ($J_{AA'} \neq 0$ Hz). The multiplicity of the methyl resonance is perturbed not only by the adjacent alkenic proton but also by the proton on the opposite side of the ring. Expressed in the Pople notation, **4.51** is $(AX_3)_2$, but **4.50** is $AA'X_3 X'_3$. This effect is not altered at a higher field because A and A' are chemically equivalent.

4.9.3 Shift Reagents

Sometimes proton spectra are second order even at 500 MHz or higher (aside from the AA' case, which is always second order). In addition, some institutions still have available only iron-core, 60 MHz spectrometers, which produce largely second-order proton spectra. These spectra may be clarified somewhat by the use of *paramagnetic shift reagents*. These molecules contain unpaired electron spins and form Lewis acid–base complexes with dissolved substrates. The unpaired spin exerts a strong paramagnetic shielding effect (hence, to higher frequency or downfield) on nuclei close to it. The effect drops off rapidly with distance, so that those nuclei in the substrate that are closest to the site of acid–base binding are affected more. Consequently, the shift to higher frequency varies from proton to proton within the substrate and leads to greater separation of peaks. Two common shift reagents contain lanthanide elements: *tris*(dipivalomethanato)europium(III)·2(pyridine) [called Eu(dpm)$_3$ without pyridine] and 1,1,1,2,2,3,3-heptafluoro-7,7-dimethyloctanedionatoeuropium(III) [Eu(fod)$_3$]. Shift reagents are available with numerous rare earths, as well as with other elements. Almost all organic functional groups that are Lewis bases have been found to respond to these reagents. When the shift reagent is chiral, it can complex with the enantiomers and generate separate resonances from which enantiomeric ratios may be obtained.

4.9.4 Isotope Satellites

Spectral analysis can sometimes be facilitated by taking advantage of dilute spins present in the molecule. Earlier in this chapter, cyclopropene (**4.5**) was mentioned as an example of an A_2X_2 spectrum, and, in Section 4.5, the vicinal coupling between the protons on the double bond (J_{AA}) was quoted as being 1.3 Hz. How was such a coupling constant between two chemically equivalent protons measured? Its small value prohibits the use of deuterium, as J_{HD} would be only 0.2 Hz for cyclopropene. For 1.1% of the molecules, the double-bond spin system is H—^{12}C—^{13}C—H. The

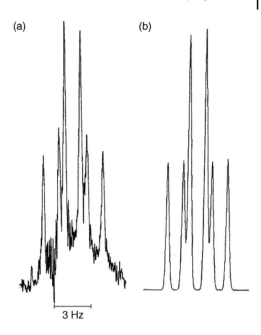

Figure 4.12 The 90 MHz ^{1}H spectrum of cyclopropene, showing the observed (a) and calculated (b) high-frequency ^{13}C satellite of the alkenic protons. Source: Lambert et al. 1970 [3]. Copyright 1970. Reproduced with permission from American Chemical Society.

proton on ^{13}C in such molecules resonates at almost the same position as in the molecules with no ^{13}C. The proton on ^{12}C resonates at almost the same position as the molecules with no ^{13}C. The large one-bond ^{13}C—^{1}H coupling produces multiplets, called *satellites*, on either side of the centerband and separated from it by about ½J(CH). Small isotope effects can shift the center of the satellites. The separation of each satellite from the centerband serves as an effective chemical shift difference, so the H—H coupling between H—^{12}C and H—^{13}C is present in the satellite. Figure 4.12 shows the satellite spectrum of the alkenic protons of cyclopropene. The satellite is a doublet of triplets, since the alkenic proton on ^{13}C is coupled to the other alkenic proton and to the two methylene protons. In this way, ^3J(HC=CH) can be measured between the normally equivalent hydrogens in cyclopropane. Other dilute spins produce satellite spectra that are commonly observed in proton spectra, including ^{15}N, ^{29}Si, ^{77}Se, ^{111}Cd, ^{113}Cd, ^{117}Sn, ^{119}Sn, ^{125}Te, ^{195}Pt, and ^{199}Hg.

The most general and effective method for analyzing complex proton spectra involves the use of two dimensions, as described in Chapter 6. Even this method, however, has limitations imposed on it by the presence of second-order relationships.

4.10 Tables of Coupling Constants

Tables 4.1–4.5 summarize values of coupling constants by class of structure, extracted from the references found at the end of the chapter. Further examples may be obtained by examining these references.

Table 4.1 One-bond coupling constants.

$^{13}C-^{1}H$	CH_3CH_3	125
	$(CH_3)_4Si$	118
	CH_3Li	98
	$(CH_3)_3N$	132
	CH_3CN	136
	$(CH_3)_2S$	138
	CH_3OH	142
	CH_3F	149
	CH_3Cl	150
	CH_2Cl_2	177
	$CHCl_3$	208
	Cyclohexane	125
	Cyclobutane	136
	Cyclopropane	162
	Tetrahydrofuran (α, β)	145, 133
	Norbornane (C1)	142
	Bicyclo[1.1.1]pentane (C1)	164
	Cyclohexene (C1)	157
	Cyclopropene (C1)	226
	Benzene	159
	1,3-cyclopentadiene (C2)	170
	CH_2=CHBr	197
	Acetaldehyde (CHO)	172
	Pyridine (α, β, γ)	177, 157, 160
	Allene	168
	$CH_3C\equiv CH$	248
	$(CH_3)_2C^+H$	164
	HC≡N	269
	Formaldehyde	222
	Formamide	191
$^{13}C-^{19}F$	CH_2F_2	235
	CF_3I	345
	C_6F_6	362
$^{13}C-^{31}P$	CH_3PH_2	9.3
	$(CH_3)_3P$	−13.6
	$(CH_3)_4P^+\ I^-$	56
$^{13}C-^{15}N$	CH_3NH_2	−4.5
	$C_6H_5NH_2$	−11.4
	$CH_3(CO)NH_2$	−14.8
	$CH_3C\equiv N$	−17.5
	Pyridine	+0.62
	$CH_3HC=N-OH$ (E, Z)	−4.0, −2.3

Table 4.1 (Continued)

$^{15}N-^{1}H$	CH_3NH_2	−64.5
	$CH_3(CO)NH_2$	−89
	Pyridinium	−90.5
	$HC\equiv N^+H$	−134
	$(C_6H_5)_2C=NH$	−51.2
$^{15}N-^{15}N$	Azoxybenzene	12.5
	Phenylhydrazine	6.7
$^{15}N-^{31}P$	$C_6H_5NHP(CH_3)_2$	53.0
	$C_6H_5NH(PO)(CH_3)_2$	−0.5
	$[(CH_3)_2N]_3P=O$	−26.9
$^{13}C-^{13}C$	CH_3CH_3	35
	$CH_3(CO)CH_3$	40
	CH_3CO_2H	57
	$CH_2=CH_2$	68
	$CH\equiv CH$	171
$^{31}P-^{1}H$	$C_6H_5(C_6H_5CH_2)(PO)H$	474
$^{31}P-^{31}P$	$(CH_3)_2P-P(CH_3)_2$	−179.7
	$(CH_3)_2(PS)(PS)(CH_3)_2$	18.7

Table 4.2 Geminal proton–proton (H—C—H) coupling constants (Hz).

CH_4	−12.4
$(CH_3)_4Si$	−14.1
$C_6H_5CH_3$	−14.4
$CH_3(CO)CH_3$	−14.9
CH_3CN	−16.9
$CH_2(CN)_2$	−20.4
CH_3OH	−10.8
CH_3Cl	−10.8
CH_3Br	−10.2
CH_3F	−9.6
CH_3I	−9.2
CH_2Cl_2	−7.5
Cyclohexane	−12.6
Cyclopropane	−4.3
Aziridine	+1.5
Oxirane	+5.5
$CH_2=CH_2$	+2.3
$CH_2=O$	+40.22
$CH_2=NOH$	9.95
$CH_2=CHF$	−3.2

(Continued)

Table 4.2 (Continued)

$CH_2=CHNO_2$	−2.0
$CH_2=CHOCH_3$	−2.0
$CH_2=CHBr$	−1.8
$CH_2=CHCl$	−1.4
$CH_2=CHCH_3$	2.08
$CH_2=CHCO_2H$	1.7
$CH_2=CHC_6H_5$	1.08
$CH_2=CHCN$	0.91
$CH_2=CHLi$	7.1
$CH_2=C=C(CH_3)_2$	−9.0

Table 4.3 Vicinal proton–proton (H—C—C—H) coupling constants (Hz).

CH_3CH_3	8.0
$CH_3CH_2C_6H_5$	7.62
CH_3CH_2CN	7.60
CH_3CH_2Cl	7.23
$(CH_3CH_2)_3N$	7.13
CH_3CH_2OAc	7.12
$(CH_3CH_2)_2O$	6.97
CH_3CH_2Li	8.90
$(CH_3)_2CHCl$	6.4
$ClCH_2CH_2Cl$ (neat)	5.9
$Cl_2CHCHCl_2$ (neat)	3.06
Cyclopropane (cis, trans)	8.97, 5.58
Oxirane (cis, trans)	4.45, 3.10
Aziridine (cis, trans)	6.0, 3.1
Cyclobutane (cis, trans)	10.4, 4.9
Cyclopentane (cis, trans)	7.9, 6.3
Tetrahydrofuran (α–β: cis, trans)	7.94, 6.14
Cyclopentene (3–4: cis, trans)	9.36, 5.72
Cyclohexane (av.: cis, trans)	3.73, 8.07
Cyclohexane (ax–ax)	12.5
Cyclohexane (eq–eq and ax–eq)	3.7
Piperidine (av. α–β: cis, trans)	3.77, 7.88
Oxane (av. α–β: cis, trans)	3.87, 7.41
Cyclohexanone (av. α–β: cis, trans)	5.01, 8.61
Cyclohexene (3–4: cis, trans)	2.95, 8.94
$CH_2=CH_2$ (cis, trans)	11.5, 19.0
$CH_2=CHLi$ (cis, trans)	19.3, 23.9
$CH_2=CHCN$ (cis, trans)	11.75, 17.92

Table 4.3 (Continued)

$CH_2=CHC_6H_5$ (cis, trans)	11.48, 18.59
$CH_2=CHCO_2H$ (cis, trans)	10.2, 17.2
$CH_2=CHCH_3$ (cis, trans)	10.02, 16.81
$CH_2=CHCl$ (cis, trans)	7.4, 14.8
$CH_2=CHOCH_3$ (cis, trans)	7.0, 14.1
$ClHC=CHCl$ (cis, trans)	5.2, 12.2
Cyclopropene (1–2)	1.3
Cyclobutene (1–2)	2.85
Cyclopentene (1–2)	5.3
Cyclohexene (1–2)	8.8
Benzene	7.54
C_6H_5Li (2–3)	6.73
$C_6H_5CH_3$ (2–3)	7.64
$C_6H_5CO_2CH_3$ (2–3)	7.86
C_6H_5Cl (2–3)	8.05
$C_6H_5OCH_3$ (2–3)	8.30
$C_6H_5NO_2$ (2–3)	8.36
$C_6H_5N(CH_3)_2$ (2–3)	8.40
Naphthalene (1–2, 2–3)	8.28, 6.85
Furan (2–3, 3–4)	1.75, 3.3
Pyrrole (2–3, 3–4)	2.6, 3.4
Pyridine (2–3, 3–4)	4.88, 7.67

Table 4.4 Carbon coupling constants other than $^1J(^{13}C-^1H)$ (Hz).

CH_3CH_3	−4.8
CH_3CH_2Cl	2.6
$Cl_2CH-CHCl_2$	+1.2
Cyclopropane (^2J)	−2.6
$(CH_3)_2CHCH_2CH(CH_3)_2$	5.
$(CH_3)_2C=O$	5.9
$CH_3(CO)H$	26.7
$CH_3CH=C(CH_3)_2$	4.8
$CH_2=CH_2$	−2.4
$CHCl=CHCl$ (cis, trans)	16.0, 0.8
$CH_2=CHBr$ (cis, trans)	−8.5, +7.5
Benzene [$^2J(CH)$, $^3J(CH)$]	+1.0, +7.4
$CH_3C≡CH$ (CH_3, ≡CH)	−10.6, +50.8
CF_3CF_3	46.0
$CH_3(CO)F$	59.7
$Cl_2C=CF_2$	44.2

(Continued)

Table 4.4 (Continued)

CH_3CH_3	34.6
CH_3CH_2OH	37.7
CH_3CHO	39.4
$CH_3C\equiv N$	56.5
$CH_3CO_2C_2H_5$	58.8
$CH_2=CH_2$	67.2
$CH_2=CHCN$	74.1
C_6H_5CN (ipso)	80.3
$C_6H_5NO_2$ (1,2)	55.4
$HC\equiv CH$	170.6
$(CH_3CH_2)_3P$	+14.1
$(CH_3CH_2)_4P^+\ Br^-$	−4.3
$(CH_3O)_3P$	+10.05
$(CH_3O)_3P=O$	−5.8
$(CH_3)_3P=S$	+56.1
$CH_3(CH_3O)_2P=O$	+142.2

Table 4.5 Nitrogen-15 coupling constants beyond one bond (Hz).

CH_3NH_2	−1.0
Pyrrole (HNCH)	−4.52
Pyridine (NCH)	−10.76
Pyridinium (HNCH)	−3.01
$(CH_3)_2NCHO$ (CH_3, CHO)	+1.1, −15.6
H—C≡N	8.7
$H_2N(CO)CH_3$	1.3
Pyrrole (HNCCH)	−5.39
Pyridine (NCCH)	−1.53
Pyridinium (HNCCH)	−3.98
$CH_3C\equiv N$	−1.7
$CH_3CH_2CH_2NH_2$	1.2
CH_3CONH_2	9.5
$CH_3C\equiv N$	3.0
Pyridine (NCC)	+2.53
Pyridinium (HNCC)	+2.01
Aniline (NCC)	−2.68
Pyrrole (HNCC)	−3.92
$CH_3CH_2CH_2NH_2$	1.4
Pyridine (NCCC)	−3.85
Pyridinium (HNCCC)	−5.30
Aniline (NCCC)	−1.29

Problems

4.1 Characterize the indicated protons as (1) homotopic, enantiotopic, or diastereotopic and (2) magnetically equivalent or nonequivalent. In parts (h) and (i), the Cr(CO)$_3$ ligand remains on one side of the benzene ring.

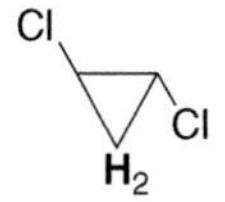

(a) cyclopropane with Cl, Cl, H$_2$

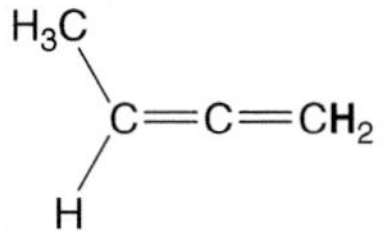

(b) H$_3$C, H on C=C=CH$_2$

(c) cyclopentene with H, H and –CH(CH$_3$)(C$_2$H$_5$) substituent

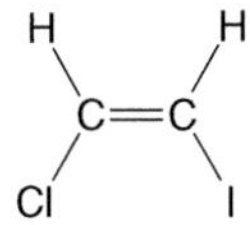

(d) H, H on C=C with Cl and I

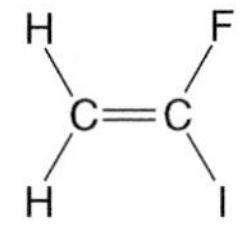

(e) H, H on C=C with F and I

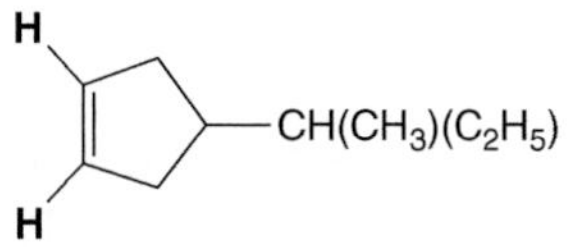

(f) epoxide with H$_3$C, H$_3$C on one carbon and H, CH(CH$_3$)$_2$ on the other (Two answers—one for each geminal pair of methyls)

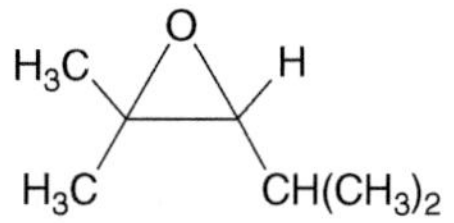

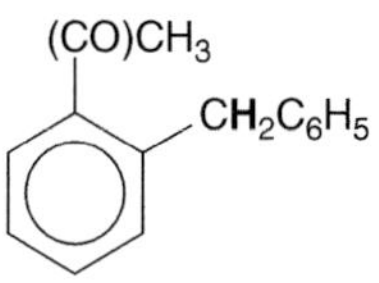

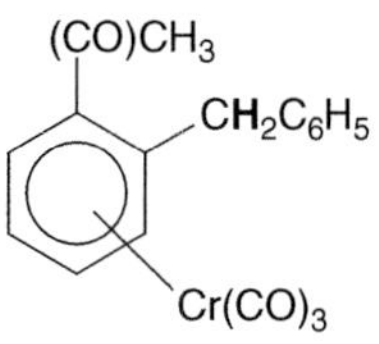

(g) benzene ring with (CO)CH$_3$ and CH$_2$C$_6$H$_5$ (ortho)

(h) benzene ring with (CO)CH$_3$, CH$_2$C$_6$H$_5$, and Cr(CO)$_3$

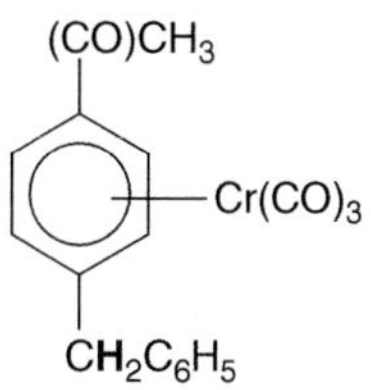

(i) benzene ring with (CO)CH$_3$ and CH$_2$C$_6$H$_5$ (para), with Cr(CO)$_3$

4.2 (a) In the molecule illustrated below, are the protons on the double bond homotopic, enantiotopic, or diastereotopic? Explain.

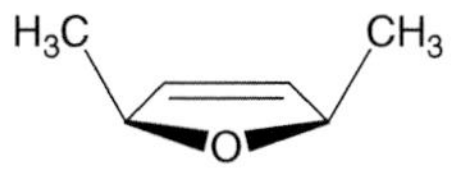

H$_3$C — [furan-like ring with O] — CH$_3$

(b) Answer as in (a) for the following molecule.

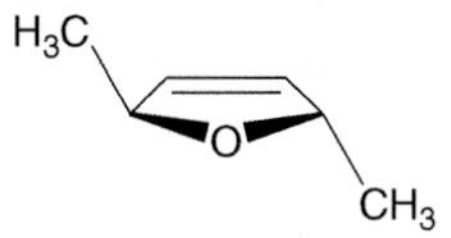

H$_3$C — [furan-like ring with O] — CH$_3$

4.3 The 3′ proton of the illustrated indene dimer exhibits the 600 MHz spectrum depicted below. Construct the tree diagram for this proton resonance. Measure the coupling constants, assign them to protons pairs in the structure, and rationalize their magnitudes in terms of structure and stereochemistry.

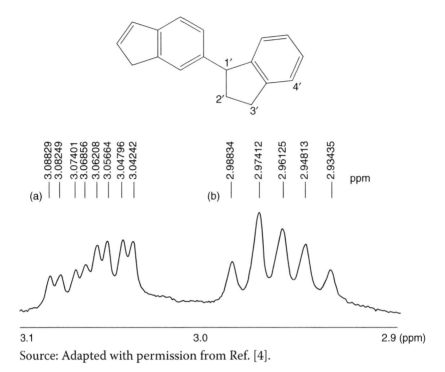

4.4 What is the spin notation for each of the following molecules (AX, AMX, AA'XX', etc.)? Consider only major isotopes.

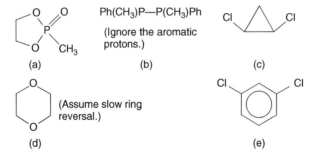

4.5 Write out the rotamers of 2-chloroethanol ($ClCH_2CH_2OH$). What is the spin notation at slow rotation for each rotamer and at fast rotation for the average?

4.6 Consider the following ^{1}H-decoupled ^{31}P spectrum of the platinum complex with the illustrated structure (the resonances of the anion are omitted). Explain all the peaks and give the spin notation. What should the ^{195}Pt spectrum look like?

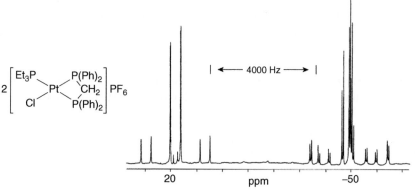

Source: Adapted from Ref. [5]. Copyright 1994 American Chemical Society. Reprinted by permission of the American Chemical Society.

4.7 Several binary structures between phosphorus and sulfur are possible, including the three shown below.

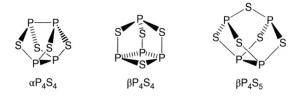

(a) What are the spin systems for these molecules?
(b) The ^{31}P spectra for these three molecules are given below. Which structure corresponds to which spectrum? Explain all the splittings.

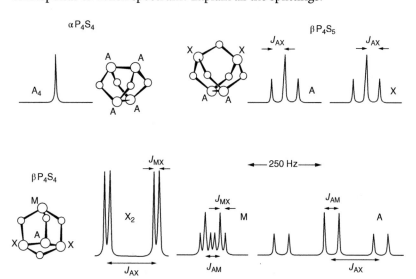

Source: Adapted from Ref. [6]. Reproduced by permission of the Oxford University Press.

4.8 Construct the stick diagram for the CH resonance of the molecule BrCH$_2$CH(CH$_3$)CH$_2$OH, in which ^3J(CH—CH$_3$) = 6 Hz and ^3J(CH—CH$_2$) = 7 Hz (for either methylene group).

4.9 The ^{1}H spectra below are of 1,2-dichlorobenzene at (a) 90 MHz and (b) 750 MHz. Examine each spectrum separately. Is it first or second order? Explain. What effect does higher field have on the structure of the spectrum?

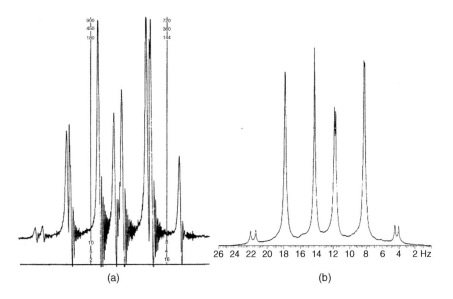

4.10 Eliminating four moles of HBr from the molecule below should give the indicated cyclopropane. The $^1J(^{13}C-^1H)$ for the bridge CH$_2$ group in the isolated product was measured to be 142 Hz. Explain in terms of product structures.

4.11 There are two isomers of thiane 1-oxide, (a) and (b). The observed geminal coupling constant between the α protons is −13.7 Hz in one isomer and −11.7 Hz in the other. Which coupling belongs to which isomer and why?

4.12 The ^{1}H spectrum of 1,3-dioxane (below) at slow ring reversal contains three multiplets with the following geminal couplings: −6.1, −11.2, and −12.9 Hz. Without reference to any chemical shift data, assign the resonances.

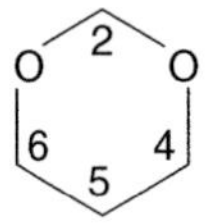

4.13 Does the angular methyl group in *trans*-decalins (a) or in *cis*-decalins (b) have the larger line width? Explain.

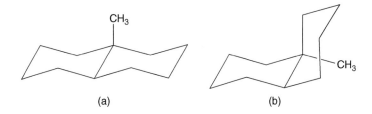

4.14 In cycloheptatriene (a), J_{23} is 5.3 Hz, whereas in the bistrifluoromethyl derivative (b), J_{23} is 6.9 Hz. Explain.

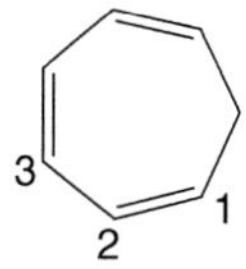

(a) 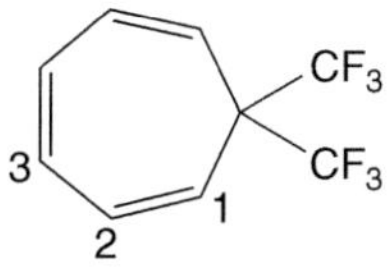   (b)

4.15 Explain the following couplings in terms of structure and mechanism.

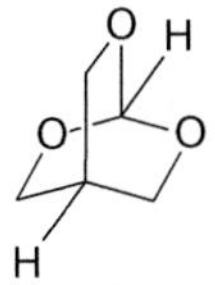

(a) $^5J = 1.7$ Hz

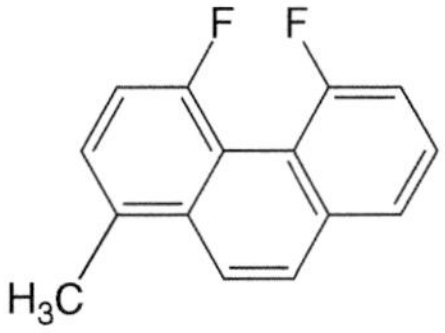

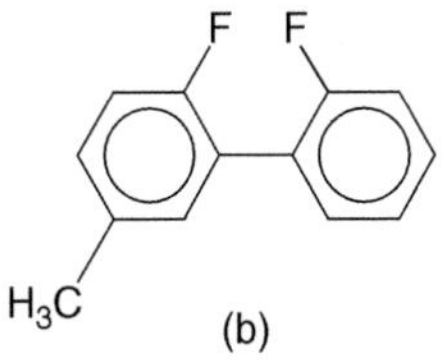

(b) $^5J = 16.5$ Hz

(c) $^5J = 170$ Hz

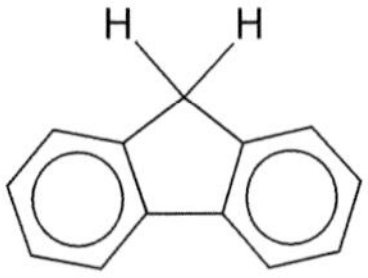

(d) $^2J = -22.3$ Hz

4.16 The following four 300 MHz ^{1}H spectra are of lutidines (dimethylpyridines). From the chemical shifts and the coupling patterns, deduce the placement of methyl groups on each molecule. Assume the spacings are first order.

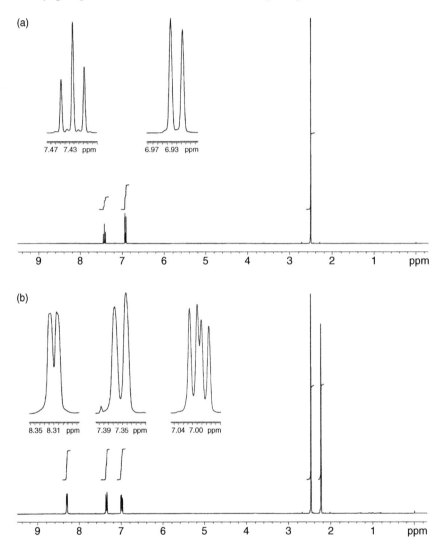

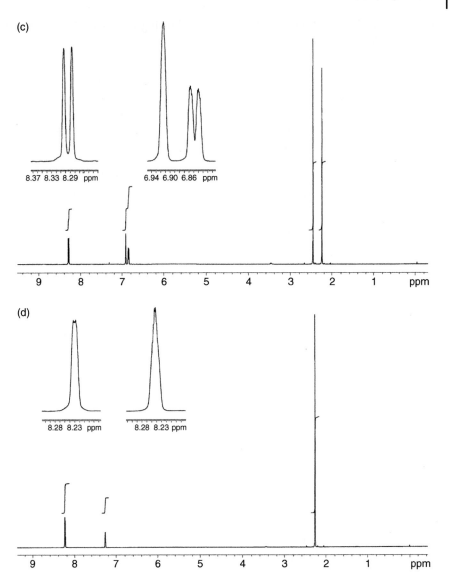

4.17 Proceed as in Problem 4.16 with the following four 300 MHz ^{1}H spectra of dichlorophenols.

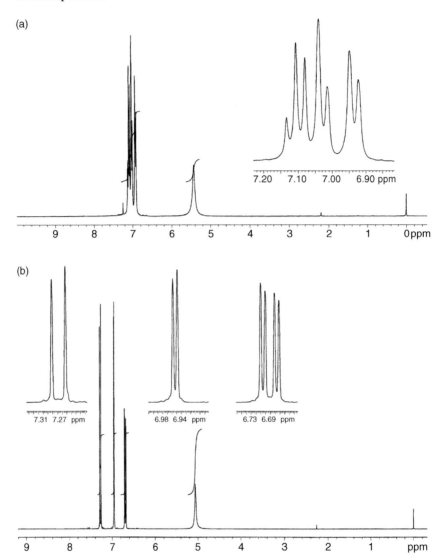

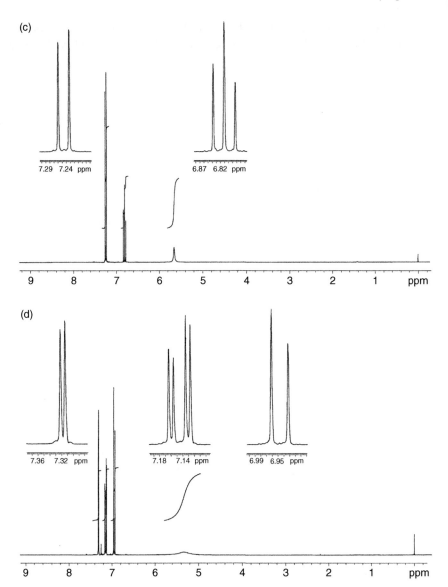

4.18 Is the 400 MHz ^{1}H spectrum below of the cis or the trans isomer of dimethyl 1,2-cyclopropanedicarboxylate (conditions: CDCl$_3$, 25 °C, 400 MHz)? Explain.

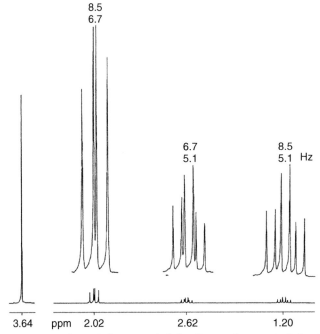

Source: Adapted from Ref. [7]. Copyright 1993 John Wiley & Sons, Ltd. Reprinted by permission of John Wiley & Sons, Ltd.

4.19 The ^{1}H spectrum of 2-hydroxy-5-isopropyl-2-methylcyclohexanone has a $^3J_{56} = 3$ Hz in benzene-d_6 but 11 Hz in CD$_3$OD. Explain.

4.20 Analyze the following ^{1}H spectrum of the illustrated thujic ester. The CH$_3$ resonances are not shown. Assign the resonances to specific protons and give approximate coupling constants. Explain your chemical-shift assignments.

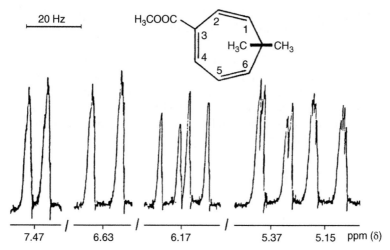

4.21 Deduce the structure (with relative stereochemistry) of the compound $C_6H_{12}O_6$ having the following 300 MHz ^{1}H NMR spectrum in D_2O [the peaks at δ 2.9 are from the aqueous reference, 3-(trimethylsilyl)propionic acid]. Hydroxyl resonances are not shown. The triplet (a) at δ 4.04 has an integral of 1 and $J = 2.8$ Hz. The second-order triplet (b) at δ 3.61 has integral 2 and $J = 9.6$ Hz. The second-order doublet of doublets (c) at δ 3.52 has integral 2 and $J = 2.8$, 9.6 Hz. The triplet (d) at δ 3.26 has integral 1 and $J = 9.6$ Hz. The ^{13}C NMR spectrum shows four resonances, all between δ 71 and 75.

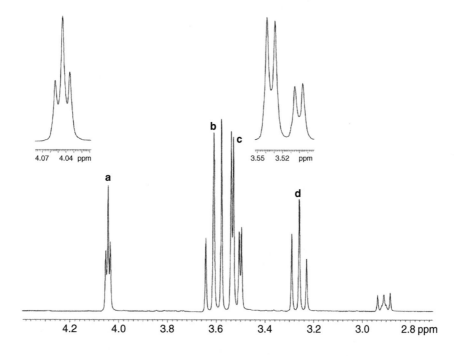

4.22 The covalent and oligomeric nature of organolithium compounds has been demonstrated by examining the spectra of compounds fully labeled with ^{13}C or ^{6}Li.

(a) The following ^{7}Li{^{1}H} spectrum [the symbols {/} signify double irradiation at the frequency of the nucleus within the brackets, in this case of ^{1}H] of $(Li^{13}CMe_3)_x$ [the oligomer of *tert*-butyllithium labeled with ^{13}C at the quaternary carbon] is a 1:3:3:1 quartet with $^{1}J(^{7}Li-^{13}C) = 14.3$ Hz. What can you conclude about the number of nearest neighbor *tert*-butyl groups to lithium in solution? Explain.

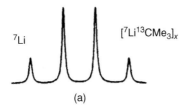
(a)

(b) The ^{13}C{^{1}H} spectrum of $(^{6}Li^{13}CMe_3)_x$ at $-88\,^\circ$C in cyclopentene is a 1:3:6:7:6:3:1 septet (recall that ^{6}Li has a spin of 1), with $^{1}J(^{6}Li-^{13}C) = 5.4$ Hz. How many nearest neighbor lithiums are indicated? Explain.

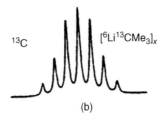
(b)

(c) Suggest a structure for *tert*-BuLi under these conditions. Explain.
(d) Above $-5\,^\circ$C, the septet is replaced by a nonet (nine lines) with $^{1}J = 4.1$ Hz, as in the following spectrum. Explain in terms of your structure.

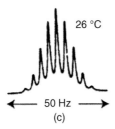
(c)

Source: Adapted from Ref. [9]. Copyright 1986 American Chemical Society. Reprinted by permission of the American Chemical Society.

4.23 Explain the following couplings in terms of structure and mechanism.

(a) $^5J = 1.7$ Hz

(b) $^5J = 16.5$ Hz

(c) $^5J = 170$ Hz

(d) $^2J = -22.3$ Hz

References

4.1 Anet, F.A.L. (1961). *Can. J. Chem.* 39: 2267.
4.2 Becker, E.D. (1980). *High Resolution NMR*, 2e. Orlando, FL: Academic Press.
4.3 Lambert, J.B., Jovanovich, A.P., and Oliver, W.L. Jr., (1970). *J. Phys. Chem.* 74: 2221.
4.4 Spiteller, P., Spiteller, M., and Jovanovich, J. (2002). *Magn. Reson. Chem.* 40: 372.
4.5 Berry, D.E. (1994). *J. Chem. Educ.* 71: 899–902.
4.6 Hore, P.J. (1993). *Nuclear Magnetic Resonance*, 30. Oxford: Oxford University Press.
4.7 Breitmaier, E. (1993). *Structure Elucidation by NMR in Organic Chemistry*. Chichester: Wiley.
4.8 Günther, H. et al. (1974). *Org. Magn. Reson.* 6: 388.
4.9 Thomas, R.D., Clarke, M.T., Jensen, R.M., and Young, T.C. (1986). *Organometallics* 5: 1851.

Further Reading

Coupling (General)

Ando, I. and Webb, G.A. (1983). *Theory of NMR Parameters*. London: Academic Press.

Magnetic Equivalence

Jennings, W.B. (1975). *Chem. Rev.* 75: 307.
Mislow, K. and Raban, M. (1966). *Top. Stereochem.* 1: 1.
Pirkle, W.H. and Hoover, D.J. (1982). *Top. Stereochem.* 13: 263.

One-Bond Couplings

Goldstein, J.H., Watts, V.S., and Rattet, L.S. (1971). *Prog. Nucl. Magn. Reson. Spectrosc.* 8: 103.
Jameson, C.J. and Gutowsky, H.S. (1969). *J. Chem. Phys.* 51: 2790.
McFarlane, W. (1969). *Q. Rev. Chem. Soc.* 23: 187.

Geminal, Vicinal, and Long-Range ^{1}H—^{1}H Couplings

Barfield, M. and Charkrabarti, B. (1969). *Chem. Rev.* 69: 757.
Barfield, M., Spear, R.J., and Sternhell, S. (1976). *Chem. Rev.* 76: 593.
Bothner-By, A.A. (1965). *Adv. Magn. Reson.* 1: 195.
Bystrov, V.F. (1972). *Russ. Chem. Rev.* 41: 281.
Hilton, J. and Sutcliffe, L.H. (1975). *Prog. Nucl. Magn. Reson. Spectrosc.* 10: 27.
Sternhell, S. (1964). *Rev. Pure Appl. Chem.* 14: 15.
Sternhell, S. (1969). *Q. Rev. Chem. Soc.* 23: 236.

Carbon-13 Couplings

Ewing, D.F. (1975). *Annu. Rep. NMR Spectrosc.* 6A: 389.
Hansen, P.E. (1978). *Org. Magn. Reson.* 11: 215.
Hansen, P.E. (1981). *Annu. Rep. NMR Spectrosc.* 11A: 65.
Hansen, P.E. (1981). *Prog. Nucl. Magn. Reson. Spectrosc.* 14: 175.
Hansen, P.E. (1981). *Org. Magn. Reson.* 15: 102.
Krivdin, L.B. and Della, E.W. (1991). *Prog. Nucl. Magn. Reson. Spectrosc.* 23: 301.
Krivdin, L.B. and Kalabia, G.A. (1989). *Prog. Nucl. Magn. Reson. Spectrosc.* 21: 293.
Levy, G.C., Lichter, R.L., and Nelson, G.L. (1980). *Carbon-13 Nuclear Magnetic Resonance Spectroscopy*, 2e. New York: Wiley.

Marshall, J.L. (1983). *Carbon–Carbon and Carbon–Proton NMR Couplings.* Deerfield Beach, FL: Wiley-VCH.
Marshall, J.L., Müller, D.E., Conn, S.A. et al. (1974). *Acc. Chem. Res.* 7: 333.
Parella, T. and Espinosa, F. (2013). *Prog. Nucl. Magn. Reson. Spectrosc.* 73: 17.
Stothers, J.B. (1973). *Carbon-13 NMR Spectroscopy.* New York: Academic Press.
Wasylishen, R.E. (1977). *Annu. Rep. NMR Spectrosc.* 7: 118.
Wray, V. (1979). *Prog. Nucl. Magn. Reson. Spectrosc.* 13: 177.
Wray, V. and Hansen, P.E. (1981). *Annu. Rep. NMR Spectrosc.* 11A: 99.

Fluorine-19 Couplings

Emsley, J.M., Phillips, L., and Wray, V. (1977). *Prog. Nucl. Magn. Reson. Spectrosc.* 10: 82.

Phosphorus-31 Couplings

Finer, E.G. and Harris, R.K. (1970). *Prog. Nucl. Magn. Reson. Spectrosc.* 6: 61.

Shift Reagents

Gribnau, M.C.M., Keijzers, C.P., and De Boer, E. (1985). *Magn. Reson. Rev.* 10: 161.
Hofer, O. (1976). *Top. Stereochem.* 9: 111.
Inagaki, F. and Miyazawa, T. (1981). *Prog. Nucl. Magn. Reson. Spectrosc.* 14: 67.
Mayo, B.C. (1973). *Chem. Soc. Rev.* 2: 49.
Morrill, T.C. (ed.) (1986). *Lanthanide Shift Reagents in Stereochemical Analysis.* Deerfield Beach, FL: Wiley-VCH.
Sullivan, G.R. (1978). *Top. Stereochem.* 10: 287.
Wenzel, T.J. (1987). *NMR Shift Reagents.* Boca Raton, FL: CRC Press.

Spectral Analysis

Abraham, R.J. (1971). *The Analysis of High Resolution NMR Spectra.* Amsterdam: Elsevier Science Inc.
Diehl, P., Kellerhals, H., and Lustig, E. (1972). *NMR Basic Princ. Prog.* 6: 1.
Garbisch, E.W. Jr., (1968). *J. Chem. Educ.* 45, 311, 403, 481.
Günther, H. (1972). *Angew. Chem. Int. Ed. Engl.* 11: 861.
Haigh, C.W. (1971). *Annu. Rep. NMR Spectrosc.* 4: 311.
Hoffman, R.A., Forsén, S., and Gestblom, B. (1971). *NMR Basic Princ. Prog.* 5: 1.
Hoye, T.R., Hanson, P.R., and Vyvyan, J.R. (1994). *J. Org. Chem.* 59: 4096–4103.

Manatt, S.L. (2002). *Magn. Reson. Chem.* 40: 317.

Roberts, J.D. (1961). *An Introduction to the Analysis of Spin–Spin Splitting in High-Resolution Nuclear Magnetic Resonance Spectra*. New York: W. A. Benjamin.

Wiberg, K.B. and Nist, B.J. (1962). *The Interpretation of NMR Spectra*. New York: W. A. Benjamin.

5

Further Topics in One-Dimensional NMR Spectroscopy

Although the chemical shift and the coupling constant are the two fundamental measurable quantities in NMR spectroscopy, several other phenomena may be studied in a single NMR time dimension. In this chapter, we first examine the processes of spin–lattice and spin–spin relaxation, whereby a system moves toward spin equilibrium (Sections 2.3 and 5.1). Relaxation times or rates provide another important measurable quantity related to both structural and dynamic factors. Second, we explore in greater detail structural changes that occur on the NMR time scale (Sections 2.8 and 5.2). The temporal dependence of chemical shifts and coupling constants influences both line shapes and intensities and can be used to generate rate constants for reactions. Third, we describe the family of experiments that utilize a second irradiation frequency, B_2 (Section 5.3). Double irradiation can simplify spectra, perturb intensities, and provide information about structure and rate processes. Finally, we expand on the technique of using multiple pulses, often of varied duration rather than only a single 90° pulse, sometimes separated by specific time periods, and even of nonrectalinear shape. Multiple pulses may be used to improve sensitivity, simplify spectral patterns, measure relaxation times and coupling constants, draw structural conclusions, and improve the accuracy of pulse timing and definition (Sections 5.4–5.8).

5.1 Spin–Lattice and Spin–Spin Relaxation

Application of the B_1 field at the resonance frequency results in energy absorption and the conversion of some $+\frac{1}{2}$ spins into $-\frac{1}{2}$ spins, so that magnetization in the z direction (M_z) decreases. Spin–lattice, or longitudinal, relaxation returns the system to equilibrium along the z axis with time constant T_1 and rate constant R_1 ($=1/T_1$). Such relaxation occurs because of the presence of natural magnetic fields in the sample that fluctuate at the Larmor frequency. When the frequencies match, excess spin energy can flow into the molecular surroundings, sometimes called the *lattice*, and $-\frac{1}{2}$ spins can return to the $+\frac{1}{2}$ state.

5.1.1 Causes of Relaxation

The major source of these magnetic fields is nearby magnetic nuclei in motion. Like the classic model of a charge moving in a circle, a magnetic dipole in motion creates a magnetic field, whose frequency depends on the rate of motion and on the magnetic moment

Nuclear Magnetic Resonance Spectroscopy: An Introduction to Principles, Applications, and Experimental Methods,
Second Edition. Joseph B. Lambert, Eugene P. Mazzola, and Clark D. Ridge.
© 2019 John Wiley & Sons Ltd. Published 2019 by John Wiley & Sons Ltd.

of the dipole. For appropriate values of these parameters, the resulting magnetic field can fluctuate at the same frequency as the resonance (Larmor) frequency of the nucleus in question, permitting energy to flow from excited spins to the lattice. Such a process is called *dipole–dipole relaxation* (T_1(DD)), because it involves interaction of the resonating nuclear magnetic dipole with the dipole of the nucleus in motion that gives rise to the fluctuating field of the lattice. The resulting relaxation time depends on nuclear properties of both resonating and moving nuclei, on the distance between them, and on the rate of motion of the moving nucleus. Mathematically, the dependence of relaxation on these factors takes the form of Eq. (5.1a)

$$R_1(\text{DD}) = \frac{1}{T_1(\text{DD})} = n\gamma_C^2\gamma_H^2\hbar^2 r_{CH}^{-6}\tau_c \tag{5.1a}$$

for the case of ^{13}C relaxed by protons in motion or the form of Eq. (5.1b)

$$\frac{1}{T_1(\text{DD})} = \frac{3}{2}n\gamma_H^4\hbar^2 r_{HH}^{-6}\tau_c \tag{5.1b}$$

for protons relaxed by protons. As usual, the nuclear properties are represented by the gyromagnetic ratios. The symbol n stands for the number of protons that are nearest neighbors to the resonating nucleus and hence are most effective at relaxing it. The rapid falloff with distance is indicated by the inverse sixth power of the distance r_{CH} (r_{HH}) to the nearest-neighbor C—H (H—H). The motional properties of the protons are described by the effective correlation time τ_c, which is the time required for the molecule to rotate 1 rad and is typically in the nanosecond-to-picosecond range for organic molecules in solution.

Thus, carbon relaxation is faster (and the relaxation time is shorter) when there are more attached protons, when the internuclear C—H or H—H distance is smaller, and when the rotation rate in solution decreases. A quaternary carbon has a long relaxation time because it lacks an attached proton and because the distance r_{CH} to other protons is large. The ratio of the carbon relaxation time of methinyl to methylene to methyl is 6 : 3 : 2 (equivalent to 1 : ½ : ⅓), due to differences in the number of attached protons, other things being equal. Because the rate of molecular tumbling in solution slows as molecular size increases, larger molecules relax more rapidly. Thus, cholesteryl chloride relaxes more rapidly than phenanthrene, which relaxes more rapidly than acetone. Equation (5.1a) is an approximation to a more complete equation and represents what is called the *extreme narrowing limit* for smaller molecules. Because the frequency of motion of the moving nuclear magnet must match the resonance frequency of the excited nuclear magnet, dipolar relaxation becomes ineffective for both rapidly moving small molecules and slowly moving large molecules. Many molecules of interest to biochemists fall into the latter category, to which Eq. (5.1a) does not apply. Rapid internal rotation of methyl groups in small molecules also can reduce the effectiveness of dipole–dipole relaxation. The optimal correlation times (τ_c) for dipolar relaxation lie in the range of about 10^{-7}–10^{-11} s (the inverse of the resonance frequency). Because the resonance frequency depends on the value of B_0, this range also depends on B_0.

When dipolar relaxation is slow, other mechanisms of relaxation become important. Fluctuating magnetic fields also can arise from (i) interruption of the motion of rapidly tumbling small molecules or rapidly rotating groups within a molecule (*spin rotation relaxation*), (ii) tumbling of molecules with anisotropic chemical shielding at high fields, (iii) scalar coupling constants that fluctuate through chemical exchange or through quadrupolar interactions, (iv) tumbling of paramagnetic molecules (unpaired electrons have very large magnetic dipoles), and (v) tumbling of quadrupolar nuclei. In the absence of quadrupolar nuclei or paramagnetic species, these last two mechanisms often are unimportant. A major exception is the relaxation of methyl (CH_3) and trifluoromethyl carbons by spin rotation. At higher temperatures relaxation by dipolar interactions becomes less effective but relaxation by spin rotation becomes more effective, so these mechanisms may be distinguished by measuring T_1 at multiple temperatures.

5.1.2 Measurement of Relaxation Time

The actual value of T_1 must be known at least approximately in order to decide how long to wait between pulses for the system to return to equilibrium (the delay time). In addition, $T_1(DD)$ offers both structural information, because of its dependence on r_{CH}, and dynamic information, because of its dependence on τ_c. For these reasons, convenient methods have been developed for measuring T_1, the commonest of which is called *inversion recovery*. The strategy is to create a nonequilibrium distribution of spins and then to follow their return to equilibrium as a first-order rate process. Inverting the spins through the application of a 180° pulse creates a maximum deviation from equilibrium (Figure 5.1b). If a very short amount of time τ is allowed to pass (Figure 5.1c) and a 90° pulse is applied to move the spins into the xy plane for observation, the nuclear magnets are aligned along the $-y$ axis (Figure 5.1d), and an inverted peak is obtained. During time τ, some T_1 relaxation occurs (Figures 5.1b,c). The z magnetization at the end of time τ (Figure 5.1c) is smaller than at the beginning (Figure 5.1b). Consequently, the peak produced after the 90° pulse is smaller than if the 180° and 90° pulses had been combined initially as a 270° pulse, i.e. if $\tau = 0$. The inversion recovery pulse sequence is summarized as (180° – τ – 90° – Acquire) and is an example of a simple multipulse sequence. In this shorthand the term "Acquire" refers to the period of time during which the free induction decay is recorded (the acquisition time).

Further such experiments with increasingly longer values of τ result in greater relaxation between the 180° and 90° pulses. After the 90° pulse (Figure 5.1d), the resulting

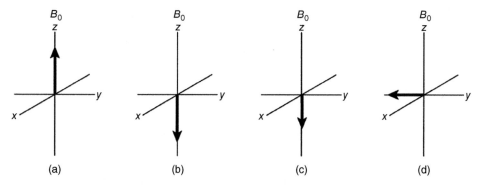

Figure 5.1 The inversion recovery experiment. (a) Initial condition. (b) After 180° pulse. (c) After τ period. (d) After final 90° pulse.

peak evolves from negative through zero to positive as τ increases, until complete relaxation occurs with very long values of τ. Figure 5.2 shows a stack of such experiments for the carbons of chlorobenzene.

Because the carbon ipso to chlorine (C-1) has no directly attached proton, much longer values of τ are needed for its inverted peak to turn over. Relaxation for C-1 is not complete even by $\tau = 80$ s. The intensity I measured at a series of times τ follows an exponential decay according to the first-order kinetics of Eq. (5.2)

$$I = I_0(1 - 2e^{-\tau/T_1}) \tag{5.2}$$

in which I_0 is the equilibrium intensity (measured, for example, initially or after a very large value of τ) and the factor of 2 arises because magnetization recovery begins from a fully inverted condition. A plot of the natural logarithm of $(I_0 - I)$ vs τ gives a straight line with a slope of $-1/T_1$. An estimate for the spin–lattice relaxation time may be obtained through knowledge of the time τ at which the intensity passes through a null ($\tau(\text{null}) = 8$ s for C-4 in Figure 5.2). At the null time, $I = 0$, so that T_1 corresponds to $\tau(\text{null})/\ln 2$, or $1.443\tau(\text{null})$. Such an estimate might be useful, for instance, in deciding how long to wait between repetitive pulses, but it should never be considered a rigorous measurement of T_1.

5.1.3 Transverse Relaxation

Relaxation in the xy plane, or spin–spin (transverse) relaxation (T_2), might be expected to be identical to T_1, because movement of the magnetization from the xy plane

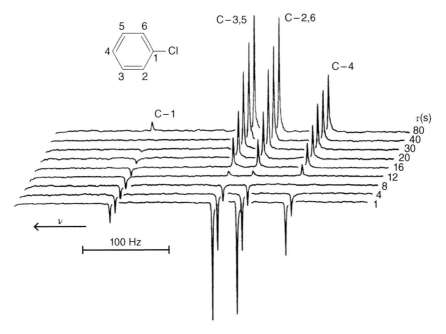

Figure 5.2 A stack plot for the inversion recovery experiment of the ^{13}C resonances of chlorobenzene at 25 MHz. The time τ in the pulse sequence $180° - \tau - 90°$ is given in seconds at the right.
Source: From Harris 1983 [1]. Reproduced with permission of Addison Wesley Longman, Ltd.

back onto the z axis restores z magnetization at the same rate as it depletes xy magnetization. There are, however, other mechanisms of xy relaxation that do not affect z magnetization. We already saw in Section 2.3 that inhomogeneity of the B_0 magnetic field randomizes phases in the xy plane and hastens xy relaxation. As a result, T_2 is expected to be less than or equal to T_1. In addition, xy (T_2) relaxation can occur when two nuclei mutually exchange their spins, one going from $+\frac{1}{2}$ to $-\frac{1}{2}$ and the other from $-\frac{1}{2}$ to $+\frac{1}{2}$. This spin–spin, double-flip, or flip-flop mechanism is most significant in large molecules. The process can result in *spin diffusion*. The excitation of a specific proton changes the magnetization of surrounding protons as flip–flop interactions spread through the molecule. The interpretation of the spectra of large molecules such as proteins must take such processes into consideration.

5.1.4 Structural Ramifications

Proton spin–lattice relaxation times depend on the distance between the resonating nucleus and the nearest-neighbor protons. The closer the neighbors are, the faster the relaxation and the shorter the T_1. The two isomers **5.1a** and **5.1b**

(Bz = Ph(C=O)) may be distinguished by their proton relaxation times. In **5.1a**, H_1 is axial and close to the 3 and 5 axial protons, resulting in a T_1 of 2.0 s. In **5.1b**, H_1 is equatorial and has more distant nearest neighbors, resulting in a T_1 of 4.1 s. In this way, the structure of these anomers may be distinguished. The remaining values of T_1 may be interpreted in a similar fashion. For example, H_2 in isomer **5.1a** has only the H_4 axial proton as a nearest neighbor, so its T_1 is a relatively long 3.6 s. In **5.1b**, H_2 has both the axial H_4 and the vicinal H_1 as nearest neighbors, so T_1 is shorter, viz, 2.1 s.

5.1.5 Anisotropic Motion

When a molecule is rigid and rotates equally well in any direction (isotropically), all the carbon relaxation times (after adjustment for the number of attached protons) should be about the same. The nonspherical shape of a molecule, however, frequently leads to preferential rotation in solution around one or more axes (anisotropic rotation). For example, toluene prefers to rotate around the long axis that includes the methyl, ipso, and para carbons, so that less mass is in motion. On average, these carbons (and their attached protons) move less in solution than do the ortho and meta carbons, because atoms on the axis of rotation remain stationary during rotation. The more rapidly moving ortho and meta carbons thus have shorter effective correlation times τ_c and hence, by Eq. (5.1a), a longer T_1. The actual values are shown in structure **5.2**.

$$\underset{\textbf{5.2}}{\text{18}\;\langle\bigcirc\rangle\;\overset{23\quad 23}{\underset{}{}}\;\overset{58}{}\;\underset{}{T_1\,(s)}\;-\text{CH}_3}$$

The longer value for the ipso carbon arises because it lacks a directly bonded proton and because r_{CH} in Eq. (5.1a) is very large.

5.1.6 Segmental Motion

When molecules are not rigid, the more rapidly moving pieces relax more slowly because their τ_c is shorter. Thus, in decane (**5.3**)

$$\begin{array}{l}\text{CH}_3\text{CH}_2\text{CH}_2\text{CH}_2\text{CH}_2\text{CH}_2\text{CH}_2\text{CH}_2\text{CH}_2\text{CH}_3\\ nT_1\quad 26.1\;\;13.2\;\;11.4\;\;10.0\;\;8.8\end{array}$$

5.3

the methyl carbon relaxes most slowly, followed by the ethyl carbon, and so on, to the fifth carbon in the middle of the chain. Structure **5.3** gives the values of nT_1 (n is the number of attached protons), so that the figures may be compared for all carbons without considering any substitution patterns. These values reflect the relative rates of motion of each carbon.

5.1.7 Partially Relaxed Spectra

The inversion recovery experiment used to measure T_1 also may be exploited to simplify spectra. In Figure 5.2, the spectrum for $\tau = 40$ s lacks a resonance for the ipso carbon (C-1). Similarly, for a τ of about 10 s, all the other ring carbons are nulled, and only the negative peak for C-1 is obtained. Such partially relaxed spectra can be used not only to obtain partial spectra in this fashion, but also to eliminate specific peaks. When deuterated water (D$_2$O) is used as the solvent, the residual HOD peak is undesirable. An inversion recovery experiment can reveal the value of τ for which the water peak is nulled. The rest of the protons will have positive or negative intensities at that τ, depending on whether they relax more rapidly or more slowly than water. The experiment may be refined by applying the 180° pulse selectively only at the resonance position of water. Selection of τ for nulling of this peak then produces a spectrum that lacks the water peak, but otherwise is quite normal for the remaining resonances. Such a procedure is an example of *peak suppression* or *solvent suppression*.

5.1.8 Quadrupolar Relaxation

The dominant mode of spin–lattice relaxation for nuclei with spins greater than ½ results from the quadrupolar nature of such nuclei. These nuclei are considered to have an ellipsoidal rather than a spherical shape. When $I = 1$, as for ^{14}N or ^{2}H, there are three

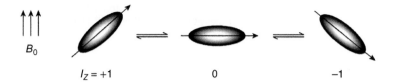

Figure 5.3 The spin states for a nucleus with $I = 1$.

stable orientations in the magnetic field: parallel, orthogonal, and antiparallel, as shown in Figure 5.3. When these ellipsoidal nuclei tumble in solution within an unsymmetrical electron cloud of the molecule, they produce a fluctuating electric field that can bring about relaxation.

The mechanism is different from dipole–dipole relaxation in two ways. First, it does not require a second nucleus in motion; the quadrupolar nucleus creates its own fluctuating field by moving in the unsymmetrical electron cloud. Second, because the mechanism is extremely effective when the quadrupole moment of the nucleus is large, T_1 can become very short (milliseconds or less). In such cases, the uncertainty principle applies, whereby the product of ΔE (the spread of energies of the spin states, as measured by the line width $\Delta \nu$) and Δt (the lifetime of the spin state, as measured by the relaxation time) must remain constant ($\Delta E \Delta t \sim$ Planck's constant). Thus, when the relaxation time is very short, the line width becomes very large. Nuclei with large quadrupole moments often exhibit very large line widths—for example, about 20 000 Hz for the ^{35}Cl resonance of CCl_4. The common nuclides ^{17}O and ^{14}N have smaller quadrupolar moments and exhibit sharper resonances, typically tens of Hz. The small quadrupole moment of deuterium results in quite sharp peaks, usually one or a few Hz. The line width also depends on the symmetry of the molecule, which controls how unsymmetrical the electron cloud is. Systems with π electrons are more unsymmetrical and give broader lines (as in amides and pyridines for ^{14}N). Spherical or tetrahedral systems have no quadrupolar relaxation, since the electron cloud is symmetrical. Such systems exhibit very sharp line widths, like those of spin-½ nuclei (^{14}N in $^+NH_4$, ^{6}Li or ^{7}Li in Li$^+$, ^{10}B in $^-BH_4$, ^{35}Cl in Cl$^-$ or $^-ClO_4$, and ^{33}S in SO_4^{2-}).

Very important to the organic chemist is the effect of quadrupolar nuclei on the resonances of nearby protons. When quadrupolar relaxation is extremely rapid, a neighboring nucleus experiences only the average spin environment of the quadrupolar nucleus, so that no spin coupling is observed. Hence, protons in chloromethane produce a sharp singlet, even though ^{35}Cl and ^{37}Cl have spins of 3/2 and exist in four spin states. Chemists have come to think of the halogens (other than fluorine) as being nonmagnetic, although they appear so only because of their rapid quadrupolar relaxation. At the other extreme, deuterium has a weak quadrupole moment and possesses only s electrons, so that neighboring protons exhibit normal couplings to ^{2}H. Thus, nitromethane with one deuterium (CH_2DNO_2) shows a 1 : 1 : 1 triplet, because the protons are influenced by the three spin states ($+1$, 0, and -1) of deuterium (in analogy to Figure 5.3). Nitromethane with two deuteriums (CHD_2NO_2) shows a 1 : 2 : 3 : 2 : 1 quintet from coupling to the various combinations of the three spin states (++; +0, 0+; +−, 00, −+; −0, 0−; −−). This quintet is often observed in deuterated solvents such as acetone-d_6, acetonitrile-d_3, or nitromethane-d_3, because incomplete deuteration results in an impurity with a CHD_2 group.

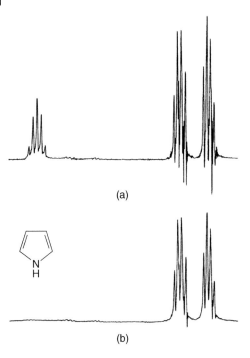

Figure 5.4 The 90 MHz proton spectrum of pyrrole with (a) and without (b) ^{14}N decoupling.

The ^{14}N nucleus falls between these extremes. In highly unsymmetrical cases, such as the interior nitrogen in biuret, $NH_2(CO)NH(CO)NH_2$, quadrupolar relaxation is rapid enough to produce the average singlet for the attached proton. The protons of the ammonium ion, on the other hand, give a sharp 1 : 1 : 1 triplet with full coupling between ^{1}H and ^{14}N, since quadrupolar relaxation is absent. When relaxation is at an intermediate rate, it is possible to observe three broadened peaks, one broadened average peak, or broadening to the point of invisibility. Irradiation at the ^{14}N frequency removes the ^{14}N–^{1}H coupling interaction (Section 5.3), so that ^{14}N appears to be nonmagnetic. Figure 5.4 shows the normal spectrum of pyrrole at the bottom, containing only the AA′BB′ set from the CH protons and no visible NH resonance, because the line is extremely broad. Irradiation at the ^{14}N frequency decouples the NH proton from ^{14}N and results in a quintet NH resonance from coupling to the four CH protons.

5.2 Reactions on the NMR Time Scale

NMR is an excellent tool for following the kinetics of an irreversible reaction traditionally through the disappearance or appearance of peaks over periods of minutes to hours. The spectrum is recorded repeatedly at specific intervals, and rate constants are calculated from changes in peak intensities. Thus, the procedure is a classical kinetic method, performed on the *laboratory time scale*. The molecular changes take place on a time scale much longer than the pulse or acquisition times of the NMR experiment. In addition, and more importantly, NMR has the unique capability for study of the kinetics of reactions that occur at equilibrium, usually with activation energies in the range from 4.5 to 25 mol^{-1} (Section 2.8), a range that corresponds to rates in the range from 10^0 to 10^4 s^{-1}. This *NMR time scale* refers to the rough equivalence of the reaction rate in s^{-1} to the frequency spacing in Hz ($\Delta\nu$) between the exchanging nuclei.

A series of spectra for the interchange of axial and equatorial protons in cyclohexane-d_{11} as a function of temperature is illustrated in Figure 2.31. When the interchange of two such chemical environments occurs much faster than the frequency differences between the two sites, the result is a single peak, reflecting the average environment (*fast exchange*). Keep in mind that these exchanges occur reversibly, and the system remains at equilibrium. When the interchange is slower than the frequency differences, the NMR result is two distinct peaks (*slow exchange*). When the interchange is comparable to the frequency differences, broad peaks typically result. The reaction then is said to occur on the NMR time scale. Both fast and slow exchange sometimes may be reached by altering the temperature of the experiment. Intermolecular reactions, such as the acid-catalyzed interchange of protons, also may be studied, as in the case of the hydroxy proton of methanol (Figure 2.30). The following examples are intramolecular.

5.2.1 Hindered Rotation

Normally, rotation around single bonds has a barrier below 5 kcal mol^{-1} and occurs faster than the NMR time scale. Rotation around the double bond of alkenes, on the other hand, has a barrier that is normally above 50 kcal mol^{-1} and is slow on the NMR time scale. There are numerous examples of intermediate bond orders, however, whose rotation occurs within the NMR time scale. Hindered rotation about the C—N bond in amides such as *N*,*N*-dimethylformamide (**5.4**)

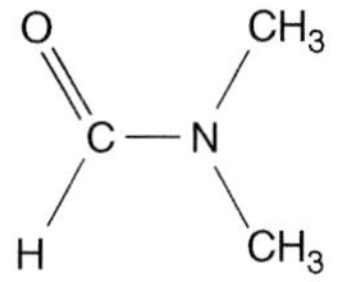

5.4

provides a classic example of site exchange. At room temperature, exchange is slow and two methyl resonances are observed, whereas, above 100 °C, exchange is fast and a single resonance is observed. The measured barrier is about 22 kcal mol^{-1}.

Hindered rotation occurs on the NMR time scale for numerous other systems with partial double bonds, including carbamates, thioamides, enamines, nitrosamines, alkyl nitrites, diazoketones, aminoboranes, and aromatic aldehydes. Formal double bonds can exhibit free rotation when alternative resonance structures suggest partial single bonding. The calicene **5.5**,

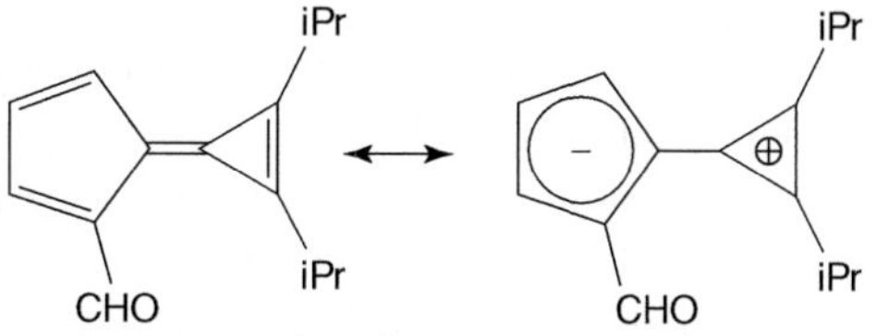

5.5

for example, has a barrier to rotation about the central bond of only 20 kcal mol^{-1}.

Steric congestion can raise the barrier about single bonds enough to bring it into the NMR range. Rotation about the single bond in the biphenyl **5.6**

is raised to a measurable 13 kcal mol^{-1} by the presence of the ortho substituents, which also provide diastereotopic methylene protons as the dynamic probe. Hindered rotation about an sp^3—sp^3 bond can sometimes be observed when at least one of the carbons is quaternary. Thus, at −150 °C, the *tert*-butyl group in *tert*-butylcyclopentane (**5.7**) gives two resonances in the ratio of 2 : 1, since two of the methyl groups are different from the third (**5.7a**).

Hindered rotation has frequently been observed in halogenated alkanes. The increased barrier probably arises from a combination of steric and electrostatic interactions. 2,2,3,3-Tetrachlorobutane (**5.8**)

exhibits a 2 : 1 doublet below −40 °C from anti and gauche rotamers that are rotating slowly on the NMR time scale.

When both atoms that constitute a single bond possess nonbonding electron pairs, the barrier often is in the observable range. The high barrier may be due to electrostatic interactions or repulsions between lone pairs. For example, the barrier to rotation about the sulfur–sulfur bond in dibenzyl disulfide ($C_6H_5CH_2S$—$SCH_2C_6H_5$) is 7 kcal mol^{-1}. Similar high barriers have been observed in hydrazines (N—N), sulfenamides (S—N), and aminophosphines (N—P).

5.2.2 Ring Reversal

Axial–equatorial interconversion through ring reversal has been studied in a wide variety of systems in addition to cyclohexane, including heterocycles such as piperidine (**5.9**),

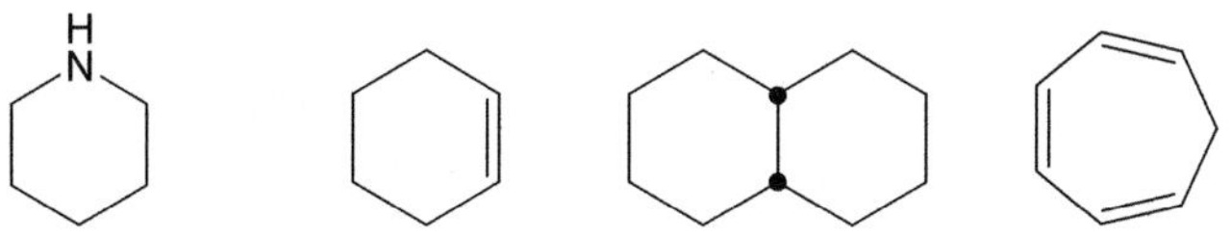

5.9   **5.10**   **5.11**   **5.12**

unsaturated rings such as cyclohexene (**5.10**), fused rings like *cis*-decalin (**5.11**), and rings of other than six members, such as cycloheptatriene (**5.12**). Cyclooctane and other eight-membered rings have been examined extensively. The pentadecadeutero derivative of the parent compound exhibits dynamic behavior below −100 °C, with a free energy of activation of 7.7 kcal mol^{-1}. The dominate conformation appears to be the boat–chair (**5.13**).

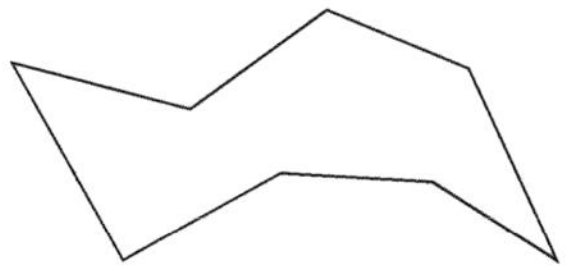

5.13

Cyclooctatetraene (**5.14**)

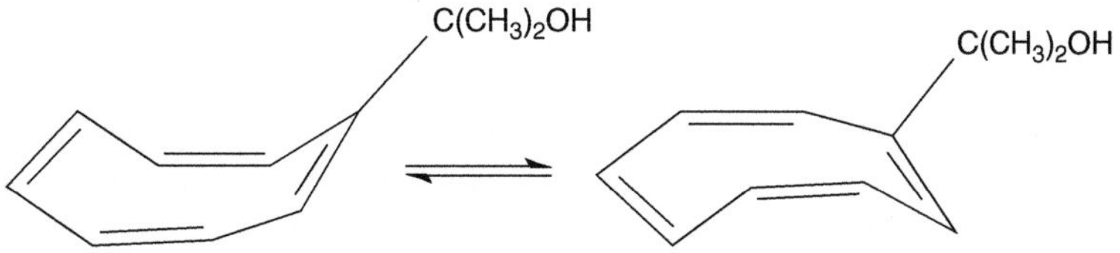

5.14

undergoes a boat–boat ring reversal. The methyl groups on the side chain provide the diastereotopic probe and reveal a barrier of 14.7 kcal mol^{-1}. The favored transition state is a planar form with alternating single and double bonds.

5.2.3 Atomic Inversion

Trisubstituted atoms with a lone pair, such as amines, may undergo the process of pyramidal atomic inversion on the NMR time scale. The resonances of the two methyls in the aziridine **5.15**

5.15

become equivalent at elevated temperatures through rapid nitrogen inversion. This barrier is particularly high (18 kcal mol^{-1}) because of angle strain in the three-membered ring, which is higher in the transition state than in the ground state. The effect is observed to a lesser extent in azetidines (**5.16**, 9 kcal mol^{-1}) and in strained bicyclic systems such as **5.17** (10 kcal mol^{-1}).

The inversion barrier may be raised when nitrogen is attached to highly electronegative elements. This substitution increases the s character of the ground-state lone pair. Since the transition-state lone pair must remain p-hybridized, the barrier is higher, as in N-chloropyrrolidine (**5.18**).

When neither ring strain nor electronegative substituents are present, barriers are low, but still often measurable, as in N-methylazacycloheptane (**5.19**, 7 kcal mol^{-1}) and 2-(diethylamino)propane, $(CH_3CH_2)_2NCH(CH_3)_2$ (6.4 kcal mol^{-1}). In the latter case, the transition state is considered to be a mix of nitrogen inversion and C—N bond rotation.

Inversion barriers for elements in lower rows of the periodic table generally are above the NMR range. Thus chiral phosphines and sulfoxides are isolable. Barriers must be brought into the observable NMR range by substitution with electropositive elements, as in the diphosphine $CH_3(C_6H_5)P-P(C_6H_5)CH_3$, whose barrier of 26 kcal mol^{-1} compares with 32 kcal mol^{-1} in $CH_3(C_6H_5)(C_6H_5CH_2)P$. The former is within the NMR range and the latter is not. The barrier in phosphole **5.20**

is lowered because the transition state is aromatic. Its barrier of 16 kcal mol^{-1} compares with 36 kcal mol^{-1} in a saturated analogue, **5.21**.

5.2.4 Valence Tautomerizations and Bond Shifts

The barriers to many valence tautomerizations fall into the NMR range. A classic example is the Cope rearrangement of 3,4-homotropilidine (**5.22**).

5.22

At low temperatures, the spectrum has the features expected for the five functionally distinct types of protons (disregarding diastereotopic differences). At higher temperatures, the Cope rearrangement becomes fast on the NMR time scale, and only three types of resonances are observed (14 kcal mol^{-1} for the 1,3,5,7-tetramethyl derivative). When a third bridge is added, as in barbaralone (**5.23**),

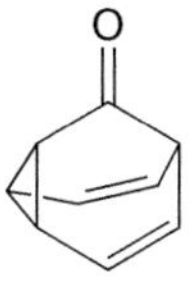

5.23 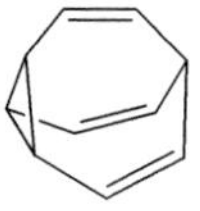   **5.24**

steric requirements of the rearrangement are improved, and the barrier is lowered to 9.6 kcal mol^{-1}. When the third bridge is an ethylenic group, the molecule is bullvalene (**5.24**). All three bridges are identical, and a sequence of Cope rearrangements renders all protons (or carbons) equivalent. Indeed, the complex spectrum at room temperature becomes a singlet above 180 °C (12.8 kcal mol^{-1}). Molecules that undergo rapid valence tautomerizations often are said to be *fluxional*.

Cyclooctatetraene offers another example of fluxional behavior. In an operation distinct from boat–boat ring reversal depicted in **5.14**, the locations of the single and double bonds are switched via the antiaromatic transition state **5.25b**.

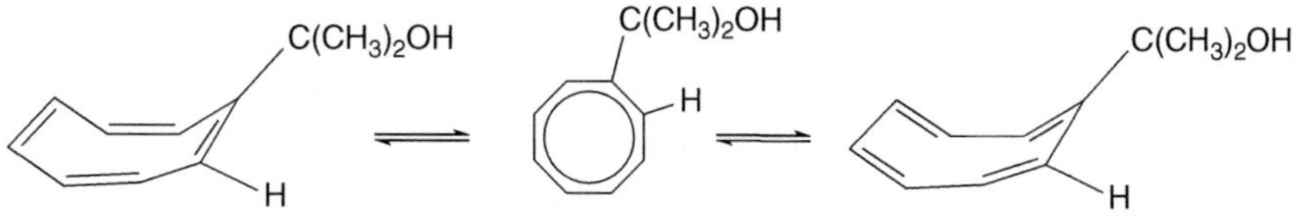

5.25a   **5.25b**   **5.25c**

The transition state to ring reversal in **5.14** has alternating single and double bonds. The proton adjacent to the substituent is different in the bond-shift isomers **5.25a** and **5.25c**. The barrier to bond switching was determined from the conversion of the proton resonance from two peaks to one (17.1 kcal mol^{-1}). The barrier to bond switching is higher than that to ring reversal because of antiaromatic destabilization that is present in the equal-bond-length transition state **5.25b**. Rearrangements of carbocations also may be studied by NMR methods. The norbornyl cation (**5.26**) may undergo 3,2- and 6,2-hydride shifts, as well as Wagner–Meerwein (W–M) rearrangements. The sum of these processes renders all protons equivalent, so that the complex spectrum below −80 °C becomes a singlet at room temperature. The slow process appears to be the 3,2-hydride shift, whose barrier was measured to be 11 kcal mol^{-1}.

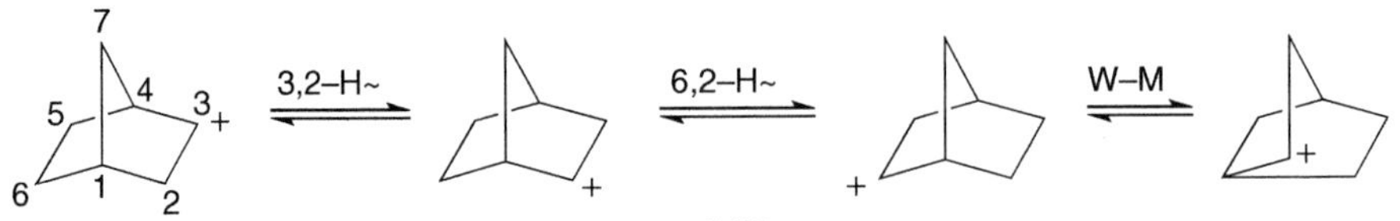

Many examples of fluxional organometallic species have been investigated. Tetramethylalleneiron tetracarbonyl (**5.27**)

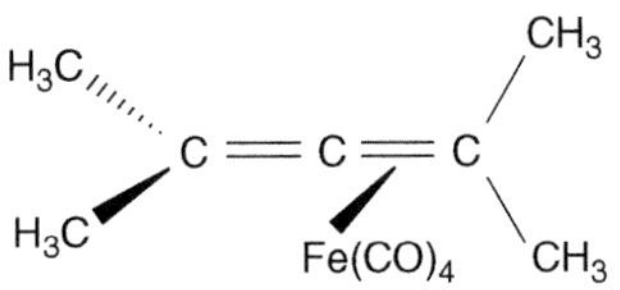

exhibits three distinct methyl resonances in the ratio 1:1:2 at −60 °C, in agreement with the structure depicted. Above room temperature, however, the spectrum becomes a singlet (9 kcal mol^{-1}) as the Fe(CO)$_4$ unit circulates about the allenic π-electron structure by moving orthogonally from one alkenic unit to the other.

In cyclooctatetreneiron tricarbonyl (**5.28**),

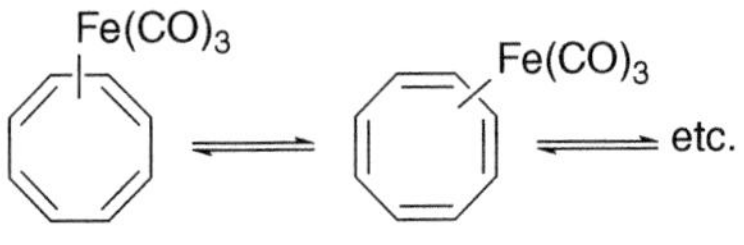

the spectrum below −150 °C indicates four protons on carbons bound to iron and four on carbons not bound to iron, consistent with the η^4 structure shown. Above −100 °C, all the protons converge to a singlet as the iron atom moves around the ring as shown. A bond shift occurs with each 45° movement of the iron atom. Eight such operations result in complete averaging of the ring protons or carbons.

A series of 1,5-sigmatropic shifts occurs in triphenyl-(7-cycloheptatrienyl)tin (**5.29**).

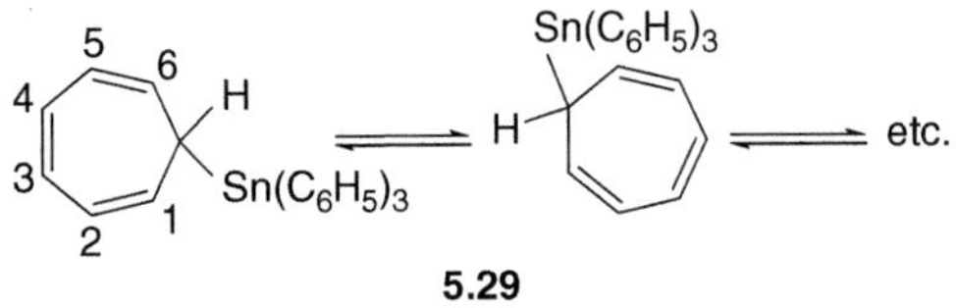

At 0 °C, the spectrum indicates that bond shifts are slow on the NMR time scale, but, at 100 °C, all of the ring protons are equivalent. That the migration is a 1,5 shift to the

3 or 4 positions (rather than a 1,2 or 1,3 shift) was demonstrated by double-irradiation experiments (saturation transfer; see shortly).

5.2.5 Quantification

For the simple case of two equally populated sites that do not exhibit coupling (such as cyclohexane-d_{11} in Figure 2.31 or the amide **5.4**), the rate constant (k_c) at the point of maximum peak broadening (the coalescence temperature T_c, approximately $-60\,°C$ in Figure 2.31) is $\pi \Delta v/\sqrt{2}$, in which Δv is the distance in Hz between the two peaks at slow exchange. The free energy of activation then may be calculated as $\Delta G_c^{\ddagger} = 2.3 RT_c [10.32 + \log(T_c/k_c)]$. This result is extremely accurate and certainly easy to obtain, but the equation is limited in its application. For the two-site exchange between coupled nuclei, the rate constant at T_c is $\pi(\Delta v^2 + 6J^2)^{1/2}/\sqrt{2}$.

To include unequal populations, more complex coupling patterns, and more than two exchange sites, it is necessary to use computer programs such as DNMR3, which can simulate the entire line shape at several temperatures. Such a procedure generates Arrhenius plots from which enthalpic and entropic activation parameters may be obtained. The procedure is more elegant and more comprehensive, but it is more susceptible to systematic errors involving inherent line widths and peak spacings, than is the coalescence temperature method. Consequently, it is always a good idea to use both line-shape fitting and coalescence temperature methods, when possible, as a cross check.

The proportionality between k_c and Δv ($k_c = \pi \Delta v/\sqrt{2}$) implies that the rate constant is dependent on the field strength (B_0). Thus, a change in field from 300 to 600 MHz alters the rate constant at T_c. The practical result is that T_c changes. Since the slow exchange peaks are farther apart at 600 MHz, a higher temperature is required to achieve coalescence than at 300 MHz. At a given field strength, two nuclides such as ^{1}H and ^{13}C have different values of Δv for analogous functionalities and achieve coalescence at different temperatures. Since Δv is usually larger for ^{13}C than for ^{1}H, the ^{13}C coalescence temperature often is much higher than the ^{1}H coalescence temperature, e.g. for the methyl carbons and hydrogens of N,N-dimethylformamide (**5.4**), even though a single rate process is involved.

5.2.6 Magnetization Transfer and Spin Locking

Alternative procedures not requiring peak coalescence have been developed to expand the kinetic dynamic range of NMR spectroscopy. In many cases, coalescence and fast exchange are never attained. The system may exchange too slowly on the NMR time scale at the highest available temperatures (as determined by the temperature range of the spectrometer, volatility of the solvent, or stability of the sample). An alternative technique, called *saturation transfer* or *magnetization transfer*, can provide rate constants without peak coalescence, that is, solely at the slow exchange limit. Continuous, selective irradiation of one slow-exchange peak may partially saturate the other peak. Some of the nuclei from the first site turn into nuclei of the second type by the exchange process. The intensity of the second peak then is reduced because the newly transformed nuclei already had been saturated in their previous identity. This reduction in intensity is related to the rate constant of interchange and the relaxation time. Saturation transfer is observed for rates in the range from 10^{-3} to 10^1 s^{-1}, which extends the NMR range based on line-shape coalescence (10^0–10^4 s^{-1}) on the slow-exchange end by about 3 orders of magnitude. In addition to expanding the dynamic range of NMR kinetics, this method permits the easy identification of exchanging partners. For example, in the cycloheptatrienyltin **5.29**, saturation of the 7 proton resonance (geminal to tin) at $-10\,°C$ (below the coalescence temperature) brings about a decrease in the intensity of the 3,4 proton resonance, indicative of a 1,5 shift. A 1,2 shift would have saturated the 1,6 resonance, and a 1,3 shift would have saturated the 2,5 resonance. The two-dimensional version of this experiment is termed *EXchange SpectroscopY* (EXSY) and is discussed in Chapter 6.

Rates that are fast on the NMR line-shape time scale (when peaks fail to decoalesce at high temperatures) sometimes may be measured by observation at a different resonance frequency. Normally, nuclear spins precess around the B_0 field at their Larmor frequency. Application of the usual 90° pulse in the x direction places the spins in the xy plane, along the y axis (Figure 2.15a). Continuous B_1 irradiation along the y axis (not a pulse) forces magnetization to precess around that axis (called *spin locking*, as in the cross-polarization experiment of Section 2.9). The spins are said to be locked onto the y axis. Because the spins are precessing at a lower frequency (γB_1, rather than γB_0), they are sensitive to a different range of rate processes, one corresponding to about 10^2–10^6 s^{-1}, which extends the NMR range on the fast-exchange end by about 2 orders of magnitude. Rates are obtained by comparing the relaxation time when the system is spin locked ($T_{1\rho}$) with the usual spin–lattice relaxation time (T_1) and analyzing any differences.

Through line-shape, saturation transfer, and spin–lock methods, the entire range of rates accessible to NMR is about 10^{-3}–10^6 s^{-1}, 10 orders of magnitude. Thus, NMR has become an important method for studying the kinetics of reactions at equilibrium over a very large dynamic range.

5.3 Multiple Resonance

Special effects may be routinely and elegantly created by using sources of radiofrequency energy in addition to the observation frequency ($v_1 = \gamma B_1$) ($\gamma = \gamma/2\pi$). The technique is called *multiple irradiation* or *multiple resonance* and requires the presence of a second transmitter coil in the sample probe to provide the new irradiating frequency $v_2 = \gamma B_2$. When the second frequency is applied, the experiment, which is widely available on modern spectrometers, is termed *double resonance* or *double irradiation*. Less often, a third frequency ($v_3 = \gamma B_3$) also is provided, to create a *triple-resonance* experiment. We already have seen several examples of double irradiation experiments, including the removal of proton couplings from ^{13}C (Figure 2.25), the elimination of solvent peaks by peak suppression (Section 5.11), the sharpening of NH resonances by irradiation of ^{14}N (Figure 5.4), and the study of rate processes by saturation transfer (Section 5.2).

5.3.1 Spin Decoupling

One of the oldest and most generally applicable double-resonance experiments is the irradiation of one proton resonance (H_X) and observation of the effects on the AX coupling (J_{AX}) present in another proton resonance (H_A). The traditional and intuitive explanation for the resulting spectral simplification, known as *spin decoupling*, is that the irradiation shuttles the X protons between the $+½$ and $-½$ spin states so rapidly that the A protons no longer have a distinguishable independent existence. As a result, the A resonance collapses to a singlet. This explanation, however, is inadequate in that it fails to account for phenomena at weak decoupling fields (spin tickling) and even some phenomena at very strong decoupling fields.

The actual experiment involves getting the coupled nuclei to precess about orthogonal axes. The magnitude of the coupling interaction between two spins is expressed by the scalar, or dot, product between their magnetic moments and is proportional to the expression $J\mu_1 \cdot \mu_2 = J\mu_1\mu_2\cos\phi$. The quantity ϕ is the angle between the vectors (the axis of precession of the nuclei). So long as both sets of nuclei precess around the same (z) axis, ϕ is zero, $\cos 0° = 1$, and full coupling is observed. The geometrical relationship between the spins may be altered by subjecting one of them to a B_2 field. Imagine observing ^{13}C nuclei as they precess around the z axis at the frequency B_1. When the attached protons are subjected to a strong B_2 field along the x axis, they will precess around that axis. The angle ϕ between the ^{13}C and ^1H nuclear vectors then is 90°, as they respectively precess around the z and x axes. As a result, their spin–spin interaction goes to zero because the dot product is zero ($\cos 90° = 0$). The nuclei are then said to be decoupled.

Spin decoupling has been useful in identifying coupled pairs of nuclei. Figure 5.5 provides such an example for the molecule ethyl *trans*-crotonate (ethyl *trans*-but-2-enoate).

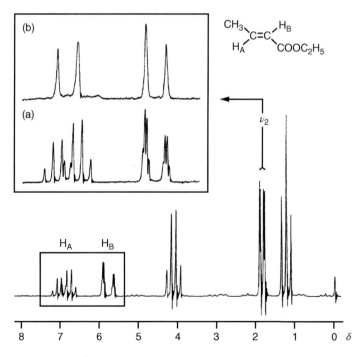

Figure 5.5 The ^1H spectrum of ethyl *trans*-crotonate. The inset contains an expansion of the alkenic range (a) without and (b) with decoupling of the methyl resonance at δ 1.8. Source: Günther 1992 [2]. Reproduced with permission of John Wiley & Sons.

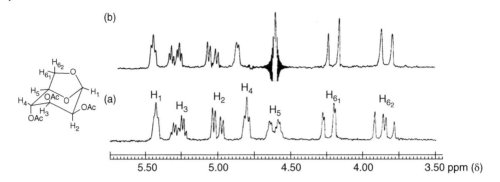

Figure 5.6 The 100 MHz ^{1}H spectrum of mannosan triacetate in CDCl$_3$ without decoupling (a) and with double irradiation at δ 4.62 (b). Source: Courtesy of Varian Associates.

The alkenic protons split each other, and both are split by the allylic methyl group to form an ABX$_3$ spin system. Irradiation at the methyl resonance frequency produces the upper spectrum in the inset for the alkenic protons, which have become a simple AB quartet. A more complex example is illustrated in Figure 5.6. The bicyclic sugar mannosan triacetate, whose structure is given on the left of the figure, has a nearly first-order spectrum with numerous coupling partners. Irradiation of H$_5$ (δ 4.62) produces simplification of the resonances of its vicinal partners H$_4$, H$_{6/1}$, and H$_{6/2}$, as well as its long-range zigzag partner H$_3$.

5.3.2 Difference Decoupling

With complex molecules, it is useful to record the difference between coupled and decoupled spectra. Features that are not affected by decoupling are subtracted out and do not appear. Figure 5.7 shows the ^{1}H spectrum of 1-dehydrotestosterone. The complex region between δ 0.9 and 1.1 contains the resonances of four protons. A comparison of the coupled (Figure 5.7a in the inset) and decoupled (Figure 5.7b) spectra from irradiation of the 6α resonance shows little change as the result of double irradiation. The *difference decoupling spectrum* (Figure 5.7c) is the result of subtracting (a) from (b). Unaffected overlapping peaks are gone. The original resonances of the affected protons are observed as negative peaks with coupling, and the simpler decoupled resonances of the same protons are present as positive peaks. The resonances must be due to the 7α protons. The procedure provides coupling relationships when spectral overlap is a serious problem. This and other simple spin-decoupling experiments have been entirely superseded by two-dimensional experiments (Chapter 6).

5.3.3 Classes of Multiple Resonance Experiments

Experiments in which both the irradiated and the observed nuclei are protons are called *homonuclear double resonance* experiments and are represented by the notation ^{1}H{^{1}H}. The irradiated nucleus is denoted by braces. When the observed and irradiated nuclei are different nuclides, as in proton-decoupled ^{13}C spectra, the experiment is a *heteronuclear double resonance* experiment and is denoted, for example, ^{13}C{^{1}H}, or ^{13}C{^{1}H}{^{31}P} for a triple resonance experiment.

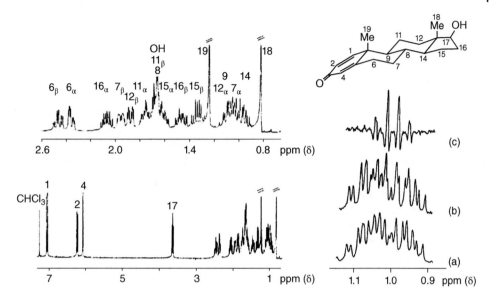

Figure 5.7 The 400 MHz ^{1}H spectrum of 1-dehydrotestosterone. The complete spectrum and an expansion of the low-frequency region are given on the left. On the right are shown (a) the coupled spectrum for the δ 0.9–1.1 region; (b) the same region, decoupled from the 6α proton; and (c) the difference spectrum obtained by subtracting (b) from (a). Source: Hall and Sanders 1980 [3]. Copyright 1980. Reproduced with permission of American Chemical Society.

Double resonance experiments also may be classified according to the intensity or bandwidth of the irradiating frequency. If irradiation is intended to cover only a portion of the resonance frequencies, the technique is known as *selective irradiation* or *selective decoupling*. The decoupling shown in Figures 5.5 and 5.6, the peak suppression described in Section 5.1, and the magnetization transfer discussed in Section 5.2 are three examples of selective double irradiation. In the two decoupling experiments, only couplings to the selectively irradiated proton are removed. Nonirradiated resonances can exhibit a small movement in frequency, called the *Bloch–Siegert shift*, which is related to the intensity of the B_2 field and the distance between the observed and irradiated frequencies. An examination of Figure 5.6 reveals several such shifts, found by comparing the relative positions of the resonances in the upper and lower spectra. When all frequencies of a specific nuclide are irradiated, the experiment is termed *nonselective irradiation* or *broadband decoupling*. Figure 2.25 illustrates the ^{13}C spectrum of 3-hydroxybutyric acid both with and without broadband proton double irradiation. The invention of this technique was instrumental in the development of ^{13}C NMR spectroscopy as a routine tool. To cover all the ^{1}H frequencies, B_2 was modulated with white noise, so the technique often was called *noise decoupling*.

5.3.4 Off-resonance Decoupling

The broadband decoupling experiment removes coupling patterns that could indicate the number of protons attached to a given carbon atom. The *off-resonance decoupling* method was developed to retain this information and still provide some of the advantages of the decoupling experiment. Irradiation above or below the usual 10-ppm range of ^{1}H frequencies leaves residual coupling given by the approximate formula $J_{res} = 2\pi J(\Delta v)/\gamma B_2$, in which J is the normal coupling, γ is the gyromagnetic ratio of the irradiated nucleus, and Δv is the difference between the decoupler frequency and the resonance frequency of a proton coupled to a specific carbon. Because carbon multiplicities remain intact, this technique is useful for determining, with minimal peak overlap, whether carbons are methyl (quartet), methylene (triplet), methine (doublet), or quaternary (singlet). If methylene protons are diastereotopic, methylene carbons can appear as two doublets. The outer peaks of the off-resonance decoupled triplets and quartets usually are weaker than one might expect from the binomial coefficients. As a result, doublets and quartets sometimes are difficult to distinguish. Figure 5.8 shows the spectrum of vinyl acetate with full decoupling and with off-resonance decoupling. In complex molecules, peak overlap and ambiguities with regard to quartets often make assignments by this technique difficult. Therefore, it has been superseded by the editing experiments described in Section 5.5.

In early spin-decoupling experiments, the irradiation frequency was left on continuously while the experimenter observed the resonating nuclei. There are two significant

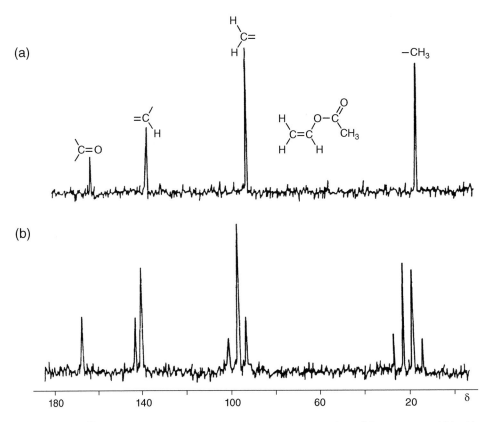

Figure 5.8 The ^{13}C spectrum of vinyl acetate (a) with complete decoupling of the protons and (b) with off-resonance decoupling of the protons. Source: Günther 1992 [4]. Reproduced with permission of John Wiley & Sons.

problems with this method. First, the application of RF energy at the decoupling frequency generates heat. As B_0 fields increased from 60 to above 900 MHz, higher decoupling intensities were required. The resultant heating was unacceptable for biological samples and for many delicate organic or inorganic samples. Second, with higher field strengths, it became increasingly more difficult for B_2 to cover the entire range of ^{1}H frequencies, which had been about 600 Hz at 60 MHz, but became 5000 Hz at 500 MHz.

To overcome these problems of heteronuclear decoupling, modern methods replaced continuous irradiation with a series of pulses that eliminate the effects of coupling. In a ^{13}C{^{1}H} experiment (Figure 5.9 for two spins, ^{13}C–^{1}H), a 90° B_1 pulse applied to the observed ^{13}C nuclei along the x direction moves magnetization from carbon coupled to either spin-up or spin-down protons into the xy plane along the y axis ((Figure 5.9a) → (Figure 5.9b)). The reference frequency is considered to coincide with the y axis and be midway between the frequencies of the carbons associated with the spin-up (β) and spin-down (α) protons. The two carbon vectors then diverge in the xy plane after the 90° pulse, one becoming faster and the other slower than the carrier frequency (Figure 5.9c). After time τ, a 180° proton pulse (the B_2 of the decoupling experiment) switches the locations of the vectors. The slower moving vector that was dropping behind the carrier frequency now is replaced by the faster moving vector (and the faster moving vector by the slower moving vector), so that both carbon vectors start to move back toward the y axis (Figure 5.9d). After an equal second period τ, the two vectors coincide on the y axis. At this time, only one frequency or peak occurs, and coupling to the protons disappears (Figure 5.9e). The process is repeated during acquisition at a rate (in Hz) that is faster than the coupling constant, so that the effects of coupling are removed. In this way, decoupling can be achieved with short pulses

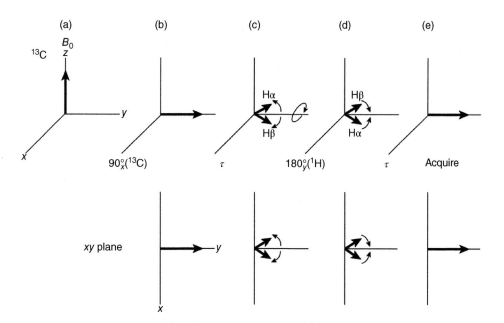

Figure 5.9 Pulse sequence to remove heteronuclear coupling.

during acquisition rather than with a continuous, high-intensity field during the entire experiment.

In practice, the method is limited because the 180° pulse must be very accurate and because the B_2 field is inhomogeneous. Refinements of this experiment have been achieved by replacing the 180° pulse with several pulses (*composite pulses*) and by cycling their order (*phase cycling*) so as to cancel out the inaccuracies (Section 5.8). Some successful such methods include MLEV-16 (for *Malcolm LEVitt*, the name of one of the developers), with 16 phase cycles and, in particular, WALTZ-16, which achieves full decoupling across a much wider range than the original continuous method and with a fraction of the power. The source of the name WALTZ is provided in Section 5.8.

5.4 The Nuclear Overhauser Effect

Dipole–dipole relaxation occurs when two nuclei are located close together and are moving at an appropriate relative rate (Section 5.1). Irradiation of one of these nuclei with a B_2 field alters the Boltzmann population distribution of the other nucleus and therefore perturbs the intensity of its resonance. No J coupling need be present between the nuclei. The original phenomenon was discovered by Overhauser, but between nuclei and unpaired electrons. The Overhauser effect when both spins are of nuclei was observed first by Anet and Bourne and is of more interest to the chemist. It has great structural utility, because the dipole–dipole mechanism for relaxation depends on the distance between the two spins (Eq. (5.1)).

5.4.1 Origin

The physical basis of the *Nuclear Overhauser Effect* (NOE) is illustrated in Figure 5.10. On the left are the states for two spins (A and X) in the absence of double irradiation. Effects of J are irrelevant and are ignored. The diagram represents an expansion of Figure 2.4a for one spin, with β standing for $+\frac{1}{2}$ and α for $-\frac{1}{2}$. There are four spin states: when

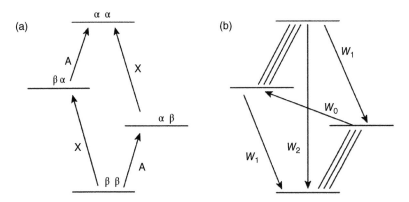

Figure 5.10 (a) The spin states for a normal two-spin (AX) system. (b) The spin states for an AX system when the frequency of A is doubly irradiated. Arrows denote spin excitations.

both spins are β, when the first spin (A) is β and the second (X) is α, when the first is α and the second is β, and when both are α. There are two A-type transitions (when the A spin flips from β to α)—for example, ββ to αβ—and there are two X-type transitions (when the X spin flips from β to α)—for example, αβ to αα. When $J = 0$, the two A transitions coincide, as do the two X transitions. Because chemical shifts are very small in comparison with the Larmor frequency, the αβ and βα states are almost degenerate. Their difference has been exaggerated to emphasize the different chemical shifts.

The normal intensities of the A and X resonances are determined by the difference between the populations of the upper and lower spin states in a spin transition – for instance, between αβ and αα for one X transition. In the NOE experiment, one resonance frequency (A) is doubly irradiated, and intensity perturbations are monitored at the other resonance frequency (X). When the A resonance is irradiated, as is represented by the multiple parallel lines in the right-hand diagram of Figure 5.10, the population difference between the spin states connecting an A transition decreases through partial saturation. Compared with the normal situation on the left, the populations of αα and αβ (the upper states) have increased, while those of βα and ββ (the lower states) have decreased. Dipolar relaxation from αα to ββ, labeled W_2 in the figure, can help restore the system to equilibrium. The new equilibrium present during irradiation of A thus can carry spins along the route βα → αα → ββ → αβ, depleting βα and augmenting αβ. Depleting βα enhances the population difference for one X transition (ββ → βα), while augmenting αβ enhances the population difference for the other X transition (αβ → αα). This enhanced polarization of nuclear spin states means that the X intensity is higher in the new equilibrium during the irradiation of A. To a first approximation, the populations of αα and ββ are constant, as they are simultaneously augmented by one transition and depleted by another. Normal relaxation of X nuclei, labeled W_1 in the figure (αα → αβ or βα → ββ), does not alter the X intensity. These processes are unchanged from the diagram on the left.

Relaxation from αβ to βα (called W_0 in Figure 5.10) also can move the irradiated system back toward equilibrium. This relaxation mechanism, however, would result in a *decrease* in intensity of X, since it depletes ββ and augments αα, (ββ → αβ → βα → αα) with αβ and βα constant. For liquids and relatively small molecules, $W_0 \ll W_2$, so that enhanced intensities are expected. The frequencies of W_2 are in the MHz range (represented by the large distance between the ββ and αα levels in the figure), whereas those of W_0 are much smaller, in the kHz or Hz range (represented by the small, but exaggerated, distance between the αβ and βα levels). Small molecules tumbling in solution produce fields in the MHz range and hence can provide W_2 relaxation. On the other hand, large molecules tumbling in the Hz or kHz range can provide W_0 relaxation.

5.4.2 Observation

Double irradiation of A in molecules of molecular weight up to c. 3000 Da thus enhances the X intensity, provided that the two nuclei are close enough for W_2 relaxation to dominate (less than about 5 Å). This circumstance corresponds to what we previously called the extreme narrowing limit. For larger molecules—certainly those with molecular weights over 5000—W_0 dominates, and reductions in peak intensity or inverse peaks occur. At some intermediate size (1000–3000), the effect disappears as the crossover between regimes occurs. The change in intensity (denoted by the Greek letter η (eta)) thus depends on the difference between the W_2 and W_0 relaxation rates, in comparison with the total relaxation rates, as given by Eq. (5.3).

$$\eta = \frac{\gamma_{irr}}{\gamma_{obs}} \left(\frac{W_2 - W_0}{W_0 + 2W_1 + W_2} \right) \quad (5.3)$$

The factor W_1 carries a coefficient of 2 because there are two such modes. The effect is observed by comparing intensities I in the presence of double irradiation with those I_0 in its absence via Eq. (5.5).

$$\eta = \frac{(I - I_0)}{I_0} \quad (5.4)$$

For small molecules (the extreme narrowing limit), the maximum increment in intensity, η_{max}, is $\gamma_{irr}/2\gamma_{obs}$ (this factor requires assigning values to the relaxation rates W in Eq. (5.3)), so that an initial intensity of unity ($I_0 = 1.0$) increases up to ($1 + \eta_{max}$). In our example, A is irradiated ("irr") and X observed ("obs"). The maximum enhanced intensity, obtained by rearrangement of Eq. (5.4), is given by Eq. (5.5).

$$I_{max}(NOE) = I_0 \left(1 + \frac{\gamma_{irr}}{2\gamma_{obs}} \right) \quad (5.5)$$

The increase is almost always less than the maximum, because nondipolar relaxation mechanisms are present and because the observed nucleus is relaxed by nuclei other than the irradiated nucleus.

Whenever the two nuclei are the same nuclide, e.g. both protons, the gyromagnetic ratios in Eq. (5.5) cancel, η_{max} becomes 0.5, and the maximum intensity enhancement ($1 + \eta_{max}$) is a factor of 1.5, or 50%. For the common case of broadband ^{1}H irradiation with observation of ^{13}C, [^{13}C{^{1}H}], η_{max} is 1.988, so the enhancement is a factor of up to 2.988, or about 200%. Other maximum Overhauser enhancement factors ($1 + \eta_{max}$) include 2.24 for ^{31}P{^{1}H}, 3.33 for ^{195}Pt{^{1}H}, and 3.39 for ^{207}Pb{^{1}H}.

Certain nuclei have negative gyromagnetic ratios, so that, in the extreme narrowing limit, η_{max} becomes negative and a negative peak can result. For irradiation of ^{1}H and observation of ^{15}N [^{15}N{^{1}H}], η_{max} is −4.94. The maximum negative intensity is thus 3.94 times that of the original peak, or an increase of 294% ((3.94 − 1.00) × 100), but as an inverse peak. If dipolar relaxation is only partial, the ^{15}N{^{1}H} NOE can result in decreased intensity or even a completely nulled resonance. Silicon-29 also has a negative gyromagnetic ratio, so similar complications ensue. For the ^{29}Si{^{1}H} experiment, $\eta_{max} = -2.52$. The maximum enhancement factor ($1 + \eta_{max}$) is then −1.52, which results in an inverted intensity with an increase of 52% over the unirradiated case. For ^{119}Sn{^{1}H} ($\eta_{max} = -1.34$), there is actually a net loss in intensity. The maximum enhancement factor is −0.34, representing a 66% loss in intensity of the negative peak, compared with the peak at the unirradiated position. The NOE is entirely independent of spectral changes that arise from the collapse of spin multiplets through spin decoupling. The NOE does not require that nuclei A and X be spin coupled—only that they be mutually relaxed through a dipolar mechanism.

Large molecules, such as proteins or nucleic acids, with molecular weights over about 3000–5000 are dominated by W_0 relaxation. Since the other terms (W_2 and W_1) in Eq. (5.3) are small, the value of η_{max} becomes −1 for the homonuclear proton case. Such a situation can result in a loss of signal. Consequently, for large molecules, transient rather than steady-state NOEs often are studied. For example, the buildup (or loss) of signal from the NOE can provide interproton distances. By observing many such relationships, the structures of large biomolecules may be determined quantitatively in a process that rivals X-ray crystallography, but applies to the liquid state (Nobel Prize, 2002).

At the crossover between the extreme narrowing and the large-molecule limits, it is possible that W_2 and W_0 are comparable in magnitude, so that, by Eq. (5.3), the NOE goes to zero. The spectroscopist may improve the situation somewhat by changing the solvent or the temperature in order to alter τ_c. Viscosity, in addition to molecular size, can affect the tumbling rates and hence the rate of dipolar relaxation. Thus, viscous media can lower nuclear Overhauser enhancements.

In the traditional NOE experiment, the spectrum is recorded twice, with and without the NOE. Figure 5.11 illustrates the relative timing for the heteronuclear case of a ^{13}C pulse (B_1), a ^{1}H double irradiation field (B_2), and acquisition of the ^{13}C signal (not to scale) in order to carry out these two experiments. In the original experiment with continuous broadband decoupling (Figure 5.11a), the B_2 field is turned on and left on. It must be on during acquisition to ensure decoupling, but also during the recovery time, when relaxation occurs and the NOE builds up. The power can be lower during times other than acquisition (*power-gated decoupling*). This experiment results in both decoupling and the Overhauser effect and provides the quantity I in Eq. (5.4). By gating the decoupler off during the recovery period, as in (Figure 5.11b), but keeping it on during acquisition, the spectroscopist obtains decoupling, but no NOE, providing the quantity I_0 in Eq. (5.4). Without irradiation during the recovery period, there is insufficient time for the NOE to build up, and unperturbed intensities are obtained. In practice, the double-resonance frequency is not actually turned off, but is moved far off resonance. A comparison of the intensity I in experiment (a) with the intensity I_0 in experiment (b) provides the NOE via Eq. (5.4). Figure 5.11c illustrates an alternative procedure, in which the B_2 field is gated off during acquisition, but is on during the recovery period.

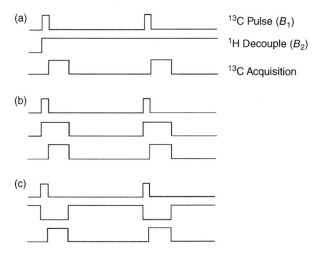

Figure 5.11 (a) Observation of ^{13}C with continuous double irradiation of ^{1}H (decoupling and NOE). (b) Double irradiation applied during acquisition, but gated off during the wait period (decoupling, no NOE). (c) Double irradiation applied only during the wait period (NOE, no decoupling). The pulse widths are not to scale. The scheme is shown for two cycles.

Such an experiment provides no decoupling, but generates the NOE, so it is useful for measuring ^{1}H–^{13}C couplings with enhanced intensity.

5.4.3 Difference NOE

For the homonuclear proton NOE experiment (^{1}H{^{1}H}) that parallels experiments (a) and (b) in Figure 5.11, it has traditionally been supposed that the NOE (in percentage, 100η) must exceed about 5% to be accepted as experimentally significant. The *difference NOE experiment*, however, can measure enhancements reliably to below 1%. In this procedure, spectra obtained by the methods analogous to Figure 5.11a,b are alternatively recorded and subtracted. Unaffected resonances disappear, and NOEs are signified by residual peaks.

Figure 5.12 illustrates the difference NOE spectrum for a portion of the ^{1}H spectrum of progesterone (**5.30**),

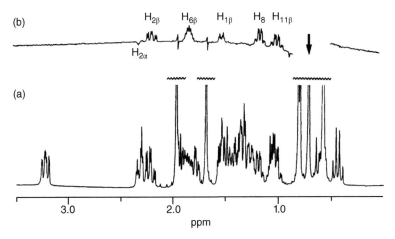

in which the resonance of the 19 methyl group has been irradiated (arrow). The unirradiated spectrum is given at the bottom, the difference spectrum at the top. Enhancements are seen by difference for five nearby protons. In general, for molecules in the extreme

Figure 5.12 The 400 MHz ^{1}H spectrum (in part) of progesterone (a) without double irradiation and (b) with irradiation of the CH_3–19 resonance displayed as a difference spectrum. Source: Sanders and Hunter 1993 [5]. Reproduced with permission of Oxford University Press.

narrowing limit, the NOE difference experiment is preferred to the direct experiment. Proton H$_{2\alpha}$ (the equatorial 2 proton) is not close to the 19 methyl group, but its resonances show a small negative NOE. This finding is the result of a three-spin effect. (A is relaxed by B and B by C.) Irradiation at A increases the Boltzmann population for B and enhances the intensity of B. By spin diffusion (Section 5.1), this enhanced intensity of B has the opposite effect on C, decreasing the Boltzmann population and the intensity. As a result, C appears as a negative peak in the difference NOE spectrum. In this example, A is Me-19, B is H$_{2\beta}$, and C is H$_{2\alpha}$. The process occurs most commonly with very large molecules.

5.4.4 Applications

The NOE experiment has three distinct uses. For heteronuclear examples, the foremost use is the increase in sensitivity, which combines with the collapse of multiplets through decoupling to provide the standard ^{13}C spectrum composed of a singlet for each carbon. Because most carbons are relaxed almost entirely by their attached protons, the NOE commonly attains a maximum value of about 200%. Quaternary carbons, with more distant nearest neighbors, do not enjoy this large enhancement.

Second, interpreting ^{13}C spin–lattice relaxation routinely requires a quantitative assessment of the dipolar component, $T_1(DD)$. Because the NOE results from dipolar relaxation, its size is related to the dipolar percentage of overall relaxation. If the maximum, or full, NOE for ^{13}C{^{1}H} of 200% is observed, then $T_1(obs) = T_1(DD)$. When other relaxation mechanisms contribute to ^{13}C relaxation, the enhancement is less than 200%. The dipolar relaxation for ^{13}C{^{1}H} then may be calculated from the expression $T_1(DD) = \eta T_1(obs)/1.988$, in which η is the observed NOE and 1.988 is the maximum NOE (η_{max}). It is possible then to discuss $T_1(DD)$ in terms of structure, according to Eq. (5.1).

In the third application, the dependence of the NOE on internuclear distances can be exploited to determine structure, stereochemistry, and conformation. Enhancements are expected when nuclei are close together. The adenosine derivative **5.31**

5.31

(2′,3′-isopropylidene adenosine) can exist in the conformation shown with the purine ring lying over the sugar ring (syn) or in an extended form with the proton on C8 lying over the sugar ring (anti). Saturation of the H1′ resonance brings about a 23%

enhancement of the H8 resonance, and saturation of H2′ produces an enhancement of H8 of 5% or less. Thus, H8 must be positioned most closely to H1′, as in the syn form shown. Structural and stereochemical distinctions frequently are made possible by determining the relative orientations of protons. The synthetic penicillin derivative **5.32**

5.32a **5.32b**

could have the spiro sulfur heterocycle oriented either as shown in (a) or with the sulfur atom and $(CH)_{10}$ switched, as in (b). Irradiation of the methyl protons brings about an enhancement of H10 as well as of H3 and clearly demonstrates that the stereochemistry is that in **5.32a**.

5.4.5 Limitations

Despite the considerable advantages of the NOE experiment, its limitations must be appreciated. First, three-spin effects, or spin diffusion, may cause misleading intensity perturbations when the third spin is not close to the irradiated nucleus ($H_{2\alpha}$ in Figure 5.12). Second, the size of the molecule can cause NOE effects that are positive, negative, or null. Third, nuclei with negative gyromagnetic ratios can give diminished positive peaks, no peak, or negative peaks with diminished or enhanced intensity. Fourth, chemical exchange can cause an intensity perturbation analogous to the three-spin effect. Irradiation of a nucleus can lead to intensity changes at another nucleus, which can alter its chemical identity through a dynamic exchange such as a bond rotation or valence tautomerization. The NOE can then be observed for the product nucleus, provided chemical exchange is faster than relaxation of the NOE effects. Fifth, closely coupled systems can give complex results because of mixing of spin states. Sixth, unintentional paramagnetic impurities can alter the NOE through intermolecular dipole–dipole relaxation. All these considerations must be taken into account in interpreting NOE experiments. Despite its limitations, the NOE is a very important tool for enhancing intensities and elucidating structures.

5.5 Spectral Editing

For deducing the structure of organic molecules, one of the most useful pieces of information is a compilation of the substitution pattern of all the carbons – that is, a census of which carbons are methyl, methylene, methine, or quaternary. We have already seen (Figure 5.8) that the off-resonance decoupling procedure provides such information, although with less-than-ideal results. Through the choice of appropriate pulses and timing, the chemist may accomplish the same task by eliminating some of the resonances from the spectrum or by altering their polarization. Such an experiment is called *spectral editing* and includes solvent suppression for example.

5.5.1 The Spin–Echo Experiment

Most spectral editing procedures are based on the *spin echo* experiment devised by Hahn, Carr, Purcell, Meiboom, and Gill in the 1950s, largely to measure spin–spin relaxation times (T_2). An example of this experiment was given in Figure 5.9, in which a 180° pulse brought vectors from spin–spin interactions back together on the y axis as an echo. Such a procedure also refocuses dispersion in the chemical shift caused by magnetic inhomogeneity in the following fashion. As shown in Figure 5.13, in the absence of J, a resonance (Figure 5.13b) fans out over a range of frequencies (Figure 5.13c), because not every nucleus of a given type has exactly the same resonance frequency in an inhomogeneous field. The 180° pulse refocuses all the magnetization back onto the y axis after time 2τ, as in Figure 5.13e. Chemical-shift differences also may be eliminated in this fashion. Repetition of the 180° pulse every 2τ produces a train of peaks whose intensities die off with time constant T_2. This relaxation time provides a measure of spin–spin interactions alone, free from the usually dominating effects of field inhomogeneity. The notation T_2^* sometimes is used to denote transverse relaxation that includes the effects of inhomogeneity.

5.5.2 The Attached Proton Test

Although developed to measure T_2, this pulse sequence is able to improve resolution or eliminate coupling constants or chemical shifts after a single cycle. Moreover, it may be modified to achieve other effects. To obtain information about how many protons are attached to a carbon, the coupling information must be manipulated in a fashion different from that used in Figure 5.5. This is a double-resonance procedure, with pulses applied at both ^{13}C (B_1) and ^1H (B_2) frequencies (Figure 5.14 for a methine group, $^{13}C-^1H$, with the reference frequency set at the ^{13}C resonance). The protons are subjected to a 180° pulse (B_2) at the same time that the carbons are subjected to their 180° pulse (B_1). The two pulses cancel each other, but the vectors from spin–spin

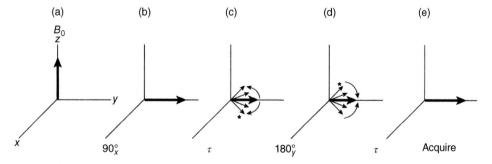

Figure 5.13 Spin echo experiment to eliminate the effects of B_1 inhomogeneity.

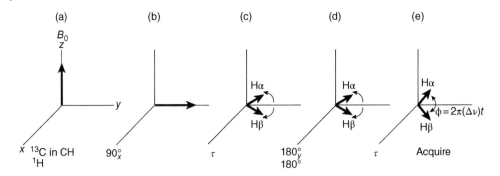

Figure 5.14 Pulse sequence that allows spin vectors to evolve to an arbitrary frequency separation ϕ.

coupling continue to diverge, as in Figure 5.14d. The cancelation occurs in the following fashion. Just as the ^{13}C spins are rotated by the 180° ^{13}C pulse between (c) and (d), the signs of the ^{1}H spins are reversed by the 180° ^{1}H pulse. At point (c), the $+\frac{1}{2}$ protons are precessing around the $+z$ axis and the $-\frac{1}{2}$ protons around the $-z$ axis, as in Figure 2.10. The 180° ^{1}H pulse (around either the x or the y axis) switches these identities. The nuclei that were precessing around the $+z$ axis ($+\frac{1}{2}$) are now precessing around the $-z$ axis ($-\frac{1}{2}$) and vice versa. Consequently, the identities of the protons have all been switched. Consider, for example, the faster moving ^{13}C vector, which may have been associated with the $+\frac{1}{2}$ protons (H$_\beta$). After the 180° ^{13}C rotation, the vector would start catching up to the y axis in the absence of the 180° ^{1}H pulse (as was the case in Figure 5.9d). In the presence of the pulse, however, this vector is now associated with $-\frac{1}{2}$ protons (H$_\alpha$) and hence is still dropping behind the carrier frequency, as shown in Figure 5.14d. Thus, the effects of the two 180° pulses (^{13}C and ^{1}H) on the vectors derived from coupling cancel out, but those on inhomogeneity do not. The net effect is to achieve an improvement in homogeneity, while at the same time controlling the angle of divergence between the vectors that arise from spin–spin splitting. After the second τ period (total time $t = 2\tau$), these vectors have further diverged to an arbitrary angle ϕ, which is dependent on the difference in their frequencies ($\Delta v = J$) and on the total time since the initial 90° pulse, that is, $\phi = (\Delta\omega)t = 2\pi(\Delta v)t = 2\pi J(2\tau) = 4\pi J\tau$.

As an aside, in a homonuclear decoupling experiment such as ^{1}H{^{1}H}, a 180° pulse that follows the initial 90° pulse by a time τ has the same effect as the pair of 180° pulses in Figure 5.14. The homonuclear sequence (90° – τ – 180° – τ – Acquire) results in refocusing of field inhomogeneities, but continued divergence of the two vectors. The 180° nonselective pulse not only rotates the directions of the vectors for the observed nucleus in the manner of Figure 5.14c, but also rotates all of the spins of the irradiated nucleus from above the xy plane to below it and vice versa, thus flipping the spins. For example, the pulse rotates the faster moving vector for the observed nucleus around the y axis. Because of the switch of spins of the irradiated nucleus, it becomes the more slowly moving vector and hence continues to move away from the y axis. After time 2τ, the angle between the vectors is $\phi = 2\pi(\Delta v)(2\tau)$.

Returning to the spectral editing experiment begun in Figure 5.14, let us set the time τ to the specific value of $[2J(^{13}\text{C}-^1\text{H})]^{-1}$ (J is the coupling between the carbon and hydrogen in the methine group (Figure 5.15)). The vectors diverge during one period τ until they are 180° apart, as in Figure 5.15d, since $\phi = 2\pi J(2J)^{-1} = \pi$. After the full pulse

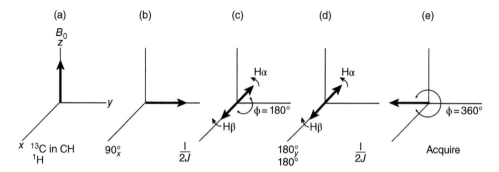

Figure 5.15 Pulse sequence for spectral editing of a methine (CH) resonance.

sequence ($\tau = 2\pi$), the angle between the vectors is $4\pi J(2J)^{-1}$, or 360°, as in Figure 5.15e. If the spectrum is sampled at this time, the result is a negative singlet, because the spins are all aligned along the negative y direction.

If the same experiment is carried out for a carbon attached to two protons (CH_2, Figure 5.16), the middle peak of the triplet remains on the y axis (coincident with the reference frequency, like the $+y$ vector in Figure 5.16c), and the diverging smaller peaks now differ by $\Delta v = 2J$ (the distance between the outer peaks of the triplet). The value of $\phi = 2\pi(\Delta v)t$ after τ then is $2\pi(2J)(\tau)$, so that, for $\tau = (2J)^{-1}$, the angle is $4\pi J(2J)^{-1}$, or 2π, as for the $-y$ vector in Figure 5.16c,d. After 2τ, $\phi = 4\pi$, so that both vectors are coincident with the positive y axis, as in Figure 5.16e. Consequently, we get a positive peak for methylene carbons and a negative peak for methine protons. Quaternary carbons, of course, always give a positive peak, because, being unsplit, they remain on the positive y axis throughout these pulses. The value of Δv for the four peaks of a methyl carbon is either J (for the middle two peaks) or $3J$ (for the outer two peaks). These separations result in refocusing all vectors onto the negative y axis after 2τ and hence producing a negative peak.

Figure 5.17 illustrates the result of the complete editing experiment for cholesteryl acetate, which gives negative peaks for CH and CH_3 resonances and positive peaks for C and CH_2. Proton irradiation during acquisition provides decoupling. This experiment affords a visual identification of the substitution pattern of all carbons, and has been called *J modulation* or the *Attached Proton Test* (APT). It exists in many variants.

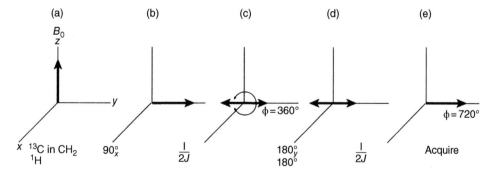

Figure 5.16 Pulse sequence for spectral editing of a methylene (CH_2) resonance.

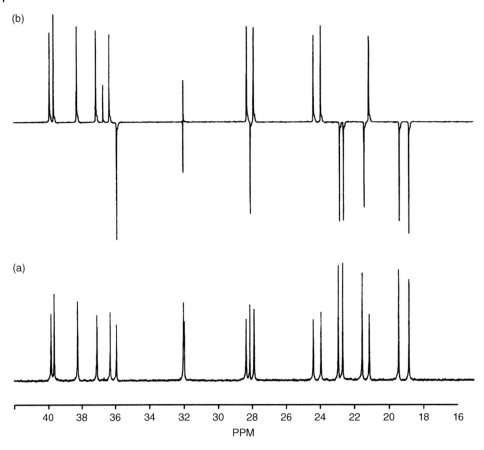

Figure 5.17 (a) The normal proton-decoupled ^{13}C spectrum of cholesteryl acetate. (b) The attached proton test (APT), phased so that CH_2 and quaternary carbons are positive and CH and CH_3 carbons are negative. Source: Derome 1987 [6]. Reproduced with permission of Elsevier.

5.5.3 The DEPT Sequence

The procedure illustrated in Figure 5.17 does not distinguish between methine and methyl carbons, so alternative editing procedures have been developed that can provide separate spectra for each substitution pattern. Figure 5.18 illustrates the full set of spectra for the trisaccharide gentamycin, using the DEPT pulse sequence (defined in greater detail in the next section). The DEPT experiment often is presented alternatively as three, rather than four, spectra: (i) the fully decoupled spectrum with all carbons as positive singlets, (ii) a spectrum with only CH carbons as positive singlets, and (iii) a spectrum with CH_3 and CH carbons positive and CH_2 carbons negative (quaternary carbons then are identified by difference from the complete spectrum). The various DEPT experiments probably are the most commonly used experiments today for ascertaining carbon substitution patterns, because (i) they depend less on the exact value of J than does the aforementioned APT experiment, (ii) they provide signal enhancement (Section 5.6), and (iii) they easily distinguish CH and CH_3 groups. An

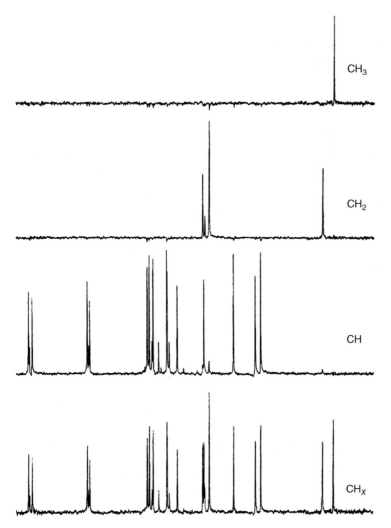

Figure 5.18 Spectral editing of the 75.6 MHz ^{13}C spectrum of the trisaccharide gentamycin by the DEPT sequence. The bottom spectrum contains resonances of all carbons with attached protons, and the ascending spectra are respectively of the methine, methylene, and methyl carbons. Source: Courtesy of Bruker Instruments, Inc.

edited ^{13}C spectrum is a standard, and sometimes necessary, part of the structural analysis of complex organic molecules.

5.6 Sensitivity Enhancement

Some important nuclei, including ^{13}C and ^{15}N, have low natural abundances and sensitivities. Pulse sequences have been devised to improve the observability of these nuclei when they are coupled to another nucleus of high receptivity, usually a proton. Pulses are applied in such a way that the favorable population of the sensitive nucleus S is transferred to the insensitive nucleus I.

5.6.1 The INEPT sequence

A common sequence developed by Raymond Freeman for this purpose is called INEPT, for *Insensitive Nuclei Enhanced by Polarization Transfer*, as follows:

$$^1\text{H(S)} \quad 90^\circ_x - 1/4J - 180^\circ_y - 1/4J - 90^\circ_y$$
$$^{13}\text{C(I)} \qquad\qquad\qquad\quad 180^\circ - 1/4J - 90^\circ_x - \text{Acquire}$$

The pulses are closely related to the spin echo experiment in Figure 5.14, with $\tau = (4J)^{-1}$ chosen to leave the ^{1}H and ^{13}C spin vectors 180° apart, or *antiphase*, after 2τ ($\phi = 2\pi J \cdot 2 \cdot (4J)^{-1} = \pi$). The additional 90° pulses after 2τ are necessary to place the vectors on the appropriate axes.

The results of the pulses are illustrated in Figure 5.19 for the case of two spins, for example, ^{13}C–^{1}H (an isolated methinyl group). The first set of pulses is applied to the sensitive nucleus (^{1}H) to prepare it in the antiphase arrangement. The first 90° pulse moves the proton magnetization into the xy plane (Figure 5.19b). The simultaneous 180° pulses on both the proton and the carbon nuclei remove the effects of inhomogeneity, but allow the proton vectors to continue to diverge, as in Figure 5.14. After $\tau = (4J)^{-1}$, the protons are 90° apart (Figure 5.19c), and, after the second $(4J)^{-1}$ period, they are 180° apart (Figure 5.19d). The 90° pulse along the y direction rotates the proton vectors back onto the z axis (Figure 5.19e). Whereas in (a) the protons associated with both carbon spin up and carbon spin down are pointed in the $+z$ direction, in (b) the protons associated with carbon spin $+½$, or β, are pointed along the $+z$ but the protons associated with carbon spin $-½$, or α, are pointed along the $-z$ direction (or the reverse, depending on the sign of the ^{13}C–^{1}H coupling constant). This situation is termed *antiphase*.

The spin energy diagram after these proton pulses is compared with that for the normal two-spin system at the beginning of the sequence in Figure 5.20. The normal diagram on the left shows that the Boltzmann distributions result in more intense ^{1}H resonances (βα → αα and ββ → αβ) than ^{13}C resonances (αβ → αα and ββ → βα, as represented by the greater vertical length of the arrows for the ^{1}H transitions than for the ^{13}C transitions. Each arrow goes from a lower to a higher state and hence represents absorption (a positive peak).

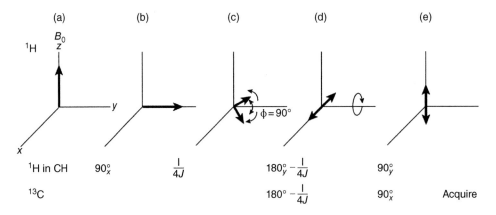

Figure 5.19 Pulse sequence for the INEPT experiment, showing the effects on the ^{1}H spin vectors.

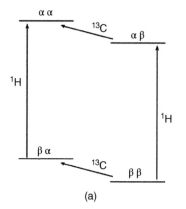 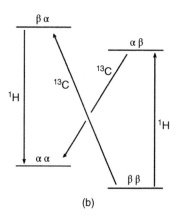

Figure 5.20 Spin states for a two-spin (^{13}C—^{1}H) system, normally (a) and after the INEPT pulse sequence (b).

The antiphase INEPT arrangement of ^{1}H spin vectors on the right of Figure 5.20 means that two ^{1}H energy levels (αα and βα) are interchanged, so that the ^{1}H spin flip (βα → αα) gives a negative peak, while the other spin flip (ββ → αβ) still gives a positive peak. Thus, one ^{1}H signal is positive and the other is negative (the spectral result of the antiphase relationship). An examination of the carbon transitions in the INEPT diagram indicates that their Boltzmann distributions have increased to protonlike proportions (look at the vertical lengths of the arrows, although the representation is not proportional). In this fashion, protons have transferred polarization to carbons. According to the spin energy diagram at the right of the figure, the carbon vectors also are antiphase, since the carbon transition associated with proton spins $+\frac{1}{2}$, or β, is absorptive and must be pointed along the $+z$ direction, whereas the carbon transition associated with proton spins $-\frac{1}{2}$, or α, is emissive and must be pointed along the $-z$ direction. The situation for carbons is identical to that for protons in Figure 5.19e. The final carbon pulse, which is 90° along the x axis, places the antiphase vectors along the y direction for observation. Because of the antiphase relationship, one carbon transition (ββ to βα) is positive (absorption) and one (αβ to αα) is negative (emission).

The INEPT sequence results in enhanced signals for the insensitive I nuclei, half of which give negative and half positive peaks for a CH group, such as pyridine at the top of Figure 5.21. For comparison, the figure includes the normal spectrum at the bottom and the spectrum in the middle with gated irradiation in order to obtain the NOE without decoupling. The INEPT spectrum clearly achieves a greater enhancement of sensitivity than does that produced with NOE alone. The maximum increment in intensity is $(1+|\gamma_S/\gamma_I|)$ (absolute value of γ_S/γ_I) for INEPT, but that for the NOE $(1+\eta_{max})$ is only $(1+\gamma_S/2\gamma_I)$ (Eq. (5.5)) and can be positive or negative. The maximum enhanced intensity available from the INEPT experiment, analogous to that obtained from Eq. (5.5) for the NOE experiment, is given by Eq. (5.6).

$$I_{max}(\text{INEPT}) = I_0 \left|\frac{\gamma_{irr}}{\gamma_{obs}}\right| \qquad (5.6)$$

For ^{13}C{^{1}H}, maximum increased intensities ($I/I_0 = 1+\eta_{max}$) are 3.98 for INEPT and 2.99 for NOE. When the gyromagnetic ratio of the insensitive nucleus is negative,

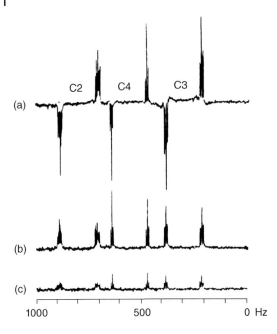

Figure 5.21 The proton-coupled ^{13}C spectrum of pyridine, (a) with INEPT, (b) with NOE only, and (c) unenhanced, all on the same scale. Source: Morris and Freeman 1979 [7]. Copyright 1979. Reproduced with permission of American Chemical Society.

INEPT has an even greater advantage because of the subtractive factor present in the NOE expression. For ^{15}N{^{1}H}, the INEPT and NOE factors are 9.87 and −3.94, respectively; for ^{29}Si{^{1}H}, 5.03 and −1.52; and for ^{119}Sn{^{1}H}, 2.68 and −0.34. Clearly, INEPT is significantly more effective in each case and is always positive.

5.6.2 Refocused INEPT

There is one apparent drawback to the INEPT experiment. Decoupling of the −1 : 1 pattern for each CH resonance would lead to precise cancelation and hence a null signal. As methylene triplets give −1 : 0 : 1 INEPT intensities and methyl quartets give −1 : −1 : 1 : 1 intensities, both also would give null signals on decoupling. The *refocused* INEPT pulse sequence was designed to get around this problem and permit decoupling by repeating the INEPT pulses a second time in the following fashion:

^{1}H(S) $90°_x - 1/4J - 180°_y - 1/4J - 90°_y - 1/4J - 180° - 1/4J$ − Decouple
^{13}C(I) $180° - 1/4J - 90°_x - 1/4J - 180°_y - 1/4J$ − Acquire

The second refocusing period again is a spin echo in which chemical shifts are focused by the 180° pulses. The spin roles, however, are reversed in the second set, so that I magnetization is refocused back to two positive peaks for the CH case. The decoupling of protons during carbon acquisition thus does not result in the cancelation of any peaks. The spectrum that is obtained contains decoupled peaks with enhanced intensity. Figure 5.22 compares the various experiments for chloroform.

5.6.3 Spectral Editing with Refocused INEPT

The value of $(2J)^{-1}$ for the total period between the last 90° pulse and acquisition (sometimes called Δ_2 to distinguish it from the period Δ_1 or 2τ between the first and last

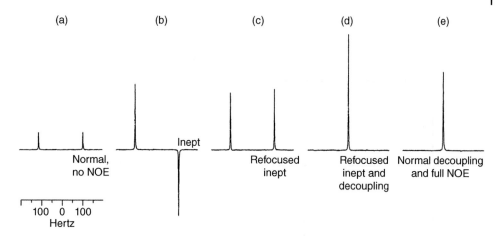

Figure 5.22 The ^{13}C spectrum of chloroform (a) without double irradiation, (b) with ^{1}H irradiation to achieve the INEPT enhancement, (c) with ^{1}H irradiation to achieve refocused INEPT enhancement, (d) with ^{1}H irradiation to achieve refocused INEPT enhancement and decoupling, and (e) with normal decoupling to achieve only the NOE. Source: Derome 1987 [8]. Reproduced with permission of Elsevier.

90° pulses) is appropriate only for the methine fragment CH. For methylene (CH$_2$) and methyl (CH$_3$) groups, the vectors do not refocus, so that decoupling would still result in a canceled signal. Alternative values of Δ_2, however, can lead to improved refocusing, with Δ_2 for an arbitrary CH$_n$ fragment given by Eq. (5.7).

$$\Delta_2 = (1/\pi J)\sin^{-1}\left(\frac{1}{\sqrt{n}}\right) \tag{5.7}$$

The respective optimum values of Δ_2 for CH, CH$_2$, and CH$_3$ are $(2J)^{-1}$, $(4J)^{-1}$, and $\approx(5J)^{-1}$. Thus $\approx(3.3J)^{-1}$ represents a compromise value that yields enhanced, but not optimal, intensities for all substitution patterns under decoupling conditions. In the absence of decoupling, phase differences within a collection of CH, CH$_2$, and CH$_3$ resonances would result in peak distortions.

Because the choice of $\Delta_2 = (2J)^{-1}$ for the decoupled, refocused INEPT experiment leads to completely refocused doublets, but to antiphase triplets and quartets, this particular experiment with decoupling produces a subspectrum that contains only methinyl resonances. Values of Δ_2 also can be selected to optimize the intensities of methylene and methyl resonances. The idea can be depicted graphically by defining an imaginary angle $\theta = \pi J \Delta_2$. Signal intensities then are found to be proportional to $\sin \theta$ for CH, $\sin(2\theta)$ (or $2\sin \theta \cos \theta$) for CH$_2$, and $3\sin \theta \cos^2 \theta$ for CH$_3$. Thus, when $\theta = \pi/2$ (and $\Delta_2 = (2J)^{-1}$), the CH signal is optimized and the other signals go to zero. For all other values of θ, the spectrum contains varying proportions of all substitution types. Figure 5.23 illustrates the experiment for three values of θ ($\pi/4$, $\pi/2$, and $3\pi/4$) that correspond to $\Delta_2 = 3(4J)^{-1}$, $(2J)^{-1}$, and $(4J)^{-1}$, respectively. Linear combinations of these spectra can lead to edited spectra that contain only methylene or only methyl resonances. Figure 5.24 is a plot of the signal intensities for the three types of carbon as a function of the angle $\theta = \pi J \Delta_2$. The spectra shown were taken as cuts at $\theta = 135°$, $90°$, and $45°$.

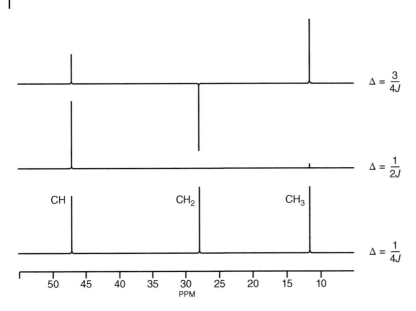

Figure 5.23 Intensities of the carbon resonances of an imaginary molecule containing one CH, one CH_2, and one CH_3 under varying values of Δ_2 in the refocused INEPT experiment. Source: Derome 1987 [9]. Reproduced with permission of Elsevier.

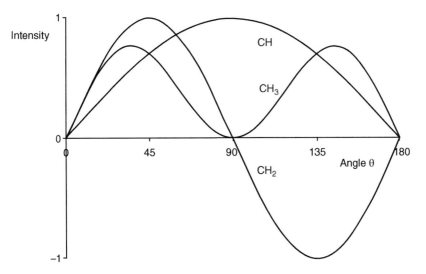

Figure 5.24 Variation of signal intensities for CH, CH_2, and CH_3 as a function of $\theta = \pi J \Delta_2$ in the refocused INEPT experiment. Source: Claridge 1999 [10]. Reproduced with permission of Elsevier.

5.6.4 DEPT Revisited

A comparison of the preceding three INEPT spectra allows the multiplicity of all protonated carbon resonances to be determined, albeit with intensities that are not optimized. Section 5.5 describes the APT, which does not distinguish CH from CH_3.

The DEPT sequence provides an editing technique that suffers from the drawbacks of neither of these other methods and moreover is less sensitive to experimental imperfections, such as the exact value of J. Already mentioned as the method of choice for spectral editing, DEPT is similar to refocused INEPT. There are a pair of τ periods ($=(2J)^{-1}$) followed in the proton channel by a single variable pulse θ, an angle corresponding to that previously defined ($\pi J \Delta_2$, in which Δ_2 corresponds to two τ periods):

$$S \quad 90^\circ_x - 1/2J - 180^\circ_y - 1/2J - \theta_y \quad - 1/2J - \text{Decouple}$$
$$I \qquad\qquad 90^\circ_x - 1/2J - 180^\circ_x - 1/2J - \text{Acquire}$$

The DEPT and refocused INEPT sequences begin in a similar fashion, with the 90° ^{1}H pulse generating proton magnetization that evolves under the influence of coupling to carbon. Whereas the time τ between the first and second proton pulses is $(4J)^{-1}$ for refocused INEPT, it is $(2J)^{-1}$ for DEPT, as it was for APT. The second (180°) ^{1}H pulse refocuses proton chemical shifts. The simultaneous initial 90° ^{13}C pulse generates carbon magnetization and brings about a situation that cannot be followed by the vector model we have used throughout this textbook. As both proton and carbon magnetizations, linked by the C—H coupling, are evolving together, the phenomenon is termed *Multiple Quantum Coherence* (MQC), or, more specifically, *Heteronuclear Multiple Quantum Coherence* (HMQC). In essence, the proton and carbon magnetizations have become pooled. The MQC continues to evolve during the second $(2J)^{-1}$ period. The final proton pulse, of duration θ, converts the MQC to single quantum carbon coherence. MQC cannot be observed, as it induces no signal in the detection coil, so it must be transformed back into *single quantum coherence*. The final $(2J)^{-1}$ period allows the development of carbon magnetization, with a dependence on the number of attached protons (CH, CH_2, or CH_3) determined by the value of θ.

As with refocused INEPT, the modulation of θ, now a pulse length, results in a series of edited spectra such as those in Figure 5.18. One of the most common sets of experiments uses the angles 45°, 90°, and 135°. The DEPT-45 spectrum contains resonances of all types except quaternary, DEPT-90 contains only CH, and DEPT-135 contains CH/CH_3 positive and CH_2 negative, analogous to the plot in Figure 5.24, readily permitting an assignment of each type of substitution. Spectral subtraction with some loss of signal is required to obtain the fully edited spectra illustrated in Figure 5.18. The term "distortionless" was applied because the initial set of pulses (up to the first 2τ) results, not in a combination of positive and negative peaks, but rather in positive 1 : 1 doublets, 1 : 2 : 1 triplets, and 1 : 3 : 3 : 1 quartets in the absence of decoupling.

The INEPT and DEPT sequences assume that coupling between the I and S nuclei is dominant, so that other couplings must be negligibly small. For one-bond ^{13}C–^{1}H couplings, this assumption holds, as all ^{1}H–^{1}H couplings are much smaller. If polarization is to be transferred from two- or three-bond ^{13}C–^{1}H couplings, however, the homonuclear couplings no longer are small in comparison. This situation is more likely to occur when attempts are being made to transfer polarization from protons to silicon, nitrogen, or phosphorus. Because Si–H, N–H, and P–H bonds are relatively uncommon (compared with C–H), recourse must be made to longer range coupling constants, with attendant difficulties.

5.7 Carbon Connectivity

The one-bond $^{13}C-^{13}C$ coupling potentially contains a wealth of structural information, as it specifies and characterizes carbon–carbon linkages. Unfortunately, only 1 in about 10 000 pairs of carbon atoms contains two ^{13}C atoms and hence displays a $^{13}C-^{13}C$ coupling in the ^{13}C spectrum. These resonances can be detected as very low intensity satellites on either side of the centerband that is derived from molecules containing only isolated ^{13}C atoms. For bonded pairs of ^{13}C atoms, ^1J is about 30–50 Hz, and the satellites are separated from the centerband by half that amount. Coupling also may be present over two or three bonds (^2J, ^3J) in the range of about 0–15 Hz. Not only are these satellites low in intensity and possibly obscured by the centerband, but, in addition, spinning sidebands, impurities, and other resonances may get in the way.

The pulse sequence INADEQUATE (*Incredible Natural Abundance DoublE QUAntum Transfer Experiment*) was developed by Freeman to suppress the usual (single-quantum) resonances and exhibit only the satellite (double-quantum) resonances. The pulse sequence is $90°_x - \tau - 180°_y - \tau - 90°_x - \Delta - 90°_\phi$. The homonuclear 180° pulse refocuses field inhomogeneities, but allows the vectors from different $^{13}C-^{13}C$ coupling arrangements to continue to diverge (Section 5.5). If the carrier frequency coincides with the centerband of a carbon resonance, the spins represented in the centerband remain on the y axis after the first 90° pulse. The delay time τ is set to $(4J)^{-1}$, so that the vectors for the two satellites from the coupled $^{13}C-^{13}C$ system diverge by 180° after 2τ $(= 2\pi(\Delta v)t = 2\pi J(2/4J) = \pi)$ and lie respectively on the $+x$ and $-x$ axes. The second $90°_x$ pulse then rotates the centerband spins to the $-z$, axis, but leaves the satellite spins aligned along the x axis. Thus, the centerband signal is not available for detection in the xy plane, but the satellites are.

This pulse sequence is another example of MQC. After the second 90° pulse, the coupled pairs of ^{13}C nuclei evolve together. Note that each ^{13}C pair is an isolated AX system because of the natural abundance of ^{13}C. During the period Δ, homonuclear double quantum coherence evolves as the sum of the Larmor frequencies of the two coupled spins. In the two-dimensional variant (Chapter 6), the constant period Δ becomes a variable period. The final 90° pulse reconverts multiple to single quantum coherence for observation. The phase of the final 90° pulse ($90°_\phi$) is cycled through a series of directions represented by ϕ ($+x, +y, -x, -y$). The vector diagrams used throughout this book illustrate only the coherence of the spins of a single nucleus. Spins in the xy plane are said to be coherent when they have an ordered relationship between their phases, so that they all precess around one axis and can be depicted by a vector along that direction. Spins rotating with random phase in the xy plane are said to be incoherent. The simultaneous coherence of two spins, as created in the INADEQUATE experiment, is not well depicted by the vector diagrams, so the reason for the final ^{13}C 90° pulse is not well represented.

Figure 5.25 contains the INADEQUATE spectrum for piperidine. The double-quantum (satellite) peaks are antiphase, so each $^{13}C-^{13}C$ coupling constant is represented by a pair of peaks—one up, one down (+1 :−1). The spectrum for C4 of piperidine thus contains two such doublets: a large one for $^1J_{34}$ and a small one for $^2J_{24}$. For C3, there are two large doublets, because the one-bond couplings $^1J_{23}$ and $^1J_{34}$ to

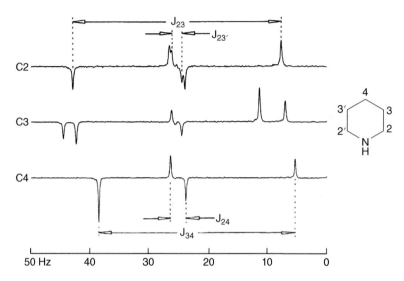

Figure 5.25 The one-dimensional INADEQUATE spectrum for the carbons of piperidine. Source: Bax et al. 1980 [11]. Copyright 1980. Reproduced with permission of American Chemical Society.

the adjacent carbons are slightly different. There also is a small $^3J_{23'}$ between C2 and the nonadjacent C3. The spectrum for C2 shows $^1J_{23}$, $^2J_{24}$, and $^3J_{23'}$.

Although more distant couplings are observable, the most important are the one-bond couplings, which vary slightly for every carbon–carbon bond. Thus, a match of $^1J(^{13}\text{C}-^{13}\text{C})$ for any two carbons strongly suggests that they are bonded to each other. Even in complex molecules, there is sufficient variability of couplings that INADEQUATE can be used to map the complete connectivity of the carbon framework, provided that it is not broken by a heteroatom. The major drawback to the INADEQUATE experiment is its extremely low sensitivity, as it uses only 0.01% of the carbons in the molecule. The two-dimensional version is discussed in Section 6.4.

5.8 Phase Cycling, Composite Pulses, and Shaped Pulses

5.8.1 Phase Cycling

We have used 90° and 180° pulses extensively to carry out a variety of experiments. In each case, it is important that the length of the pulse provide the desired angle of rotation accurately. Various artifacts can arise because of imperfections in the pulses. Figure 5.26 illustrates the effect on the inversion recovery experiment ($180°_x - \tau - 90°_x$ Figure 5.1) used to determine T_1, but with the initial inverting pulse not quite 180°. The magnetization after the pulse is slightly off the z axis (Figure 5.26b), so there is a small amount of transverse (xy) magnetization present at the start of the τ period. Only the y component is shown in (Figure 5.26b). After the period τ, the z magnetization has decreased through T_1 relaxation, and the component of magnetization in the xy plane caused by the pulse imperfection persists (Figure 5.26c). Following the final 90° pulse, the z magnetization is moved into the xy plane for detection (Figure 5.26d). The pulse imperfection in the drawing causes a reduction in intensity, but the spectral phase also

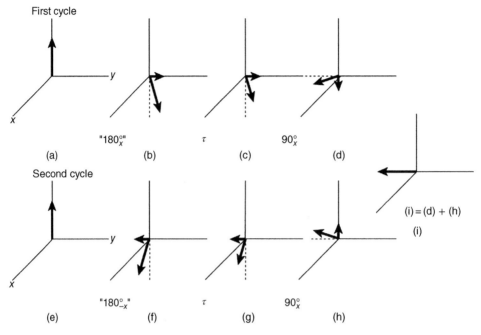

Figure 5.26 Phase cycling in the inversion recovery experiment.

can be altered. Almost certainly, there would be errors in the 90° pulse as well, but these are not under consideration here.

Such errors may be eliminated largely by alternating the relative phase of the 180° pulse. The result of an inversion in which the 180° rotation is carried out counterclockwise instead of clockwise about the x axis (−x, or −180°) is illustrated in Figure 5.26f. The unwanted transverse magnetization now appears along the −y axis. After time τ (Figure 5.26g) and the final 90° pulse (Figure 5.26h), the imperfection is still present, but now has the opposite effect on the z magnetization from that in Figure 5.26d. When the two results are added, as in Figure 5.26i, the effect of the imperfection cancels out. The pulse therefore is alternated between x and −x. Such a procedure is called *phase cycling*, a technique that permeates modern NMR spectroscopy.

Phase cycling has improved procedures for broadband heteronuclear decoupling. As described in Section 5.3, modern methods use repeated 180° pulses rather than continuous irradiation. Imperfections in the 180° pulse would accumulate and render the method unworkable. Consequently, phase-cycling procedures have been developed to cancel out the imperfections. The most successful to date is the WALTZ method of Freeman, which uses the sequence $90°_x$, $180°_{-x}$, $270°_x$ in place of the 180° pulse (90 − 180 + 270 = 180), with significant cancelation of imperfections. The expanded WALTZ-16 sequence cycles through various orders of the simple pulses and achieves a very effective decoupling result. The sequence 90°, −180°, 270° was given the shorthand notation $1\bar{2}3$ (1 for 90°, $\bar{2}$ for −180° in which the bar denotes the negative direction, and 3 for 270°). The implied rhythm of the sequence suggested its name.

A third example of phase cycling is used to place the reference frequency in the middle of the spectrum, instead of off to one side. As described heretofore, the NMR experiment is sensitive only to the difference $\Delta\omega$ between a signal and the reference frequency. This situation necessitates placing the reference frequency to one side of all the resonances, so that there is no confusion of two signals that are respectively at a higher and a lower frequency than the reference frequency by the exact same amount ($+\Delta\omega$ and $-\Delta\omega$). Such sideband detection, however, always contains signals from noise on the signal-free side of the reference. Placement of the reference in the middle of the spectrum avoids this unnecessary noise, but requires a method for distinguishing between signals with $+\Delta\omega$ and those with $-\Delta\omega$. *Quadrature detection* accomplishes this task by splitting the signal in two and detecting it twice, using reference signals with the same frequency, but 90° out of phase. Signals with the same absolute value of $\Delta\omega$, but opposite signs, are distinguished in the experiment (in terms of obtaining θ by knowing both $\sin\theta$ and $\cos\theta$, which are 90° out of phase). Systematic errors, however, can arise if the two reference frequencies are not exactly 90° out of phase. The resulting signal artifacts, called *quad images*, can appear as low intensity peaks. The CYCLically Ordered Phase Selection (CYCLOPS) sequence involves four steps that move the 90° pulse and the axis of detection from $+x$ to $+y$ to $-x$ to $-y$ and change the way the two receiver channels are added, with the result that imperfections in the phase difference cancel out.

Phase cycling not only can remove artifacts from pulse or phase imperfections, but also can assist in the selection of coherence pathways. The inversion recovery experiment can be described with a slightly different vocabulary to illustrate this process. When spins are aligned entirely along the z axis, the order of coherence is said to be zero (phases around the xy plane are random). An exact 90° pulse creates maximum single quantum coherence by lining the spins up along, for example, the y direction. Phase cycling in the inversion recovery experiment (Figure 5.26) removes undesired single quantum coherence (transverse or xy magnetization) and leaves coherence of order zero until the end of the τ period, at which time the final 90° pulse creates single quantum coherence. In this way, phase cycling selects the desired degree of coherence. Double quantum coherence, involving the relationship between two spins, is not well illustrated by these vector diagrams. The INADEQUATE experiment involves the selection of double over single quantum coherence (elimination of the centerband and retention of the satellites), in part through phase cycling in the final 90° pulses, whose subscript ϕ refers to a sequence of pulses with different phases.

5.8.2 Composite Pulses

Imperfections in pulses also may be corrected by using *composite pulses* instead of single pulses. The 180° pulse that inverts longitudinal magnetization for the measurement of T_1 or other purposes may be replaced by the series $90°_x$, $180°_y$, $90°_x$, which results in the same net 180° pulse angle, but reduces the error from as much as 20% to as little as 1%. As Figure 5.27 shows, the 180° pulse compensates for whatever imperfection existed in the 90° pulse. Normally, 180° is taken as double the optimized 90° pulse, so errors in one are present in the other. The three components of the WALTZ-16 method ($90°_x$, $180°_{-x}$, $270°_x$) also constitute a composite pulse for $180°_x$.

5.8.3 Shaped Pulses

For the most part, pulses have been generated by applying RF energy equally over the entire frequency range, with a short duration on the order of microseconds. Such excitations are sometimes referred to as *hard pulses*, in distinction to pulses that require

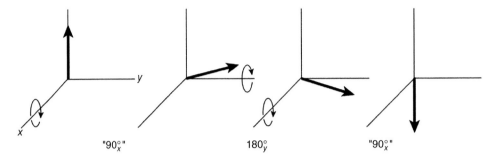

Figure 5.27 A composite pulse equivalent to a single 180° pulse.

selective excitation, i.e. excitation over a restricted frequency range. Selective excitation has been mentioned on several occasions. It is useful, for example, in the saturation transfer experiment (Section 5.2) and in the suppression of specific unwanted peaks (Section 5.1). Frequency selection within two-dimensional spectra (Chapter 6) results in a reduction in dimensionality, so that effects at a single frequency can be examined in detail. A one-dimensional cut of a two-dimensional spectrum offers the twin advantages of reduced experimental time and decreased storage needs.

The procedure for producing a selective pulse is to reduce the RF power (B_1), so that the effective frequency range also is reduced. To counter the reduction in power and still achieve the required tip angle, the duration of the pulse is increased, typically into the millisecond range. The simplest such *soft pulses* would have rectangular shapes, i.e. from zero intensity instantly up to full intensity for a period of milliseconds and then back to zero intensity, similar to the shapes of the hard pulses in our vector diagrams. Unfortunately, such a pulse shape generates wiggles, or feet, on the signal (Figure 5.28a). By a process called with apodization (meaning "no-feet-ization"), these wiggles may be removed by smoothing off the edges of the peaks (Figure 5.28b).

Such excitations have been called *shaped pulses*, and a considerable effort has been expended in an attempt to optimize their shapes. A simple Gaussian shape is a considerable improvement over a rectangular pulse, but is not entirely effective in achieving an optimal peak shape. The use of more elaborate mathematical functions improves the

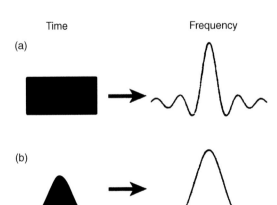

Figure 5.28 (a) The result of Fourier transformation of a low-power rectangular pulse. (b) The result for a shaped Gaussian pulse. Source: Claridge 1999 [12]. Reproduced with permission of Elsevier.

shape of the signal, although with increasing loss of intensity. The BURP (*Band selective, Uniform Response, Pure phase*) family utilizes an exponentially dependent sinusoidal series of Gaussians with considerable success in a variety of situations (EBURP for 90°, REBURP for 180°).

An early alternative to soft pulses was the *Delays Alternating with Nutation for Tailored Excitation* (DANTE) experiment, which used a sequence of short, hard pulses of angle $\alpha \ll 90°$ along the x axis, followed by a fixed delay τ to achieve selective excitation. Thus, the pulse sequence is $(\alpha_x - \tau)_n$. Nuclei that are on resonance are driven eventually to the y axis and hence are selected, whereas those more removed from the frequency range are not affected. The sequence of hard pulses can achieve a result similar to that of soft pulses and even can be shaped by modulating the duration of the pulse lengths, but DANTE pulses lead to spectral artifacts not created by soft pulses, such as unwanted sidebands.

Problems

5.1 Give the spectral notation (AB, ABX, etc.) for the following substituted ethanes, first at slow C—C rotation, then at fast rotation. Draw all stable conformations. The spectral notation for each frozen form gives the slow-rotation answer. Then imagine free rotation about the C—C bond. The identity of certain protons may average for the fast-rotation answer.
(a) CH_3CCl_3
(b) CH_3CHCl_2
(c) CH_3CH_2Cl
(d) $CHCl_2CH_2Cl$

5.2 Ring reversal in 7-methoxy-7,12-dihydropleiadene (below) can be frozen out at $-20\,°C$. Resonances from two conformations are observed, in the ratio $2:1$. When the high-frequency (low-field) part of the 12-CH_2 AB quartet in the minor isomer is doubly irradiated, the intensity of the 7-methine proton is enhanced by 27%. Double irradiation of the same proton in the major isomer has no effect on the spectrum. What are the two conformational isomers and which is more abundant?

5.3 Permethyltitanocene reacts with an excess of nitrogen below $-10\,°C$ to form a $1:1$ complex:

$$[C_5(CH_3)_5]_2 Ti + N_2 \rightleftharpoons [C_5(CH_3)_5]_2 TiN_2.$$

The methyl resonance of the complex is a sharp singlet above $-50\,°C$. Below $-72\,°C$, the resonance splits reversibly into two peaks of not quite the same intensity. If the nitrogen molecule is doubly labeled with ^{15}N, the ^1H-decoupled ^{15}N spectrum contains a singlet and an AX quartet $[J(^{15}N-^{15}N)=7\,Hz]$ of not quite the same overall intensity at low temperatures. Explain these observations in terms of structures.

5.4 (a) The resonance of the methylene protons of $C_6H_5CH_2SCHClC_6H_5$ in $CDCl_3$ is an AB quartet at room temperature. Why?
(b) The AB spectrum coalesces at high temperatures to an A_2 singlet with a $\Delta G^{\ddagger}$ of 15.5 kcal mol^{-1}. The rate is independent of concentration in the range 0.0190–0.267 M. Explain in terms of a mechanism.

5.5 The ^1H spectrum of the following molecule contains resonances from two isomers at room temperature.

(a) The spectrum of isomer A contains the resonance of H2' at δ 6.42, and that of isomer B contains the same resonance in the region δ 7–8. What can you say about the conformations of A and B?
(b) The sample crystallizes only as isomer A. Dissolution of these crystals, however, produces the spectra of both isomers, A and B. What can you say about the barrier for the equilibrium A ⇌ B? When CH_3 on the double bond is replaced by CMe_2OH, crystallization still produces only A, but redissolution of the crystals also yields A only and none of B. What can you say about the A ⇌ B barrier in this compound?

5.6 No coupling is observed between CH_3 and ^{14}N in acetonitrile ($CH_3-C\equiv N$), but there is a coupling in the corresponding isonitrile ($CH_3-N\equiv C$). Explain. This is not a distance effect. The phenomenon is general for nitriles and isonitriles.

5.7 Comment on the following ^{14}N line widths (Hz):

$^+NMe_4$	<0.5	$MeNO_2$	14
		$\underset{N}{\overset{H}{\diagdown}}\!\!\bigcirc$	
Me_3N	77		172
Aniline ($C_6H_5NH_2$)	1300		

5.8 The inversion-recovery (180° – τ – 90°) spectral stack for the aromatic carbons of *m*-xylene is given below. Assign the resonances and explain the order of T_1 (look at the nulls).

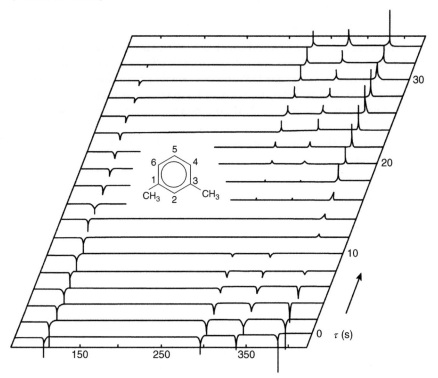

Source: Reprinted with permission from Ref. [13].

5.9 1-Decanol has the following carbon T_1 values (s). Explain the order.

$$CH_3-CH_2-CH_2-CH_2-(CH_2)_5-CH_2-OH$$
3.1   2.2   1.6   1.1   0.8 – 0.83   0.65

5.10 In ribo-C-nucleosides, the base is attached to C1′ by carbon. The α and β forms (C1′ epimers) may be distinguished by T_1 studies.

(a) Consider the following proton T_1 (s) data.

	H1'	H3'	H5'	H5''
Isomer 1	1.60	1.31	0.45	0.45
Isomer 2	3.33	1.37	0.40	0.40

Which isomer (1 or 2) is α and which is β? Why are the T_1 values for H1' different for isomers 1 and 2, but the values for H3', H5', and H5'' are about the same? Why are the T_1 values for H5' and H5'' smaller than the other values? Use the equation for dipolar relaxation (Eq. (5.1)) in your reasoning.

(b) Suggest another (not T_1) NMR method for distinguishing these α and β forms.

5.11 (a) Consider the following molecule **A**, in which rotation is rapid around all C—C bonds.

What is the spin system when R = R'? Are the protons within a single methylene group homotopic, enantiotopic, diastereotopic, or magnetically nonequivalent? More than one category may apply.

(b) Answer the same question when R ≠ R'.

(c) The R and R' groups were chosen to be a donor (D, 9-anthracyl) or an acceptor (A, 3,5-dinitrophenyl). Three molecules can be constructed, in which both R and R' are D (D–D), both R and R' are A (A–A), and R = D when R' = A (D–A). The ^{1}H spectra of these three molecules, as well as the spectra of the model compounds containing only a single A or D and the spectrum of a solution containing equal amounts of D–D and A–A, are given below. Explain the splitting patterns of the A–A molecule in the top, right spectrum and of the D–D molecule in the spectrum directly below that of A–A. All spectra were measured in C_6D_6.

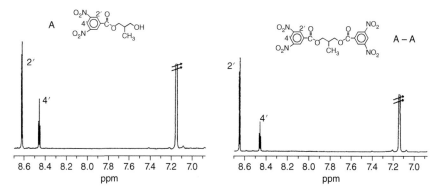

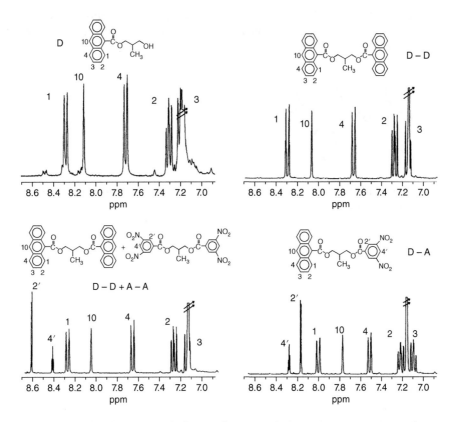

(d) Explain why there is no difference between the aromatic resonances of A–A and of A (top, left), nor between those of D–D and D (directly below that of A).

(e) Explain why the spectrum of D–A (bottom, left), however, is quite different from those of A–A, D–D, A, and D. What is the purpose of spectrum with equal amounts of D–D and A–A (bottom, left)? What mechanism(s) of interaction between the D and A moieties is (are) eliminated from consideration by these observations?

(f) There are one trans (anti) (**B**) and two gauche (**C, D**) conformations around the C—C bonds. For the two C—C bonds, there can be trans–trans and various gauche–trans and gauche–gauche arrangements. The trans–trans conformer, for example, resembles E.

$$\text{E}$$

The bonds labeled 1 and 2 in the following table are different for D–A, but the same for A–A or D–D.

	J(AX) (Hz)	J(BX) (Hz)
A–A	6.52	5.62
D–D	6.46	5.55
(A–D)-1	3.60	8.21
(A–D)-2	4.09	8.23

Rotation is fast on the NMR time scale. Couplings were measured at 298 K in C_6D_6 between CHMe (H_X) and CH_2 (H_A and H_B). What conformational conclusion may be drawn from these numbers? Explain.

(g) NOE experiments were carried out on D–A. Irradiation of H10 (see the above spectrum of D–A) enhanced the resonances of H4, H2′, and H4′. Irradiation of H4′ enhanced the resonances of H1, H4, and H10. Explain.

Source: Reproduced with permission from Ref. [14]. Copyright 1998. American Chemical Society.

5.12 (a) Trimethylsilylation of N-(triisopropylsilyl)indole gave a single product in which the ^1H spectrum contained four doublets and one doublet of doublets (ignoring long-range couplings). What structures are compatible and incompatible with these observations? Explain.

(b) Double irradiation of the trimethylsilyl ^1H resonance increased the intensity of two of the doublets. Irradiation of the isopropyl septet increased the intensity of the other two doublets. What is the structure of the product? Explain.

5.13 The 75 MHz ^{13}C spectrum of the drug N-propyl-3,4-methylenedioxyamphetamine hydrochloride is given below with WALTZ-16 decoupling (lower) and with the APT (upper). Assign the carbons in the full spectrum, using the edited spectrum and your knowledge of α, β, and γ substituent effects.

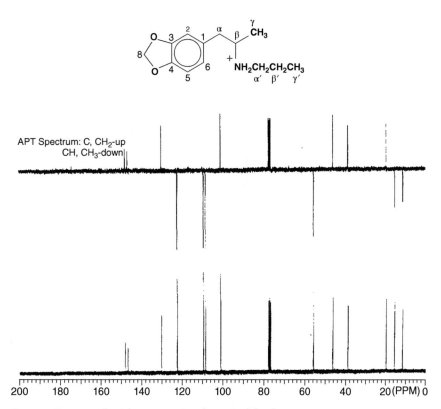

Source: Reprinted with permission from Ref. [15].

5.14 Hydroformulation of myrtenol (**A**) was supposed to give the aldehyde **B** but instead produced another product **C** with the bicyclic portion of the molecule entirely entact.

The 400 MHz ^{1}H and various difference NOE spectra are provided below, with the resonances from the protons in the bicycle assigned (the multiplet at δ 2.2 comes from H7$_{eq}$ and H4α). The ^{13}C spectrum contained nine peaks in the region δ 20–45, plus peaks at δ 70.06 and 105.31. See structure **C** for the numbering system. The target of NOE irradiation is indicated in each spectrum by a deep valley. Proton 4b also showed a strong NOE with irradiation of the proton at δ 5.4. Use all this information to prove the structure of the product, including stereochemistry.

224 | Nuclear Magnetic Resonance Spectroscopy

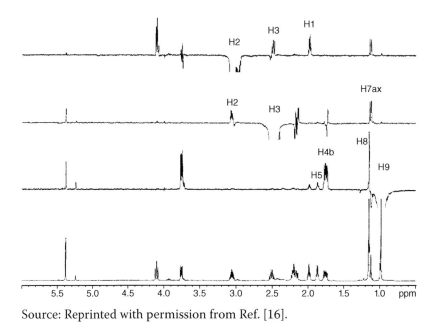

Source: Reprinted with permission from Ref. [16].

5.15 The following 500 MHz ¹H spectrum is of the taxane structure (**A**) illustrated below.

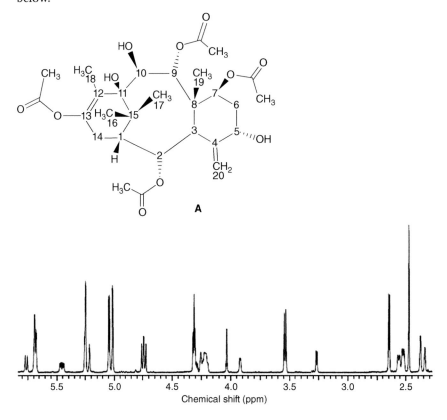

Note the following important spectral characteristics.

1) The methyl resonances and those of H1 and H6 are off scale to low frequency.
2) The middle ring has eight members. Do not expect cyclohexane-like couplings.
3) The OH protons at δ 2.47 and 2.65 are exchanging slowly on the NMR time scale, so that you may expect to see vicinal couplings.
4) H1 resonates at δ 1.89. By spin decoupling, it is coupled to the proton at δ 5.68.
5) The protons at δ 2.36 and 2.55 are coupling to each other ($J = 19$ Hz).
6) The OH proton at position 5 is a doublet off scale to low frequency.

Assign peaks to H2, H3, H5, H7, H9, H10, H14 (two protons), H20 (two protons), OH at position 10, and OH and position 11.

Source: Reprinted with permission from Ref. [17].

5.16 Glycidol has the following 300 MHz ^{1}H spectrum.

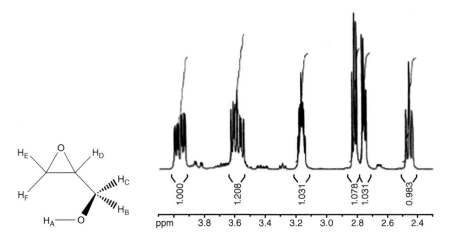

The positions of the ^{1}H resonances are as follows: δ 2.53, 2.76, 2.82, 3.16, 3.60, 3.95. The ^{13}C spectrum contains peaks at δ 44.9, 52.8, and 62.5. This experiment was carried out in extremely dry CDCl$_3$, so that coupling of the hydroxy proton H$_A$ appears in the spectrum (no fast exchange or broadening). From the following observations, assign all the ^{1}H and ^{13}C peaks.

(a) The peak at δ 2.53 disappeared on shaking the sample with D$_2$O, and the peaks at δ 3.60 and 3.95 broadened.
(b) The DEPT experiment showed that the carbon at δ 52.8 is methinyl.
(c) Heteronuclear ^{1}H{^{13}C} (actually 2D *HETeronuclear chemical-shift CORrelation* (HETCOR), Chapter 6) double irradiation of the ^{13}C resonance at δ 52.8 decoupled the proton at δ 3.16.
(d) Homonuclear ^{1}H{^{1}H} (actually 2D *COrrelation SpectroscopY* (COSY), Chapter 6) of the ^{1}H resonance at δ 2.53 decoupled the peaks at δ 3.60 and 3.95.

(e) Heteronuclear $^1H\{^{13}C\}$ (actually 2D HETCOR, Chapter 6) double irradiation of the ^{13}C resonance at δ 62.5 decoupled the protons at δ 3.60 and 3.95.
(f) The resonance at δ 2.82 is a triplet with $J = 5.0$ Hz.
(g) The resonance at δ 2.76 is a doublet of doublets with $J = 2.7$ and 5.0 Hz.
(h) The resonances at δ 3.60 and 3.95 share a coupling of 12.7 Hz.
(i) Theoretical calculations indicated that the favored conformation is as shown above.
(j) The resonance at δ 3.60 has a slightly larger coupling with the resonance at δ 2.53 than does the resonance at δ 3.95, as measured from the ddd spacings centered at δ 3.60 and 3.95 (not clear from the broaden triplet at δ 2.53).

In your answer, interpret the couplings of 5.0 and 12.7 Hz. If any protons are enantiotopic or diastereotopic, identify them as such and explain why.

Source: Reproduced with permission from Ref. [18].

5.17 The 500 MHz ^1H spectra and selective nuclear Overhauser experiments given below are of the two γ-butyrolactones **A** and **B**, a functionality found in about 10% of all natural products. The lactone ring may have either a cis- or a trans-fusion to the attached ring. The structures are illustrated without that stereochemistry. In the spectra, the Greek letter α indicates that the proton is down and β that it is up. Each set of spectra contains an entirely assigned ^1H spectrum followed by a series of 1D difference NOE experiments. Assign the stereochemistry of each molecule.

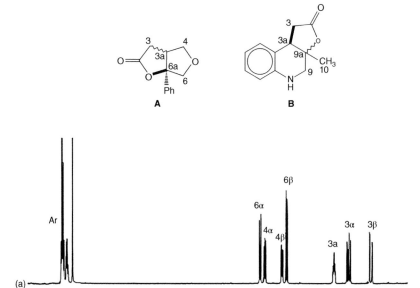

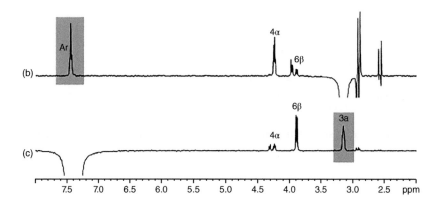

Selective NOEs of **A**: (a) normal ^{1}H spectrum, (b) irradiation of H3a, (c) irradiation of the aromatic protons.

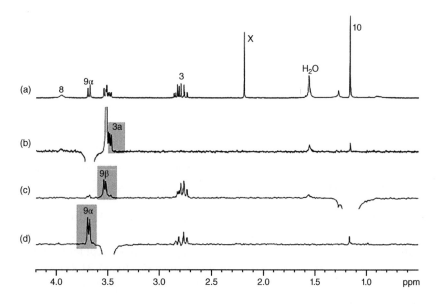

Selective NOEs of **B**: (a) normal ^{1}H spectrum, (b), irradiation of H9α, (c) irradiation of H10, (d) irradiation of H3a.
Source: Reproduced with permission from Ref. [19].

5.18 The cannabinoid receptor that gives marijuana its psychopharmacological properties also serves as the receptor for the molecule anandamide (from the Sanskrit for *bliss*), claimed

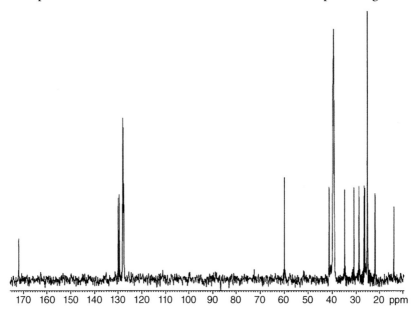

to be present in small amounts in chocolate. From the ^{13}C spectrum given below, assign as many peaks as possible. The large peak at δ 40 is from the solvent, DMSO-d_6, and the peak at δ 25 is the superposition of four peaks. Relaxation times (T_1 in s) were measured for the saturated carbons (except for the four superimposed peaks at δ 25): δ 60.8 (0.74 s), 42.3 (0.65), 35.7 (0.58), 31.8 (1.8), 29.6 (1.5), 27.5 (1.5), 27.1 (0.68), 22.9 (2.6), 14.8 (3.5). These values will be useful in distinguishing C16–C20.

Source: Reproduced with permission from Ref. [20].

5.19 The following two isomers were prepared in a synthetic project.

The 300 MHz ^{1}H spectra are given below for the two isomers with the label (a) (aromatic region only). The R groups are aliphatic and off scale, as are the NH resonances. The spectra labeled (b) are the difference NOE spectra for the two isomers. Assign the isomers. Comment on chemical shift and coupling constants as well as the NOEs. In the process, assign the H2, H3, H5, and H6 resonances on the trisubstituted rings and the ortho, meta, and para resonances on the monosubstituted phenyl groups.

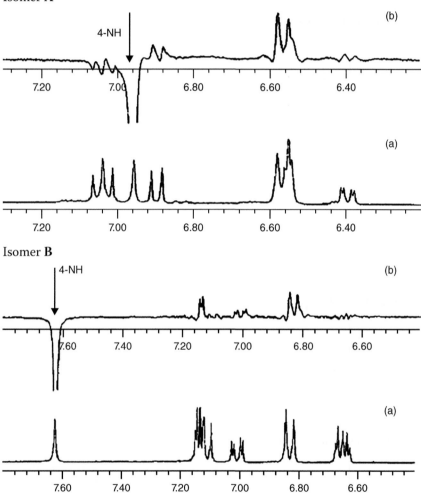

Source: Reproduced with permission from Ref. [21].

5.20 Draw out the spin vectors in the rotating coordinate system for each step in the following pulse sequence:

$$90°_x - (1/4J) - 180°_x - (1/4J) - \text{Acquire}$$

(all pulses are applied to the observed nucleus). Imagine that your observed nucleus is a ^{13}C atom attached to two protons, with the reference frequency fixed at the Larmor frequency of the carbon. What conclusion can you draw about the appearance of the observed peak at the end of the pulse sequence?

Equilibrium state

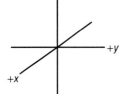

After the $90°_x$ pulse

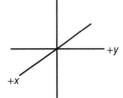

After the 1/4J fixed delay

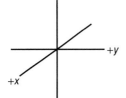

After the $180°_x$

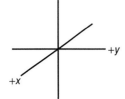

After the second 1/4J fixed delay

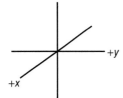

References

5.1 Harris, R.K. (1983). *Nuclear Magnetic Resonance Spectroscopy*, 82. London: Pitman Publishing, Ltd.
5.2 Günther, H. (1992). *NMR Spectroscopy*, 2e, 46. Chichester: Wiley.
5.3 Hall, L.D. and Sanders, J.K.M. (1980). *J. Am. Chem. Soc.* 102: 5703.
5.4 Günther, H. (1992). *NMR Spectroscopy*, 2e, 270. Chichester: Wiley.
5.5 Sanders, J.K.M. and Hunter, B.K. (1993). *Modern NMR Spectroscopy*, 2e, 191. Oxford: Oxford University Press.
5.6 Derome, A.E. (1987). *Modern NMR Techniques for Chemical Research*, 261. Oxford: Pergamon Press.
5.7 Morris, G.A. and Freeman, R. (1979). *J. Am. Chem. Soc.* 101: 760.
5.8 Derome, A.E. (1987). *Modern NMR Techniques for Chemistry Research*, 137. Oxford: Pergamon Press.
5.9 Derome, A.E. (1987). *Modern NMR Techniques for Chemical Research*, 143. Oxford: Pergamon Press.
5.10 Claridge, T.D.W. (1999). *High-Resolution NMR Techniques in Organic Chemistry*, 138. Amsterdam: Pergamon Press.
5.11 Bax, A., Freeman, R., and Kempsell, S.P. (1980). *J. Am. Chem. Soc.* 102: 4849.
5.12 Claridge, T.D.W. (1999). *High-Resolution NMR Techniques in Organic Chemistry*, 349. Amsterdam: Pergamon Press.
5.13 Bremser, W., Hill, H.D.W., and Freeman, R. (1971). *Messtechnik* 79: 14.
5.14 Heaton, N.J., Bello, P., Herradón, B. et al. (1998). *J. Am. Chem. Soc.* 120: 9636.
5.15 Dal Cason, T.A., Meyers, J.A., and Lankin, D.C. (1997). *Forensic Sci. Int.* 86: 19.
5.16 Shi, Q.W., Sauriol, F., Park, Y. et al. (1999). *Magn. Reson. Chem.* 37: 127.
5.17 Sirol, S., Gorricon, J.-P., Kalck, P. et al. (2005). *Magn. Reson. Chem* 43: 799.
5.18 Helms, E., Arpaia, N., and Widener, M. (2007). *J. Educ. Chem.* 84: 1329.
5.19 Xie, X., Tschan, S., and Glorius, F. (2007). *Magn. Reson. Chem.* 45: 384–385.
5.20 Bonechi, G., Brizzi, A., Brizzi, V. et al. (2001). *Magn. Resonan. Chem.* 39: 433–535.
5.21 Katritzky, A.R., Akhmedov, N.G., Wang, M. et al. (2004). *Magn. Reson. Chem.* 42: 652.

Further Reading

Relaxation Phenomena

General

Bakhmutov, V. (2004). *Practical NMR Relaxation for Chemists*. New York: Wiley.
Murali, N. and Krishnan, V.V. (2003). A primer for nuclear magnetic relaxation in liquids. *Concepts Magn. Reson. Part A* 17: 86.
Weiss, G.H. and Ferretti, J.A. (1988). *Prog. Nucl. Magn. Reson. Spectrosc.* 20: 317.
Wink, D.J. (1989). *J. Chem. Edu.* 66: 810.
Wright, D.A., Axelson, D.E., and Levy, G.C. (1979). *Magn. Reson. Rev.* 3: 103.

Carbon-13 Relaxation

Craik, D.J. and Levy, G.C. (1983). *Top. Carbon-13 Spectrosc.* 4: 241.
Lyerla, J.R. Jr., and Levy, G.C. (1974). *Top. Carbon-13 Spectrosc.* 1: 79.
Wehrli, F.W. (1976). *Top. Carbon-13 Spectrosc.* 2: 343.

Nuclear Overhauser Effect

Bell, R.A. and Saunders, J.K. (1973). *Top. Stereochem.* 7: 1.
Kövér, K.E. and Batta, G. (1987). *Prog. Nucl. Magn. Reson. Spectrosc.* 19: 223.
Neuhaus, D. and Williamson, M.P. (2000). *The Nuclear Overhauser Effect in Structural and Conformational Analysis*, 2e. New York: Wiley–VCH.
Noggle, J.H. and Schirmer, R.E. (1971). *The Nuclear Overhauser Effect*. New York: Academic Press.
Saunders, J.K. and Easton, J.W. (1976). *Determ. Org. Struct. Phys. Meth.* 6: 271.
Vögeli, B. (2014). The nuclear overhauser effect from a quantitative perspective. *Progr. Nucl. Magn. Reson. Spectrosc.* 78: 1.

Reactions on the NMR Time Scale

General

Bain, A.D. (2008). *Annu. Rep. NMR Spectrosc.* 63: 23.
Binsch, G. (1968). *Top. Stereochem.* 3: 97.
Binsch, G. and Kessler, H., *Angew. Chem. Int. Ed. Engl.*, 19, *411* (1980).
Casarini, D., Lunazzi, L., and Mazzanti, A. (2010). *Eur. J. Org. Chem.* 2035.
Jackman, L.M. and Cotton, F.A. (ed.) (1975). *Dynamic Nuclear Magnetic Resonance Spectroscopy*. New York: Academic Press.
Kaplan, J.I. and Fraenkel, G. (1980). *NMR of Chemically Exchanging Systems*. New York: Academic Press.
Kolehmainen, E. (2003). *Annu. Rep. NMR Spectrosc.* 49: 1.
Mann, B.E. (1977). *Prog. Nucl. Magn. Reson. Spectrosc.* 11: 95.
Ōki, M. (ed.) (1985). *Applications of Dynamic NMR Spectroscopy to Organic Chemistry*. Deerfield Beach, FL: Wiley–VCH.
Sändstrom, J. (1982). *Dynamic NMR Spectroscopy*. London: Academic Press.
Steigel, A. (1978). , NMR Basic Principles and Progress, vol. 15, 1.

Hindered Rotation

Bushweller, C.H., Lambert, J.B., and Takeuchi, Y. (ed.) (1992). *Acyclic Organonitrogen Stereodynamics*, 1–55. New York: Wiley–VCH.
Kessler, H., *Angew. Chem. Int. Ed. Engl.*, 9, *219* (1970).

Martin, M.L., Sun, X.Y., and Martin, G.J. (1985). *Annu. Rep. NMR Spectrosc.* 16: 187.
Nelsen, S.F. (1992). *Acyclic Organonitrogen Stereodynamics* (ed. J.B. Lambert and Y. Takeuchi), 89–121. New York: Wiley–VCH.
Ōki, M. (1983). *Top. Stereochem.* 14: 1.
Pinto, B.M. (1992). *Acyclic Organonitrogen Stereodynamics* (ed. J.B. Lambert and Y. Takeuchi), 149–175. New York: Wiley–VCH.
Raban, M. and Kost, D. (1992). *Acyclic Organonitrogen Stereodynamics* (ed. J.B. Lambert and Y. Takeuchi), 57–88. New York: Wiley–VCH.
Stewart, W.E. and Siddall, T.H. (1970). *Chem. Rev.* 70: 517.

Ring Reversal and Cyclic Systems

Booth, H. (1969). *Prog. Nucl. Magn. Reson. Spectrosc.* 5: 149.
Eliel, E.L. and Pietrusiewicz, K.M. (1979). *Top. Carbon-13 Spectrosc.* 3: 171.
Günther, H. and Jikeli, G., *Angew. Chem., Int. Ed. Engl.*, 16, 599 (1977).
Lambert, J.B. and Featherman, S.I. (1975). *Chem. Rev.* 75: 611.
Marchand, A.P. (1982). *Stereochemical Applications of NMR Studies in Rigid Bicyclic Systems*. Deerfield Beach, FL: Wiley–VCH.
Riddell, F.G. (1980). *The Conformational Analysis of Heterocyclic Compounds*. London: Academic Press.

Atomic Inversion

Delpuech, J.J. (1992). *Cyclic Organonitrogen Stereodynamics* (ed. J.B. Lambert and Y. Takeuchi), 169–252. New York: Wiley–VCH.
Jennings, W.B. and Boyd, D.R. (1992). *Cyclic Organonitrogen Stereodynamics* (ed. J.B. Lambert and Y. Takeuchi), 105–158. New York: Wiley–VCH.
Lambert, J.B. (1971). *Top. Stereochem.* 6: 19.
Rauk, A., Allen, L.C., and Mislow, K. (1970). *Angew. Chem. Int. Ed. Engl.* 9: 400.

Organometallics

Mann, B.E. (1982). *Annu. Rep. NMR Spectrosc.* 12: 263.
(a) Orrell, K.G. and Šik, V. (1987). *Annu. Rep. NMR Spectrosc.* 19: 79; (b) Orrell, K.G. and Šik, V. (1993). *Annu. Rep. NMR Spectrosc.* 27: 103; (c) Orrell, G. and Šik, V. (1999). *Annu. Rep. NMR Spectrosc.* 37: 1.
Vrieze, K. and Vanleeuwen, P.W.N.M. (1971). *Progr. Inorg. Chem.* 14: 1.

Rates from Relaxation Times

Lambert, J.B., Nienhuis, R.J., and Keepers, J.W. (1981). *Angew. Chem. Int. Ed. Engl.* 20: 487.

Multiple Irradiation

General

Castanar, L. and Parella, T. (2015). *Magn. Reson. Chem.* 53: 399.
Dalton, L.R. (1972). *Magn. Reson. Rev.* 1: 301.
Hoffman, R.A. and Forsén, S. (1966). *Prog. Nucl. Magn. Reson. Spectrosc.* 1: 15.
Kowalewski, V.J. (1969). *Prog. Nucl. Magn. Reson. Spectrosc.* 5: 1.
McFarlane, W. (1971). *Determ. Org. Struct. Phys. Meth.* 4: 150.
McFarlane, W. and Rycroft, D.S. (1985). *Annu. Rep. NMR Spectrosc.* 16: 293.
Micher, R.L. (1972). *Magn. Reson. Rev.* 1: 225.
von Philipsborn, W., *Angew. Chem. Int. Ed. Engl.*, 10, *472* (1971).

Difference Spectroscopy

Sanders, J.K.M. and Merck, J.D. (1982). *Prog. Nucl. Magn. Reson. Spectrosc.* 15: 353.

Selective Excitation

Freeman, R. (1991). *Chem. Rev.* 91: 1397.

Broadband Decoupling

Levitt, M.H., Freeman, R., and Frenkiel, T. (1983). *Adv. Magn. Reson.* 11: 47.
Shaka, A.J. and Keeler, J. (1987). *Prog. Nucl. Magn. Reson. Spectrosc.* 19: 47.

One-Dimensional Multipulse Methods

General

R. Benn and H. Günther, *Angew. Chem. Int. Ed. Engl.*, 22, 350 (1983).
Ernst, R.R. and Bodenhausen, G. (1990). *Principles of Nuclear Magnetic Resonance in One and Two Dimensions*. Oxford: Oxford University Press.
Morris, G.A. (1986). *Magn. Reson. Chem.* 24: 371.
Nakanishi, K. (1990). *One-Dimensional and Two-Dimensional NMR Spectra by Modern Pulse Techniques*. Mill Valley, CA: University Science Books.
Turner, C.J. (1984). *Prog. Nucl. Magn. Reson. Spectrosc.* 16: 311.
Turner, D.L. (1989). *Annu. Rep. NMR Spectrosc.* 21: 161.

Composite Pulses

Levitt, M.H. (1986). *Prog. Nucl. Magn. Reson. Spectrosc.* 18: 61.

Multiple Quantum Methods

Bodenhausen, G. (1986). *Prog. Nucl. Magn. Reson. Spectrosc.* 14: 137.
Norwood, T.J. (1992). *Prog. Nucl. Magn. Reson. Spectrosc.* 24: 295.

6

Two-Dimensional NMR Spectroscopy

NMR spectroscopy always has been multidimensional. In addition to frequency and intensity, which serve as the coordinates of the standard one-dimensional (1D) spectrum, reaction rates and relaxation times have provided further dimensions, often presented in the form of stacked plots (Figures 1.31 and 5.2). The second dimension of modern NMR spectroscopy, however, refers to an additional frequency axis. This concept was suggested first in a lecture by Jean Jeener in 1971 and reached wide application in the 1980s, when instrumentation caught up with theory. We can think of the first frequency dimension as the traditional characterization of nuclei in terms of chemical shifts and coupling constants. By introducing a second frequency dimension, we examine magnetic interactions between nuclei through structural connectivity, spatial proximity, or kinetic interchange.

6.1 Proton–Proton Correlation Through *J* Coupling

In the single-pulse experiments considered up to this point, a 90° pulse is followed by a period (t) during which the free-induction decay (FID) is acquired (Figure 6.1a). Fourier transformation of the time-dependent magnetic information into a frequency dimension provides the familiar spectrum of δ values, henceforth called a 1D spectrum.

If the 90° acquisition pulse is preceded by another 90° pulse (Figure 6.1b), useful relationships between spins can evolve. This two-pulse experiment has been named *COrrelation SpectroscopY* (COSY). Figure 6.2 illustrates what happens in terms of magnetization vectors. Consider a sample that contains only one type of nucleus without any coupling partners, for example, the ^{1}H spectrum of chloroform or tetramethylsilane. Although the final result of this particular experiment may seem trivial or even pointless at first, it will take on fuller meaning when we introduce relationships with other nuclei. The isolated nucleus of Figure 6.2 begins with the net magnetization **M** aligned along the z-axis (Figure 6.2a). The magnetization realigns along the y-axis after application of the 90° pulse (Figure 6.2b). If the coordinate system rotates at the reference frequency and the nucleus resonates at a slightly higher frequency, the spin vector picture begins to evolve. After a short amount of time, the vector **M** moves to a new position in the *xy* plane, for example, as in Figure 6.2c. We ignore longitudinal relaxation (T_1) to simplify the drawings. The evolving magnetization vector may be decomposed geometrically into a *y* component ($M_y = M \cos \omega t_1$) and an *x* component ($M_x = M \sin \omega t_1$), in which ω is

Nuclear Magnetic Resonance Spectroscopy: An Introduction to Principles, Applications, and Experimental Methods,
Second Edition. Joseph B. Lambert, Eugene P. Mazzola, and Clark D. Ridge.
© 2019 John Wiley & Sons Ltd. Published 2019 by John Wiley & Sons Ltd.

The one-dimensional experiment

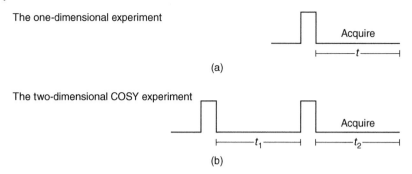

(a)

The two-dimensional COSY experiment

(b)

Figure 6.1 The pulse arrangements for a single cycle of one-dimensional NMR spectroscopy (a) and for two-dimensional NMR correlation spectroscopy (COSY) (b). In this diagram, each pulse is 90°. Data are acquired during the time t in the one-dimensional experiment and during t_2 in the two-dimensional experiment.

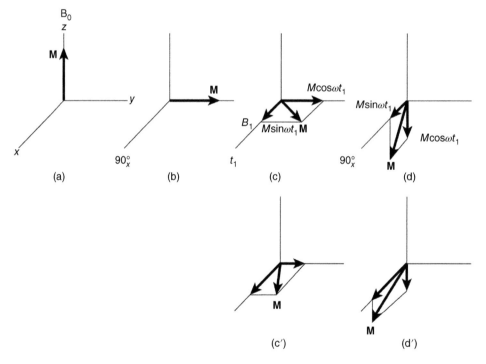

Figure 6.2 The pulse sequence for the COSY experiment. (a) Initial condition. For (c'), the magnetization M is allowed to evolve a longer time from (b) before the final $90°_x$ pulse is applied to give (d') than was the case from (b) for (c), which would give (d) after the final $90°_x$ pulse.

the difference between the frequency of the reference and that of the resonating nucleus and t_1 is the time elapsed since the 90° pulse.

If, at this point, the second 90° pulse of Figure 6.1b is applied, again along the x-axis, the result is different for the two magnetization components illustrated (Figure 6.2d). The x component is unaffected, but the y component is transferred to the negative z-axis. If magnetization is detected in the xy plane, only M_x remains. This quantity appears as

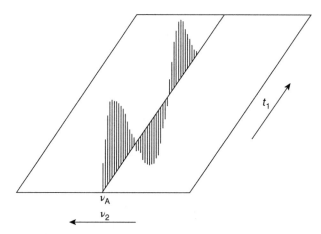

Figure 6.3 The solid line slanting upward at frequency v_A in the horizontal plane serves as the baseline for a series of ^{1}H spectra of chloroform, according to the COSY pulse sequence for a series of values of t_1. Each peak results from one cycle of 90°–t_1–90°-acquire, followed by Fourier transformation during t_2 (Figure 6.1) to give frequency v_A on the axis labeled v_2 (corresponding to the time domain t_2). The period t_1 is ramped up after each cycle. Fourier transformation in the t_1 dimension has not been carried out.

a FID during the time t_2 after the second pulse. Fourier transformation of the FID as a function of t_2 yields a signal at the resonance frequency (v_A). The intensity of this signal is determined by the length of the time period t_1 ($M \sin \omega t_1$). The subscripts are necessary to distinguish the evolution period t_1 from the acquisition period t_2 (Figure 6.1b). If t_1 is relatively short, M_x (= $M \sin \omega t_1$) is small, little x magnetization has developed (Figure 6.2d), and the resulting peak is small. A slightly longer value of t_1 yields a larger x component (Figure 6.2c′,d′) as $M \sin \omega t_1$ grows. Note that the spin population (the z, or longitudinal, magnetization) is inverted in Figures 6.2d,d′.

Figure 6.3 shows the result of a whole series of such experiments, with a buildup of M_x (= $M \sin \omega t_1$) as t_1 increases, reaching a maximum when the spin vector M is lined up along the x-axis. The peak height then decreases as the vector moves to the left of the x-axis, reaching zero intensity when it is lined up along the negative y-axis. As it passes behind the y-axis, the intensity becomes negative, attaining a negative maximum when the vector is aligned along the $-x$-axis. This negative maximum would be slightly smaller than the initial positive maximum, because of T_2 relaxation. It is clear from Figure 6.3 that this family of experiments generates a sine curve when M_x is plotted as a function of t_1. Only a cycle and a half are illustrated. Frequency (obtained from the Fourier transformation of t_2 to give v_2) is along the horizontal dimension (all the peaks are at v_A, intensity is along the vertical dimension, and time t_1 is along the more or less diagonal dimension (in Cartesian terms, these respectively are the x, z, and y coordinates in Figure 6.3). The set of data generated by stepping t_1 in this fashion in fact constitutes a FID that also may be Fourier transformed. Because the frequency ω represented by the sine curve in t_1 is the same as the frequency from the initial Fourier transformation in t_2, the result of the second Fourier transformation is a single peak at the coordinates (v_A, v_A) when plotted in two frequency dimensions (Figure 6.4a). This is the trivial result previously alluded to. If the molecule contains two uncoupled nuclear sets, as for example in methyl acetate (CH$_3$(C=O)OCH$_3$), two peaks appear along the

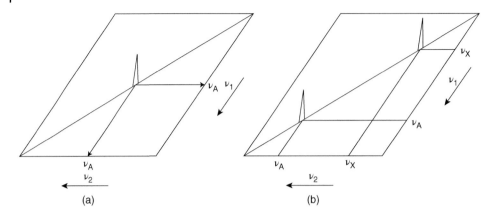

Figure 6.4 (a) The result of the COSY experiment after double Fourier transformation for a single isolated nucleus such as that in Figure 6.3. (b) The result of the COSY experiment for two uncoupled nuclei.

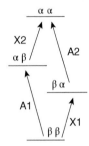

Figure 6.5 The energy diagram for an AX spin system.

diagonal, as in Figure 6.4b, with peaks, respectively, at (v_A, v_A) and (v_X, v_X). These peaks are necessarily on the diagonal of the two-dimensional (2D) representation.

The utility of the experiment just discussed becomes evident when two coupled nuclei are treated in the indicated fashion. Profound complications arise when the two nuclei are coupled. Figure 6.5 illustrates the possible spin states for nuclei A and X, as for example the alkenic protons in β-chloroacrylic acid, ClCH=CHCO$_2$H (the carboxyl hydrogen is ignored). The 1D AX spectrum contains four peaks, due to scalar (J) coupling. The diagram is intended to indicate the four different frequencies, from the highest (A1) to the lowest (X2). It is useful first to consider population perturbations during an old-style, 1D selective-decoupling continuous-wave (CW) experiment. Irradiation, for example, of only transition A1 tends to bring the αβ and ββ states closer together in population. Consequently, there is a direct effect on the intensities of the connected transitions X1 and X2, which propagates as a secondary effect on the intensity of the A2 transition. With respect to A1, X2 is called a *progressive transition* (a transition that goes on to a higher spin state), X1 is called a *regressive transition*, and A2 is called a *parallel transition*.

In the pulse experiment, energy absorption at frequency A1 has similar effects, which bring about magnetization or population transfer. The first 90° pulse serves to label all the magnetization with the 1D frequencies during period t_1: A1, A2, X1, and X2. Figure 6.2d,d′ represent a starting point at which the second 90° pulse of Figure 6.1b is applied. Let us set the reference frequency of the rotating frame at frequency A1. The magnetization analogous to that in Figure 6.2 had the frequency A1 during the period t_1. The second 90° pulse moves any z magnetization into the xy plane, where it can precess at any of the allowed frequencies: A1, A2, X1, or X2. Thus the magnetization that had frequency A1 during t_1 can be transferred to A2, X1, or X2 during t_2, and some magnetization remains at A1 during the second time period. Magnetization with frequency A1 during both time periods appears on the diagonal in the 2D representation analogous to Figure 6.4b, at position (A1, A1). The transferred magnetization, however, appears as *cross peaks* off the diagonal, at positions (A1, A2), (A1, X1), and (A1, X2). Each of the four resonances of Figure 6.5 undergoes analogous operations to generate three more diagonal peaks at (A2, A2), (X1, X1), and (X2, X2), plus nine more off-diagonal peaks at, for example, (A2, X1) and (X2, A1).

The 16 peaks (four on the diagonal and 12 off the diagonal) from the 2D experiment are illustrated in Figure 6.6 for β-chloroacrylic acid (the carboxyl resonance is omitted). The *stacked representation* contains several hundred complete 1D experiments, each representing a different value of t_1. The closely packed horizontal lines are barely distinguishable. On the diagonal from the lower left to the upper right (as drawn into the plots in Figure 6.4) are the four peaks that arise directly from resonance without magnetization transfer, that is, the components of magnetization that possess the same frequency in t_1 and t_2, such as (A1, A1). These four peaks along the diagonal constitute the normal four-peak 1D spectrum.

All the peaks off the diagonal represent magnetization transfer by scalar (J) coupling. For example, a transfer between the parallel transitions A1 and A2 is found as symmetrical peaks just above and below the diagonal at the lower left. One peak represents a transfer from A1 to A2, the other from A2 to A1. Because of this reciprocal relationship, all off-diagonal peaks appear in pairs reflected across the diagonal. Normally, the off-diagonal peaks between parallel transitions are more of a nuisance than useful, and they can be reduced or deleted by special techniques. The important information results from magnetization transfer between the A and the X nuclei, whose peaks are the

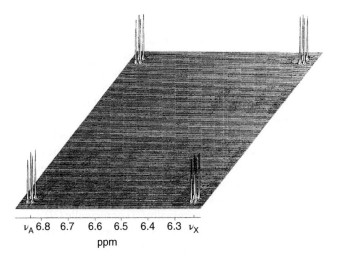

Figure 6.6 The stacked representation of the COSY experiment for the two coupled nuclei of β-chloroacrylic acid. Source: Reproduced with permission from Ref. [6].

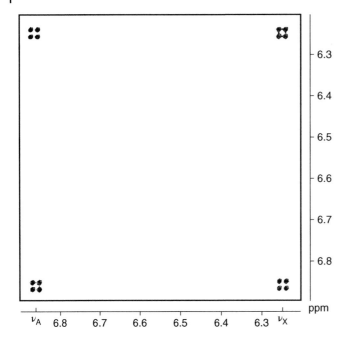

Figure 6.7 The contour representation of the COSY experiment for two coupled nuclei of β-chloroacrylic acid. Source: Reproduced with permission from Ref. [7].

clusters in the upper left and lower right of Figure 6.6, representing all possible transfers between the A and X transitions: A1 to X1, X2 to A1, and so on — eight in total, including the mirror-image pairs (A to X and X to A) on either side of the diagonal.

Figure 6.7 is the alternative *contour representation* of the same data, in which the distracting baselines are removed and only the peak bases remain, as if the spectator is viewing the spectrum from directly above it. By convention, the original diagonal is from lower left to upper right and the depiction necessarily is square. The Jeener experiment commonly is given the quasi acronym COSY. Since most 2D experiments involve spectral correlations, the name is not really apt. Alternative terms, such as 90° COSY, COSY90, H,H-COSY, or homonuclear HCOSY, have gone by the wayside through public acceptance of the general term COSY. The experiment itself has become an essential part of the analysis of complex proton spectra in order to determine which nuclei are coupled to each other, the 2D analogue of the 1D decoupling experiment.

Figure 6.8 is the COSY experiment for the indicated annulene. The 1D spectrum is shown along both the horizontal at the top and the vertical axes on the left, and the resonances are labeled α,β, and A through F. The aromatic protons that are ortho and meta to the ring fusion provide an isolated spin system, and their coupling is represented by the cross peak labeled α,β. The presence of a cross peak normally indicates that the protons giving the connected resonances on the diagonal are geminally (^2J) or vicinally (^3J) coupled. Long-range couplings usually do not provide significant cross peaks. Exceptions, however, can be expected, since long-range couplings can be large (Section 4.6).

The COSY analysis of the rest of the spectrum in Figure 6.8 assigns the remaining peaks and confirms the structure. Only protons A and F are split by a single neighbor,

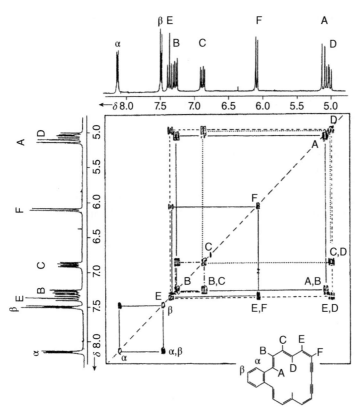

Figure 6.8 The COSY experiment for the illustrated annulene. Source: Reproduced with permission from Ref. [8].

to give doublets. The trans coupling of A with B should be larger than the *cis* coupling between E and F. The two doublets then may be assigned as F (with the smaller coupling) at δ 6.1 and A (with the larger coupling) at δ 5.2. It usually is essential in a COSY analysis to be able to make an initial assignment through traditional considerations of chemical shifts and coupling constants (Chapters 3 and 4). The COSY analysis then consists of moving from this known diagonal peak to a cross peak, and then back to the diagonal for the assignment of a new peak. Only A and F have single cross peaks (one coupling partner). All remaining resonances in the large ring have two cross peaks (two coupling partners), which provide the means for assignment. We can start with either A or F. Dropping down from A leads to the cross peak A,B, and moving horizontally to the left leads to a new diagonal peak and assignment of proton B. This horizontal path passes through another cross peak, which must be between B and its other coupling partner, C. Moving up from the cross peak at B,C then leads to a new diagonal peak and assignment of proton C. Moving horizontally to the right leads to the new cross peak C,D and returning upward to the diagonal assigns proton D. Dropping back down from D and passing through C,D leads to the other cross peak from D, labeled E,D. Returning to the diagonal to the left assigns proton E and passes through the other cross peak from E, labeled E,F. Returning upward from E,F to the diagonal completes the assignment with proton F.

Nuclear Magnetic Resonance Spectroscopy

A group at IBM has provided a useful example of a more complex COSY analysis with the tripeptide Pro–Leu–Gly (**6.1**).

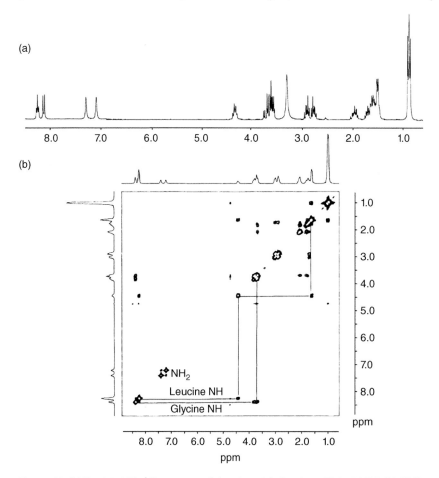

6.1

The three carbonyl groups disrupt vicinal connectivities, so the molecule consists of four independent spin systems: proline, leucine, glycine, and the terminal amide. The 1D ^{1}H spectrum is given in Figure 6.9a without any assignments. The high-frequency (low field) peaks at δ 7.0–8.3 are from the protons on nitrogen, as there are no aromatic hydrogens,

Figure 6.9 (a) The 300 MHz ^{1}H spectrum of the tripeptide Pro–Leu–Gly in DMSO. (b) COSY spectrum of Pro–Leu–Gly with connectivities of the NH protons. Source: Courtesy of IBM Instruments, Inc.

and the broad peak at δ 3.3 is from the solvent HOD. Figure 6.9b contains the COSY spectrum with connectivities drawn in for the amide resonances. The nonequivalent terminal NH$_2$ resonances are assigned immediately as δ 7.0 and 7.2 because they have no external connectivities and hence no cross peaks other than between themselves. The Gly NH proton is assigned at δ 8.2 because it is a triplet (next to a CH$_2$) and has only the single connectivity with the CH$_2$ group at δ 3.6 (completing the Gly portion of the spectrum). The remaining NH resonance, at δ 8.1, from Leu is a doublet (next to a CH) and has a cross peak with the resonance at δ 4.3, which has other connectivities. There is no third NH resonance, so the Pro NH must be quadrupolar broadened, or is exchanging with HOD.

The spectrum of Figure 6.10a completes the COSY analysis of the Leu portion and confirms the fact that the NH resonating at δ 8.1 is part of Leu rather than Pro. The expected Leu connectivity is NH → CH → CH$_2$ → CH → (CH$_3$)$_2$. Cross peaks with the following connectivities (starting with NH) are observed: δ 8.1(d) → 4.3(q or dd) → 1.5(m) → 0.9(dd). Apparently, two of the proton resonances coincide, most likely those from CH$_2$ and the isopropyl CH. The CH$_3$ resonance, as expected, is at the lowest frequency and cannot be from any Pro group. Its higher multiplicity (dd, Figure 6.9a) arises because the two methyl groups are diastereotopic due to the chiral center to which the butyl group is attached.

The spectrum of Figure 6.10b shows the Pro connectivity. The highest-frequency resonance (δ 3.7) should be from the CH group adjacent (α) to the carbonyl group. The entire resonance at δ 3.7 is an overlap of this Pro CH (the higher-frequency portion) with the Gly CH$_2$ (the lower frequency portion). The Pro CH has two cross peaks with the diastereotopic β protons at δ 1.7 and 1.9, which are mutually coupled and have their own cross peak. Unfortunately, the γ protons are nearby (δ 1.6), but their cross peak with the δ protons at δ 2.8 completes the assignment of the spectrum. The fully assigned 1D spectrum and structure are given in Figure 6.11.

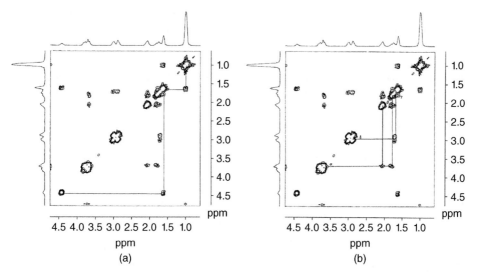

Figure 6.10 (a) Connectivity within the leucine portion of Pro–Leu–Gly by COSY. (b) Connectivity within the proline portion of Pro–Leu–Gly by COSY. Source: Courtesy of IBM Instruments, Inc.

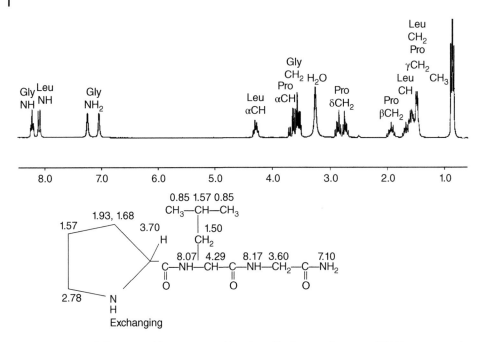

Figure 6.11 The fully assigned ¹H spectrum of Pro–Leu–Gly. Source: Courtesy of IBM Instruments, Inc.

False peaks and lack of symmetry around the diagonal can arise in the COSY experiment for several reasons. First, differences in digital resolution in the two periods, t_1 and t_2, may prevent perfect symmetry. Second, incorrect pulse lengths or, third, incomplete transverse relaxation during the delay time can create false cross peaks. Fourth, there may be effects from longitudinal relaxation. Any magnetization in the z direction does not precess during t_1 and therefore is rotated by the second 90° pulse into the position recognized as $v_1 = 0$ (the position of the reference frequency). Signals thus occur at $v_1 = 0$ and at any value of v_2 associated with a resonance, resulting in a stream of lines, called *axial peaks* or t_1 *noise*, in the 2D plot. Fifth and finally, folding can occur in two dimensions and can give rise to off-position diagonal peaks and even cross peaks. All these artifacts may be minimized by optimizing pulse lengths, by allowing sufficient time for transverse relaxation, by using phase cycling, or by employing symmetrization. Axial peaks may be suppressed largely by alternating +90° and −90° for the second pulse, thus canceling z magnetization. The more complex CYCLOPS procedure suppresses axial peaks and eliminates other artifacts, such as quad images (Section 5.8). *Symmetrization* is a procedure for imposing bilateral symmetry around the diagonal. Most artifacts are conveniently eliminated by this procedure, but not all. For example, if two resonances have streams of t_1 noise, a point on one stream can occur at the precise mirror position (with respect to the diagonal) of a point on the other stream. Although the streams are largely eliminated, the two peaks at the symmetrical positions are retained and appear as handsome cross peaks. Usually common sense can reject them. Experimental procedures for optimizing the COSY experiment are discussed in Chapter 7.

There are many variants of the standard COSY experiment that either improve on its basic aims or provide new information. We shall consider several of them without appreciable attention to the details of the pulse sequences.

6.1.1 COSY45

The large size of diagonal peaks sometimes can be a deterrent to understanding the significance of nearby cross peaks. The problem is aggravated by the presence of cross peaks from parallel transitions (Figure 6.5). The COSY45 experiment reduces the intensities of both the diagonal peaks and the cross peaks from parallel transitions. Figure 6.12 compares the COSY90 and COSY45 experiments for 2,3-dibromopropionic acid ($CH_2BrCHBrCO_2H$). The COSY45 experiment clarifies cluttered regions close to the diagonal and also provides information on the signs of coupling constants. The name derives from alteration of the second pulse length: $90°-t_1-45°-t_2$ (acquire).

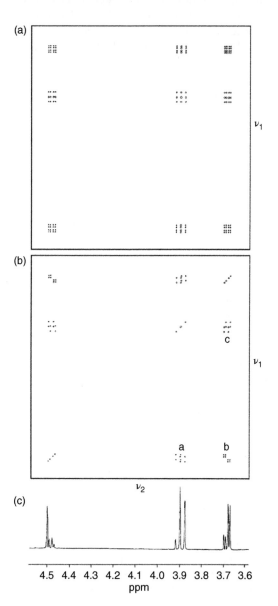

Figure 6.12 The COSY90 (a) and COSY45 (b) spectra of 2,3-dibromopropionic acid. (c) The 1D spectrum is a cross section through the COSY45 spectrum. Source: Reproduced with permission from Ref. [9].

The use of the smaller tip angle restricts the magnetization transfer between nuclei, but the effect is larger for the directly connected parallel transitions than for the more remotely connected progressive and regressive transitions. The diagonal peaks are clarified by suppression of the progressive cross peaks. Inevitably, however, there is loss of signal. A tip angle of 60° (COSY60) may be used as a compromise, but any gains in sensitivity occur at the expense of clarification of the diagonal.

Examination of the cross peaks in the COSY45 spectrum of Figure 6.12 reveals overall appearances different from those in normal COSY spectra. Rather than possessing the usual squarish or rectangular shape, many of the cross peaks have taken on a decided tilt. The direction of tilt is related to the relative signs of coupling constants. In an AMX system, for example, the A, M off-diagonal peak is caused by magnetization transfer through J_{AM}, referred to as the *active coupling*. Its tilt, however, depends on whether couplings of A and M with the third nucleus, X, have the same or opposite signs. A tilt with a positive slope (parallel to the diagonal), for example, results if the two *passive couplings*, J_{AX} and J_{MX}, have the same sign. A tilt with a negative slope (orthogonal to the diagonal) results if they have opposite signs.

COSY cross peaks are caused predominantly by either geminal (HCH) or vicinal (HCCH) couplings. Because connectivity inferences are based largely on vicinal couplings, it would be useful to be able to distinguish these two classes. As described in Chapter 4, vicinal couplings are, in general, positive, and geminal couplings (at least on saturated carbons) are negative. Consequently, the two classes can, in principle, be distinguished by the slope of the tilt in the COSY45 spectrum, as illustrated in Figure 6.12. This spectrum is closer to ABX than AMX but nonetheless shows the expected off-diagonal COSY peaks. The resonances of the diastereotopic CH_2 protons are found at δ 3.67 and 3.89, and the resonance of the methine proton is at δ 4.49, shifted to higher frequency (lower field) by attachment of the carbon to two electron-withdrawing substituents (Br and CO_2H). The off-diagonal peaks that have been labeled a and b result from the active coupling of either methylene proton with the methine proton: a positive, vicinal coupling. The passive couplings for these cross peaks are the geminal coupling to the diastereotopic partner and the vicinal coupling to the other methylene proton. As these couplings have opposite sign, the tilt has a negative slope. The off-diagonal peak labeled c results from active coupling between the diastereotopic methylene protons. The passive couplings for this cross peak are the two vicinal couplings between the diastereotopic protons and their vicinal neighbor. Because both passive couplings are positive, the tilt has a positive slope. In this fashion, c is spotted as a cross peak between geminal protons.

6.1.2 Long-Range COSY (LRCOSY or Delayed COSY)

The normal assumption in the COSY experiment is that two- or three-bond (geminal or vicinal) couplings provide the dominant magnetization transfer to create cross peaks. Information from longer range couplings, however, also can be useful. Introducing fixed delays Δ during the evolution and detection periods ($90°-t_1-\Delta-90°-\Delta-t_2$ (acquire)) enhances magnetization transfer from small couplings at the expense of large couplings. Figure 6.13 compares the COSY and LRCOSY experiments for a polynuclear aromatic compound. In the COSY spectrum, cross peaks occur only between ortho neighbors (1,2 and 3,4). In the LRCOSY spectrum, additional cross peaks arise between

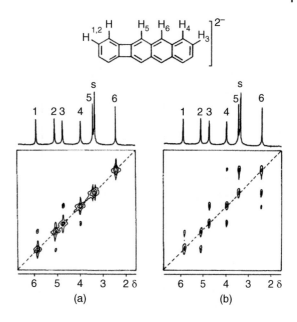

Figure 6.13 The 400 MHz COSY (a) and LRCOSY (b) spectra of naphthobiphenylene dianion. (The signal S is from solvent.) Source: Reproduced with permission from Ref. [10].

peri neighbors (5,6 and 4,6). Information on the connectivity between fused aromatic rings thus becomes available in the LRCOSY case.

6.1.3 Phase-Sensitive COSY (ϕ-COSY)

Fourier transformation involves building up a signal from the sum of sine and cosine curves. Every point in the spectrum has both sine and cosine contributions, which are 90° out of phase. These contributions sometimes are called, respectively, the imaginary and real terms and lead mathematically to *dispersion-mode* and *absorption-mode* spectra. An in-phase, or absorption, signal has the familiar form of a positive peak. A dispersion signal, commonly used for electron spin resonance spectra, has a sideways S shape with a portion below and a portion above the baseline (Figure 2.11). Such a signal produces both negative and positive maxima for a given peak and a value of zero at the resonance frequency as the sign changes. NMR experiments normally are tuned to the absorption mode by the process of phasing, but the two signals also may be combined mathematically to produce what is called a *magnitude*, or *absolute-value, spectrum*.

Many of the COSY experiments we have examined thus far used magnitude representations because of phase differences between various peaks in the pure modes. Both magnetization that is not transferred (and thus appears on the diagonal) and magnetization that is transferred to parallel transitions undergo no phase shift. Other cross peaks, however, exhibit phase shifts. A transfer between progressive transitions (A1 to X2 in Figure 6.5) shifts the phase −90°, and a transfer between regressive transitions (A1 to X1) shifts it +90°. Because absorption and dispersion modes differ by 90°, phasing the diagonal peak to absorption results in dispersive cross peaks, or vice versa. Moreover, cross peaks from progressive and regressive transitions are always out of phase by 180°. If one cross peak represents positive absorption, then the other represents negative absorption; or if one cross peak begins a dispersive signal negatively, the other begins positively.

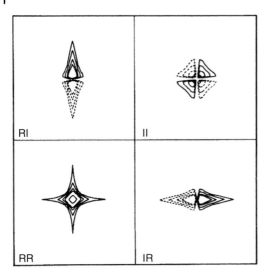

Figure 6.14 The four types of 2D phase quadrants, corresponding to frequency modes that are real–real (RR), imaginary–real (IR), real–imaginary (RI), and imaginary–imaginary (II). Source: Reproduced with permission from Ref. [11].

The magnitude, or absolute-value, mode is used to eliminate all phase differences and produce absorption-like peaks. The resulting peaks tend to be broad and often are distorted. In small molecules with little peak overlap, there may be no problem with the use of magnitude spectra, but larger molecules such as proteins, polysaccharides, or polynucleotides may produce unacceptable overlap. The phase-sensitive COSY experiment then can tune the cross peaks to a pure absorption (real) mode. This experiment not only provides enhanced resolution but also enables coupling constants to be read more easily from the cross peaks when the data are highly digitized.

Because 2D methods involve two time domains, the transformations in both t_1 and t_2 generate real and imaginary components. As a result, the phase-sensitive 2D signal has four modes rather than two. These phase modes, or quadrants, correspond to the four situations in which (i) both frequency signals are real, (ii) both are imaginary, or (iii and iv) one is real while the other is imaginary. Figure 6.14 illustrates the four modes. The real–real (RR) mode produces the familiar peak with a contour shaped like a four-pointed star at the base. Figure 6.15 illustrates what the COSY spectrum for two spins looks like when the diagonal and parallel components are tuned dispersively (both imaginary) and the progressive and regressive cross peaks are tuned absorptively (both RR, but 180° out of phase). This common phase-sensitive representation provides straightforward identification of the cross peaks derived from coupling and hence determines J.

6.1.4 Multiple Quantum Filtration

The 1D INADEQUATE pulse sequence suppresses the centerband singlet in order to measure ^{13}C–^{13}C couplings from the satellites (Section 5.7). The procedure involves creating double quantum coherences (Section 5.8). A similar procedure may be used in 2D ^1H spectra to suppress singlets, which are single quantum coherences. Such singlets may arise from solvent or from uncoupled methyl resonances, both of which can constitute major impediments in locating highly split resonances in a complex spectrum. In a *Double Quantum Filtered COSY* (DQF-COSY) experiment, an extra 90° pulse is added after

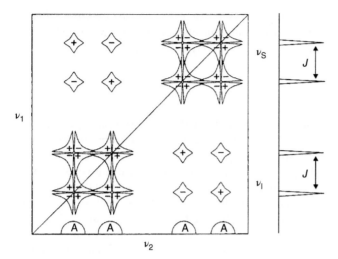

Figure 6.15 Phase-sensitive COSY diagram for two spins, with the diagonal peaks in dispersion mode and the cross peaks in antiphase absorption mode. The 1D spectrum is on the right. Source: Reproduced with permission from Ref. [12].

the second 90° COSY pulse, and phase cycling converts multiple quantum coherences into observable magnetizations. The resulting 2D spectrum lacks all singlets along the diagonal. For example, the spectrum of lysine, $^+NH_3CH(CH_2CH_2CH_2CH_2NH_2)CO_2^-$, in Figure 6.16a has no solvent (HOD) peak, which was suppressed as a single quantum coherence. An important feature of the phase-sensitive DQF-COSY experiment is that double quantum filtration allows both diagonal and cross peaks to be tuned into pure absorption at the same time. This feature reduces the size of all the diagonal signals and permits cross peaks close to the diagonal to be observed. The only disadvantage of

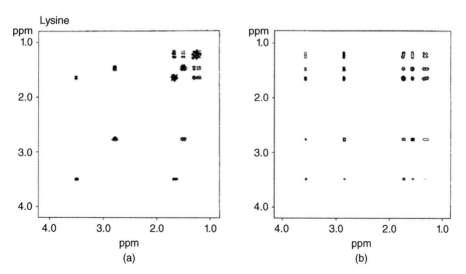

Figure 6.16 The (a) DQF-COSY and (b) TOCSY spectra of lysine. Source: Reproduced with permission from Ref. [13].

DQF-COSY is a reduction in sensitivity by a factor of 2. The *Triple Quantum Filtered COSY* (TQF-COSY) experiment removes both singlets and AB or AX quartets, providing even greater spectral simplification. It is rarely used because of a concomitant increased loss of sensitivity.

6.1.5 TOtal Correlation SpectroscopY (TOCSY)

In the standard COSY experiment, the connectivity within an entire spin system, such as that in a butyl group ($CH_3CH_2CH_2CH_2-$), must be mapped out from proton to proton via a series of cross peaks. By spin locking the protons during the second COSY pulse, the chemical shifts of all the protons may be brought essentially into equivalence. Recall that resonance frequencies of protons and carbons are made equal through cross polarization for solids by achieving the Hartmann–Hahn condition (Section 1.9). In the 2D variant of this experiment, the initial 90° pulse and the period t_1 occur as usual, but the second pulse locks the magnetization along the y-axis so that all protons have the spin lock frequency. All coupled spins within a spin system then become closely coupled to each other, and magnetization is transferred from one spin to all the other members, even in the absence of J couplings. The spectrum in Figure 6.16b shows the TOtal Correlation SpectroscopY (TOCSY) experiment for lysine. The methylene group at the lowest frequency (upper right corner) exhibits four TOCSY cross peaks, one with each of the other three methylene groups and one with the methine proton. The TOCSY experiment, a variation of which is called the *HOmonuclear HArtmann-HAhn* or HOHAHA, experiment, has particular advantages for large molecules, including enhanced sensitivity and, if desired, the phasing of both diagonal and cross peaks to the absorption mode. The process of identifying resonances within specific amino acid or nucleotide residues is considerably simplified by this procedure. Each residue can be expected to exhibit cross peaks among all its protons and none with protons of other residues.

6.1.6 Relayed COSY

An alternative, but less general, method for displaying extended levels of connectivity is provided by *Relayed Coherence Transfer* (RCT). The typical COSY experiment for an AMX system with $J_{AX} = 0$ produces cross peaks between A and M and between M and X. It is not unusual for the key diagonal peak for M to be coincident with a resonance from another spin system, making it difficult to follow the connectivity path. (Recall, for example, the Leu portion of the COSY spectrum of Pro–Leu–Gly in Figure 6.10.) The RCT experiment generates a cross peak between the A,M and M,X cross peaks, eliminating the ambiguity. The result is shown diagrammatically in Figure 6.17 for AMX and A′M′X′ systems whose M and M′ resonances coincide. The COSY experiment contains the expected four cross peaks. The RCT experiment contains two additional cross peaks, connecting the cross peaks of the individual spin systems. The two new cross peaks are labeled (A,M,X) and (A′,M′,X′). The connectivity of AMX and of A′M′X′ is then rendered unambiguous.

6.1.7 J-Resolved Spectroscopy

In Chapter 5, we saw how the spin echo experiment can isolate or remove characteristics of chemical shifts or coupling constants. Spin echoes may be used in two

Figure 6.17 Diagram of (a) COSY and (b) relayed coherence transfer (RCT) experiments for two three-spin systems (AMX and A'M'X') whose M and M' portions overlap. Source: Adapted with permission from Ref. [14].

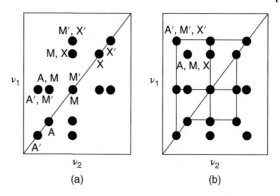

dimensions to generate one frequency dimension representing chemical shifts and another representing coupling constants. The sequence $90°-\frac{1}{2}t_1-180°-\frac{1}{2}t_1-t_2$, for example, uses the 180° pulse to refocus chemical shifts during t_1. The result for a glucose derivative is shown in Figure 6.18. The ^{1}H frequencies are found on the horizontal axis (v_2 in δ), with the normal 1D spectrum displayed at the top (Figure 6.18a). The vertical axis ("f_1" = v_1) contains only proton–proton coupled multiplets, each centered about a zero frequency point, i.e. all multiplets occur at the same chemical shift in $v_1 = 0$. Thus, the multiplet at the highest frequency (lowest field) from H-3 is a quartet, seen with further splitting when viewed from the vertical axis. By taking a projection at an

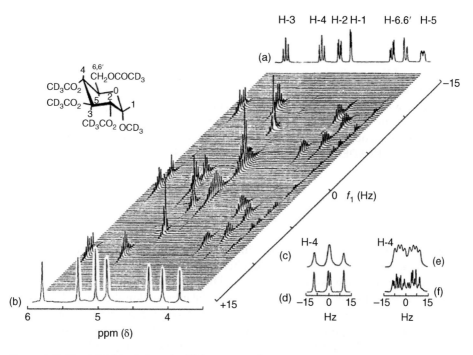

Figure 6.18 The 270 MHz 2D J-resolved ^{1}H spectrum of 2,3,4,6-tetrakis-O-trideuteroacetyl-α-D-glucopyranoside. Source: Reproduced with permission from Ref. [15].

angle (45°) that causes each of the members of the individual multiplets to overlap when viewed from the horizontal axis, as at the bottom (Figure 6.18b), a display is obtained that, in essence, is a proton–proton decoupled proton spectrum. Resonances devoid of any couplings are present at each frequency. This projection is a novel way to examine ^{1}H spectra, although it has not seen widespread use because it reveals no connectivities.

The pulse sequence, as a variant of the spin echo experiment, also refocuses the spread of frequencies caused by field inhomogeneity so that some improvement in resolution is obtained. The inset at the lower right of Figure 6.18 shows the normal 1D spectra of H-4 and H-5 at the top (Figure 6.18c,e) and the unrotated projection of the 2D J-resolved spectra at the bottom (Figure 6.18d,f, extracted from the projected spectrum (Figure 6.18a) at the top of the 2D display). The much higher resolution of the 2D resonances is clearly evident. Thus, the procedure is an effective way to measure J accurately, particularly when J is poorly resolved in the 1D spectrum. The experiment fails for closely coupled nuclei (second-order spectra).

In addition to resolving small couplings that may be absent in the 1D spectrum, the J-resolved procedure can be used to distinguish homonuclear from heteronuclear couplings. The vertical axis in the figure displays couplings only between spins that were affected by the 180° pulse. Hence, only ^{1}H–^{1}H couplings appear on that axis. Couplings to heteronuclei are not phase modulated and, consequently, appear as spacings along the horizontal axis. In this way, ^{1}H–^{19}F and ^{1}H–^{31}P couplings may be distinguished from ^{1}H–^{1}H couplings, which are removed in the rotated spectrum such as that shown in (b) at the bottom of the figure.

6.1.8 COSY for Other Nuclides

The basic COSY experiment can be carried out for any spin-½ nucleus that is 100% abundant. In addition to ^{1}H–^{1}H COSY, the procedure thus is applicable to ^{19}F–^{19}F (F,F-COSY) and to ^{31}P–^{31}P (P,P-COSY), respectively, in organofluorine and organophosphorus compounds. When the nuclide is less than 100% abundant, as for Li,Li-COSY, the uncoupled centerband must be separated from the coupled satellites, usually by the 2D INADEQUATE procedure (Section 6.4).

6.2 Proton–Heteronucleus Correlation

Cross peaks in the standard COSY experiment are generated through magnetization transfer that arises from scalar (J) coupling between protons. Coupling from a proton to a different nuclide, such as carbon-13, should be able to generate a similar response. Analogous cross peaks then would provide very useful information about which carbons are bonded to which protons. Thus, the assignment of a proton resonance would automatically lead to the assignment of the resonance of the carbon to which it is bonded, and vice versa. This field has seen considerable development recently, and there now are several pulse sequences commonly used to explore connectivity between protons and carbon or other heteronuclides.

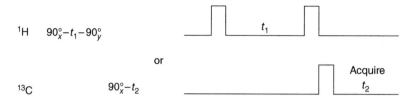

Figure 6.19 The pulse sequence for the HETCOR experiment.

6.2.1 HETCOR

The simplest 2D sequence that includes magnetization transfer proton to carbon takes the form given in Figure 6.19. The pulse sequence is reminiscent of the 1D INEPT sequence, and manipulation of magnetization is much the same as in Figure 5.19, but without the 180° pulse. The initial 90° ^{1}H pulse generates y magnetization. For the simplest case, in which one carbon is bonded to one proton (as in CHCl$_3$), ^{1}H magnetization evolves during the period t_1 according to its Larmor frequency. Two ^{1}H vectors diverge due to coupling with ^{13}C. The second 90° ^{1}H pulse generates nonequilibrium z magnetization that is transferred to ^{13}C in the manner of the INEPT experiment of Figure 5.19. The single 90° ^{13}C pulse then provides the ^{13}C FID that is acquired during t_2. The 2D spectrum then has one axis in ^{1}H frequencies (ν_1) and one in ^{13}C frequencies (ν_2).

For the simple heteronuclear AX case, the 2D spectrum contains two peaks when it is projected onto either the ^{1}H or the ^{13}C axis (the A and X portions, respectively, of the AX spectrum), as in the INEPT experiment. Thus, the ^{1}H resonances that are detected correspond to the ^{13}C satellites of the usual ^{1}H spectrum. The 2D display contains four peaks: two along a diagonal and two symmetrically off the diagonal. Moreover, the peaks are in antiphase for each nuclide, since the INEPT spectrum without decoupling generates one peak up and one peak down. Decoupling would result in summing the peaks algebraically to zero.

To bring about decoupling, another pulse and two fixed periods are added (Figure 6.20). The first (90°$_x$) ^{1}H pulse allows chemical shifts and coupling constants to evolve during t_1. The 180° ^{13}C pulse refocuses the H—C coupling constants in the ^{1}H dimension, thereby decoupling ^{1}H from ^{13}C. The fixed time Δ_1 allows the ^{1}H

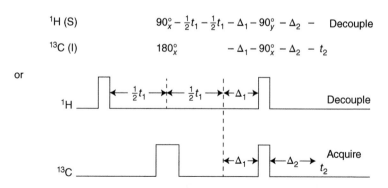

Figure 6.20 The pulse sequence for the HETCOR experiment with decoupling.

vectors to obtain the antiphase (180° out of phase) relationship illustrated in Figure 5.19e. The second (90°$_y$) ^{1}H pulse moves the vectors in antiphase relationship onto the z-axis and enables polarization to be transferred to ^{13}C, also in the antiphase relationship. The 90° ^{13}C pulse is for observation. The second fixed time, Δ_2, restores phase alignment and permits ^{13}C to be decoupled from ^{1}H during ^{13}C acquisition.

Figure 6.21 illustrates this experiment for an adamantane derivative. The ^{1}H frequencies are on the vertical axis, and the ^{13}C frequencies are on the horizontal axis. The respective 1D spectra are illustrated on the left and at the top. The 2D spectrum is composed only of cross peaks, each one relating a carbon to its directly bonded proton(s). There are no diagonal peaks (and no mirror symmetry associated with a diagonal), because two different nuclides are represented on the frequency dimensions. Quaternary carbons are invisible to the technique, as the fixed times Δ_1 and Δ_2 normally are set to values for one-bond couplings. This experiment often is a necessary component in the complete assignment of ^{1}H and ^{13}C resonances. Its name, *HETeronuclear chemical-shift CORrelation*, is abbreviated as HETCOR, but other acronyms, e.g. HSC, for *Heteronuclear Shift Correlation*, and H,C-COSY, have been used. The method may be applied to protons coupled to many other nuclei, such as ^{15}N, ^{29}Si, and ^{31}P, as well as ^{13}C.

Figure 6.21 illustrates two advantages of the HETCOR experiment. (i) The correlation between protons and carbons means that spectral assignments for one nuclide automatically lead to spectral assignments for the other. Thus, ^{13}C assignments can assist in ^{1}H assignments, and vice versa. (ii) Overlapping proton resonances often can be dispersed in the carbon dimension. Even at very high fields, proton resonances can

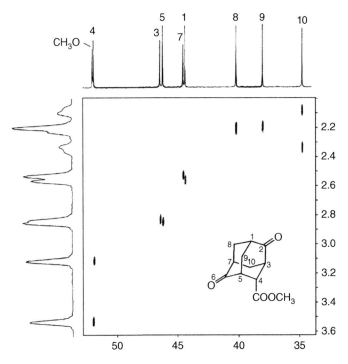

Figure 6.21 The HETCOR spectrum of 4-(methoxycarbonyl)adamantane-2,6-dione. Source: Reproduced with permission from Ref [16].

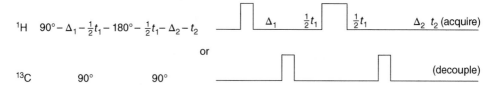

Figure 6.22 The pulse sequence for the HMQC experiment.

overlap; consider, for example, those of H8 and H9, which coincide at δ 2.2 in the figure. The presence of the two cross peaks from C8 (δ 40) and C9 (δ 38) reveals the spectral overlap in the proton dimension. Similar considerations apply to H3 and H5.

COSY cross peaks can arise from either geminal (HCH) or vicinal (HCCH) connectivities so that ambiguities can be present. Geminally related protons that are diastereotopic are attached to a common carbon and hence have HETCOR connectivities to a single ^{13}C frequency, whereas vicinally related protons are attached to different carbons and thus have HETCOR connectivities to two different ^{13}C frequencies. This advantage is not illustrated in the figure, but is useful for distinguishing geminal from vicinal relationships in COSY spectra.

6.2.2 HMQC

A major drawback to the HETCOR experiment is the low sensitivity that results from detection of the X nucleus (usually ^{13}C). The HMQC (*Heteronuclear correlation through Multiple Quantum Coherence*) experiment uses *inverse detection*, whereby ^{13}C responses are observed in the ^1H spectrum. The pulse sequence is given in Figure 6.22 and represents a transfer of coherence rather than of polarization. The initial ^1H magnetization from the 90° pulse becomes antiphase during the fixed period Δ_1 through the ^1H–^{13}C coupling constant. Multiple quantum coherence then is created by the first ^{13}C pulse. The remainder of the sequence is designed to select double or higher quantum coherence (from the ^{13}C satellites in the ^1H spectrum) over single quantum coherence (from the ^1H centerbands), in a process similar to the 1D INADEQUATE experiment in Section 5.7. The 2D representation, as shown in Figure 6.23 for camphor, still includes the ^1H–^{13}C coupling information in the ^1H dimension, although ^{13}C irradiation can be applied during the ^1H t_2 acquisition period to provide decoupling. The major difference between HETCOR and HMQC is that the acquisition period t_2 is at ^{13}C frequencies in the former experiment (Figure 6.20), but at ^1H frequencies in the latter (Figure 6.22). Consequently, the HMQC experiment is much more sensitive. Like HETCOR, HMQC can be used with heteronuclei other than ^{13}C, of which the experiment involving ^{15}N is the most common.

6.2.3 BIRD-HMQC

The most difficult aspect of implementing the HMQC experiment is the suppression of signals from protons attached to ^{12}C (the centerband or single quantum coherences) in favor of the protons attached to ^{13}C (the satellites or double quantum coherences). The use of pulsed field gradients (PFG's, Section 6.6) is the most effective technique, but many spectrometers still lack the hardware required for their generation. Fortunately, there is an effective alternative for the suppression of centerbands by means of the *BIlinear Rotation Decoupling* (BIRD) sequence, which is outlined by the vector notation

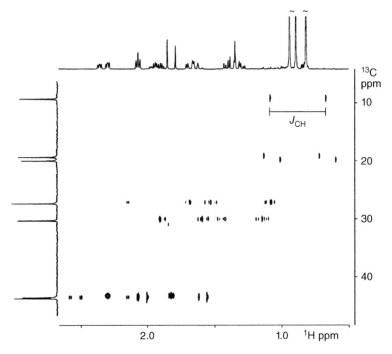

Figure 6.23 The HMQC spectrum of camphor, with ^{1}H–^{13}C couplings retained in the ^{1}H dimension. Source: Reproduced with permission from Ref. [17].

in Figure 6.24. Two sets of vectors are followed, one for the protons attached to ^{12}C and one for the protons attached to ^{13}C. The initial 90° proton pulse (Figure 6.24a,a′) along the x direction moves all magnetization onto the y-axis (Figure 6.24b,b′). (Keep in mind that eventually inverse detection HMQC pulses will be applied, with ultimate detection in the proton channel.) Protons on ^{12}C are unsplit, so their magnetization evolves as depicted in the upper set of vector diagrams. After the delay period $(2J)^{-1}$, these vectors have reached some arbitrary angle with respect to the y-axis (Figure 6.24c), according to their individual Larmor frequencies. (Only one such frequency is illustrated.) By contrast, protons on a ^{13}C evolve as two vectors (Figure 6.24c′), separated by the frequency of the one-bond coupling constant ($\Delta\omega = 2\pi\Delta\nu = 2\pi J$), as illustrated in the lower set of vector diagrams. The centers of the diagrams are maintained on the y-axis for viewing simplicity, but the chemical shifts of the vectors evolve according to their Larmor frequencies. After the delay period $(2J)^{-1}$, these vectors are separated by a 180° angle (Figure 6.24c′, Section 5.5, $\phi = 2\pi(\Delta\nu)\,t = 2\pi J(2J)^{-1} = \pi$). At this time, a 180° pulse is applied to the ^{1}H channel along the y-axis, and a 180° pulse is applied to the ^{13}C channel. The ^{1}H pulse rotates the vector for the protons attached to ^{12}C about the y-axis (Figure 6.24d), so that, after another period $(2J)^{-1}$, that vector converges onto the y-axis to create a spin echo (Figure 6.24e). The simultaneous ^{13}C pulse, however, switches the spin identities of the coupled partners for the protons attached to ^{13}C so that the vectors continue to diverge, as was discussed with regard to Figure 5.14. After the second period $(2J)^{-1}$, these vectors converge onto the $-y$-axis (Figure 6.24e′, $\phi = 360°$). Application of the second 90° pulse along the x-axis moves all the proton magnetization back to

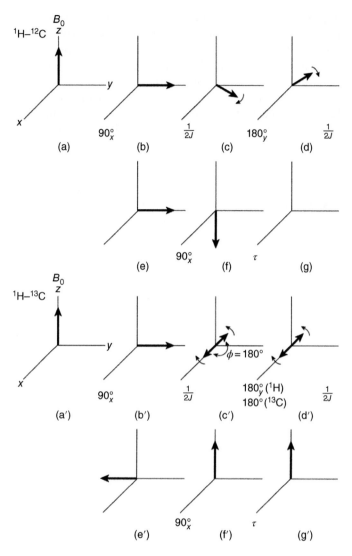

Figure 6.24 The BIRD sequence for the selection of signals from protons attached to ^{13}C over those from protons attached to ^{12}C.

the z-axis. Because they had opposite phases in Figure 6.24e,e′, the protons attached to ^{12}C now point along the −z-axis (Figure 6.24f), whereas the protons attached to ^{13}C point along the +z-axis (Figure 6.24f′). The experiment then requires a delay time τ for relaxation. The protons attached to ^{13}C are already essentially at equilibrium and remain unaffected (Figure 6.24g′), but those attached to ^{12}C are upside down and hence begin to relax back to equilibrium. The HMQC pulses are applied after a time τ that brings the magnetization for the protons attached to ^{12}C exactly to zero (Figure 6.24g). At this point, BIRD has suppressed the single quantum coherences (^{12}C–^{1}H, Figure 6.24g) and selected the multiple quantum coherences (^{13}C–^{1}H, Figure 6.24g′).

The overall experiment thus becomes BIRD–τ–HMQC–DT, in which τ is chosen to null the single quantum coherences and DT is the normal delay time between pulse repetitions. DT includes the time taken to acquire the signal, as well as a recycle time during which the signal is regenerated through relaxation. The details of the experiment require some knowledge of the relaxation times T_1 and an appropriate choice of the intervals τ and DT. The optimized result can provide effective suppression of the unwanted signals for small molecules. Larger molecules, however, can undergo a negative nuclear Overhauser effect (NOE) (Section 6.3) from the inverted signals, with loss of sensitivity.

6.2.4 HSQC

The *Heteronuclear Single Quantum Correlation* (HSQC) experiment is an alternative to HMQC that accomplishes a similar objective. The experiment generates, via an INEPT sequence, single quantum ^{13}C (or ^{15}N) coherence, which evolves and then is transferred back to the proton frequency by a second INEPT sequence, this time in reverse. The main difference from the HMQC result is that HSQC spectra do not contain ^{1}H–^{1}H couplings in the ^{13}C (v_1) dimension. As a result, HSQC cross peaks tend to have improved resolution over analogous HMQC cross peaks. HSQC is preferred when there is considerable spectral overlap.

6.2.5 COLOC

To focus on longer range H—C couplings, the fixed times Δ_1 and Δ_2 of HETCOR can be lengthened accordingly. Loss of magnetization due to transverse relaxation then reduces sensitivity significantly. The *COrrelation spectroscopy* via *LOng-range Coupling* (COLOC) pulse sequence avoids this problem by incorporating the ^{1}H evolution period t_1 inside the Δ_1 delay period. Figure 6.25 shows the COLOC spectrum for vanillin. The circled cross peaks are residues from one-bond couplings. The only long-range coupling of the methoxy group is with C-3, which indicates that methoxy is connected at that point. Other long-range couplings, however, also are seen, for example, between C-1 and H-5, C-3 and H-5, and C-2 and H-7.

The principal disadvantage of the COLOC sequence lies in the fixed nature of the evolution period. In such a pulse sequence, C—H correlations are diminished or even absent when two- and three-bond ^{1}H–^{13}C couplings are of a magnitude similar to those of ^{1}H–^{1}H couplings within a molecular fragment. This situation occurs quite commonly (Chapter 4). The FLOCK sequence (so named because it contains three BIRD sequences, Figure 6.24; also see Section 7.3.3.2) contains a variable evolution time, in which t_1 becomes progressively larger, and avoids the potential loss of C—H correlations. Although the experiment detects carbon, it is quite useful when there is signal overlap in the ^{1}H spectrum.

6.2.6 HMBC

Correlations through longer range H—C couplings offer at least two potential advantages. (i) Heteronuclei that lack an attached proton are invisible to the HETCOR/HMQC/HSQC family of experiments. Thus, carbonyl groups and tetrasubstituted carbons cannot be studied. Such carbons, however, are likely to show coupling to protons two or more bonds away. (ii) A single carbon can be correlated with several

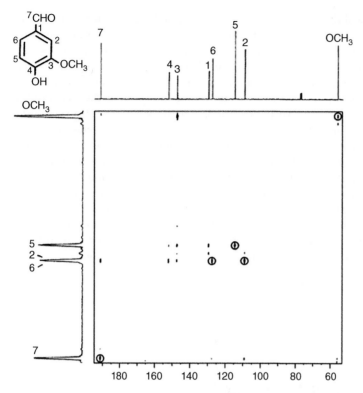

Figure 6.25 The COLOC spectrum of vanillin. Source: Reproduced with permission from Ref. [18].

neighboring protons. Hence, connectivities over heteroatoms or carbonyl groups can help define larger groupings of atoms, thereby complementing information from COSY. The COLOC and FLOCK sequences, however, suffer from the same drawback as HETCOR, namely, low sensitivity arising from direct observation of a nuclide (^{13}C, ^{15}N, and so on) with a low gyromagnetic ratio.

Proton-detected *Heteronuclear Multiple Bond Correlation* (HMBC) is designed to provide correlations between protons and heteronuclei such as carbon or nitrogen by a pulse sequence similar to that used in HMQC (Figure 6.26). A second delay time (Δ_2) is no longer necessary, because H—C couplings are not intended to be removed. The major difference from HMQC is the duration of the initial delay time Δ, which is lengthened in order to select for the smaller couplings over two or more bonds. These couplings are usually in the range 0–15 Hz (Section 4.4 for ^2J(HCC) and Section 4.5

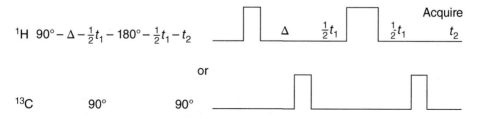

Figure 6.26 The pulse sequence for the HMBC experiment.

for ^3J(HCCC)). A typical delay Δ of $(2J)^{-1}$ then corresponds to the range 60–200 ms ($J = 2.5$–8.3 Hz), in comparison with 24 ms for HMQC. Shorter delay times tend to improve sensitivity (there is less loss of signal through relaxation), but longer times may be entirely acceptable for small molecules with relatively long relaxation times. Longer delay times occasionally permit observation of connectivities through four-bond couplings, e.g. ^4J(HCCCC), or even five (Section 4.6). Interpreting the magnitudes of H—C couplings over two to five bonds entails all the subtleties of analogously interpreting H—H couplings and includes consideration of inductive effects, zigzag pathways, π bonding, Karplus stereochemical considerations, and so on (Sections 4.5 and 4.6). Methyl groups often exhibit the most intense HMBC correlations, because of the multiplicative effect of simultaneous detection by three protons and because free rotation provides almost no stereochemical restrictions that can reduce couplings.

Figure 6.27 provides the HMBC spectrum of the illustrated heterocycle. As with HET-COR, the spectrum contains only cross peaks. The carbonyl carbon C8 apparently couples with H2, H5, H6, H6′, H7, and H7′, as it exhibits cross peaks with all these protons, representing all possible two- and three-bond couplings. Three further points are worth noting. First, the proton pairs on C4, C6, and C7 are diastereotopic. Separate peaks are observed for each such connectivity, e.g. that between C8-H7 and C8-H7′. Second, some

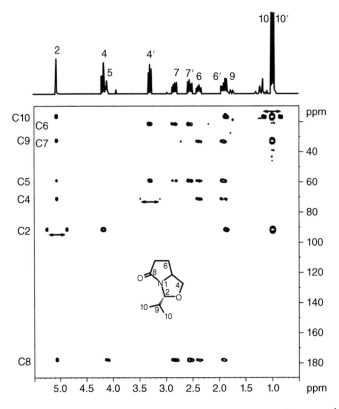

Figure 6.27 The HMBC spectrum of the illustrated heterocycle. The 1D ^{1}H spectrum is given at the top. The vertical axis corresponds to ^{13}C frequencies, the horizontal axis to ^{1}H frequencies. One-bond correlations are indicated by peaks connected by doubly headed arrows. Source: Reproduced with permission from Ref. [19].

one-bond couplings break through the selection process, because of accidental phase coincidence. Such couplings show up as doublets (the one-bond H—C coupling survives because HMBC does not include carbon irradiation during t_2) and are noted in the spectrum by doubly headed arrows. The cross peaks are obvious, because of the large magnitude of $^1J(^{13}C-^1H)$ and because they appear at the coincidence of C and H chemical shifts for the CH fragment, such as C2/H2. They can be filtered out by a more complex pulse scheme. Third, suppression of single coherence ($^1H-^{12}C$) signals is the primary task of the experiment, and the process is significantly enhanced by the use of PFG's (Section 6.6).

6.2.7 Heteronuclear Relay Coherence Transfer

The relay COSY experiment (RCT) may be adapted to the HETCOR context. The result of such an experiment is illustrated in Figure 6.28 for the illustrated acetal of acrolein, along with the normal HETCOR experiment ("HSC" in the figure). The following HETCOR cross peaks are present: H_C/C_2 at the bottom left, H_B/C_1 and H_A/C_1 in the middle, and H_D/C_3 at the upper right. This experiment shows that C_3 is bonded to H_D, C_2 is bonded to H_C, and C_1 is bonded to both H_A and H_B. The H–H–C relay coherence transfer experiment is depicted at the upper right. In a general fragment $H_X-C_X-C_Y-H_Y$, HETCOR peaks occur for H_X/C_X and H_Y/C_Y, respectively, defining the H_X-C_X and H_Y-C_Y fragments. In the RCT experiment, off-diagonal peaks also

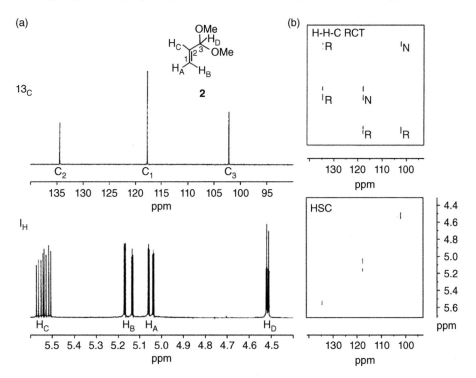

Figure 6.28 (a) The ^{13}C and ^1H spectra of the dimethyl acetal of acrolein. (b) The normal HETCOR (labeled HSC, bottom) and the H–H–C relay coherence transfer (top) experiments for the same molecule. Source: Reproduced with permission from Ref. [20].

occur for H_X/C_Y and H_Y/C_X, thereby defining the larger H_X–C_X–C_Y–H_Y fragment. The normal HETCOR cross peaks are labeled N (that for H_C/C_2 is missing). The additional cross peaks resulting from relayed connectivity between given CH pieces are labeled R. Thus, the relayed (R) peak at the upper left indicates that the H_D/C_3 and H_C/C_2 pairs are connected (H_D–C_3–C_2–H_C). The relayed peak in the middle left indicates that the $H_{A/B}/C_1$ and H_C/C_2 pairs are connected ($H_{A/B}$–C_1–C_2–H_C). The other two peaks labeled R (bottom, middle, and right) are mirror images of the former relayed peaks across the diagonal because relay occurs symmetrically ($H_D/C_3 \to H_C/C_2$ is the same as $H_C/C_2 \to H_D/C_3$). This particular experiment is called H–H–C RCT because the pulses involve ^{1}H signals twice and ^{13}C signals once in that order. Other heteronuclear relay experiments can involve a different order (H–C–H) or different nuclei (H–H–N).

6.3 Proton–Proton Correlation Through Space or Chemical Exchange

In the original depiction of the 2D experiment in Figure 6.2, the magnetization vector was resolved into sine and cosine components during t_1. The sine component was followed through the second 90° pulse and into the t_2 domain to create the COSY sequence. We ignored the cosine component, which was placed along the $-z$-axis after the second 90° pulse and hence was unobservable. There are mechanisms other than scalar coupling for transferring magnetization, and these methods can alter z magnetization and hence change the cosine component. Irradiation at the frequency of one proton can transfer magnetization to nearby protons through dipolar interactions (the NOE). The clear effect of this technique on z magnetization is reflected in spectral intensity perturbations in one dimension (Section 5.4). Altering the chemical identity of a nucleus through chemical exchange similarly affects z magnetization. A nucleus resonating at one frequency becomes a nucleus resonating at a different frequency (Section 5.2). The population (z magnetization) thus decreases at the first frequency and increases at the second.

After the second 90° pulse in Figure 6.2d, both the NOE and the chemical exchange mechanisms can modulate the cosine component of the magnetization along the z-axis. The frequency of modulation is the resonance frequency of the magnetization transfer partner, either from dipolar relaxation or chemical exchange. After a suitable fixed period of time (τ_m, the mixing period), during which this modulation is optimized, the cosine component may be moved to the xy plane by a third 90° pulse and be detected along the y-axis during a t_2 acquisition period. Thus, the complete experiment is 90°–t_1–90°–τ_m–90°–t_2 (acquire). Because the frequency of magnetization of some nuclei during t_1 moves to another value during τ_m and is observed at the new frequency during t_2, the 2D representation of this experiment exhibits cross peaks. When the cross peaks derive from magnetization transfer through dipolar relaxation, the 2D experiment is called *NOE SpectroscopY* (NOESY). When they derive from chemical exchange, the experiment is called *EXchange SpectroscopY* (EXSY).

The duration of the fixed time τ_m depends on the relaxation time T_1, the rate of chemical exchange, and the rate of NOE buildup. In the case of the NOESY experiment, valuable information can be ascertained about the distance between various protons within a molecule. Figure 6.29 illustrates the NOESY spectrum for a complex heterocycle. As with COSY, the 1D spectrum is found along the diagonal. Cross peaks occur when two

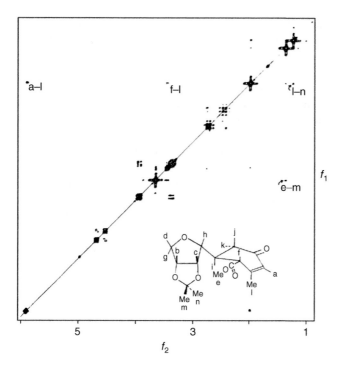

Figure 6.29 The ¹H NOESY spectrum for the indicated compound. Source: Courtesy of Bruker Instruments, Inc.

protons are close to each other. Thus, methyl group l shows an expected cross peak with the adjacent alkenic proton a (upper left). Additional cross peaks of methyl l indicate its closeness to the methinyl proton f and the acetal methyl n. The ester methyl e is close to the other acetal methyl m. The NOESY experiment can provide both structural and conformational information. In practice, cross peaks become unobservable when the proton–proton distance exceeds about 5 Å.

At least three factors complicate the analysis of NOESY spectra. First, COSY signals may be present from scalar couplings and may interfere with interpretations intended to be based entirely on interproton distances. Vicinal couplings, for example, are largest when the coupled nuclei are farthest apart, a situation that occurs in the antiperiplanar or trans geometry. COSY signals may be reduced through phase cycling or by statistical variation of τ_m by about 20%. (The NOESY signals grow monotonically, but the COSY signals are sinusoidal and cancel out.) In the phase-sensitive NOESY experiment, NOESY cross peaks due to positive NOE's may be distinguished from COSY cross peaks because they have opposite phases. Weak NOESY cross peaks, however, may be canceled in this experiment when breakthrough COSY cross peaks happen to be of similar intensity. Also, COSY signals are not distinguished from NOESY signals that are due to negative NOE's.

Second, in small molecules, the NOE builds up slowly and attains a theoretical maximum of only 50%, as noted earlier in the 1D context (Section 5.4). Because a single proton may be relaxed by several neighboring protons, the actual maximum normally is much less than 50%. (Of course, the same problem exists in the 1D NOE experiment.) Moreover, as the molecular size increases and behavior departs from the extreme narrowing limit, the maximum NOE decreases to zero and becomes negative. Thus, particularly for medium-sized molecules, the NOESY experiment may fail. For larger molecules, whose relaxation is dominated by the W_0 term, not only is the maximum NOE −100% rather than +50% but also the NOE buildup occurs more rapidly. The NOESY experiment thus has been of particular utility in the analysis of the structure and conformation of large molecules such as proteins and polynucleotides.

Third, in addition to its transfer directly from one proton to an adjacent proton, magnetization may be transferred by spin diffusion. In this mechanism, already described for the 1D experiment (Section 5.4), magnetization is transferred through the NOE from one spin to a nearby second spin and then from the second to a third spin that is close to the second spin, but not necessarily to the first one. These multistep transfers can produce NOESY cross peaks between protons that are not close together. Spin diffusion can occur even through two or more intermediate spins, but the process becomes increasingly less efficient. The direct transfer of magnetization and its transfer by spin diffusion sometimes may be distinguished by examining the NOE buildup rate, as illustrated in Figure 6.30. In this hypothetical plot of the NOE intensity as a function of the mixing time τ_m for the system D—A–B–C, AA is the intensity of the diagonal peak, and the other lines represent intensities of cross peaks. The NOE between two close protons (A and B, separated by 2 Å in the model) rises most rapidly. Protons A and D are 4 Å apart and show a very small NOE. Protons A and C also are 4 Å apart (2 Å between A and B and another 2 Å between B and C), but the intensity of the AC cross peak rises steadily through spin diffusion with B as the intermediary. The model shows that spin diffusion provides the major contribution to the AC cross peak for a distance of 4 Å, but the buildup is slower than for the direct transfer that gives rise to the AB cross peak.

The rotating-frame NOESY experiment (ROESY) provides some advantages for small and medium-sized, as well as large, molecules. The pulse sequence for ROESY (previously called CAMELSPIN) is similar to TOCSY or HOHAHA, although the period of spin locking is chosen to optimize magnetization transfer through the NOE (via dipolar interactions) rather than through scalar couplings. Whereas the NOE decreases to zero and becomes negative as the mean correlation time τ_c for molecular rotation increases (larger molecules move more slowly), in the rotating frame the maximum NOE remains positive and even increases from 50% to 67.5% (Figure 6.31). In addition to enhancing the signal, the ROESY experiment decreases spin diffusion,

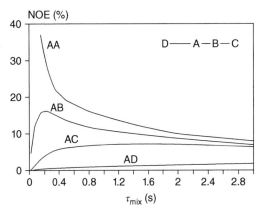

Figure 6.30 Peak intensities calculated for a hypothetical NOESY experiment involving four nuclei, D–A–B–C, with D 4 Å from A and B 2 Å from A and C. The curve labeled AA is for the diagonal peak, and the remaining curves are for the various cross peaks. Source: Reproduced with permission from Ref. [21].

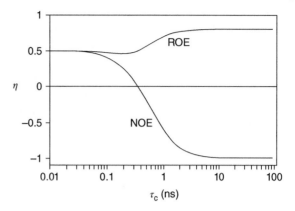

Figure 6.31 The enhancement factor η as a function of the effective correlation time τ_c for the standard nuclear Overhauser experiment (NOE) and for the spin-lock, or rotating-coordinate, variant (ROE). The curves were calculated for an interproton distance of 2.0 Å and a spectrometer frequency of 500 MHz. Source: Reproduced with permission from Ref. [22].

offering advantages for large molecules. Just as COSY artifacts may be present in the NOESY spectrum, so can TOCSY artifacts be present in the ROESY spectrum. Steps must be taken to remove them, as provided by the T-ROESY variant. The use of a weak static spin-lock pulse can reduce the TOCSY peaks. Positive ROESY cross peaks also can be distinguished easily from negative TOCSY peaks in a phase-sensitive ROESY experiment. As with COSY/NOESY, there is the possibility of canceling signals if TOCSY artifacts have intensities similar to those of the desired ROESY cross peaks.

When magnetization is transferred via chemical exchange in the EXSY experiment, it may be necessary to perform several preliminary experiments to optimize the value of τ_m, which should be approximately $1/k$. Figure 6.32 illustrates the EXSY experiment from an early example by Ernst, in which the diagonal peaks run nontraditionally from upper left to lower right. At fast exchange, the 1D ^{1}H spectrum of the heptamethylbenzenium ion contains only one methyl resonance, as the methyl group moves around the ring. At slow exchange, there are distinct resonances for the four types of methyls labeled on the left in the figure. The EXSY experiment shows which methyls interchange with which. One can imagine 1,2, 1,3, or 1,4 shifts, but the EXSY experiment agrees only with the 1,2 mechanism. Each off-diagonal peak indicates magnetization transfer between two diagonal peaks. Thus, the 1 methyls have a cross peak only with (and hence

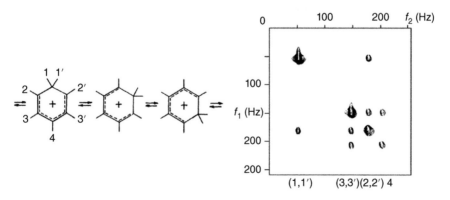

Figure 6.32 The ^{1}H EXSY spectrum for the heptamethylbenzenium ion. Source: Meier and Ernst (1979) [23]. Copyright 1979. Reproduced with permission from American Chemical Society.

exchange only with) the 2 methyls, the 2 methyls exchange with the 1 and 3 methyls, the 3 methyls exchange with the 2 and 4 methyls, and the 4 methyls exchange only with the 3 methyls. This is the pattern expected for 1,2 shifts.

The intensities of the cross peaks depend on the rate constant for exchange. For the case of exchange between equally populated sites lacking spin–spin coupling, e.g. the two methyls of N,N-dimethylformamide [H(CO)N(CH$_3$)$_2$], the rate constant k is related to the mixing time τ_m, the intensity of the cross peak I_c, and the intensity of the diagonal peaks I_d by the formula of Eq. (6.1).

$$\frac{I_d}{I_c} \sim \frac{(1 - k\tau_m)}{k\tau_m} \quad (6.1)$$

Rearranging this expression algebraically gives the rate constant k from Eq. (6.2).

$$k \sim \frac{1}{[\tau_m(I_d/I_c + 1)]} \quad (6.2)$$

Since the pulse sequence is the same for EXSY and NOESY, cross peaks in the NOESY (or ROESY) experiment might be mistaken for EXSY cross peaks, or vice versa. They can be distinguished in the phase-sensitive experiment, since EXSY and NOESY/ROESY peaks have opposite phases.

6.4 Carbon–Carbon Correlation

The 1D INADEQUATE experiment provides a method for measuring ^{13}C–^{13}C coupling constants and for determining carbon–carbon connectivity by establishing coupling magnitudes that are common to two carbon atoms (Section 5.7). In practice, application of the method to solving connectivity problems is restricted not only by the inherently low sensitivity of detecting two dilute nuclei but also by the similarity of many ^{13}C–^{13}C couplings. Duddeck and Dietrich have pointed out that all the one-bond carbon–carbon couplings in cyclooctanol fall into the narrow region from 34.2 to 34.5 Hz, except for C$_1$–C$_2$, which is 37.5 Hz. The second problem may be largely alleviated by translating the experiment into two dimensions. The original INADEQUATE experiment (Section 5.7) can be adapted directly to two dimensions by incrementing the fixed time Δ as the t_1 domain: $90°_x - 1/4J_{CC} - 180°_y - 1/4J_{CC} - 90°_x - t_1 - 90°_\phi - t_2$ (acquire). The period t_1 is used to encode the double quantum frequency domain. The resulting 2D display contains a horizontal axis in ν_2 (the normal ^{13}C frequencies) and a vertical axis that is a double quantum domain represented by the sum of the frequencies of coupled ^{13}C nuclei ($\nu_1 = \nu_A + \nu_X$). The latter frequencies are referenced to a transmitter frequency at zero.

Figure 6.33 illustrates the 2D INADEQUATE spectrum of menthol (**6.2**).

6.2

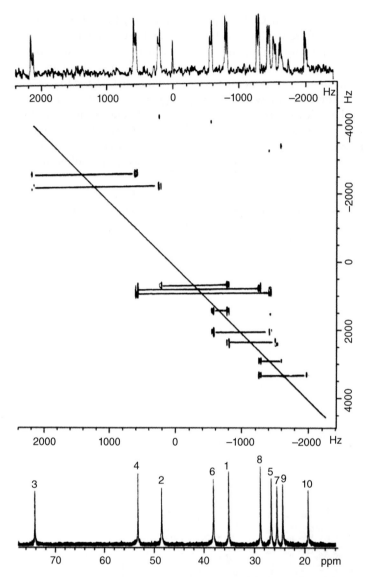

Figure 6.33 The 2D INADEQUATE spectrum of menthol, with the ^{1}H-decoupled ^{13}C spectrum. Source: Reproduced with permission from Ref. [24].

The experiment also has been called C,C–COSY, as the cross peaks represent connectivity between two carbons. There are no diagonal peaks (which would arise from ^{13}C nuclei with ^{12}C neighbors) because the experiment removes single quantum signals. The diagonal usually is drawn in, as in the figure, and may appear either as in the figure or from lower left to upper right. The normal proton-decoupled ^{13}C spectrum is shown at the bottom. At the top, the 2D procedure permits recovery of the carbon-coupled ^{13}C spectrum through a projection of the v_2 dimension.

To obtain connectivity from a 2D INADEQUATE experiment, a single assignment is made and the remainder of the structure is mapped, much as with COSY. Only a gap caused by the presence of a heteroatom, C–X–C, prevents mapping the entire skeleton. For menthol, the oxygen-substituted C-3 resonates at the highest frequency (far left) and serves as the starting point. Horizontal lines are drawn between coupled carbons in the 2D spectrum, passing through the diagonal at their midpoints. There are two cross peaks at the C-3 frequency, corresponding to connectivities to C-2 and to C-4. Of these, the secondary C-2 should be at lower frequency (higher field). The connectivity then may be mapped in the following fashion: C-2 → C-1 → C-6 (and from C-1 to the C-7 methyl) → C-5 → C-4 → C-8 (and from C-4 to the original C-3) → C-9 and C-10. The 2D INADEQUATE procedure also is applicable to concatenations of other coupled, dilute nuclides, such as ^{29}Si/^{29}Si, ^{11}B/^{11}B, and ^{6}Li/^{6}Li in organosilicon, organoboron, and organolithium systems, respectively.

The major disadvantage of this experiment is its extremely low sensitivity. A variation is INEPT-INADEQUATE (sometimes called ADEQUATE), which uses proton observation and PFG's over pathways such as H–C–C. The resulting spectrum resembles that of 2D INADEQUATE, but can contain peaks only when at least one of the paired carbons has an attached proton. Although such technical refinements ameliorate the problem of sensitivity, this family of experiments has not been widely used.

6.5 Higher Dimensions

The enormous complexity of spectra of large biomolecules such as proteins, polynucleotides, and polysaccharides has led to the development of three- and four-dimensional experiments. Two independently incremented evolution periods (t_1 and t_2), in conjunction with three separate Fourier transformations and of an acquisition period t_3, result in a cube of data with three frequency coordinates.

Figure 6.34, from a study by van de Ven, illustrates the complexity of the 2D NOESY spectrum of a DNA-binding protein of phage Pf3, consisting of 78 amino acids. The

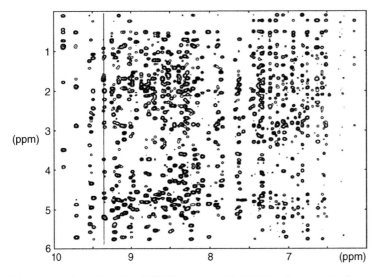

Figure 6.34 A portion of the NOESY spectrum of a DNA-binding protein of phage Pf3 containing cross peaks between NH and aliphatic protons. Source: Reproduced with permission from Ref. [25].

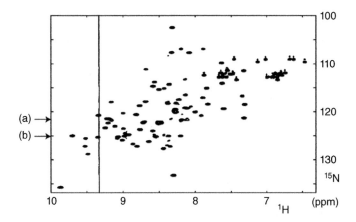

Figure 6.35 The ^{1}H/^{15}N HMQC spectrum of ^{15}N-labeled Pf3. The arrows are explained in Figure 6.38. Source: Reproduced with permission from Ref. [26].

vertical line at δ 9.35 highlights the problems at a single resonance position in the NH region. The NH proton in a given peptide unit —CHR′—CO—NH—CHR—CO could have one cross peak with its own CHR proton and another with the neighboring CHR′ protein, but the NOESY spectrum contains more than a dozen cross peaks at the one frequency of δ 9.35. Thus, more than one NH must be generating cross peaks at that frequency.

The nitrogen HMQC experiment provides information about the connectivity of nitrogens and their attached protons. For proteins, the use of HMQC normally requires isotopic enrichment of ^{15}N, which is obtained by growing an organism in a medium containing a single nitrogen source, such as ^{15}NH$_4$Cl. (Similarly, ^{13}C enrichment may be obtained from a medium containing ^{13}C-labeled glucose.) The normal 2D HMQC spectrum (^{15}N vs ^{1}H) for this same protein is given in Figure 6.35, and two connectivities are seen at the ^{1}H frequency of δ 9.35 (vertical line). Accordingly, there are two NH resonances (or more if there are coincidences) at δ 9.35.

The 3D experiment takes the 2D experiments in Figures 6.34 and 6.35 into an additional dimension. The 3D procedure illustrated in Figure 6.36 labels each NOESY peak with the ^{15}N frequency through the HMQC method, thus combining NOESY and HMQC data. The pulses and time delays constitute the standard ^{1}H NOESY sequence through the third 90° ^{1}H pulse. The pulses and delays thereafter make up the standard HMQC sequence, which ends with inverse detection of ^{15}N at the ^{1}H frequencies in t_3. The totality of data requires a cube for its representation, as illustrated diagrammatically in Figure 6.37, in which the flat dimensions are the NOESY data in ^{1}H frequencies and the vertical axis provides the ^{15}N frequencies from HMQC. In practice, horizontal planes (single ^{15}N frequencies) are selected for analysis, as in Figure 6.38 for δ 120.7 and 124.9 (arrows labeled a and b in Figure 6.35). The vertical lines at δ 9.35 each show two dominant cross peaks for the NH NOE's to the inter- and intraresidue CHR. Note

Figure 6.36 The pulse sequence for the 3D NOESY/HMQC experiment.

^{1}H: $90° - t_1 - 90° - \tau_m - 90° - 180° - t_3$ (Acquire)

^{15}N: $180° - \Delta - 90° - t_2 - 90° - \Delta$ − Decouple

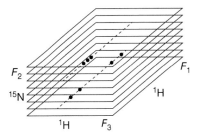

Figure 6.37 Diagram of the 3D NOESY–(^{1}H/^{15}N)HMQC spectrum of Pf3 in three frequency dimensions: F_1, F_2, and F_3 (v_1, v_2, and v_3). Source: Reproduced with permission from Ref. [27].

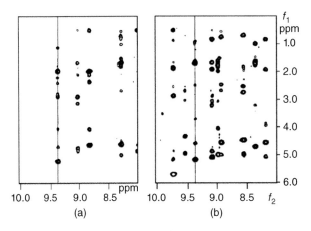

Figure 6.38 Two v_1–v_3 (F_1–F_3) planes taken from the 3D NOESY–(^{1}H/^{15}N)HMQC spectrum of Pf3, corresponding to the frequencies indicated by arrows in Figure 6.35. Source: Reproduced with permission from Ref. [28].

that both ^{15}N frequencies show a cross peak for δ 9.35 at a CHR frequency of δ 5.2 so that the question of overlap in Figure 6.35 is resolved.

This type of heteronuclear 3D experiment is called NOESY–HMQC. (Figure 6.37, with two ^{1}H dimensions and one ^{15}N dimension.) Most 3D experiments use high-sensitivity methods and displays that are particularly effective for large molecules. Thus, COSY is not often used, but TOCSY–HMQC is a useful method for separating ^{1}H–^{1}H coupling connectivities into a ^{13}C or ^{15}N dimension. The homonuclear 3D experiment NOESY–TOCSY, in which all three dimensions are ^{1}H, separates through-space connectivities from coupling connectivities by the pulse sequence $90°-t_1-90°-\tau_m-t_2-$(spin lock)$-t_3$ (acquire). The three dimensions of each may represent a different nuclide, as in ^{1}H/^{13}C/^{15}N, to provide a 3D variant of the HETCOR experiment. The nuclides usually are selected to explore specific connectivities in biomolecules. The H–N–CO experiment looks at the connection ^{1}H–^{15}N—(C)—^{13}C=O in the peptide unit —NH—CHR—CO— and requires double labeling of ^{15}N and ^{13}C to provide sufficient sensitivity in proteins. The 3D cross peaks connect the HN proton in the first dimension, the HN nitrogen in the second, and the intraresidue carbonyl carbon in the third. An analogue in nucleotide analysis is the H–C–P experiment, in which the third dimension is ^{31}P. Numerous variations of these triple resonance experiments exist. In particularly complex cases, a fourth time domain t_4 may be introduced to produce 4D experiments.

When through-bond connectivity experiments are combined with the spatial information from buildup rates of NOESY cross peaks, proton–proton distances can be obtained by comparison with known bond lengths. The result can be a complete three-dimensional (3D) structure of large molecules. Such solution-phase structures complement solid-phase information from X-ray crystallography. In this way, NMR spectroscopy has become a structural tool for obtaining detailed molecular geometries of complex molecules in solution.

6.6 Pulsed Field Gradients

Field inhomogeneity has been mentioned as the primary contributor to transverse (xy) relaxation (T_2) (Sections 1.3 and 5.1). Transverse magnetization arises when the phases of individual magnetic vectors become coherent (Figure 1.11) rather than random (Figure 1.10). In a perfectly homogeneous field, this coherence would relax only through spin–spin interactions. In an inhomogeneous field, however, the existence of slightly unequal Larmor frequencies permits vectors to move faster or more slowly than the average, thereby randomizing their phases and destroying transverse magnetization, as described in Section 1.3.

There are several situations in which transverse magnetization is unwanted and may be eliminated by the application of a PFG, also called a *gradient pulse*. An example of a PFG is presented in Figure 6.39. The gradient is along the direction (z) of the B_0 field. When a PFG is applied, nuclei with different positions in the sample (different z coordinates) resonate at different frequencies. Such spatial encoding of frequency information is the fundamental principle of *Magnetic Resonance Imaging* (MRI). In the present context, PFG's may be viewed as a method for inducing transverse relaxation very quickly by a rapid dephasing of the spins. In a typical 2D pulse sequence, a delay time is necessary between repetitions of the pulse sequence in order for relaxation to occur. If repetition occurs before transverse magnetization has relaxed to zero, sensitivity is reduced and artifacts may occur in the 2D spectrum. Consequently, the application of a PFG at the beginning of the sequence reduces or avoids these problems.

For the NOESY experiment, the following pulse sequence may be used: G1–90°–t_1– 90°–τ_m, G2–90°–t_2 (acquire). The first PFG (G1) destroys residual transverse magnetization from previous pulses by dephasing the magnetic vectors. The second PFG (G2) is applied during the mixing period τ_m. Only the effects on longitudinal (z) magnetization are of interest during this period. The second PFG helps to eliminate false cross peaks that can arise from pre-existing transverse magnetization.

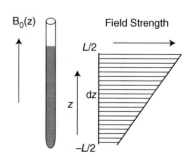

Figure 6.39 Diagram of a B_0 field gradient along the z direction. Source: Reproduced with permission from Ref. [29].

PFGs also may be used to remove unwanted resonances. One of the most successful methods for solvent suppression destroys solvent magnetization by a PFG, but retains all other resonances through the sequence G–S–G–t(acquire), which is reminiscence of solvent suppression through partial relaxation (Section 5.1). In this sequence (called WATERGATE for *WATER suppression by GrAdient-Tailored Excitation*), the S pulse is chosen to invert all resonances except the solvent peak, which is left at zero magnetization, as in partially relaxed spectra (Section 5.1). The two identical gradients mimic a spin-echo process, whereby the dephasing of the first PFG is undone by the second PFG through the process of *rephasing*. Normally, the result is a *gradient echo*. Such an echo occurs for all resonances except solvent. Because the middle pulse eliminates solvent magnetization, the final PFG cannot rephase solvent resonance. All other resonances, however, are rephased by the second gradient pulse.

In addition to dephasing transverse magnetization, PFG's are used to select a coherence order. The use of phase cycling to select a coherence order inevitably involves multiple scans, by which pulse sequences move through 4, 16, or 64 variations with switching, for example, of x, $-x$, y, and $-y$. A full exploitation of phase cycling thus is time consuming. The development of zero, single, or double quantum coherence depends on the rate of various dephasing processes. Proper use of PFG's permits the selection of a coherence order without the repetitive scans of phase cycling. For example, in the inverse-detection HMQC experiment, the single quantum coherence signal for protons attached to ^{12}C (or ^{14}N) must be suppressed while selecting the multiple quantum coherence signal for protons attached to ^{13}C (or ^{15}N). Through phase cycling, this selection is achieved by measuring the difference between two strong signals. The PFG method selects and measures the small difference signal directly in a single scan.

PFG procedures have been developed to implement most of the 1D and 2D experiments discussed in Chapters 5 and 6. In particular, the procedures may be used for INADEQUATE, all common 2D experiments, and spectral editing. A PFG combination of DEPT and HMQC results in editing of proton spectra according to carbon substitution patterns. A PFG-based multiple quantum filtration leads to evolution of double, triple, or quadruple quantum coherence providing proton spectra containing only CH, CH_2, CH_3 resonances, respectively. Broadband decoupling removes coupling to ^{13}C, while proton–proton couplings remain. Figure 6.40 illustrates the result for brucine (**6.3**).

6.3

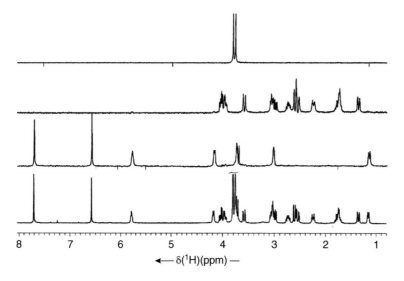

Figure 6.40 Editing of the ^{1}H spectrum of brucine (**6.3**) into subspectra for CH_3, CH_2, and CH (top to bottom). The complete ^{1}H spectrum is given at the bottom. Source: Reproduced with permission from Ref. [30].

PFG's may be used to optimize the NOE experiment. Although the 2D NOESY experiment is useful in analyzing spatial relationships in large molecules such as proteins, the enhancements are weaker for smaller molecules. The 1D NOE experiment thus may be more appropriate for small molecules. The difference experiment described in Section 5.4 was instrumental in lowering the limit for considering a ^{1}H NOE to be significant from about 5% to about 1%. The experiment, however, suffers from problems arising from incomplete subtraction between the irradiated and unirradiated spectra. Such *subtraction artifacts* limit the difference NOE method. The use of pairs of PFG's can yield Overhauser enhancements without difference methods. This procedure has been called *excitation sculpting* and involves a pulse sequence of the general type G1–S–G1–G2–S–G2, in which S represents a selective inversion pulse or sequence of pulses (to produce, for example, the NOE). For the selected spins, the two identical G1 pulses act in opposition and hence refocus the magnetization to produce a gradient echo. Spins outside the selected frequency range absorb the cumulative effect of G1+G1 and are fully dephased, reminiscent of the WATERGATE procedure. Hence, after a single gradient echo, all resonances have been eliminated, except those in the selected range of the pulse S. Because S may not have ideal phase properties, the gradient echo is repeated a second time (G2/G2) with a gradient of different magnitude, to avoid accidental refocusing of unwanted dephased magnetization. The resulting sequence has been called the *Double PFG Spin Echo* (DPFGSE) experiment. In practice, the sequence is preceded by a nonselective 90° pulse. All nuclei move into the *xy* plane, and the gradient pulses then dephase all resonances except those selected by the S pulse.

When the double PFG's are followed by a mixing time τ_m that permits the development of the NOE's that arise due to irradiation by the pulse S, the only resonances that develop are those coming from the NOE. All others have been dephased. The

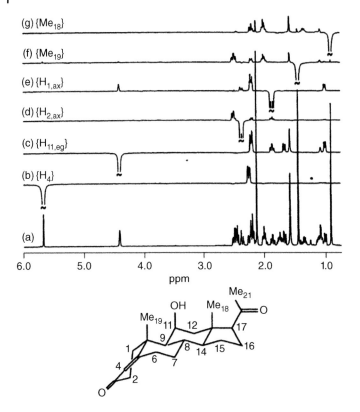

Figure 6.41 The double pulse field gradient spin echo (DPFGSE) NOE experiment for 11β-hydroxyprogesterone: (a) the unirradiated spectrum and (b)–(g) spectra with irradiation at selected frequencies. Source: Reproduced with permission from Ref. [31].

sequence then continues with another PFG to eliminate transverse magnetization and finishes with a 90° pulse for acquisition. The complete pulse sequence is thus 90°–G1–S–G1–G2–S–G2–90°–τ_m–G3–90°–t(acquire), in which all 90° pulses are along the x-axis, S is a selective pulse for the target (irradiated) nucleus, τ_m is set to optimize the NOE's, and G3 eliminates transverse magnetization. Figure 6.41 shows the results of this experiment with 11β-hydroxyprogesterone (compare Figure 5.12). The bottom spectrum (a) contains the unirradiated spectrum, and the ascending spectra contain the DPFGSE NOE experiments on a series of target nuclei. These spectra are taken directly, not by a difference technique. Only Overhauser-enhanced resonances are observed. The technique easily extends the limit of significance for ^{1}H NOE's to 0.1% and has been used to observe enhancements as small as 0.02%.

Excitation sculpting also can be used for solvent suppression in the DPFGSE version of WATERGATE (G1–S–G1–G2–S–G2–t (acquire)), in which the solvent peak is selected for dephasing during S and all other resonances are refocused. Figure 6.42 illustrates the removal of the solvent resonance for 2 mM sucrose in 9 : 1 H_2O/D_2O.

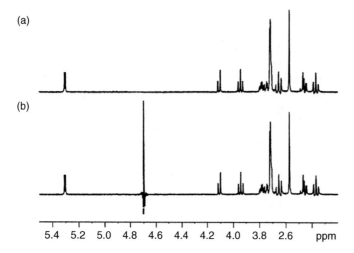

Figure 6.42 (b) Water suppression with excitation sculpting on 2 mM sucrose in 9:1 H_2O/D_2O. (a) The residual solvent peak has been eliminated by further processing. Source: Reproduced with permission from Ref. [32].

6.7 Diffusion-Ordered Spectroscopy

Entirely different from all previously described experiments is a family based on molecular diffusion. Magnetic properties of molecules have a fundamental connection with their motional properties. For example, most relaxation mechanisms arise from rotational motion that generates fluctuating magnetic fields (Section 5.1). In an inhomogeneous field such as provided by gradients (Figure 6.39), translational motion can move a molecule from one value of the magnetic field to another one. Consequently, the diffusional coefficient of a molecule determines in part how a molecule responds to magnetic field gradients. Molecular diffusion depends on many properties of the environment, such as temperature and viscosity, which normally are constant within a liquid. Thus, differences in diffusional properties between molecules are primarily a function of molecular size and shape. Multiple components in solution can diffuse at different rates and respond to the gradient in different fashions. Methods have been developed to exploit these differences in order to use NMR to separate components according to their diffusion properties. Such experiments as a group have been termed *Diffusion Ordered SpectroscopY* or DOSY.

The intensity I_A of the signal for component A in such a mixture can be expressed as Eq. (6.3),

$$I_A = I_A(0) \exp(-D_A Z) \tag{6.3}$$

in which $I_A(0)$ is the intensity of the signal at zero gradient (the normal 1D signal), I_A is the intensity of the signal in the presence of a gradient of formula Z, and D_A is the diffusion coefficient of component A. The intensities of other components of the mixture differ only as the result of their diffusion coefficients D_i, as all components experience the same gradient Z. There are many different DOSY experiments, and the art of the experiment is the nature of the function Z. A series of experiments is recorded with variation of the amplitude of the gradient: the stronger the gradient, the smaller the signal intensity. After Fourier transformation of the FID, the experiment generates a series of peak intensities as a function of the gradient Z. A 2D array is created similar to that in Figure 6.3, with two important differences. First, although the horizontal axis still is the normal 1D frequency dimension, the vertical axis is the gradient amplitude. Second, instead of generating a sinusoidal array of intensities along the y-axis, the intensities decrease exponentially. This unrecorded intermediate stage analogous to Figure 6.3 would have a series of peaks, one series for each component along the x-axis, whose intensities decay from the bottom of the spectrum (zero gradient) to a diminishingly small value at the top (large gradient).

Because these simple exponential decays are not sinusoidal, the final step is not a second Fourier transformation but rather a process that has been termed a *DOSY transformation*, whereby the data are transformed into a diffusion domain instead of a frequency domain. There are numerous mathematical procedures for carrying out the DOSY transformation, including exponential fits, maximum entropy, and multivariate analysis. The result resembles a traditional 2D display in which the horizontal axis is the normal ^{1}H frequency display and the vertical axis is the diffusion constant.

Figure 6.43 illustrates the result for the DOSY experiment of a three-component system that contains the solvent tetrahydrofuran (THF), dioctyl phthalate (DOP), and the polymer polyvinyl chloride (PVC). The normal 1D spectrum is recorded at the top. The 2D spectrum contains cross peaks located according to frequency along the horizontal

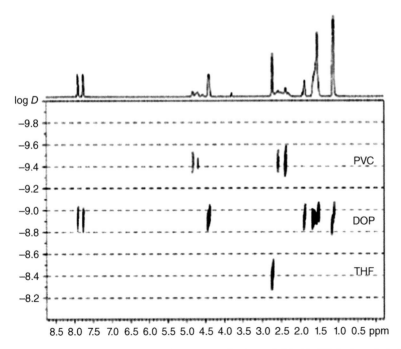

Figure 6.43 The DOSY spectrum of a mixture of polyvinyl chloride (PVC), dioctyl phthalate (DOP), and tetrahydrofuran (THF). Source: Reproduced with permission from Ref. [33].

axis and diffusion along the vertical axis. Each component diffuses at a different rate and consequently provides a series of cross peaks at different levels on the diffusion axis. More rapidly diffusing components are found lower on the plot. It is clear from the vertical positions that THF diffuses most rapidly and PVC most slowly, as expected for their molecular sizes. The cross peaks in the middle are identified as those from DOP by the presence of aromatic resonances. Each horizontal cut may be projected onto the vertical axis to create the equivalent of a chromatogram, in which each component is represented by a single peak on the gradient axis according to its diffusion coefficient, with an intensity appropriate to its molar representation in the mixture. Ideally, each distinct component of a mixture would generate a separate spectrum on the vertical axis. In essence, DOSY results in a chromatographic-like separation, with the important differences that there is no physical separation of components and no sample preparation beyond normal NMR sample preparation. The ability for NMR to separate components according to molecular size has led to a host of new applications.

6.8   Summary of 2D Methods

A bewildering array of 2D methods is available to the NMR spectroscopist today. This chapter has described a number of the most widely used such experiments. A routine structural assignment begins with recording the 1D ^{1}H and ^{13}C spectra. Many resonances may be assigned according to the principles outlined in Chapters 3 and 4 on chemical shifts and coupling constants. Normally, recourse is made to 2D methods only if this traditional approach is insufficient. Some type of spectral editing for determining the number of protons attached to each carbon, such as DEPT, is helpful in completing the ^{13}C assignments (Section 5.5). The HETCOR or HSQC experiment then provides correlations between the ^{13}C and ^{1}H resonances, and the PFG-based multiple quantum filtration (Figure 6.40) may help fill out the ^{1}H assignments.

Further 2D methods are necessary if the structure is not deduced in the process of making spectral assignments for hypothetical or expected structures. The COSY experiment lays the groundwork for building structures through ^{1}H–^{1}H connectivities based on J couplings. For small molecules, there may not be enough vicinal or geminal couplings for the method to be useful. For molecules of medium complexity, COSY may be sufficient to provide the entire structure by confirming expected ^{1}H–^{1}H connectivities based on vicinal and geminal couplings. The analogous experiment based on long-range couplings (LRCOSY) may be necessary to assign connections between molecular pieces that do not involve vicinal protons, for example, between two rings, for substituents on a ring, or over a heteroatom or carbonyl group.

Additional 2D experiments may be necessary for larger molecules. As peaks accumulate along the diagonal, the COSY45 or DQF-COSY experiment may be used to simplify that region and uncover cross peaks that are close to the diagonal. For even larger molecules, the TOCSY or relayed COSY experiments may be necessary. Further connectivities between protons and carbons may be explored through longer range couplings (HMBC) in order to define structural regions around quaternary carbons. If ^{1}H–^{1}H couplings need to be measured, either the J-resolved method or DQF-COSY may be carried out.

The NOESY experiment provides information about the proximity of protons and hence is used primarily for distinguishing structures that have clear stereochemical differences. For larger molecules, the ROESY experiment may offer some advantages because of its lower tendency to exhibit spin diffusion. The related EXSY experiment is used only when chemical exchange is being investigated, to supplement information from one dimension.

The 2D INADEQUATE or INEPT–INADEQUATE experiment requires additional spectrometer time. It is usually an experiment of last resort, although specific structures may be particularly amenable to this technique, for example, when there are several quaternary carbons that prevent COSY analysis.

Kupĉe and Freeman have developed an experiment to implement many of these methods with one super sequence, which they call PANACEA, for *Parallel Acquisition Nuclear magnetic resonance All-in-one Combination of Experimental Application*. This single experiment generates the 1D ^{13}C, 2D INADEQUATE, HSQC, and HMBC spectra, in a run of about nine hours on 10 mg of material of several hundred daltons molecular weight. Their improved version, called fast-PANACEA, ameliorates some time constraints.

Problems

For a large selection of relatively straightforward 2D spectra, see Problem 6.1. The remaining problems involve molecules of medium-to-high complexity, although none is so complex as to require 3D methods.

6.1 The spectra in this problem provide relatively routine practice. For each unknown, the 300 MHz ^{1}H, 75 MHz ^{13}C, COSY, HETCOR, and some DEPT spectra are provided. In the DEPT spectra, methine and methyl carbons give positive peaks and methylene carbons negative peaks at the top. Only methine carbons are in the middle, and a full spectrum of all carbons is at the bottom.

(a) $C_{11}H_{16}$

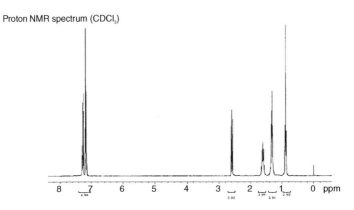

DEPT spectrum

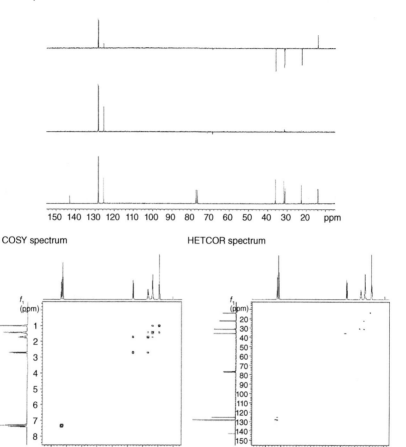

(b) C_8H_{14}

Proton NMR spectrum (CDCl$_3$)

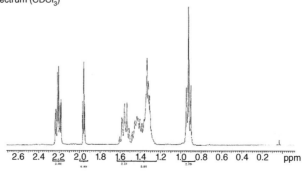

Carbon-13 NMR spectrum

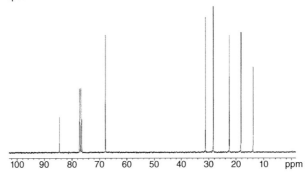

COSY spectrum

HETCOR spectrum

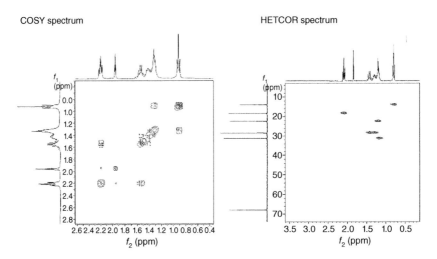

(c) C_5H_9Cl

Proton NMR spectrum (CDCl$_3$)

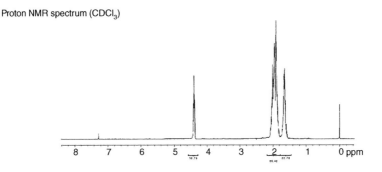

Two-Dimensional NMR Spectroscopy | 283

Carbon-13 NMR spectrum

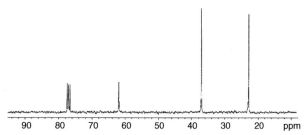

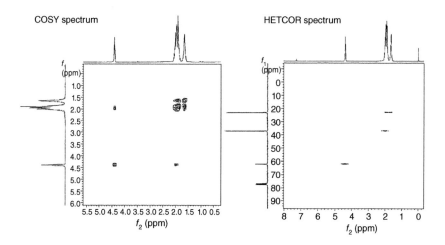

COSY spectrum

HETCOR spectrum

(d) $C_7H_8O_2$

Proton NMR spectrum $(CDCl_3)$

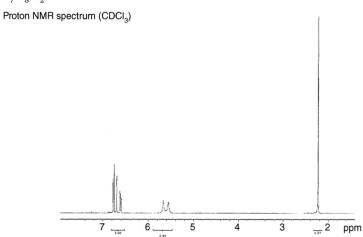

DEPT spectrum

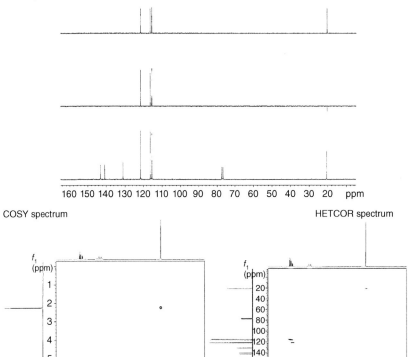

COSY spectrum

HETCOR spectrum

(e) $C_6H_{10}O$

Proton NMR spectrum (CDCl$_3$)

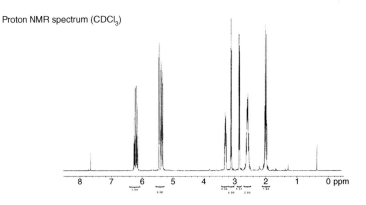

Two-Dimensional NMR Spectroscopy | 285

DEPT spectrum

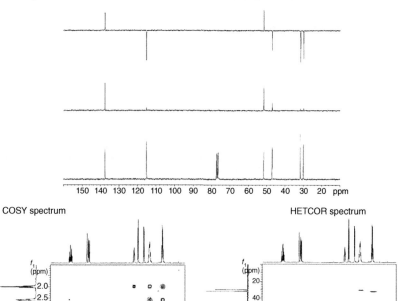

COSY spectrum

HETCOR spectrum

(f) $C_7H_8O_2$

Proton NMR spectrum (CDCl$_3$)

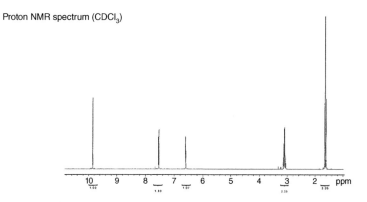

Nuclear Magnetic Resonance Spectroscopy

Carbon-13 NMR spectrum

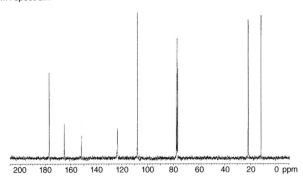

COSY spectrum

HETCOR spectrum

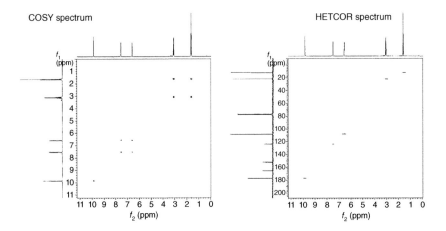

(g) C_6H_7NO

Proton NMR spectrum ($CDCl_3$)

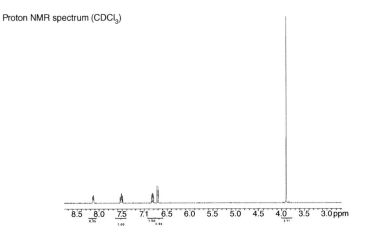

Two-Dimensional NMR Spectroscopy

Carbon-13 NMR spectrum

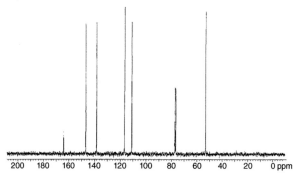

COSY spectrum

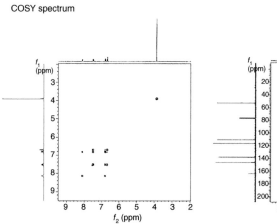

HETCOR spectrum

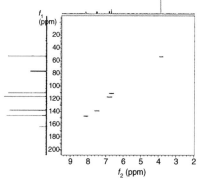

(h) $C_{10}H_{16}O$

Proton NMR spectrum (CDCl$_3$)

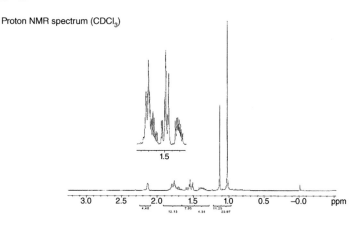

288 | Nuclear Magnetic Resonance Spectroscopy

DEPT spectrum

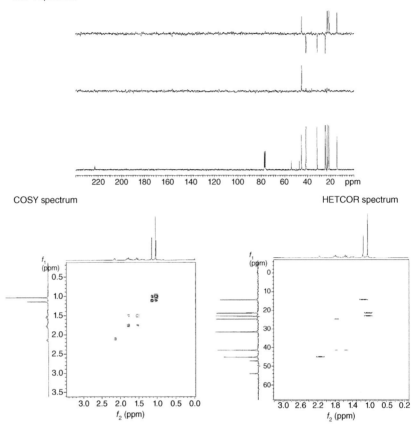

COSY spectrum

HETCOR spectrum

(i) $C_{10}H_{16}O$

Proton NMR spectrum ($CDCl_3$)

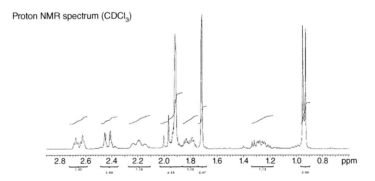

Two-Dimensional NMR Spectroscopy

DEPT spectrum

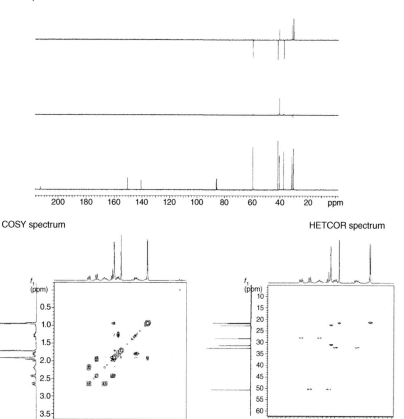

COSY spectrum

HETCOR spectrum

(j) $C_{15}H_{18}$

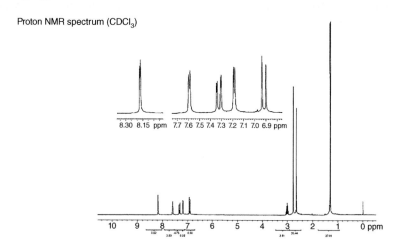

Proton NMR spectrum (CDCl$_3$)

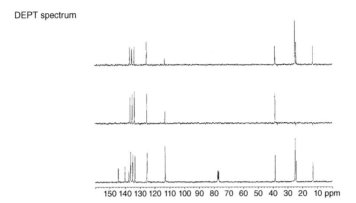

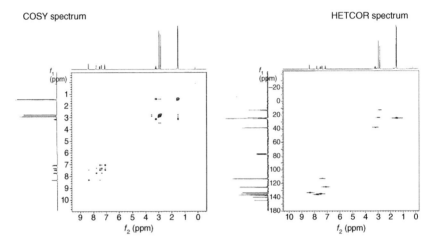

6.2 The 300 MHz COSY spectrum below is of a molecule with the formula $C_{14}H_{20}O_2$. The 1D spectrum is given on either edge. In addition to the illustrated resonances, the ^1H spectrum contains a broad singlet at δ 7.3 with integral 5. What is the structure? Show your reasoning.

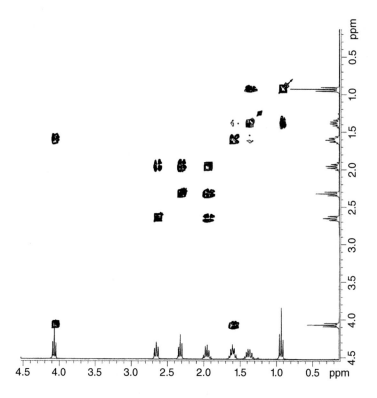

Source: Branz et al. 1995 [1]. Copyright 1995. Reproduced with permission from the American Chemical Society.

6.3 Below are the 1D ^{1}H and 2D COSY spectra of quinoline, whose structure is shown.

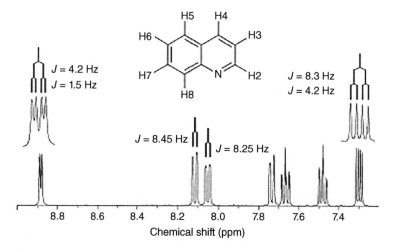

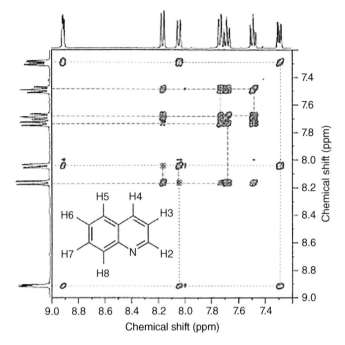

(a) Assign what resonances you can from the 1D spectrum. Explain each assignment in terms of chemical shifts and coupling constants.
(b) Assign all remaining protons from the COSY spectrum, and explain the 1D splitting patterns of these protons.

Source: Reproduced with permission from Ref. [2].

6.4 Trimerization of indole-5-carboxylic acid gives one of the following two isomers:

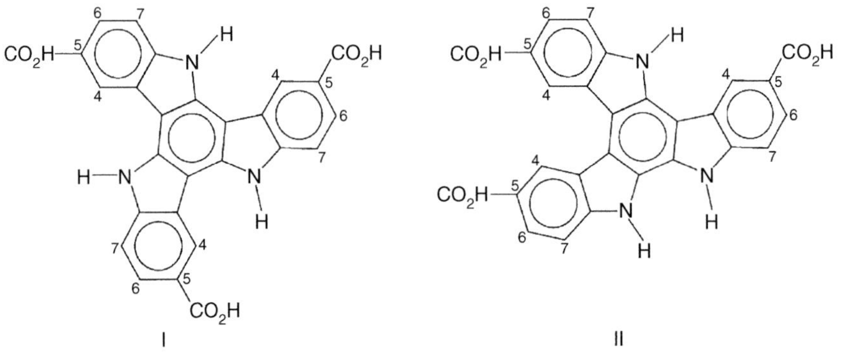

Shown below are the 360.1 MHz ^{1}H spectrum of the product and the COSY spectrum (with a blowup of the δ 7.8–8.2 region) in DMSO-d_6. The signal marked with an asterisk is an impurity. The ^{1}H signals are labeled A–N. Signals A–D were removed with the addition of D_2O. Signal A is a broad peak at the base of B. In the 1D NOE experiment, irradiation of B affected F/G and N, irradiation of C and D affected M and L, respectively, and irradiation of F/G affected B.

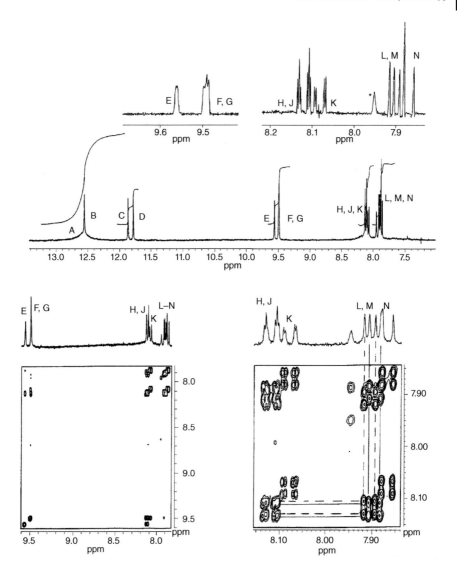

(a) From the overall appearance of the spectrum, is the trimer I or II? Explain.
(b) Using peak multiplicities, the NOE experiments, and the COSY spectrum, assign all the resonances. Discuss your reasoning in a step-by-step fashion. You should end up with an assignment of peaks A through N to specific protons.

Source: Mackintosh et al. 1994 [3]. Reproduced with permission from John Wiley & Sons.

6.5 The following 500 MHz ^{1}H spectrum is of the illustrated sugar derivative (extracted from a bean). Hydroxy resonances are omitted.

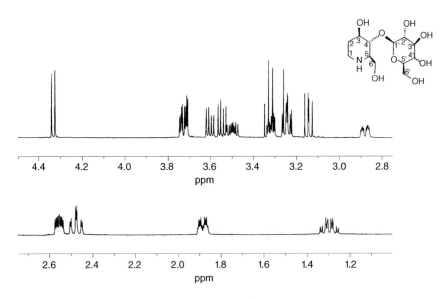

(a) The complete COSY spectrum, including the assignment of one resonance, is as follows. Complete the assignment for protons in the sugar ring.

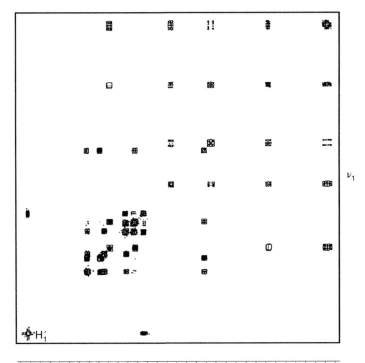

(b) The following is a slightly expanded version of the low-frequency portion of the COSY spectrum, again with the assignment of one resonance. Complete the assignment for the protons of the piperidine ring. First, it is advisable to draw out the chair conformation.

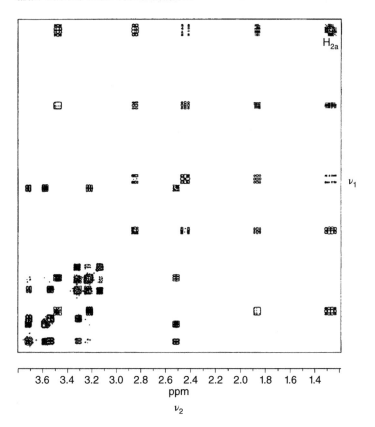

Source: Reproduced with permission from Ref. [4].

6.6 With only the following 2D INADEQUATE spectrum, derive the structure for the skeleton of isomontanolide. There are two overlapping resonances at δ 78, so you must work around ambiguities at that chemical shift. Also, there are substituents on the carbons in the range δ 60–80 that are not defined by the experiment.

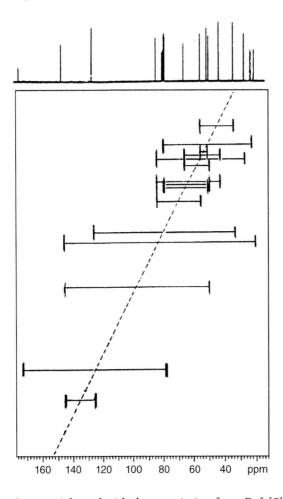

Source: Adapted with the permission from Ref. [5]

6.7 The molecule with structure **A** gave the following DQF-COSY (sugar portion only) in DMSO-d_6.

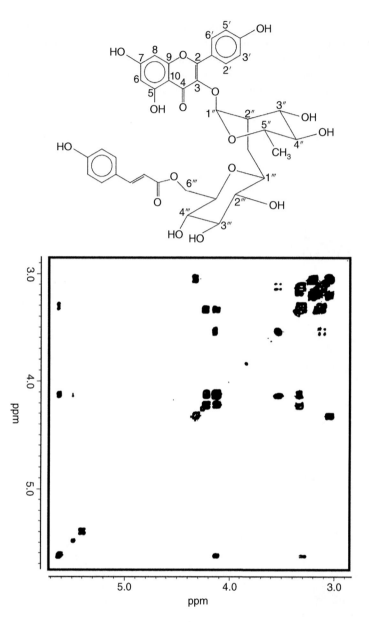

(a) Assign the sugar protons as fully as possible, and explain their positions.
(b) Assign the ¹H resonance at δ 12.56, and explain its high-frequency position.
(c) The ¹H spectrum exhibited vicinal couplings of $^3J = 9.4$ Hz for H—O—C(3″)—H and of $^3J = 3.1$ Hz for H—O—C(2‴)—H. Explain in terms of medium effects and conformations.

6.8 The isomeric 4,5-cyclopropanocholestan-3-ols **A** and **B** are expected to be inhibitors of cholesterol oxidase. (Protons pointing up are labeled β and protons pointing down are α.) The undepicted remainder of the structure is the cholestane skeleton and is not relevant to this problem. The DQF-COSY and NOESY spectra for both isomers are given below.

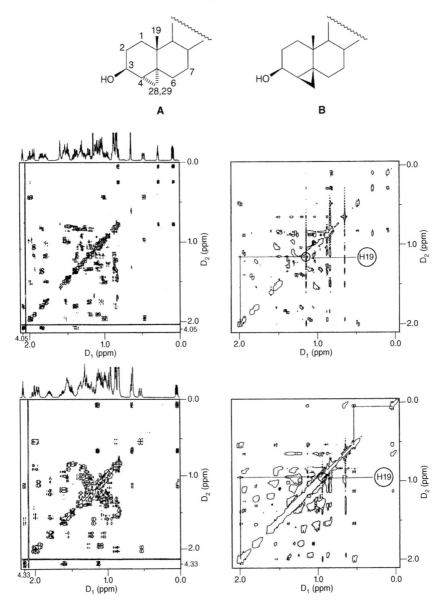

(a) Starting with H(3α) in the DQF-COSY spectrum of one isomer, assign the cyclopropane protons (H4, H28, H29). [This approach works only for one isomer; see part (b).] Note the discontinuity in the DQF-COSY axes between δ 2 and 4. Assign cross peaks for H4/H28, H4/H29, and H28/H29.
(b) The approach fails for the other isomer. Why? By analogy with your assignments in (a), assign the cyclopropane resonances and cross peaks anyway.
(c) In both isomers, there is a third low-frequency peak in addition to the cyclopropane resonances (δ 0.5). It has a large $J = 13.5$ Hz, with a partner at about δ 2 in both isomers. Identify the DQF-COSY cross peak between the $J = 13.5$ Hz partners in both isomers. Assign these resonances and explain the low-frequency position for the one partner.
(d) The NOESY spectra for both isomers reveal that H19 (the methyl group) is close to the δ 2 proton, from (c). The proton to which the δ 2 proton is coupled ($J = 13.5$ Hz) shows a NOESY cross peak with one of the other low-frequency cyclopropane protons in only one isomer. These cross peaks are indicated on the spectra. With your peak assignments and these NOESY data, assign the isomers to the spectra. The expected conformations are given below. Show the NOESY relationships.

(1a)   (1b)

6.9 (a) The ^{1}H and ^{13}C spectra of *trans*-10-chlorodecal-2-one (**A**) are given in the following HETCOR spectrum. The conformation is rigid, so each geminal proton pair exhibits distinct axial and equatorial resonances. Identify the C6, C7, and C8 resonances *as a group rather than individually*, and explain.

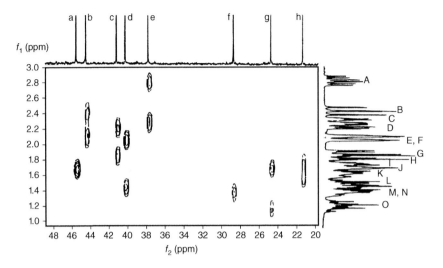

(b) Now assign H9 and C9. Explain your logic.
(c) What ¹H signals are expected to occur at the highest frequency (lowest field)? Explain.
(d) From the following COSY spectrum and the previous HETCOR spectrum, assign all the protons and carbons in the A ring. *Caution!* In this spectrum, the normal axial–equatorial relationship is reversed, as H1a and H3a, respectively, occur at a *higher* frequency than H1e and H3e.

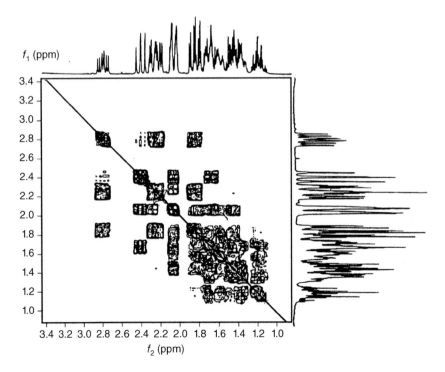

6.10 (a) The controlled substance methaqualone has the formula $C_{16}H_{14}N_2O$ (MW 250). A closely related analogue with MW 264 started to appear as a replacement in the illegal drug market. From their infrared spectra, both molecules contain a carbonyl group (C=O, at 1705 cm^{-1}). What structural fragments can you deduce from the 300 MHz ^{1}H spectra of methaqualone and its analogue given below?

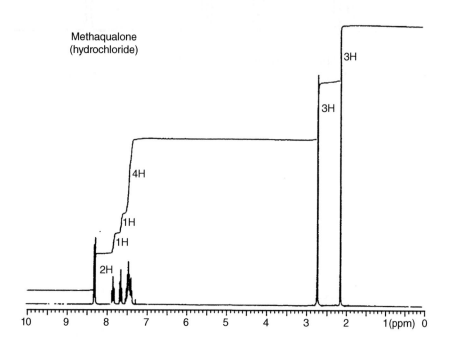

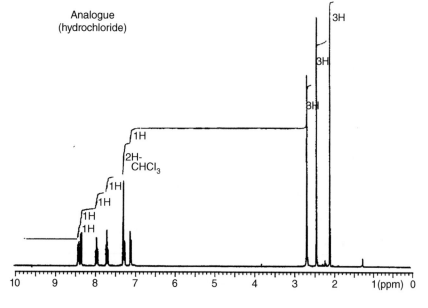

(b) From the following expansion of the high-frequency portion of the spectrum of the analogue, deduce the substructures responsible for these resonances.

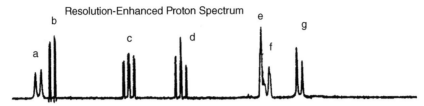

Resolution-Enhanced Proton Spectrum

(c) Both the parent and the analogue contain a pyrimidine ring (**A**) that is unsaturated and substituted at all positions. Now assemble the entire analogue molecule.

(d) Assign all the cross peaks in the following COSY spectrum for the analogue molecule.

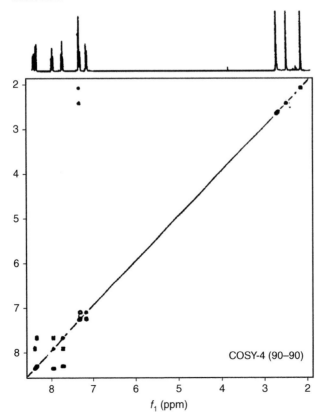

(e) From the following NOESY spectrum, complete any unresolved aspects of the structure of the analogue molecule.

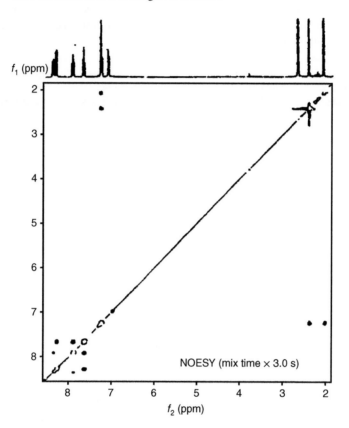

(f) From the structure of the analogue molecule and earlier spectral data, deduce the structure of methaqualone.

6.11 Upjohn scientists isolated a potent inhibitor of the cholesteryl ester transfer protein U-106305. The high-resolution mass spectrum indicated that the formula was $C_{28}H_{41}NO$, and the ^{13}C spectrum had 27 distinct resonances (including one pair of equivalent carbons at δ 20.02). DEPT spectra indicated the multiplicities given in the following table.

^{13}C chemical shift and multiplicities	^1H chemical shift and coupling constants	^{13}C chemical shift and multiplicities	^1H chemical shift and coupling constants
7.6 (T)	0.07, 0.09 (dt: 8.43, 4.85)	20.0 (D)	1.00 (m)
7.6 (T)	0.12, 0.16 (dt: 8.39, 4.90)	20.02 (Q)	0.90 (d: 6.8)
8.0 (T)	0.08 (not first order)	20.7 (D)	1.29 (m)
11.4 (T)	0.32, 0.34 (dt: 8.20, 4.77)	21.8 (D)	0.68 (m)
13.4 (T)	0.65 (dt: 8.59, 4.87)	22.4 (D)	0.94 (m)
14.8 (T)	0.34, 0.43 (dt: 8.33, 4.60)	24.0 (D)	1.01 (m)
14.8 (D)	0.63 (m)	28.5 (D)	1.77 (h: 6.8)
17.9 (D)	0.57 (dq: 13.27, 4.93)	46.7 (T)	3.20 (d: 6.8)
18.0 (D)	0.58 (m)	120.0 (D)	5.91 (d: 15.2)
18.2 (D)	0.49 (m)	130.4 (D)	4.98 (dd: 15.5, 7.2)
18.2 (D)	0.51 (m)	131.0 (D)	4.98 (dd: 15.5, 7.7)
18.4 (D)	0.53 (m)	148.8 (D)	6.24 (dd: 15.2, 9.8)
18.41 (Q)	1.02 (d: 6.0)	166.0 (S)	
18.8 (D)	0.60 (m)		

(a) The infrared spectrum showed intense bands at 1630 and 1558 cm^{-1}. What functional group is suggested? What NMR peak confirms this assignment?
(b) What substructures are suggested by the ^{13}C peaks at δ 120–150?
(c) The HETCOR spectrum gave full ^1H assignments (preceding table). From the ^1H resonances correlated with the δ 120–150 ^{13}C peaks, what else can you tell about the substructures from (b)? Use the magnitudes of J (values in parentheses), the values of the four ^1H and ^{13}C chemical shifts, and the proton multiplicities. In particular, note that δ 120 is d rather than dd.
(d) The UV–vis spectrum showed a strong band at 215 nm. From the functional groups you have already deduced, what is the chromophore?
(e) Examine the six low-frequency ^{13}C triplets. Each is correlated with a very low-frequency (high field) pair of protons (δ < 0.7). What grouping is suggested here that is present six times?
(f) Now count up your unsaturations. You should have accounted for them all. Enumerate them.
(g) The DQF-COSY spectrum was given by the authors for one substructure as follows: the integral-6 ^1H resonance at δ 0.90 (d: 6.8 Hz) was linked to the integral-1 resonance at δ 1.77 (heptet: 6.8 Hz), which was linked to the integral-2 resonance at δ 3.20 (d: 6.8 Hz). What substructure is suggested by this 2D evidence?

(h) How is this substructure linked to a previously determined functionality? Look at the chemical shifts.

(i) You now have almost all the structure. The remaining unassigned ^{13}C resonances are 12 doublets and 1 quartet. Locate these carbons (without specific assignment) on your previous substructures, and comment on the chemical shifts of the attached protons.

(j) These protons are all found in the DQF-COSY spectrum, except for the δ 1.29 resonance. Even at 600 MHz, there is severe overlap, so here are the connectivities derived from this experiment. Write out the entire structure, without some stereochemistries.

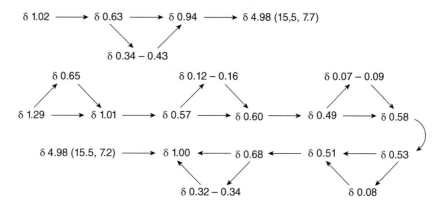

(k) Look at the available multiplicities and J values for the protons associated with the six high-frequency triplets. All are dt with J either ~8.4 or ~4.8 Hz. Give the full structure with all stereochemistries.

References

6.1 Branz, S.E., Miele, R.G., Okuda, R.K., and Straus, D.A. (1995). *J. Chem. Educ.* 72: 659–661.

6.2 Seaton, P.J., Williamson, R.T., Mitra, A., and Assarpour, A. (2002). *J. Chem. Educ.* 79: 107.

6.3 Mackintosh, J.G., Mount, A.R., and Reed, D. (1994). *Magn. Reson. Chem.* 32: 559–560.

6.4 Derome, A.E. (1987). *Modern NMR Techniques for Chemistry Research*, 257. Oxford, UK: Pergamon Press.

6.5 Budesinsky, M. and Saman, D. (1995). *Ann. Rev. NMR Spectrosc.* 30: 231–475.

6.6 Derome, A.E. (1987). *Modern NMR Techniques for Chemistry Research*, 189. Oxford, UK: Pergamon Press.

6.7 Derome, A.E. (1987). *Modern NMR Techniques for Chemistry Research*, 191. Oxford, UK: Pergamon Press.

6.8 Benn, R. and Günther, H. (1983). *Angew. Chem., Int. Ed. Engl.* 22: 350.

6.9 Derome, A.E. (1987). *Modern NMR Techniques for Chemistry Research*, 228. Oxford, UK: Pergamon Press.

6.10 Günther, H. (1995). *NMR Spectroscopy*, 2e, 300. Chichester, UK: Wiley.
6.11 Derome, A.E. (1987). *Modern NMR Techniques for Chemistry Research*, 207. Oxford, UK: Pergamon Press.
6.12 van de Ven, F.J.M. (1995). *Multidimensional NMR in Liquids*, 171. New York: Wiley-VCH.
6.13 Evans, J.N.S. (1995). *Biomolecular NMR Spectroscopy*, 428. Oxford, UK: Oxford University Press.
6.14 van de Ven, F.J.M. (1995). *Multidimensional NMR in Liquids*, 233. New York: Wiley-VCH.
6.15 Hall, L.D., Sukumar, S., and Sullivan, G.R. (1979). *J. Chem. Soc. Chem. Commun.* 292.
6.16 Duddeck, H. and Dietrich, W. (1989). *Structure Elucidation by Modern NMR*, 22. Darmstadt, Germany: Steinkopff Verlag.
6.17 Sanders, J.K.M. and Hunter, B.K. (1993). *Modern NMR Spectroscopy*, 2e, 111. Oxford, UK: Oxford University Press.
6.18 Duddeck, H. and Dietrich, W. (1989). *Structure Elucidation by Modern NMR*, 24. Darmstadt, Germany: Steinkopff Verlag.
6.19 Claridge, T.D.W. (1999). *High-Resolution NMR Techniques in Organic Chemistry*, 245. Oxford, UK: Pergamon Press.
6.20 Derome, A.E. (1987). *Modern NMR Techniques for Chemistry Research*, 257. Oxford, UK: Pergamon Press.
6.21 van de Ven, F.J.M. (1995). *Multidimensional NMR in Liquids*, 188. New York: Wiley-VCH.
6.22 van de Ven, F.J.M. (1995). *Multidimensional NMR in Liquids*, 251. New York: Wiley-VCH.
6.23 Meier, R.H. and Ernst, R.R. (1979). *J. Am. Chem. Soc.* 101: 6441.
6.24 Martin, G.E. and Zehtzer, A.S. (1988). *Two-Dimensional NMR Methods for Establishing Molecular Connectivities*, 362. New York: Wiley-VCH.
6.25 van de Ven, F.J.M. (1995). *Multidimensional NMR in Liquids*, 296. New York: Wiley-VCH.
6.26 van de Ven, F.J.M. (1995). *Multidimensional NMR in Liquids*, 297. New York: Wiley-VCH.
6.27 van de Ven, F.J.M. (1995). *Multidimensional NMR in Liquids*, 299. New York: Wiley-VCH.
6.28 van de Ven, F.J.M. (1995). *Multidimensional NMR in Liquids*, 300. New York: Wiley-VCH.
6.29 van de Ven, F.J.M. (1995). *Multidimensional NMR in Liquids*, 212. New York: Wiley-VCH.
6.30 Parella, T., Sánchez-Ferrando, F., and Virgili, A. (1995). *J. Magn. Reson. A* 117: 80.
6.31 Stott, K., Keeler, J., Van, Q.N., and Shaka, A.J. (1997). *J. Magn. Reson. A* 125: 302–324.
6.32 Claridge, T.D.W. (1999). *High-Resolution NMR Techniques in Organic Chemistry*, 365. Amsterdam: Pergamon Press.
6.33 Ahn, S., Kim, E.-H., and Lee, C. (2005). *Bull Korean Chem. Soc.* 26: 332.

Further Reading

See also general texts referenced in previous chapters

General

Atta-ur-Rahman, Choudhary, M.I., and Atia-tul-Wahab (1996). *Solving Problems with NMR Spectroscopy*. San Diego: Academic Press.

Bax, A. (1982). *Two-Dimensional Nuclear Magnetic Resonance in Liquids*. Boston: D. Reidel Publishing.

Croasmun, R.R. and Carlson, R.M.K. (ed.) (1994). *Two-Dimensional NMR Spectroscopy*, 2e. New York: Wiley-VCH.

Ernst, R.R., Bodenhausen, G., and Wokaun, A. (1987). *Principles of Nuclear Magnetic Resonance in One and Two Dimensions*. Oxford, UK: Oxford University Press.

Evans, J.N.S. (1995). *Biomolecular NMR Spectroscopy*. Oxford, UK: Oxford University Press.

H. Friebolin, *Basic One- and Two-Dimensional NMR Spectroscopy*, 4th ed. Weinheim, Germany: Wiley-VCH, 2005.

Morris, G.A. and Emsley, J.W. (ed.) (2010). *Multidimensional NMR Methods for the Solution State*. New York: Wiley.

Nakanishi, K. (1990). *One-Dimensional and Two-Dimensional NMR Spectra by Modern Pulse Techniques*. Mill Valley, CA: University Science Books.

Schraml, J. and Bellama, J.M. (1988). *Two-Dimensional NMR Spectroscopy*. New York: Wiley-Interscience.

Simpson, J.H. (2008). *Organic Structure Determination Using 2-D NMR Spectroscopy*. San Diego, CA: Academic Press.

van de Ven, F.J.M. (1995). *Multidimensional NMR in Liquids*. New York: Wiley-VCH.

COSY

Kumar, A. (1988). *Bull. Magn. Reson.* 10: 96–118.

Noda, I. (2006). *J. Mol. Struct.* 799: 2.

Long-range HETCOR

Martin, G.E. and Zektzer, A.S. (1990). *Magn. Reson. Chem.* 26: 631–652.

HMBC

Furrer, J. (2011). *Annu. Rep. NMR Spectrosc.* 74: 293.

Martin, G.E. (2002). *Annu. Rep. NMR Spectrosc.* 46: 36–100.

Schoefberger, W., Schlagnitweit, J., and Müller, N. (2011). *Annu. Rep. NMR Spectrosc.* 72: 1.

HOESY

Yemloul, M., Bouguet-Bonnet, S., Aïcha Ba, L. et al. (2008). *Magn. Reson. Chem.* 46: 939–942.

EXSY

Orrell, K.G., Ŝik, V., and Stephenson, D. (1990). *Progr. NMR Spectrosc.* 22: 141.
Perrin, C.L. and Dwyer, T.J. (1990). *Chem. Rev.* 90: 935–967.
Willem, R. (1987). *Progr. Nucl. Magn. Reson. Spectrosc.* 20: 1.

2D INADEQUATE

Buddrus, J. and Lambert, J. (2002). *Magn. Reson. Chem.* 40: 3–23.
Uhrin, D. and Rep, A. (2010). *NMR Spectrosc.* 70: 1.

Solvent Suppression

McKay, R.T. (2009). *Annu. Rep. NMR Spectrosc.* 66: 33.
Zheng, G. and Price, W.S. (2010). *Progr. Nucl. Magn. Reson. Spectrosc.* 56: 267.

Multiple Quantum Methods

Norwood, T.J. (1992). *Progr. NMR Spectrosc.* 24: 295–375.

Pulsed Field Gradients

Parella, T. (1996). *Magn. Reson. Chem.* 34: 329–347; 36, 467 (1998).
Price, W.S. (1996). *Ann. Rev. NMR Spectrosc.* 32: 51–142.

DOSY

Cohen, Y., Avram, L., and Frish, L. (ed.) (2005). *Angew. Chem. Int. Ed.* 44: 520–554.
Johnson, C.S. (1999). *Progr. NMR Spectrosc.* 34: 203–256.
M. Nilsson, ed., *Magn. Reson. Chem.*, 55 (5; Special Issue) (2017).

PANACEA

Kupĉe, E. and Freeman, R. (2008). *J. Am. Chem. Soc.* 130: 10788–10792.
Kupĉe, E. and Freeman, R. (2010). *J. Mol. Reson.* 206: 147–153.

7
Advanced Experimental Methods

7.1 Part A: One-Dimensional Techniques

The introductory experimental methods for ^{1}H and ^{13}C nuclear magnetic resonance (NMR) are presented in Chapter 2. These experiments provide information on the number of protons (Section 2.6.3) and general types of protons and carbons (Chapter 3). In addition, the analysis of ^{1}H—^{1}H coupling constants can reveal much regarding a particular proton's neighbors (Chapter 4). In the case of relatively simple organic molecules, this information, together with mass spectral and infrared spectroscopic data, can be sufficient to elucidate their structures. In most cases, however, considerably more information is required. In this chapter we will become acquainted with the performance of additional one-dimensional experiments, from which we can determine (i) the number of each type of carbon, e.g. methyl, methylene, etc. and (ii) the spatial relationships that exist between protons via the nuclear Overhauser effect (NOE) (Section 5.4). We also will see how the effective T_1 of protons is measured.

7.1.1 T_1 Measurements

In virtually every NMR experiment, the nucleus of choice is pulsed more than once. Such multipulse experiments require either knowledge, or at least a guess, of the spin–lattice relaxation times (T_1) of the nuclei under investigation (Section 2.4.6). Note that for economy of space we use the term T_1 throughout the text when we really are referring to the *effective* T_1 because, among other things, the relaxation effects of paramagnetic substances, such as molecular oxygen, are not rigorously excluded by, for example, sample degassing.

Moreover, if a particular compound is going to be more than routinely examined, it is worth investing the time to measure its ^{1}H T_1s. Knowledge of the ^{13}C T_1s also is desirable, but the time required to determine these values usually is deemed to be too long for the information gained. This situation is especially true today, since relaxation delays of the most important (2D) NMR experiments are dependent on ^{1}H, rather than ^{13}C, T_1s.

The most commonly used experiment for the measurement of T_1s is the inversion-recovery-Fourier transformation (IR-FT) method (Section 5.1): (DT–180° –τ–90°–t_a)$_n$ in which $t_a + DT > 5T_1$. In a situation similar to 90° t_p determinations, it is important that z magnetization recover fully between pulses. Practically speaking this condition means that $t_a + DT$ must be greater than $5T_1$. Equation (5.2) can be used

Nuclear Magnetic Resonance Spectroscopy: An Introduction to Principles, Applications, and Experimental Methods,
Second Edition. Joseph B. Lambert, Eugene P. Mazzola, and Clark D. Ridge.
© 2019 John Wiley & Sons Ltd. Published 2019 by John Wiley & Sons Ltd.

Table 7.1 Recovered magnetization for various values of T_1.

nT_1	I_t/I_0
1	0.264
2	0.729
3	0.900
4	0.963
5	0.987
6	0.995

to determine the amount of recovered z magnetization (I_t/I_0) for various multiples of T_1, and these values are given in Table 7.1.

Owing to the uncertainty of most ^{1}H T_1 values, the first run through the IR-FT experiment is just an approximation (like the first 90° t_p calibration attempt). A small number of τ values can be used, which range from 0.1 to 10 seconds (e.g. 0.004, 0.02, 0.1, 0.5, 2.5, and 10) and DT set to 5 seconds (for $T_1 \sim 1$ second). The stack plot will look something like Figure 5.2, in which some signals, with relatively short T_1s, quickly become positive and then plateau at longer τ values while others, with relatively long T_1s, do not reach maximum intensity even at the largest τ values. Most modern spectrometers have programs that automatically calculate T_1s from the available data (note that these T_1s may be greatly in error if the data are far from optimal). This preliminary experiment quickly reveals whether the selected range of τ values and DT was too small (relatively few positive signals and little, or none, of maximum amplitude) or too large (most of the resonances near, or at, maximum amplitude) for the T_1s of the compound in question.

In order to determine T_1s with reasonable accuracy, a plot similar to that shown in Figure 7.1 should be obtained for each nucleus. Because of the spread of T_1 values in

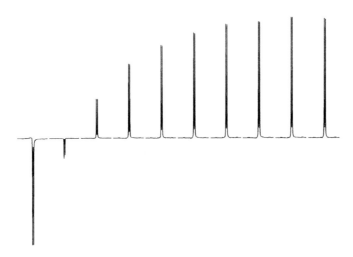

Figure 7.1 An inversion-recovery experiment. Signal intensities are shows as a function of τ.

most compounds, however, the IR-FT experiment necessarily has to be stretched to accommodate that nucleus having the longest T_1. With a better idea of the largest T_1, the IR-FT experiment can then be rerun with an appropriate range of τ values and a suitable DT. Again, modern spectrometers have T_1 programs that automatically select a range of τ values for an estimated T_1 and a DT = $5T_1$. For the example shown in Figure 7.1, $T_1 = 1$ second, then DT = 5 seconds, and the range of τ values is 0.05, 0.55, 1.05, 1.55, 2.05, 2.55, 3.05, 3.55, 4.05, and 4.55 seconds.

7.1.2 ^{13}C Spectral Editing Experiments

After preliminary ^{1}H and ^{13}C NMR experiments have been performed in a nonroutine analysis, an important process is sorting carbons in terms of their number of directly bonded hydrogens, i.e. methyl, methylene, methine, and quaternary. This knowledge can be gained from both one- and two-dimensional experiments. The principal sources of such information, however, are the one-dimensional attached proton test (APT) and distortionless enhancement by polarization transfer (DEPT) experiments (Section 5.5).

7.1.2.1 The APT Experiment

Behavior of the four types of CH$_n$ units in the APT experiment [1] is illustrated for $n = 1, 3$ in Figure 5.15 and $n = 0, 2$ in Figure 5.16. The APT pulse sequence is shown in Figure 7.2. Similar to the standard ^{13}C NMR experiment, it tends to be performed in one of two ways. The initial pulse (θ) can be set to 90° and a relaxation delay (Section 2.4.6) of c. 1 second employed. In this case, the delay time (DT) is set to zero and the second 180° pulse eliminated. We saw in Section 2.4.6, however, that the most efficient way to collect NMR data is to pulse continuously, without a DT, and with $\alpha < 90°$.

If we take the latter approach, a minor complication arises. After the $\alpha < 90°$ pulse, some magnetization remains along the z axis, which had been placed along the z axis by the first 180° pulse. In order to pulse rapidly without a DT, it is preferable to have any residual magnetization along the +z axis. This situation can be accomplished with the second 180° pulse, but small, symmetrically placed delay times (~1 ms) are necessary to refocus the magnetization vectors that move off the +y and −y axes during the second 180° pulse.

The final parameter to be selected is the delay time τ, which governs the refocusing of the carbon vectors. Two factors that slightly complicate matters are (i) the normal variation of one-bond, C—H couplings present in typical molecules and (ii) the dependency of C—H fragment behavior on hybridization of the carbon atom. To compensate for the former, τ is set to a value near the mid-point of the expected range of one-bond C—H coupling constants. The latter complication arises because vectors of sp^2 hybridized carbons dephase and refocus at faster rates than do those of sp^3 carbons. For a compound whose carbons are all sp^3 hybridized, then $^1J_{CH} \sim 125$ Hz (Table 4.1), and τ is

Figure 7.2 The APT pulse sequence.

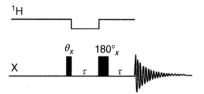

set to $1/J = 8$ ms. Likewise, if the carbons are all sp² hybridized, $^1J_{CH} \sim 160$ Hz, and τ becomes 6 ms. When both types of carbons are present, a compromise value of 7 ms is used, which corresponds to $^1J_{CH} = 140$ Hz.

In the full editing APT spectrum ($\tau = 1/J$) shown in Figure 5.17, carbons with an odd number of attached protons (CH and CH_3) are easily distinguished from those with an even number (CH_2 and quaternary, as zero is considered to be an even number). If it is necessary to differentiate carbons within either of these two groups, the APT experiment can be rerun with different values of τ. If $\tau = (2J)^{-1}$, methylene and quaternary carbons can be distinguished because the former are nulled while the latter have full intensity. Differentiation of methyl and methine carbons is less definitive. For $\tau = 2/3J$, methyl signals have approximately one-third of the intensity of the methine resonances.

The drawbacks to APT are its dependence on the magnitude of one-bond, C—H couplings and its inability to distinguish methyl from methine carbons unambiguously. A full editing experiment ($\tau = 1/J$), however, can be carried out relatively rapidly (usually half the number of scans of the parent ^{13}C NMR experiment) and is useful for characterizing compounds of known structure. Both of the above limitations are absent in the DEPT experiment, which is described in Section 7.1.2.2.

7.1.2.2 The DEPT Experiment

The DEPT experiment [2] has become the method of choice in editing procedures, and its pulse sequence is shown in Section 5.5. It gives separate subspectra that can be edited to contain only methyl, methylene, or methine carbons (Figure 5.18). The delay time $\tau = (2J)^{-1}$ is set with similar carbon hybridization considerations as the APT experiment, i.e. 4 ms for sp³ hybridized carbons, 3 ms for sp² carbons, and 3.6 ms when both are present. The main feature of the DEPT experiment is the variable θ pulse, which gives editing results somewhat reminiscent of those produced by varying the delay time τ in the APT experiment. It is, however, this dependence upon an *angle* rather than a *delay time* that makes DEPT less sensitive than APT to variations in $^1J_{CH}$ and, therefore, superior. The carbon types observed with various θ values are as follows:

1) $\theta = 45°$. All protonated carbons
2) $\theta = 90°$. Primarily methine carbons, with possible small breakthroughs of some other signals
3) $\theta = 135°$. Full spectral editing (three separate subspectra for methine, methylene, and methyl)

Additional suggested spectral parameters are the following:

1) DT = 0.5 – 1.5 seconds, which is one to three times ^{13}C $T_1(max)$, but this $T_1(max)$ is for protonated carbons only (probably methine), since quaternary carbons are not observed.
2) Steady state (dummy) pulses = 4 at a minimum (to allow the system to come to equilibrium).
3) Number of scans = multiple of 4 (for phase cycling purposes).

In addition, it is recommended that data accumulation be interleaved in blocks of 32 scans. Interleaving is common for stability purposes in lengthy experiments, and operates in the following manner. For, say, a full spectral editing experiment, instead of acquiring each of the spectra having a different angle θ in turn, one determines 32 scans for $\theta = 45°$, then 32 scans for $\theta = 90°$ (this segment is done twice), and finally 32 scans for $\theta = 135°$. This four-step cycle is repeated until sufficient scans are accumulated to produce four subspectra with acceptable S/N.

A disadvantage of DEPT is that it is a *subtraction* experiment and, therefore, much more sensitive to certain problems than typical one-dimensional techniques. One remedy, which is used in many experiments that suffer from stability problems, is the employment of *steady state* or dummy pulses. In this technique, several scans are taken as in a regular experiment, but data are not collected during what would be the normal acquisition time. This procedure allows the sample to come to equilibrium before data collection begins. Poor cancelation is generally the result of difficulties in one or more of the following areas:

1) *Lock stability.* Keep the lock power just below saturation and the lock gain in the 30% range.
2) *Sample temperature.* Perform experiments at constant temperature.
3) *Pulse calibration.* Calibration of the ^{13}C 90° and 180° pulses is critical.
4) *Incomplete cancelation in spectral subtractions.* Steady state pulses are critical.

7.1.3 NOE Experiments

The NOE provides information on the spatial proximity of nuclei (Section 5.4). NOE determinations usually are homonuclear, in the case of protons, but also can be heteronuclear, when proton signals are irradiated and those of heteronuclei are observed. NOEs occupy both an intermediate and a final position in the overall progression of structural determination. In most cases NOEs afford information on the three-dimensional structure of a molecule after its two-dimensional structure has been determined. NOEs, however, also can be used earlier in the structural elucidation process to provide answers to questions concerning *cis/trans* geometry in double-bond systems or ring substitution patterns, among others.

Classical steady-state NOE determinations involved irradiating a particular (target) signal and measuring the integrated intensities of other signals whose nuclei were believed to be spatially proximate to that nucleus whose signal was saturated (the *on-resonance* part of the experiment). Control experiments also were performed in which the irradiating frequency was positioned in a blank region of the spectrum and the integration(s) repeated (the *off-resonance* part). Some of today's NOE experiments are still carried out with on- and off-resonance components while others are selective and do not require subtraction.

Because NOE enhancements are a function of competitive relaxation, they are highly dependent on the internuclear distance(s) between the relaxed nucleus in question and nonirradiated neighbors that also can participate, as competing agents, in the overall relaxation process. For this reason, the two enhancements (A{B} and B{A}) are seldom identical and can be quite different. This statement is particularly true for CH_3—H systems for which very different results are obtained depending on which way the NOE experiment is performed. Methyl groups generally are very efficient relaxing agents for any protons that they are near, while methyl protons themselves are very little relaxed by neighboring protons. Therefore, relatively large enhancements can be observed for neighboring protons when methyl groups are irradiated, but the reverse experiment yields enhancements that are very small to nil.

7.1.3.1 The NOE Difference Experiment

Almost all modern NOE experiments are performed in the *difference* mode [3]. On- and off-resonance experiments are conducted in much the same way as before with the pulse sequence shown in Figure 7.3. On- and off-resonance spectra, or free induction decay (FIDs), when there are very strong signals like water, are stored separately in the computer memory and then subtracted to yield a spectrum that, in theory, shows signals only from those nuclei that are spatially proximate to the target proton. Typical acquisition parameters for small molecules (MW < 1000) are the following: $\alpha = 90°$ and $\tau = 5T_1$ (generally 3–5 seconds). Sufficient presaturation is important so that small NOE's have enough time to develop to the point of observation. In addition, while complete saturation of the target signal (and its subsequent disappearance from the spectrum) is ideal, this result is not always realistic in crowded spectra. A 50–75% decrease in intensity of the target signal generally indicates that a sufficient saturating power level has been attained.

NOE difference experiments are extremely sensitive to stability problems and plagued by poor cancelation. The following experimental conditions have been found to minimize subtraction problems:

1) Steady state pulses = 4.
2) Perform experiments with a non-spinning sample.
3) Perform experiments at constant temperature.
4) Keep the lock power just below saturation and the lock gain in the 20% range for optimum lock stability.
5) Number of scans = 256–1024 (512 suitable compromise) for good S/N.
6) Moderate line broadening (<2 Hz but not to cause signal overlap) reduces spectrum noise.
7) Use identical phasing for the on- and off-resonance spectra/FIDs.

When the target signal is a multiplet, modern spectrometers have programs that permit irradiation of each individual line of the multiplet. For stability purposes, four scans per resonance line can be taken per cycle before moving to the off-resonance part of the sequence. An NOE-difference spectrum for the sesquiterpene T-2 toxin (**7.1**) is shown in Figure 7.4a. The centrally located methyl-14 at δ 0.81 was irradiated and appears as a large negative signal. Signals indicative of (i) relatively large NOEs are seen for H-7B (δ 1.91) and H-13B (δ 2.80), (ii) medium-sized NOEs for H-15B (δ 4.06) and H-8 (δ 5.28), (iii) a small NOE for H-15A (δ 4.28), and a very small NOE for H-3 (δ 4.16). In addition, medium-sized signals are observed for the acyl methyls at δ 2.03 and 2.16. These resonances might appear to be indicative of moderate enhancements, but it must be remembered that they correspond to the largest signals in the normal proton spectrum and instead represent very weak NOEs. The subtraction problems discussed above are avoided with a relatively new NOE pulse sequence, which is described in Section 7.1.3.2.

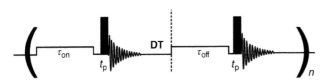

Figure 7.3 The NOE difference pulse sequence.

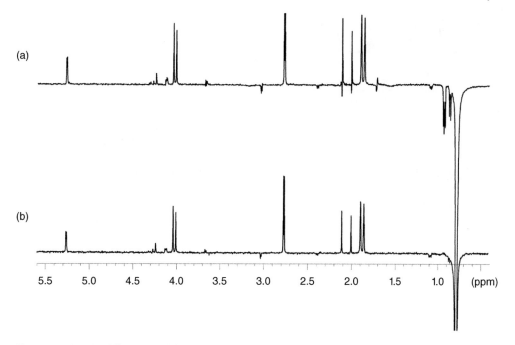

Figure 7.4 (a) NOE difference and (b) double pulsed field gradient spin echo NOE for T-2 toxin.

7.1.3.2 The Double-Pulse, Field-Gradient, Spin-Echo NOE Experiment

The double pulse field gradient spin echo-NOE (DPFGSE-NOE) sequence of Shaka is given in Figure 7.5 ([4] and Section 6.6). This experiment produces spectral results that look like a subtraction experiment, but neither spectral nor FID subtraction is involved. Critical to the DPFGSE-NOE pulse sequence is the pair of *shaped pulses* and pulsed field gradients. Modern, research-grade NMR spectrometers are able to produce finely defined pulses that are capable of irradiating extremely selective regions of a spectrum.

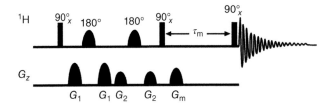

Figure 7.5 The double pulsed field gradient spin echo NOE pulse sequence. The relative strengths of gradients G_1, G_2, and G_m are 16, 7, and 13 G cm^{-1}, respectively. The mixing time is 650 ms.

After a nonselective, 90° pulse places the magnetization of all of the proton spins in the xy plane, magnetization of the target spin alone is refocused in the xy plane with a pair of selective, 180° pulses. The target magnetization is then moved to the $-z$ axis by a second non-selective, 90° pulse, and NOE's arise from this z magnetization. A DPFGSE-NOE spectrum for T-2 toxin is given in Figure 7.4b. Methyl-14 again is the target and appears as a large, negative signal. It is immediately obvious that the DPFGSE-NOE spectrum is considerably superior to the NOE-difference spectrum in Figure 7.4a.

7.2 Part B: Two-Dimensional Techniques

In Section 7.1, experimental methods are presented that permit us to determine the number of each type of carbon present in a molecule and the spatial relationships that exist between protons that are critical to the elucidation of the two- and, perhaps, three-dimensional structure of that molecule. As we saw above, it is possible that this new information combined with that obtained from chemical shifts, coupling constants, and other spectroscopic methods might be sufficient to determine a molecular structure. In most cases, however, the situation still requires that more spectral data be obtained. In this chapter we become acquainted with the performance of two-dimensional (2D) NMR experiments that provide information on *direct*, ^{1}H—^{1}H and ^{1}H—^{13}C connectivities, *longer range* ^{1}H—^{13}C connectivities, and ^{1}H—^{1}H spatial proximities.

7.2.1 Two-Dimensional NMR Data-Acquisition Parameters

Data acquisition is approached very differently in multi-dimensional as compared with 1D NMR experiments. The reason is that we are now dealing with, at minimum, a one-dimensional data matrix (for 2D NMR) and, perhaps, two or three matrices (for 3D and 4D NMR, respectively) for complex biological molecules.

7.2.1.1 Number of Data Points

To keep experimental times and data storage requirements reasonable, far fewer data points are used in each dimension than with typical 1D experiments. Reasonable digital resolution can be obtained in the *detected* dimension (f_2) for proton, homonuclear correlation, e.g. COrrelation SpectroscopY (COSY), and proton-detected, heteronuclear correlation, e.g. heteronuclear single quantum correlation (HSQC), spectra by using 1024 or 2048 data points. Due to their greater spectral widths, however, heteronuclear-detected correlation, e.g. HETeronuclear chemical-shift CORrelation (HETCOR), spectra require 2048, or even 4096, data points in f_2. Considering both computer limits and resolution requirements, 2D experiments are commonly performed with 2048 data points in each dimension.

7.2.1.2 Number of Time Increments

The number of time increments employed to create the nondetected, second dimension (f_1, Section 6.1), is just one of the factors that influences the number of data points used to describe the f_1 domain. A value of 256 increments represents a good compromise between time and resolution. It should also be remembered that this number typically doubles (to 512) for spectra presented in the absolute-value mode.

Digital resolution in f_1 (DR_1), analogous to f_2, is a function of the number of increments (ni) and the spectral width (sw_1). Its calculation is less straightforward and is considered in Section 7.2.2.3. Unlike DR_2, however, DR_1 is somewhat frustrating because, whereas sw_1 is reduced like sw_2 (the two are identical for homonuclear correlation experiments), increasing ni increases the time required to conduct the 2D experiment much more than a corresponding increase in np_2 (which has essentially no effect). If, for example, ni is increased by a factor of 2, the time of the experiment is doubled. To make matters worse, collection of 2048 increments might very well not result better, *observable* f_1 resolution. Unless the sample is very concentrated, such a large ni might effectively accomplish little more than several levels of zero filling ($n - 1$ of which do nothing to improve DR_1, Section 2.5.2). The reason is that, as the incremented time (t_1) increases, vector phase coherence (and thus xy magnetization) is lost due to magnetic field homogeneity effects and T_2 relaxation (T_2^*). There may be little xy magnetization remaining to be transferred between coupled nuclei and, therefore, little remaining to be detected by the receiver. Thus the last thousand or so increments may have essentially zero intensity. The solution to this dilemma is forward *linear prediction* (LP), which is discussed in Section 7.2.2.4. In addition, optimum resolution is achieved by reducing sw_1, as much as possible, from its value in the parent ^1H or ^{13}C spectrum.

7.2.1.3 Spectral Widths

In homonuclear correlation experiments, sw_1 is almost always set equal to sw_2. In addition, *reduced* spectral widths are used in both dimensions to improve both digital resolutions. The reduced spectral widths are set equal to the distance between the highest and lowest frequency signals plus c. 20% of this distance (c. 10% added to each end of the *reduced* sw).

7.2.1.4 Acquisition Time

Because np_2 is so much smaller than the number of points in an ordinary 1D spectrum, and sw_2 is not commensurately smaller than common 1D spectral widths, two-dimensional acquisition times are typically in the 100–300 ms range for ^1H-detected, and less than 100 ms for heteronuclear-detected, 2D experiments. Remember that, as in 1D experiments, sw_2 (in Hz) depends on the magnetic field strength and, therefore, affects the value of t_a. Similarly, t_a is normally set by the spectrometer after np_2 and sw_2 have been selected.

7.2.1.5 Transmitter Offset

When sw is reduced, it is customary for modern spectrometer programs to move the transmitter offset automatically so that it is in the middle of the reduced sw_2. The operator can verify that movement of the transmitter has occurred by noting the transmitter offset values for the original and reduced sw values.

7.2.1.6 Flip Angle

Flip angles are determined by the specific 2D experiment being conducted, and Ernst angle considerations are not an issue. The COSY family of experiments does have a variety of final pulses (Section 6.1), and they are considered in Section 7.3.1.1. Moreover, the final pulse before t_a that is delivered to the nucleus being detected is known as the *read* pulse.

7.2.1.7 Relaxation Delay

Almost all of the experiments described in this chapter, and certainly all 2D experiments, require a relaxation DT between scans. Unfortunately, many operators employ DTs that are too long for the experiments that they are performing. The recommended DTs in textbooks and spectrometer manuals are in the order of $1-3T_1$, with repetition rates (DT + t_a) for optimum sensitivity (using 90° pulses) being approximately $1.3T_1$. Of course, in a typical molecule, chemists are confronted with a range of T_1 values that are field strength dependent (Section 5.1). It is for this reason that the suggestion was made in Section 7.1.1 to determine the ^{1}H T_1s of nonroutine samples. Fortunately, even without such measurements, reasonable estimates can be made of T_1 values.

For *very small* molecules (MW < 400), ^{1}H T_1s are typically in the 1–3 seconds range, DTs in the 1–4 seconds range, and repetition rates in the 1.3–4 seconds range. Molecules of this size, however, are not usually studied except in textbooks. For *small* molecules (MW ~ 500–1000), ^{1}H T_1s are often in the 0.5–2 seconds range and DTs in the 0.4–2.5 seconds range. Repetition rates are 0.7–2.6 seconds and depend, of course, on t_a. Acquisition times, in turn, are a function of spectral width, which depends on field strength. Deciding on DTs without T_1 data would appear to be a daunting problem. Nevertheless, experience has demonstrated that DTs equal to 0.5–1 second usually suffice for the ^{1}H-detected experiments of molecules of MW < 1000. Values closer to 1 second are suitable for smaller molecules of MW < 400.

This range of DTs corresponds to a ^{1}H T_1 range of 0.5–1 second and might seem to be too small. Considering entire ranges of T_1 values to determine DTs for ^{1}H-detected experiments, however, is counterproductive. It should be remembered that methyl protons, which tend to have longer T_1s than methines and methylenes, should generally be ignored in DT deliberations. Their signals are usually much larger than those of methine and methylene protons, so they can afford to lose some signal intensity due to incomplete relaxation. Typically t_a is small, but if it exceeds 400 ms, due to a small sw, then DT can be reduced commensurately.

For the heteronucleus-detected experiments such as DEPT, HETCOR, and especially FLOCK, longer DTs are appropriate, particularly when the signals of nonprotonated nuclei are being recorded.

7.2.1.8 Receiver Gain

The receiver gain cannot be set in 2D experiments in the manner that was described for 1D experiments in Section 2.4.7. It is taken instead from the receiver gain value of the parent proton or heteronucleus experiment. Some spectrometers automatically retrieve this setting from the parent experiment while it must be manually entered in others.

7.2.1.9 Number of Scans per Time Increment

In order to minimize the overall time of a 2D experiment, one wishes to keep the number of scans per increment (ns/i) to a value that is sufficient to observe the spectrum of that particular increment. For heteronucleus detection, this number is usually large, but for protons adequate detection can often be accomplished in 1–4 scans. The minimum ns/i, however, is determined by the phase cycle (Section 5.8) of the pulse sequence being used and may be anywhere from 4 to 64 scans. As a general rule, 8 scans/i is a reasonable value for ^{1}H-detected experiments. Longer experiments that require a large ns/i, such as the study of dilute solutions (^{1}H detection) or heteronucleus detection, can make good use of interleaved acquisition with a suitable block size (as described for the DEPT experiment in Section 7.1.2.2).

In recent years, gradient versions of many of the basic 2D NMR experiments have become very popular. One of the main reasons is that the use of gradients eliminates the need for phase cycling for coherence pathway selection. Experiments involving ^{1}H detection can, therefore, often be performed with one to two transients per increment.

7.2.1.10 Steady-State Scans

Steady-state scans (Section 2.4.9) are used before the start of essentially all 2D experiments. They are particularly important in a number of pulse sequences in order to compensate for spin-lock (Subsections 7.3.1.2 and 7.3.4.2) and decoupler (Section 7.3.2) heating effects. Larger numbers of steady-state scans are employed in experiments that have either particularly long spin-lock times or X-nucleus decoupling over especially wide spectral widths.

7.2.2 Two-Dimensional NMR Data-Processing Parameters

Data processing, like acquisition, is performed differently in 2D compared to 1D NMR experiments. The principal reason is that signal truncation is a much more serious problem in 2D than 1D experiments. Zero filling also is used in 2D NMR experiments, as is linear prediction (LP).

7.2.2.1 Weighting Functions

It should not be surprising that signal (FID) truncation is a problem in the detected (t_2) domain given the short t_a (<300 ms) in that dimension. In the nondetected (t_1) domain we are dealing with *interferograms* (due to the individual signals of each spectrum at a particular chemical shift) rather than with true FID's, but their behavior with respect to truncation is similar. Because of relaxation effects, the signals at the end of an interferogram are smaller than those at its beginning, but, unless ni is unduly large, the individual interferograms do not decay to zero by the last increment. A variety of weighting functions, which have been developed so that apodization can be carried out in both the t_1 and t_2 dimensions, are pictured in Figure 7.6.

The choice of weighting function depends largely on whether the 2D data are presented in the absolute-value (magnitude) or phase-sensitive mode (Section 6.1). Absolute-value data require rather severe suppression of both the beginning and end

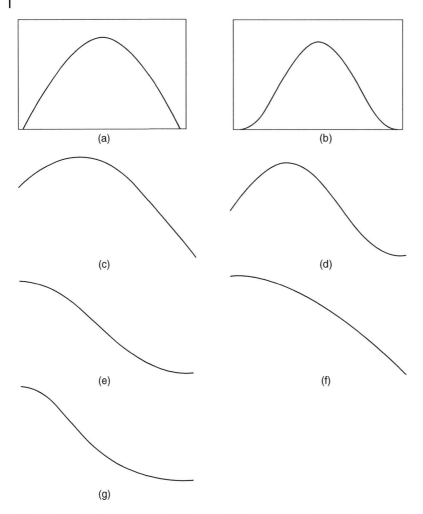

Figure 7.6 Weighting functions. (a) Sine bell. (b) Squared sine bell. (c) Shifted sine bell. (d) Shifted, squared sine bell. (e) Gaussian. (f) Cosine (90°-shifted sine bell). (g) Squared cosine (90°-shifted, squared sine bell).

parts of the FIDs. They are, therefore, usually processed with a *sine bell* function in order to eliminate the dispersive tails from the phase-twisted lineshapes. Two other choices are the *pseudo-echo* and *squared sine bell* functions that are nearly identical to each other and give results similar to sine bell weighting. They emphasize a little more the middle part of the FID and discriminate somewhat more against its beginning and end relative to the sine bell function. If sensitivity is an issue, a *shifted sine bell* function can, for example, be used in the following manner.

Most modern spectrometers have interactive software that permits the operator to vary a particular weighting parameter and immediately observe the effect of this change on the final spectrum. The first increment is generally selected as an example to see what the effects of weighting parameters are on the t_2 data. With the first FID displayed, the operator can shift the midpoint of the sine bell function toward the S/N (left) part of the FID until the best compromise is reached between sensitivity and resolution. It is important for the operator to remember that the weighting function must have a value of zero at the right end of the FID, so that it will reduce the FID to zero at both the end of (i) t_a in t_2 and (ii) the effective t_a (ni/sw$_1$) in t_1.

Data (interferograms) in the t_1 domain are handled in much the same manner as those in t_2. Appropriate weighting and Fourier transformation of the t_2 domain data result in an (f_2, t_1) data matrix that consists of horizontal rows of interferograms at the chemical shifts in the vertical (f_2) dimension. A row showing good S/N is selected and interactively weighted as described above for the first increment in t_2.

Phase-sensitive data do not have the dispersive-tail problem and require suppression of only the ends of the FIDs. They can, therefore, be processed with more gentle apodization such as a Gaussian function, but other possibilities are the *90°-shifted sine bell* (*cosine*) and the *90°-shifted, squared sine bell* (*squared cosine*) functions (Figure 7.6). If the operator has a series of compounds to be investigated, it might be worthwhile to apply interactive weighting with all three functions to determine if there is a clear advantage to any one of them.

Sine bell and Gaussian weighting functions are set up by the spectrometer software based on the acquisition parameters. Shift values for the various shifted-functions are entered interactively and then the function is readjusted so that it goes to zero at the end of t_2 and t_1.

7.2.2.2 Zero Filling
After appropriate weighting is carried out in the t_2 and t_1 domains, both data sets should be zero filled by at least a factor of 2 (one level).

7.2.2.3 Digital Resolution
We saw in Section 2.5.4 that resolution in 1D spectra of <1 Hz requires a DR of $(\delta v)/2$ or $J/2 \sim 0.5$ Hz/pt. Digital resolution likewise is an important concept in 2D NMR. If, for example, correlations from 1-Hz couplings are to be observed, one needs a DR = 1–2 Hz/pt.

Digital resolution in f_2 is a function of the number of data points (np$_2$) and the spectral width (sw$_2$) and is calculated in the same manner as 1D experiments (Section 2.5.4). If sw$_2$ = 2100 Hz and np$_2$ = 1024, then, after the FT with half of the points real and the other half imaginary (Section 2.5.1), DR$_2$ = sw$_2$/(np$_2$/2) = 2100 Hz/(1024/2) = 4.1 Hz/pt. Similar again to 1D experiments, one level of zero filling should be performed for optimum DR, so that np$_2$ = 2048. Using the one-half real points after the FT operation, DR$_2$ = 2100 Hz/(2048/2) = 2.1 Hz/pt.

As stated in Section 7.2.1.2, digital resolution in f_1 is a function of the ni and the sw$_1$. While sw$_1$ is treated in the same manner as sw$_2$, the value used for ni depends on whether the experimental data is presented in the absolute-value or the phase-sensitive mode. The situation for ni in *phase-sensitive* data is the same as np$_2$ in the f_2 domain, i.e. after the FT process, only half the points are used to describe the spectrum. If sw$_1$ = 2100 Hz and ni = 512, DR$_1$ = sw$_1$/(ni/2) = 2100 Hz/(512/2) = 8.2 Hz/pt. If one level of zero filling is carried out, then ni = 1024 and DR$_1$ = 2100 Hz/(1024/2) = 4.1 Hz/pt.

Treatment of *absolute-value* f_1 data sets is different because of the fact that *both* the absorptive (real) and dispersive (imaginary) components of the f_1 dimension are used to describe the spectrum. If ni = 512, after the FT there are 256 real and

256 imaginary points. Upon combination there is a total of 512 points. Again, if $sw_1 = 2100$ Hz and the factor of 2 is dropped from the denominator, $DR_1 = sw_1/ni = 2100$ Hz/512 = 4.1 Hz.

Optimum resolution thus is achieved by reducing sw_2 and sw_1 as much as possible from their values in the parent ^1H and ^{13}C spectra and by zero filling ni (for phase-sensitive data sets) and np_2 by a factor of 2.

7.2.2.4 Linear Prediction

Mention has been made in Section 7.2.1.2 and several other places in the text that NMR spectroscopists are engaged in seemingly never-ending battles between sensitivity and resolution and between both of these and time. Sensitivity vs time issues have existed since before the advent of FT-NMR and are well illustrated in the t_1 dimension of 2D experiments. On the one hand, there must be a sufficient number of scans per time increment to observe a spectrum, but there also must be enough increments to resolve closely situated signals. Both requirements take precious spectrometer time, and spectroscopists are forced to compromise between sensitivity and resolution in a number of 2D experiments.

Forward LP [5] represents an elegant solution to the sensitivity-resolution-time trilemma. The idea behind LP can be likened to an ideal race in which automobiles travel at a constant rate of speed. If their relative positions after 256 laps are noted, a very good estimate can be made as to what their positions will be after 1024 laps. In NMR LP, finite-length FIDs are extended by using information from the previous data points. In a time sequence of data points, the value of a particular data point, d_m, can be estimated from a linear combination (hence the name "linear prediction") of the immediately preceding data points [6] (Eq. (7.1)), in which a_1, a_2, a_3, ... are the LP coefficients (also called the LP

$$d_m = d_{m-1}a_1 + d_{m-2}a_2 + d_{m-3}a_3 + \cdots \tag{7.1}$$

prediction filter). The number of coefficients used in the LP process is known as the *order* of the LP. The number of coefficients corresponds to the number of data points that are used to predict the value of the next data point in the series.

The critical requirement for employment of LP is that the FID's must have sufficient S/N so that an accurate estimation can be made of the coefficients. This requirement presents a problem for both heteronucleus-detected 2D experiments and very dilute solutions in ^1H-detected experiments. LP, however, is less important for the former, in which protons, with their much smaller chemical shift range and, hence, intrinsically better data point resolution, constitute the f_1 dimension. Another requirement is that the number of coefficients that are used in the LP process should be *greater* than the number of signals that make up each FID that is being extended. How much greater the number of coefficients should be than the number of contributing signals depends on the manufacturer of the spectrometer.

Linear predictions are commonly performed on instruments of one manufacturer with eight coefficients while another manufacturer recommends that anywhere from 16 to 32 coefficients be used for its systems. It is very important that the correct number of coefficients be employed. Too few fails to make accurate predictions. The resulting spectra, at best, looks as if no LP has been performed and, at worst, have even poorer resolution. Use of too many coefficients results in artifacts along the f_1 axis that resemble t_1

noise. In addition, the calculations take an inordinate amount of time even if the excess number does not cause the system to shut down.

It is also important that LP not be abused. A sufficient number of experimental data points must be taken from which the FIDs can confidently be extended. A second, and greater, concern is that LP not be extended too far, e.g. 32 points predicted to 1024. Reynolds [5] has found that, as a general rule, data presented in the phase-sensitive mode can be predicted fourfold, e.g. 256 data points can be predicted to 1024, while absolute-value data can be extended twofold, 256 points to 512. For certain experiments, however, in which interferograms are composed of only one signal, 16-fold LP is performed routinely.

Because of the requirement that the number of coefficients be larger, or very much larger, than the number of signals comprising the FID's, LP usually is not appropriate for either 1D data or the incremental FID's that eventually form the f_2 dimension. It is, however, ideally suited for operating on interferogram data in the t_1 domain for several reasons. In most instances, the signals of spectra in the final time increments have nonzero intensities. If a strong weighting function is not used to drive interferograms to zero intensity, truncation wiggles are produced upon Fourier transformation (Section 2.5.3). Such apodization, however, results in line broadening, so it is really not an ideal solution. LP allows more gentle apodization functions to be employed, thereby improving spectral resolution and greatly reducing truncation errors. In addition, since t_1 interferograms are generally composed of many fewer signals than typical t_2 FID's, the above LP requirement that the number of signals be smaller that the number of coefficients generally is not violated.

It is important to realize that zero filling is complementary, and not alternative, to LP. Both techniques effectively increase t_a and, thereby, increase digital resolution (Section 2.4). LP, however, provides much greater enhancement because it extends an FID that otherwise would be truncated, while zero filling just adds zeros to an already apodized FID. This difference can be demonstrated by calculating DR for phase-sensitive data in the f_1 dimension with, and without, LP. If $sw_1 = 2100$ Hz, ni = 256, and one level of zero filling is carried out (256 points) for a total of 512 points; then $DR_1 = 2100$ Hz/(512/2) = 8.2 Hz/pt. Now if those same 256 time increments are subjected to fourfold LP (768 points) for a total of 1024 points and then zero filled by one level to 2048 points, $DR_1 = 2100$ Hz/(2048/2) = 2.1 Hz/pt. (a fourfold improvement).

LP can be employed as a data-processing method, after the fact, like zero filling. LP's maximum benefits, however, are realized if the operator plans for its use when setting up the experimental acquisition parameters. As we saw in the above example, in this approach only ¼ to ½ of the number of time increments, which would be sufficient for resolution requirements, actually are taken. The interferograms in t_1 then are subjected to a two- or fourfold LP, zero filled to a factor of 2, and processed in the normal fashion. The resulting 2D spectra have the same digital resolution in the f_1 dimension as if 2–4 ni had been used and will have been obtained in ¼ to ½ the time. If both sensitivity and resolution are problems, for a given length of spectrometer time, ni can be *reduced* and then supplemented with LP to produce increased resolution while *ns/i* is *increased* to yield greater sensitivity. LP is one of the most important 2D experimental time-saving techniques available to NMR spectroscopists. If it is not being used regularly for processing 2D data, valuable spectrometer time is being wasted. A good example of the

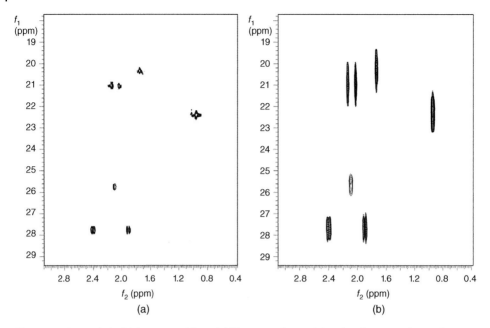

Figure 7.7 Expanded HSQC spectra with 16-fold linear prediction (a) and no linear prediction (b).

resolution-enhancing ability of LP is illustrated in Figure 7.7, which shows 2D spectra of the natural product, T-2 toxin, with and without LP.

In addition, *backward* LP can be utilized for a completely different purpose. The first few data points at the beginning of an FID occasionally become corrupted if signal acquisition starts before pulse ringdown has faded to an acceptable level. In such a situation, a large baseline roll appears in the spectrum that renders it essentially useless. This result is especially disheartening when it occurs in a ^{13}C spectrum that has been accumulated over a weekend. The problem can be remedied easily by replacing the first 5–10 data points with predicted points. Reweighting and a second FT then result in an undistorted spectrum.

7.2.3 Two-Dimensional NMR Data Display

7.2.3.1 Phasing and Zero Referencing

We saw in Section 6.1 that 2D experiments almost always are presented as contour, rather than stacked, plots. If the 1D spectrum (in the case of homonuclear experiments) or the 1D spectra (for heteronuclear experiments), from which the spectra in the f_2 dimension are derived, are phased and zero-referenced properly, both the phase and zero reference of diagonal and cross peaks along the f_2 and f_1 axes should be very close to correct. If they are not, a horizontal cross-section, or trace, through a signal in the upper right corner of the contour plot is phase corrected and checked for correct zero-referencing at the same time. Then a second cross-section through a signal in the lower left corner also is phase corrected. One cycle usually is sufficient to phase these signals in the f_2 domain. If major phase adjustments are required, however, a second cycle may be necessary.

To correct the phase and zero reference of diagonal and cross peaks along the f_1 axis, the contour plot is rotated by 90° so that f_1 is the horizontal axis. Signals at each end of the f_1 axis then are selected, phased, and checked for zero-referencing as was done for signals in the f_2 dimension.

Finally, by convention, contour plots are presented so that the detected (f_2) dimension is along the horizontal axis and the nondetected (f_1) dimension, therefore, along the vertical axis.

7.2.3.2 Symmetrization

Symmetrization is employed commonly in the display of homonuclear, absolute-value, 2D NMR data to preserve true cross peaks and discard *unsymmetrical* artifacts (Section 6.1). A particularly troublesome type of artifact, known as t_1 *noise*, is manifested as streams of signals or ridges. These ridges, associated with strong signals in a contour plot, run parallel to the f_1 axis and are located at their chemical shifts. They can interfere seriously with the observation of true cross peaks from the signals in question. Critical to its application is the requirement that digital resolution be equal in both the f_1 and f_2 dimensions. Phase-sensitive data, on the other hand, are not symmetrized because the procedure could introduce distortions in the cross peaks. Symmetrization must be undertaken with caution and is best used in comparison with the unsymmetrized data.

7.2.3.3 Use of Cross Sections in Analysis

The procedure for establishing atomic connectivities in contour plots is relatively straightforward. It should be remembered, however, that these plots can be deceiving. Threshold values, which are essential to remove baseline noise from contour plots, may be sufficiently high to eliminate smaller cross peaks. For this reason, it is highly recommended that cross sections through the individual frequencies of specific resonances be examined on the monitor and either tabulated or plotted directly. This recommendation is especially true for the longer range, proton-heteronucleus, chemical shift correlation experiments such as heteronuclear multiple bond correlation (HMBC) and FLOCK. It is further recommended that cross sections through the frequencies of both protons and the X-nuclei be analyzed for these two experiments. In the case of unknown compounds, homo- and heteronuclear, cross-section examinations are critical since dihedral angles close to 90° often result in a number of very small cross peaks.

7.3 Part C: Two-Dimensional Techniques: The Experiments

There are many ways to approach the acquisition, processing, display, and even interpretation of 2D NMR experiments. The following sections describe many of the 2D experiments that have been discussed thus far and provide guidelines for their performance. For best results, 2D experiments should be performed at constant temperature (ideally just above room temperature to decrease solvent evaporation and sample degradation) and with a nonspinning sample.

If computer speed and memory permit, 2D NMR experiments generally are planned so that a 2K (2048 points) or 4K (4096 points) FT is carried out in each dimension. In this approach, np_2 should be 1024–2048 and zero filled by one level to 2048–4096 prior to the FT2. In addition, ni should be linear predicted two to fourfold to 1024 and zero filled by one level to 2048 before FT_1. DTs of 0.5–1 second generally are sufficient. Many experiments call for the use of steady-state pulses before starting either (i) the overall experiment or (ii) the accumulation of each time increment.

A number of 2D spectra of the sesquiterpene natural product T-2 toxin (**7.1**) are collected in Section 7.3 to illustrate certain 2D experiments and the critical differences between them.

7.3.1 Homonuclear Chemical-Shift Correlation Experiments via Scalar Coupling

Experiments that correlate the chemical shifts of two nuclei of the same type (homonuclear correlation) on the basis of scalar (or spin–spin) coupling that exists between them are the most frequently performed of all the 2D procedures. Most of these experiments are adaptations of the basic COSY sequence. Another COSY-type experiment, TOtal Correlation SpectroscopY (TOCSY), is useful for observing correlations of individual protons to most, or all, of the other protons in an entire spin system.

Homonuclear contour plots are symmetrical with respect to both the spectral widths, and data point resolutions are almost always identical. LP can help greatly in making DR_1 equal to DR_2 by keeping data accumulation time within acceptable limits.

7.3.1.1 The COSY Family: COSY-90°, COSY-45°, Long-Range COSY, and DQF-COSY

7.3.1.1.1 Basic COSY

The COSY-90° experiment (which is referred to as COSY, Section 6.1, Figure 6.1, and Ref. [7]) may be said to be almost foolproof in that fairly decent results can be obtained even when the 90° pulses are not well calibrated. It is one of very few 2D experiments that is better performed in the absolute-value, rather than phase-sensitive, mode because of the mixed phasing that arises in phase-sensitive contour plots. If the cross peaks in these plots are phased to appear as absorptive signals, then the diagonal peaks are phased dispersively. Strong diagonal signals can, in turn, have long, dispersive tails (Section 7.2.2.1), which easily can obscure cross peaks located near the diagonal. Compared with other COSY experiments, phase-sensitive COSY has too many disadvantages and should not be used.

COSY experiments also have benefited greatly from the adaptation of gradients. The gradient COSY (gCOSY) pulse sequence is shown in Figure 7.8. Use of gradients is important for the following reason. In non-gradient, ^{1}H-detected experiments, more scans per time increment usually are required to satisfy the phase cycle (4–16 steps, Section 5.8) than to acquire the spectrum. gCOSY spectra are, on the other hand, routinely taken with one or two scans per increment. Another advantage of gCOSY experiments is that much shorter relaxation delay times are required. Thus, when combined with LP methods, gCOSY experiments can be carried out in a matter

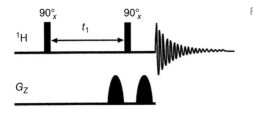

Figure 7.8 The gradient COSY pulse sequence.

of minutes. The gCOSY pulse sequence is like the COSY sequence with a pair of z gradients flanking the second 90° (read) pulse. Comparison gradient and nongradient COSY spectra of T-2 toxin are not shown here because the only substantial difference between the two experiments is the time required to accumulate spectra with equivalent S/N. The difference is about a factor of 10: 8 minutes for the gCOSY experiment vs 77 minutes for the COSY.

The following parameters are appropriate for COSY (absolute-value mode) experiments:

1) $f_1 = f_2 = -0.5$ to 9.5 ppm
2) SS = 16
3) ns/i = 1–4 (with gradients) or a multiple of 4 for the phase-cycled, nongradient version.
4) DT = 0.5–1.0 second
5) f_2 apodization: squared sine bell
6) f_1 apodization: squared sine bell
7) np = 2048, $ZF(f_2) = 2048$, ni = 512, LP = 1024, $ZF(f_1) = 2048$

7.3.1.1.2 COSY-45

The COSY-45 experiment (Section 6.1 and Figure 6.12) is essentially identical to the basic COSY experiment except that the final pulse is 45° rather than 90°. While it does suffer from a slight loss of sensitivity relative to COSY, the loss is more than compensated for by its suppression of both diagonal signals and cross peaks that are very close to the diagonal and result from parallel transitions. The decrease in intensity of these signals permits observation of more interesting cross peaks, due to progressive and regressive transitions, that happen to be located close to the diagonal. The simplified diagonal signals and tilted cross peaks of the COSY-45 spectrum of T-2 toxin are clearly visible when compared with the standard COSY spectrum in Figure 7.9.

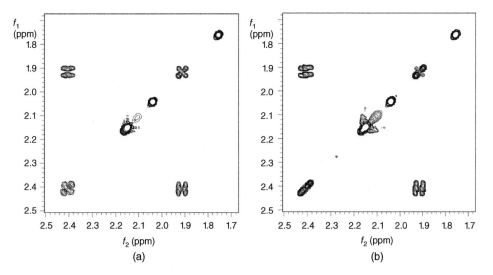

Figure 7.9 Expansions of the COSY spectrum (a) and the COSY-45 spectrum (b) for T-2 toxin.

7.3.1.1.3 Long-Range COSY

The long-range COSY (LR-COSY, Section 6.1 and Figure 6.12) experiment is, likewise, virtually identical to COSY. Two constant-time delays τ, equal to c. 200 ms, are inserted into the COSY sequence after t_1 and flanking the read pulse, making the pulse sequence $90°-t_1-\tau-90°-\tau-t_1-t_2$(acquire). These delays enhance correlations due to small couplings. Considerable digital resolution now is available in the COSY family of experiments, through increased computer memory size (larger t_1 and t_2 data sets). The resulting enhanced DR permits the observation of ^{1}H—^{1}H correlations from very small coupling constants. Many long-range correlations are, in fact, detected along with the stronger vicinal and geminal correlations in regular COSY experiments. The LR-COSY is, therefore, less of a necessity than it was previously. Interestingly, second COSY experiments sometimes are carried out in a low-resolution manner just to eliminate the longer-range correlations. High- and low-resolution COSY spectra of T-2 toxin are compared in Figure 7.10. The high-resolution COSY experiment used 512 time increments linear predicted to 1024, and the low-resolution COSY employed only 128 increments linear predicted to 256. In addition to the obviously poorer resolution in Figure 7.10b, long-range cross peaks due to (i) four-bond, W coupling between H-11 and H-7B (δ 4.34, 1.91) and H-8 and Me-16 (δ 5.28, 1.75) and (ii) homoallylic coupling between H-11 and Me-16 (δ 4.34, 1.75), among others, are clearly visible in Figure 7.10a but are noticeably absent in Figure 7.10b.

7.3.1.1.4 Double Quantum Filtered COSY

The double quantum filtered COSY experiment (DQF-COSY, Section 6.1 and Figure 6.16) is similar to COSY, but with three 90° pulses in the sequence $90°-t_1-90°-\tau-90°-t_2$ (acquire). DQF-COSY allows cross peaks close to the diagonal to be observed (like COSY-45) and also eliminates *singlet* signals from methyl groups and solvents. Use of gradients, however, is less important than for other COSY experiments because rapid pulsing (the prime advantage of gCOSY experiments) should be avoided in DQF-COSY.

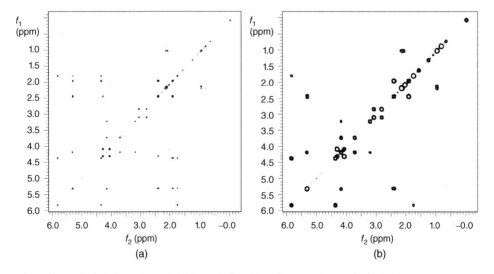

Figure 7.10 The COSY spectra with high resolution (a) and low resolution (b) for T-2 toxin.

The DQF-COSY experiment is performed in the phase-sensitive mode, but, unlike the phase-sensitive COSY experiment, both the diagonal and the cross peaks can be phased as absorptive signals. In addition, the cross peaks can be analyzed to determine the coupling constants that comprise their multiplet structures. The lower overall sensitivity of the experiment and phase-sensitive cross peaks of selected cross peaks of T-2 toxin are illustrated in Figure 7.11.

In addition to having 50% lower sensitivity compared with COSY, DQF-COSY experiments are much more sensitive to artifacts. Since DQF-COSY data are presented in the phase-sensitive mode, they are not symmetrized. Therefore, t_1 noise arising from strong signals (other than singlets) could, in principle, be a problem. It can be minimized by carefully calibrating the 90° pulse. Pulsing too rapidly also can produce artifacts so DT should be set more conservatively toward $3T_1$ when T_{1max} is not known. DQF-COSY is, therefore, not an experiment to be used regularly, like COSY, but rather in those instances when certain coupling constants have to be determined.

The following parameters are appropriate for DQF-COSY (Phase Sensitive) experiments:

1) $\tau = 4\,\mu s$
2) $f_1 = f_2 = -0.5$ to 9.5 ppm
3) SS = 16
4) ns/i = 1–8 (with gradients) or a multiple of 4 for the nongradient version.
5) DT = 0.5–1.0 second
6) f_2 apodization: 90° shifted sine bell (cosine)
7) f_1 apodization: 90° shifted sine bell (cosine)
8) np = 4096, ZF(f_2) = 8192, ni = 256, LP = 1024, ZF(f_1) = 2048

Note: Acquisition of high-resolution spectra along f_2 is highly recommended to assist in the measurement of coupling constants.

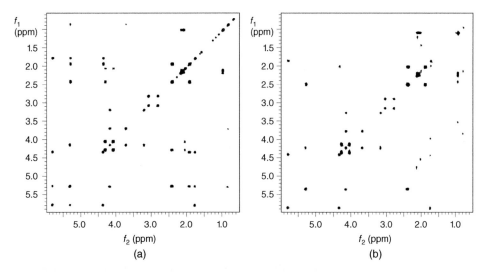

Figure 7.11 The COSY spectrum (a) and DQF-COSY spectrum (b) of T-2 toxin.

7.3.1.2 The TOCSY Experiment

The TOCSY experiment (Section 6.1, Figure 6.16, and Ref. [8]) gives information similar to relayed COSY and is well-suited for large molecules. It is performed in the phase-sensitive mode, and, like DQF-COSY, both the diagonal and cross peaks can be phased as absorptive signals. The TOCSY pulse sequence is given in Figure 7.12, and TOCSY and COSY spectra for T-2 toxin are pictured in Figure 7.13. The greater number of cross peaks in the TOCSY spectrum illustrates the further relaying of coupling information. For example, H-2 (δ 3.70) shows coupling to H-3 (δ 4.16) in both the COSY and the TOCSY spectra and also relayed coupling to OH-3 (δ 3.19) and to H-4 (δ 5.31) in the TOCSY spectrum.

The degree to which a particular proton in a spin system is coupled to its more remote neighbors increases with the length of the spin-lock mixing time. A mixing time of 20 ms, for example, produces a COSY-type of spectrum with few relays, while a mixing time of 100 ms largely transfers magnetization throughout an entire spin system. The 1D version of this experiment is especially useful and is discussed in Section 7.3.5.

The following parameters are appropriate for TOCSY and Z-TOCSY (phase-sensitive) experiments.

1) $f_1 = f_2 = -0.5$ to 9.5 ppm
2) SS = 16
3) ns/i = 2–16 (with gradients) or a multiple of 4 for the nongradient version.
4) DT = 0.5–1.0 second
5) Mixing time: 80 ms

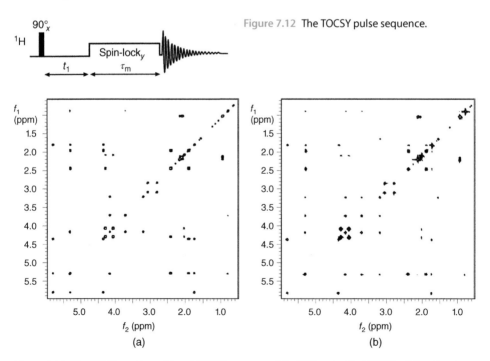

Figure 7.12 The TOCSY pulse sequence.

Figure 7.13 The COSY spectrum (a) and TOCSY spectrum (b) of T-2 toxin. The TOCSY mixing time is 80 ms.

6) f_2 apodization: Gaussian
7) f_1 apodization: Gaussian
8) np = 2048, ZF(f_2) = 4096, ni = 256, LP = 1024, ZF(f_1) = 2048
9) spin-lock mixing time = 20–100 ms

Note 1: Z-TOCSY is superior since the zero-quantum filter produces undistorted cross-peak patterns.

Note 2: A shorter mixing time (25–30 ms) gives a COSY-type spectrum with better resolution than an absolute-value COSY spectrum.

7.3.2 Direct Heteronuclear Chemical-Shift Correlation via Scalar Coupling

HETCOR experiments can be performed by detecting either protons or the X nuclei (the most common being ^{13}C, Section 6.2). The principal advantage of ^{1}H-detected experiments is their sensitivity, which is a function of the gyromagnetic ratios: $(\gamma_H/\gamma_X)3/2$. If we consider the ratios of Larmor frequencies (instead of γs) at, say, 400 MHz for ^{1}H and ^{13}C, the benefit of detecting ^{1}H rather than ^{13}C is the following: (400/100)3/2 = 8.

X-nucleus-detected experiments, however, have one advantage that can become important, since their nondetected nucleus is ^{1}H. As a rule, it is better to detect the nucleus that has the more congested spectrum, which is usually the case for protons. For certain classes of compounds, however, such as fatty acids, the proton spectra are too crowded to distinguish individual resonances. In such cases, it could be better (if sensitivity permits) to carry out an experiment that detects the X nuclei.

The two principal, ^{1}H-detected, direct (one-bond), HETCOR experiments are Heteronuclear Multiple Quantum Coherence (HMQC) and HSQC. The X-nucleus detected counterpart is HETCOR. The ^{1}H and X-nucleus spectral widths are reduced for each of these experiments. It is important to remember that the latter should be decreased to contain only the signals of *protonated* X nuclei. Quaternary carbons, for example, do not participate in these experiments, and their signals should not be included in the reduced spectral windows.

Since the X nucleus in these experiments is largely carbon, "X" refers to ^{13}C in the following discussions.

7.3.2.1 The HMQC Experiment

HMQC (Section 6.2, Figure 6.22, Ref. [9]) is now the less commonly used ^{1}H—X correlation experiment for reasons that are discussed in Section 7.3.2.2. It is a relatively robust sequence and is performed in the phase-sensitive mode. The delay time δ governs both the initial defocusing of the ^{13}C-bonded ^{1}H vectors and the final refocusing of these vectors. It is selected in a compromise manner like τ in the APT experiment (Section 7.1.2.1). Carbon decoupling also can be performed during ^{1}H signal acquisition. Such decoupling is accomplished with the GARP or WURST sequences, which are like, but superior to, WALTZ decoupling (Section 5.8).

A major difficulty with HMQC is that protons directly bonded to ^{13}C are detected in the presence of an overwhelming number of hydrogens (99%) bonded to ^{12}C. The solution to this problem of interfering ^{1}H—^{12}C magnetization can be approached in two ways. As pointed out in Section 6.2, the HMQC sequence can begin with a bilinear rotation decoupling (BIRD) pulse that is selective for protons directly bonded to ^{12}C and a nulling delay τ. Alternatively, gradients can be employed to refocus ^{1}H—^{13}C selectively while leaving ^{1}H—^{12}C magnetization defocused and, therefore, undetected.

Since the signal of any particular proton can be modulated by only its directly bonded carbon, the HMQC experiment is ideal for LP and nonuniform sampling (NUS) methods. We saw in Section 7.2.2.4 that one of the factors that influences LP is the number of signals constituting the interferogram and that the number of coefficients used in the LP process must be larger than the number of resonances that make up each FID that is being extended. Since the number of ^{13}C signals contributing to each ^{1}H interferogram is equal to one, fewer coefficients have to be used in the LP. LP for HMQC experiments is regularly extended beyond the normal fourfold limit. In addition, NUS is becoming more commonly used in direct, heteronuclear, chemical-shift correlation experiments.

The following parameters are appropriate for Gradient-Selected HMQC (Absolute-Value) experiments.

1) $f_2 = -0.5$ to 9.5 ppm
2) $f_1 = -5$ to 165 ppm
3) SS = 16
4) ns/i = 4–32
5) DT = 0.5–1.0 second
6) $^1J_{CH} = 145$ Hz
7) f_2 apodization: sine bell
8) f_1 apodization: sine bell
9) np = 2048, ZF(f_2) = 4096, ni = 512, LP = 1024, ZF(f_1) = 2048
10) $\tau = 0.3$–0.6 second if a BIRD pulse is used
11) $\Delta = (2\,^1J_{CH})^{-1} \sim 1/(2 \times 145\,\text{Hz}) = 3.5$ ms

7.3.2.2 The HSQC Experiment

The HSQC experiment (Section 6.2, Ref. [10]) is superior to HMQC in several respects and is used almost exclusively for one-bond correlation of heteronuclear chemical shifts. Probe tuning and ^{13}C 180° pulse (used twice) calibration are very important. HSQC's principal advantage over HMQC is that it has considerably better ^{13}C resolution along the f_1 axis. The reason is that HMQC cross peaks exhibit ^{1}H multiplet structure due to ^{1}H—^{1}H coupling along both the f_1 (^{13}C) and f_2 (^{1}H) axes while HSQC cross peaks are singlet in nature along f_1. This difference may not be apparent in the full HMQC and HSQC spectra of Figure 7.14a, but it can be clearly seen along the vertical dimensions of the expansion spectra of Figure 7.14b.

Like the COSY and HMQC experiments, HSQC benefits greatly from the use of gradients. Fewer scans need be taken per time increment, and the more extensive phase cycle problem is obviated. Notably absent in the gradient HSQC spectrum of T-2 toxin, shown in Figure 7.15, are the t_1 noise artifacts observed in the nongradient HSQC spectrum at the chemical shifts of the strong methyl signals (δ 0.81 and c. δ 2). HSQC is performed in the phase-sensitive mode and can incorporate a BIRD pulse for suppression of ^{1}H—^{12}C magnetization if the gradient version is not being used. Its double-INEPT pulse sequence (Section 5.6.1) is shown in Figure 7.16.

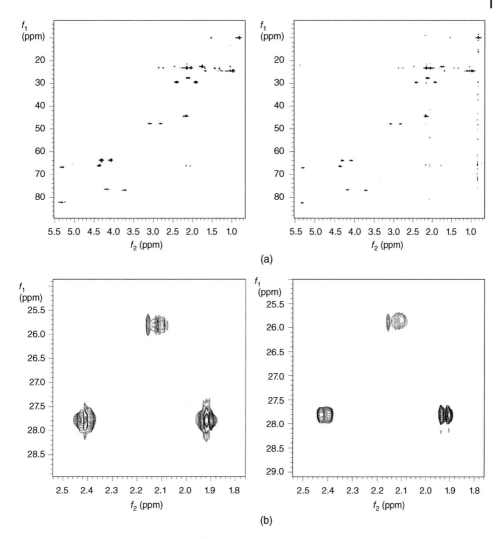

Figure 7.14 (a) The HMQC spectrum (left) and HSQC spectrum (right). (b) Expansions of HMQC spectrum (left) and HSQC spectrum (right) of T-2 toxin.

As in HMQC, a delay time Δ governs both the initial defocusing of the ^{13}C-bonded ^{1}H vectors and the final refocusing of these vectors and is, likewise, selected in a compromise manner. Carbon decoupling also can be performed during ^{1}H signal acquisition with the GARP or WURST sequences.

The same considerations for LP and NUS, which were discussed for HMQC experiments in Section 7.3.2.1, are even more pertinent to HSQC because of its greater resolution in the f_1 dimension. Remarkably, up to 16-fold LP of ^{1}H interferograms is carried out routinely. In addition, HSQC experiments can be performed with spectral editing so that methine and methyl signals are phased in one direction, generally positively, and methylene signals in the opposite direction. It is recommended that HSQC be used, rather than HMQC, unless instrumental instabilities dictate otherwise.

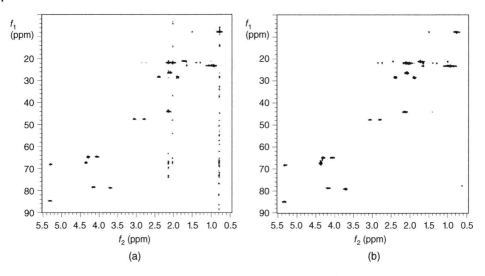

Figure 7.15 The HSQC spectrum (a) and gradient HSQC spectrum (b) of T-2 toxin.

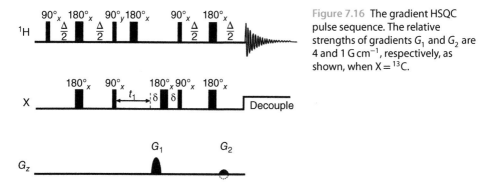

Figure 7.16 The gradient HSQC pulse sequence. The relative strengths of gradients G_1 and G_2 are 4 and 1 G cm^{-1}, respectively, as shown, when X = ^{13}C.

The following parameters are appropriate for gradient-selected HSQC (with or without ^{13}C spectral editing) experiments:

1) $f_2 = -0.5$ to 9.5 ppm
2) $f_1 = -5$ to 165 ppm
3) SS = 16
4) ns/i = 4–32
5) DT = 0.5–1.0 seconds
6) $^1J_{CH}$ = 145 Hz
7) f_2 apodization: Gaussian
8) f_1 apodization: Gaussian
9) np = 2048, ZF(f_2) = 4096, ni = 256, LP = 1024, ZF(f_1) = 2048.

7.3.2.3 The HETCOR Experiment

HETCOR (Section 6.2, Figure 6.20, and Ref. [11]) is an X-nucleus-detected experiment and, like COSY, is a relatively robust sequence. Unlike COSY, it can be performed reasonably in either the absolute-value or phase-sensitive modes, although the former gives better resolution. Since it is relatively immune to artifacts (if pulsing is not too rapid), a gradient version is largely unnecessary. Because HETCOR is a polarization-transfer experiment, the relaxation delay times are a function of the ^{1}H, and not the X-nucleus, T_1s.

There are two important delay times (Δ_1 and Δ_2) in the HETCOR experiment. Δ_1 governs the defocusing of the ^{13}C-bonded ^{1}H vectors and is set in the same compromise manner that we have seen in the APT, DEPT, and HMQC experiments. The selection of Δ_2, which controls the final refocusing of the ^{1}H-bonded ^{13}C vectors, however, is not so straightforward. A problem arises from the fact that methyl and methine carbon vectors refocus at a time of $\tfrac{1}{2}(^1J_{CH})$ whereas methylene vectors do so at $\tfrac{1}{4}(^1J_{CH})$. The latter value is commonly chosen for Δ_2 and does, in fact, represent a good compromise for the following reason. Methyl and methine carbons can have just one proton cross peak, but methylene carbons can have two when they are attached to diastereotopic protons. The intensity of many methylene cross peaks is thus halved, and this loss is compensated partially for by optimizing the refocusing time for methylene carbon vectors. In addition, a common artifact in HETCOR experiments is the appearance of small signals halfway between the correlations of diastereotopic protons bonded to the same carbon.

Proton decoupling with WALTZ decoupling (Section 5.8) typically is performed during ^{13}C signal acquisition. The following parameters are appropriate for HETCOR experiments:

1) $f_2 = -5$ to 165 ppm
2) $f_1 = -0.5$ to 9.5 ppm
3) SS = 16
4) ns/i = multiple of 4
5) DT = 1–2 seconds (1.3 ^{1}HT_1 for an optimum pulse rate)
6) $^1J_{CH}$ = 145 Hz
7) Pseudo-echo, sine bell, or squared sine bell weighting (all left-shifted to improve sensitivity)
8) np = 2048, ZF(f_2) = 4096, ni = 512, LP = 512, ZF(f_1) = 2048
9) $\Delta_1 = (2\,^1J_{CH})^{-1} \sim 1/(2 \times 145\,\text{Hz}) = 3.5$ ms
10) $\Delta_2 = (4\,^1J_{CH})^{-1} \sim 1/(4 \times 145\,\text{Hz}) = 1.7$ ms

7.3.3 Indirect Heteronuclear Chemical-Shift Correlation via Scalar Coupling

As we saw in Section 7.3.2, heteronuclear, chemical-shift correlation experiments can be performed by detecting either protons or the X nuclei. All of the comments, which were made there for *direct*, HETCOR experiments, apply equally well to their *indirect* (or longer range, i.e. two- and three-bond correlation) counterparts.

The two best, longer range, HETCOR experiments are HMBC and FLOCK. Not surprisingly, HMBC with ^{1}H detection is more commonly used than the FLOCK experiment with X-nucleus detection, because of its much greater sensitivity. As pointed out above, in those circumstances for which ^{1}H spectra are very congested, FLOCK spectra can provide a useful alternative. "X" thus refers to ^{13}C in Section 7.3.3.1 and to ^{1}H in Section 7.3.3.2.

7.3.3.1 The HMBC Experiment

HMBC (Section 6.2, Figure 6.26, and Ref. [12]) is very similar to the HMQC experiment, from which it was derived. The delay time Δ governs the defocusing of the longer range (usually 2–3 bonds) ^{13}C-bonded ^{1}H vectors and is c. 20 times longer than the corresponding delay in HMQC. It too is selected in a compromise manner and is generally set to c. 60 ms, which corresponds to $^{n}J_{CH} \sim 8$ Hz.

One of the problems with HMBC involves suppression of the signals of those protons that are directly bonded to ^{13}C. This interfering ^{1}H—^{13}C magnetization can be dealt with in the following way. A ^{13}C 90° pulse (known as a *J-filter* because it eliminates one-bond C—H couplings) can be inserted into the pulse sequence at a time equal to $\frac{1}{2}(^{1}J_{CH})$ after the initial ^{1}H 90° pulse, i.e. during the longer range delay time Δ. The $^{1}J_{CH}$ delay time allows the directly ^{13}C-bonded ^{1}H vectors to move to an antiphase orientation, and the ^{13}C 90° pulse then is applied to produce $^{1}J_{CH}$ multiple quantum coherence. This coherence is unobservable and is allowed to evolve away.

The greatest problem associated with HMBC involves the very large signals of those protons that are bonded to ^{12}C. When cancelation of the resulting ^{1}H—^{12}C magnetization is incomplete, t_1 noise ridges are produced that can interfere greatly with the analysis of HMBC spectra. A BIRD pulse cannot be used here, as in HMQC, because it also would suppress the longer range, ^{13}C-bonded ^{1}H vectors that we wish to observe. This complicating ^{1}H—^{12}C magnetization was dealt with, until recently, by phase cycling. A much better solution is the incorporation of gradients to refocus selectively longer range ^{1}H—^{13}C magnetization while leaving both direct ^{1}H—^{13}C and ^{1}H—^{12}C magnetization defocused and, thus, undetected. The gradient HMBC (gHMBC) pulse sequence is illustrated in Figure 7.17.

HMBC also is unique among 2D experiments in that it is recommended (Ad Bax, its inventor [12]) that the data be processed in the *mixed-mode*, i.e. absolute value in the t_2 dimension and phase-sensitive in t_1. This method results in a gain in sensitivity by a factor of $2^{1/2}$ relative to data processed in the absolute-value mode in both domains. This approach to data handling is related to another interesting feature of the HMBC experiment. As remarkable as the use of gradients has been in 2D experiments, they are not always the method of choice. gHMBC data are, for example, processed in the absolute-value mode in both dimensions. As we have seen above, this procedure results in a loss of sensitivity by a factor of $2^{1/2}$ with respect to mixed-mode processing. In addition, we saw that absolute-value, t_1 data can be linear predicted only twofold, while phase-sensitive data can be predicted fourfold. For equal amounts of spectrometer time, half as many time increments (with twice as many scans per time increment) are taken

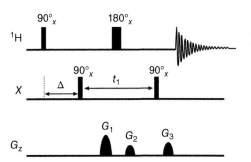

Figure 7.17 The gradient HMBC pulse sequence. The relative strengths of gradients G_1, G_2, and G_3 are 5, 3, and 4 G cm^{-1}, respectively, as shown, when X = ^{13}C.

when fourfold LP is being carried out compared to twofold LP. For this reason there is loss of sensitivity by another factor of $2^{1/2}$, for a total loss in sensitivity of a factor of 2.

A choice of HMBC vs gHMBC may be arrived at in the following manner. If a sample is fairly concentrated and displays strong signals in its ^{1}H NMR spectrum, then t_1 noise ridges could be a major problem. If this is the case, gHMBC would certainly be the method of choice. If the opposite is true, however, then sensitivity, and not t_1 noise, is the major issue. In this case, the more sensitive, nongradient HMBC might be the preferred experimental method. Figure 7.18 compares these two sequences for the first case, for which t_1 noise due to strong methyl signals in the 1- and 2-ppm regions is an obvious problem.

Because of the uncertainty concerning both the location and intensity of correlations in HMBC contour plots, cross sections should be taken through individual chemical shifts on both the ^{1}H and the ^{13}C axes.

The following parameters are appropriate for gradient-selected HMBC (absolute-value) experiments.

1) $f_2 = -0.5$ to 9.5 ppm
2) $f_1 = -5$ to 220 ppm (a high-frequency value of 200 ppm can be used if it is certain that the compound being examined has no carbonyl groups)
3) SS = 16
4) ns/i = 16–64
5) DT = 0.5–1.0 second
6) $^1J_{CH} = 145$ Hz (or 130 and 165 Hz if a two-step J-filter is used)
7) f_2 apodization: sine bell
8) f_1 apodization: sine bell
9) np = 4096, ZF(f_2) = 8192, ni = 512, LP = 1024, ZF(f_1) = 2048

The following parameters are appropriate for gradient-selected HMBC (mixed-mode processing) experiments.

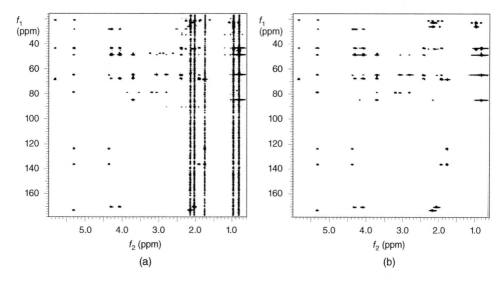

Figure 7.18 The HMBC spectrum (a) and gradient HMBC spectrum (b) of T-2 toxin.

1) $f_2 = -0.5$ to 9.5 ppm
2) $f_1 = -5$ to 220 ppm (A high-frequency value of 200 ppm can be used if it is certain that the compound being examined has no carbonyl groups.)
3) SS = 16
4) ns/i = 16–64
5) DT = 0.5–1.0 second
6) $^1J_{CH} = 145$ Hz (or 130 and 165 Hz if a two-step J-filter is used)
7) $nJ_{CH} = 8$ Hz
8) f_2 apodization: sine bell
9) f_1 apodization: Gaussian
10) np = 4096, ZF(f_2) = 8192, ni = 256, LP = 1024, ZF(f_1) = 2048

Note: This pulse sequence has better f_1 resolution and better sensitivity than the absolute-value sequence. Spectra are displayed in the absolute-value mode but processed in the phase-sensitive mode along f_1.

7.3.3.2 The FLOCK Experiment

The FLOCK experiment (Section 6.2, Ref. [13]) is an X-nucleus-detected experiment, which, despite its lower sensitivity compared with HMBC, is still useful in certain circumstances. Another longer range, ^1H—X chemical shift correlation experiment is COrrelation spectroscopy via LOng-range Coupling (COLOC), and the X nucleus in both experiments is ^{13}C.

The principal disadvantage of the COLOC sequence derives from the fact that it is a *fixed*-evolution time experiment, i.e. t_1 is incorporated into the delay time Δ_1. The major limitation with fixed-t_1 pulse sequences is that C—H correlations are considerably diminished, or completely absent, when the two- and three-bond ^1H—^{13}C coupling constants of a C—H fragment are similar in magnitude to the ^1H—^1H couplings of the same fragment. As can be seen from the compilations of coupling constants in Chapter 4, this situation is the rule rather than the exception. To minimize losing C—H correlations when $^nJ_{CH} \sim {}^nJ_{HH}$, COLOC experiments typically are conducted twice – once with $^nJ_{CH}$ optimized for, perhaps, 5-Hz couplings and again with $^nJ_{CH}$ optimized for 10-Hz coupling constants.

The FLOCK sequence (so named because it contains three BIRD pulses) of Reynolds et al. is similar to the rest of the pulse sequences that have been discussed in that it is a *variable*-evolution time experiment (t_1 becomes progressively larger). FLOCK thus avoids the potential absence of C—H correlations, which are such a problem with COLOC. Its pulse sequence is given in Figure 7.19.

Its pulse sequence is reminiscent of the HETCOR sequence with its refocusing of desired vectors during t_1 dephasing during Δ_1, polarization transfer at the end of Δ_1, and refocusing during Δ_2. The three BIRD pulses in the FLOCK experiment act in the following way. The first is selective for $^1J_{CH}$ vectors during t_1 and permits the separation of their effect from that of $^nJ_{CH}$ vectors by appropriate phase cycling. The chemical shifts of protons directly bonded to ^{13}C are rotated through a *constant* angle and thus not modulated. Similarly, $^1J_{CH}$ vectors are dephased through a variable angle that suppresses one-bond polarization transfer. Conversely, the chemical shifts of protons indirectly bonded to ^{13}C are rotated through a *variable* angle and, therefore, modulated. Likewise, $^nJ_{CH}$ vectors are refocused and thus set up to participate in longer-range polarization transfer.

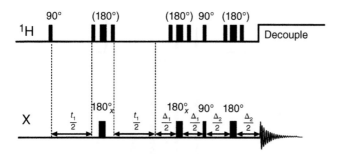

Figure 7.19 The FLOCK pulse sequence. The 180° ¹H pulses in parentheses are BIRD pulses (90°–τ–180°–τ–90°) with $\tau = (2J_{CH})^{-1}$. The relative phases of the three pulses are x, y, –x for the first BIRD pulse and x, x, –x for the second and third BIRD pulses.

The second BIRD pulse is selective for $^nJ_{CH}$ vectors during Δ_1 and acts as a simple ^{13}C 180° pulse for $^1J_{CH}$ vectors, which tend to refocus. Nonetheless, there will be some one-bond polarization transfer. The $^nJ_{CH}$ vectors, however, move to antiphase for maximum polarization transfer. In addition, the BIRD pulse refocuses magnetic field inhomogeneities for protons indirectly bonded to ^{13}C and refocuses the individual $^nJ_{CH}$ vector components (which themselves are defocusing) prior to polarization transfer.

The third BIRD pulse, during Δ_2, is also selective for $^nJ_{CH}$ vectors, which are focused from an antiphase orientation. ^{13}C nuclei that are directly bonded to protons behave in one of two ways. The $^1J_{CH}$ vectors of the majority of ^{13}C nuclei, which have experienced no polarization transfer, remain focused but are eliminated by phase cycling subtraction. On the other hand, the minority of ^{13}C nuclei, which have undergone some polarization transfer, remain at antiphase and are not observed.

In addition to a variable t_1, FLOCK has four fixed delay times. The first, DT, is a function of the ¹H, and not the X nucleus, T_1s because, like HETCOR, FLOCK is a polarization-transfer experiment. The second (τ, the delay in the BIRD pulses) is a function of the one-bond, C—H coupling and, as usual, is optimized for 140 Hz. The third and fourth, Δ_1 and Δ_2, are functions of $^nJ_{CH}$. They are set in the same compromise fashion as HMBC for $^nJ_{CH} \sim 8$ Hz. In addition, they function like the delays in HETCOR with Δ_1 optimized for proton $^nJ_{CH}$ defocusing and Δ_2 optimized for ^{13}C $^nJ_{CH}$ focusing.

Proton decoupling is typically performed during ^{13}C signal acquisition and is accomplished with WALTZ decoupling (Section 5.8). FLOCK data are presented in either the phase-sensitive or absolute-value mode. Because of the uncertainty concerning both the location and intensity of correlations in FLOCK contour plots, like HMBC spectra, cross sections should be taken through individual chemical shifts on both the ¹H and the ^{13}C axes.

As stated in Section 7.3.2, on those occasions when (i) the ¹H spectrum is more congested than the ^{13}C and (ii) sufficient sample can be dissolved (c. 20 mg for MW ~ 500) to obtain ^{13}C spectra in reasonable time, FLOCK spectra can be very useful in the elucidation of molecular structures and assignment of chemical shifts. Such a situation occurs in fatty acids, and a comparison of HMBC and FLOCK spectra of oleic acid is shown in Figure 7.20. In the HMBC spectrum (black), it is basically impossible to assign any, except the most obvious, long-range correlations between alkenic protons at δ 5.3 and methylene carbons at δ 29.5–29.8. Conversely, in the FLOCK spectrum (red),

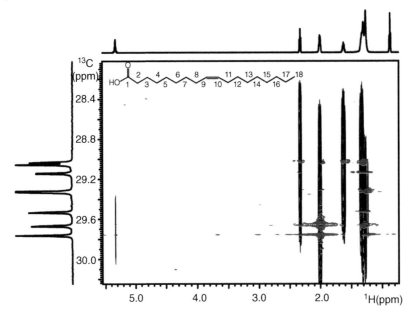

Figure 7.20 The resolution enhancement of the FLOCK experiment (red) over the proton detected HMBC (black) for oleic acid in the crowded region of the carbon dimension. The trade-off is in sensitivity. The HMBC can be up to an order of magnitude more sensitive.

many long-range correlations can be observed between the methylene carbon signals (δ 29.0–29.8) and various proton signals (δ 1.3–2.3).

The following parameters are appropriate for FLOCK experiments.

1) $f_2 = -5$ to 220 ppm (a high-frequency value of 200 ppm can be used if it is certain that the compound being examined has no carbonyl groups)
2) $f_1 = -0.5$ to 9.5 ppm
3) SS = 1
4) ns/i = multiple of 32
5) DT = 1–2 1HT_1 (0.8 second for MW < 400, 0.5 second for MW ~ 400–1000)
6) $\Delta_1 = (2\,{}^nJ_{CH})^{-1} \sim 1/(2 \times 8\,\text{Hz}) = 60$ ms
7) $\Delta^2 = (4\,{}^nJ_{CH})^{-1} \sim 1/(4 \times 8\,\text{Hz}) = 30$ ms
8) $\tau = (2\,{}^1J_{CH})^{-1} \sim 1/(2 \times 145\,\text{Hz}) = 3.5$ ms
9) Gaussian, cosine, or squared cosine weighting (for phase-sensitive data) or pseudo-echo, sine bell, or squared sine bell weighting (for absolute-value data), all left-shifted to improve sensitivity
10) np = 4096, ZF(f_2) = 8192, ni = 512, LP = 1024, ZF(f_1) = 2048.

7.3.3.3 The HSQC–TOCSY Experiment

The HSQC–TOCSY experiment gives essentially the same information as TOCSY but has the important advantage of *spectral dispersion*. When considerable spectral overlap exists in a ^{1}H NMR spectrum, TOCSY spectra may be of limited utility in establishing proton spin systems. The X-nucleus spectrum, however, almost always possesses much greater spectral dispersion, owing to its larger range of chemical shifts. It is this increased chemical-shift range that is critical to the utility of the HSQC–TOCSY experiment. Rather than being correlated with the chemical shifts of other, possibly overlapping, protons, ^{1}H signals are correlated with the chemical shifts of directly bonded X nuclei. In the HSQC–TOCSY experiment for protons and carbons, for example, magnetization is transferred from a proton to its directly bonded carbon and back again to proton, just as in a regular HSQC experiment. A TOCSY spin-lock mixing scheme is placed where acquisition would normally occur. The HSQC–TOCSY pulse sequence is pictured in Figure 7.21.

After the HSQC part of the experiment, proton magnetization is then transferred to other protons in the spin system and detected, at various mixing times, in the TOCSY part of the experiment. This process also demonstrates a limitation of the HSQC–TOCSY experiment with respect to HMBC – correlations are observed only for *protonated* X nuclei. Quaternary carbons, for instance, do not participate in HSQC–-TOCSY. A comparison of HSQC and HSQC–TOCSY spectra (Figure 7.22) shows the same extended coupling as that of COSY and TOCSY spectra in Figure 7.13. For example, C-4 (δ 84.5) shows coupling to H-4 (δ 5.31) in both the HSQC and HSQC–TOCSY spectra and also relayed coupling to H-3 (δ 4.16), H-2 (δ 3.70), and OH-3 (δ 3.19) in the HSQC–TOCSY spectrum.

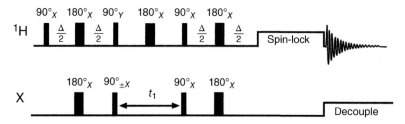

Figure 7.21 The HSQC–TOCSY pulse sequence.

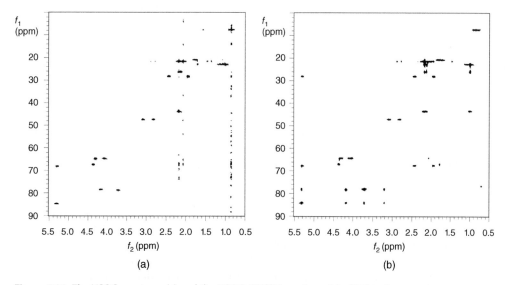

Figure 7.22 The HSQC spectrum (a) and the HSQC–TOCSY spectrum (b) of T-2 toxin.

Parameters for the HSQC–TOCSY experiment are the same as for the component HSQC and TOCSY experiments.

7.3.4 Homonuclear Chemical-Shift Correlation via Dipolar Coupling

Nuclei that undergo mutual relaxation via *dipolar coupling* are said to be dipolar coupled and give rise to the NOE. The involved nuclei also may be scalar (or spin–spin) coupled, but this question is not pertinent to the discussion (Section 5.4). NOE experiments can be both homonuclear and heteronuclear in nature, although the former, involving protons, are much more common. One-dimensional homonuclear Overhauser experiments were introduced in Section 7.1.3; their 2D versions, NOE spectroscopy (NOESY) and Rotating frame nuclear Overhauser Effect SpectroscopY (ROESY), are treated in this section.

Similar to the COSY experiments, contour plots are symmetric, with both the spectral widths and data point resolutions being almost always identical. LP can, again, help greatly in making DR_1 equal to DR_2 by keeping data accumulation time within acceptable limits. Because of the lengthy build-up nature of the NOE, the experimental times required for NOESY and ROESY experiments tend to be much longer than for corresponding 2D experiments showing spin-coupled correlations.

It also should be remembered that the results of 2D NOE experiments are averaged, i.e. an AB cross peak represents the average of the enhancement of H_B when H_A is irradiated and vice versa. Because NOE enhancements are a function of competitive relaxation, they are highly dependent on the internuclear distance(s) between the relaxed nucleus in question and nonirradiated neighbors which also can participate, as competing agents, in the overall relaxation process. For this reason, the two enhancements (A{B} and B{A}) seldom are identical and can be quite different. The resulting, averaged cross peaks can be small and difficult to detect.

The apparent absence of NOE cross peaks must, therefore, be interpreted with caution. In situations in which the presence, or especially the absence, of NOE's is critical to stereochemical decisions, e.g. whether substituents on a carbon–carbon double bond are *cis* or *trans* to one another, and the existence of cross peaks is uncertain, selective 1D NOE experiments always should be performed.

7.3.4.1 The NOESY Experiment

The NOESY experiment (Section 6.3, Ref. [14]) is an extension of the standard COSY experiment with a mixing time and a third 90° read pulse following the original two-pulse sequence. NOESY experiments should be performed in the phase-sensitive mode in order to distinguish true, *positive* NOESY cross peaks (phased positively) from both COSY-artifact and exchange spectroscopy (EXSY) cross peaks, which may be present in the spectrum but observed as negative resonances. True, *negative* NOESY cross peaks and diagonal signals also are phased negatively. Fortunately, COSY artifacts in NOESY spectra are not common. While they can be distinguished easily from positive NOESY cross peaks by their opposite phase, their real threat to the NOESY experiment lies in cancelation, complete or nearly so, of positive cross peaks. This situation cannot be tested for, and the spectroscopist should be aware of its possibility.

The successful observation of NOESY correlations can depend critically on the choice of two parameters: the mixing (Δ_m) and relaxation delay (DT) times, both of which depend on the spread of proton T_1s in a molecule. The average range of T_1s for small molecules (MW ~ 500–750) is 0.5–2 seconds. Recommended DTs are 1–2T_1 for small molecules and should be set conservatively (2–3 seconds) when the T_1s are unknown. The choice of mixing times also is important because if τ_m is too short, NOE enhancements do not have a chance to develop to detectable intensities, and NOESY cross peaks are not observed. Conversely, if τ_m is too long, the NOE enhancements disappear because of relaxation, and, again, cross peaks are absent. Compromise mixing times for a range of molecular weights can be set in the following way: 0.3–0.6 seconds for MW ~ 1000, ~1 second (c. T_1) for MW ~ 500, and 1–2 seconds for very small molecules.

Another reason for the careful selection of DT is that NOESY spectra are susceptible to artifacts from pulsing too rapidly. Problems also occur for intermediate-sized molecules (MW ~ 750–2000) in the cross-over region in which NOESY cross-peak intensities approach zero for even spatially proximate protons (Section 6.3 and Figure 6.31). The latter complication is absent in the ROESY experiment, which is discussed in Section 7.3.4.2.

Chemical and conformational exchange effects are observed in NOESY spectra, but they are not troublesome if the experiment is performed in the phase-sensitive mode. If these exchange processes occur on the same time scales that have been discussed here, the same NOESY parameters can be used for EXSY experiments (Section 6.3).

The following parameters are appropriate for NOESY (phase-sensitive) experiments:

1) $f_1 = f_2 = -0.5$ to 9.5 ppm
2) SS = 16
3) ns/i = 4–16
4) DT = 1.0–1.8 seconds (3 seconds for very small molecules)
5) Mixing time: 300–800 ms (1 second for average-sized, small molecules)
6) f_2 apodization: Gaussian
7) f_1 apodization: Gaussian
8) np = 2048, ZF(f_2) = 4096, ni = 256, LP = 1024, ZF(f_1) = 2048.

Note: Mixing time choice is critical as small molecules require long mixing times to permit the build up of reasonable NOE's. With high molecular weight (>750 Da) molecules, however, spin-diffusion occurs with long mixing times and results in erroneous apparent correlations.

7.3.4.2 The ROESY Experiment

The ROESY experiment (Section 6.3, Ref. [15]) is exactly the same as TOCSY except that the spin-lock mixing time generally is 100–200 ms compared with 20–100 ms for TOCSY. Since ROE enhancement factors do not go through zero intensity for molecules of MW ~ 750–2000 as NOE enhancements do, all ROESY cross peaks are of the same sign, and ROESY is used especially for these intermediate and large-sized molecules. The ROESY experiment is performed in the phase-sensitive mode in order to distinguish true ROESY cross peaks (phased positively) from both TOCSY-artifact and EXSY cross peaks, which may be present in the spectrum but as negative signals. Diagonal signals also are observed as negative resonances.

TOCSY artifacts are to be expected in ROESY spectra to some degree, particularly when the ROESY mixing time is close to the upper limit of the TOCSY range (c. 100 ms).

While the two types of cross peaks can be distinguished easily from each other by their opposite phases, the real threat from TOCSY signals to ROESY spectra is (like COSY signals in NOESY spectra, but more so) cross peak cancelation. TOCSY signals, however, are greatly suppressed in the transverse-ROESY (or T-ROESY) experiment. A comparison of NOESY and T-ROESY spectra of T-2 toxin is given in Figure 7.23 and shows essentially no difference, as expected, between the two experiments for this relatively small (MW = 466) molecule.

The same considerations that were presented for the selection of mixing times and relaxation delay times for NOESY experiments apply to ROESY. Suggested DTs are $1-3T_1$ for small-to-intermediate-sized molecules (MW < 1000) and generally should be in the 1–2 seconds range. Spin-lock mixing times are in the 200–600 ms range – closer to 200 ms for larger molecules and 600 ms for smaller molecules.

By analogy with NOESY experiments, ROESY parameters can be used for EXSY experiments. The following parameters are appropriate for ROESY experiments:

1) $f_1 = f_2 = -0.5$ to 9.5 ppm
2) SS = 16
3) ns/i = 4–16
4) DT = 1.0–1.8 seconds (3 seconds for very small molecules)
5) Spin-lock mixing time: 200–600 ms (1 second for average-sized, small molecules)
6) f_2 apodization: Gaussian
7) f_1 apodization: Gaussian
8) np = 2048, ZF(f_2) = 4096, ni = 256, LP = 1024, ZF(f_1) = 2048

Note: The use of ROESY rather than NOESY is strongly recommended for molecules with molecular weights >600 Da since NOESY cross peaks approach zero intensity for molecules much above this molecular weight and eventually become negative as molecular weight increases.

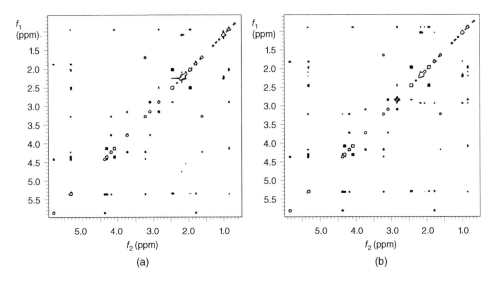

Figure 7.23 The NOESY spectrum (a) and the T-ROESY spectrum (b) for T-2 toxin.

7.3.5 1D and Advanced 2D Experiments

7.3.5.1 The 1D TOCSY Experiment

As mentioned in Section 7.3.1.2, the 1D version of the TOCSY experiment is especially useful for larger molecules that possess complicated and overlapping ^{1}H spin systems. The 2D TOCSY spectra of classes of molecules such as polysaccharides can be very difficult to interpret. 1D TOCSY experiments, however, permit the mapping of entire spin systems when the chemical shift of just one member of the system is distinct. An example is the anomeric (H-1) protons of polysaccharides, which are situated at high frequency (downfield) from the carbinol protons.

The single most important factor that has enabled 1D TOCSY experiments to become such a powerful tool is the extraordinary selectivity of modern shaped pulses (Section 5.8). These pulses for example, can irradiate the middle of three signals selectively, for which the adjacent resonances are sufficiently close that there exists very little baseline between the center and flanking signals. An example of a 1D TOCSY spectrum, which does not begin to test the selectivity of the technique, is shown in Figure 7.24 for a four-spin system of T-2 toxin composed of H-2 (δ 3.70), H-3 (δ 4.16),

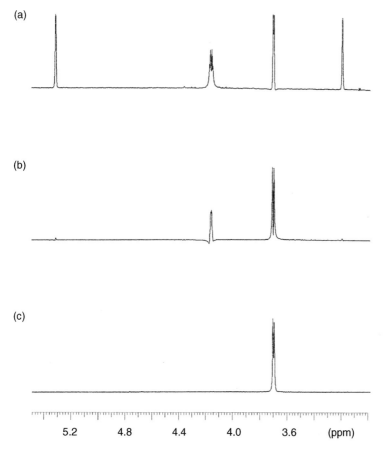

Figure 7.24 The 1D-TOCSY spectra for a four-spin system of T-2 toxin for mixing times of 0 ms (c), 120 ms (b), and 300 ms (a).

OH-3 (δ 3.19), and H-4 (δ 5.31). H-2 has been irradiated selectively, and spectra are given for the following three mixing times: (a) zero, (b) 120 ms, and (c) 300 ms. No magnetization is transferred for a mixing time of zero, and, as expected, only H-2 is found in (a). After 120 ms, magnetization has been transferred to the vicinal coupling partner (H-3), and its signal is detected. After 300 ms, magnetization has been further transferred from H-3 to OH-3 and H-4, i.e. throughout the entire spin system, and the resonances of all four protons are observed. In addition, Figure 7.25 illustrates the results of taking horizontal traces through the chemical shift of (i) H-2 (δ 3.70) for 1D (Figure 7.24) and 2D TOCSY (Figure 7.13) spectra and (ii) C-2 (δ 79.2) for the HSQC–TOCSY spectrum (Figure 7.22) of T-2 toxin. The resolution clearly is best in the 1D TOCSY trace but is largely equivalent in the other two.

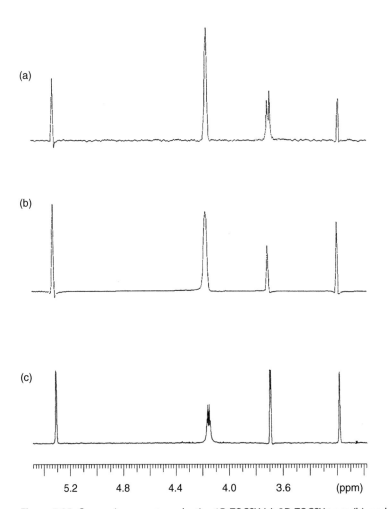

Figure 7.25 Comparison spectra: selective 1D-TOCSY (c), 2D-TOCSY trace (b), and HSQC–TOCSY trace (a).

The following parameters are appropriate for 1D TOCSY experiments:

1) $f_2 = -0.5$ to 9.5 ppm
2) SS = 16
3) ns/i = 4–64 (with gradients) or a multiple of four for the nongradient version
4) Mixing time: 0.00 s and 80 ms (or array)
5) Apodization: 0.5 Hz
6) np = 32 768
7) fn = 65 536

Note 1: If available, use of the Z-TOCSY sequence is strongly recommended.

Note 2: If a relatively long mixing time is used, a larger number of scans is needed because the initial magnetization is spread among several multiplets.

Note 3: Acquisition of an initial spectrum with a zero mixing time is recommended to ensure that a clean excitation of only the desired multiplet is achieved. Arraying the mixing time, e.g. 0.0, 0.25, 0.5, 0.75, 1.0 second, is useful as it permits the assignment of sequences of coupled protons.

7.3.5.2 The 1D NOESY and ROESY Experiments

The following parameters are appropriate for 1D NOESY and ROESY experiments.

1) $f_2 = -0.5$ to 9.5 ppm
2) SS = 16
3) ns/i = 16–256 (with gradients), see Note 2 below
4) Mixing time: 0.5 second
5) Apodization: 2-Hz line broadening
6) np = 32 768
7) fn = 65 536

Note 1: NOESY is suggested for molecules of MW < 600 Da, while ROESY is strongly recommended for those of MW > 600 Da.

Note 2: Both NOESY and ROESY measure transient NOE buildup and are relatively insensitive, thus requiring increased numbers of scans.

7.3.5.3 The Multiplicity-Edited HSQC Experiment

When dealing with any compound that contains an appreciable number of carbons, it is highly recommended that one obtain a ^{13}C NMR spectrum, especially when some degree of uncertainty exists concerning the structure of the compound in question. It is further recommended to record both DEPT and HSQC spectra (again, particularly for unknowns), so that each experiment can serve as a check on the other and better reveal spectral anomalies such as overlapping signals. If sample is very limited, however, there might not be enough material to determine a ^{13}C NMR spectrum or an edited ^{13}C spectrum.

The gradient-selected, spin-echo HSQC experiment permits one to obtain an edited HSQC spectrum in which XH and XH$_3$ groups are phased in one direction (typically positive) and XH$_2$ groups are phased oppositely (negative). This experiment suffers from two disadvantages: (i) a 15–25% decrease in sensitivity compared with the corresponding HSQC sequence, with larger molecules that have shorter T_1s exhibiting the greater reductions and (ii) potential cancelation of closely situated negative methylene

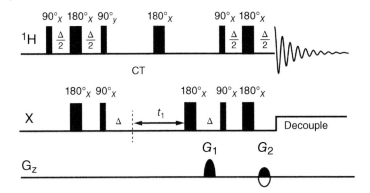

Figure 7.26 The multiplicity-edited gradient HSQC pulse sequence. The relative strengths of gradients G_1 and G_2 are 4 and 1 G cm^{-1}, respectively, as shown, when X = ^{13}C.

and positive methane or methyl signals. The experiment, however, largely eliminates the need to conduct a DEPT experiment, except in the case of complete unknowns. The multiplicity-edited HSQC experiment is basically an HSQC pulse sequence, in which two additional delay times (Δ) and a 180° X pulse are inserted around t_1, as illustrated in Figure 7.26.

Comparison HSQC spectra of T-2 toxin (**7.1**) are given in Figure 7.27. In the edited-HSQC spectrum, CH and CH$_3$ cross peaks are shown as red signals and CH$_2$ cross peaks as blue signals. Cross peaks arising from three methylene groups appear at the following C/H coordinates: δ 27.9/δ 2.41 and 1.91 (7), δ 43.8/δ 2.17 (2'), and δ 47.4/δ 3.06 and 2.80 (13). The blue peaks clearly differentiate the cross peaks from methylene 2' and 7 from the others, but those of methylene-13 are red instead of the expected blue and look like dispersive signals (Figure 2.11). The reason for their unexpected appearance as red peaks is that the one-bond C—H coupling constants for epoxides are about 175 Hz and thus approximately 40 Hz different from the average value of 140 Hz used for $^1J_{CH}$'s. By comparison, the $^1J_{CH}$ values for the sp^3 carbons are much closer to 140 Hz, ranging from 125 to 130 Hz.

The following parameters are appropriate for edited-HSQC experiments:

1) $f_2 = -0.5$ to 9.5 ppm
2) $f_1 = -5$ to 165 ppm
3) SS = 16
4) ns/i = 4–32
5) DT = 0.5–1.0 second
6) $^1J_{CH} = 145$ Hz
7) f_2 apodization: Gaussian
8) f_1 apodization: Gaussian
9) np = 2048, ZF(f_2) = 4096, ni = 256, LP = 1024, ZF(f_1) = 2048.

7.3.5.4 The H2BC Experiment

In Section 7.3.3.1, we saw that the critical delay time Δ in the HMBC experiment is set to approximately 8 Hz, which is an average value of $(2^n J_{CH})^{-1}$, with n usually equal to 2 or 3. Implicit in this relationship is the fact that the HMBC experiment cannot

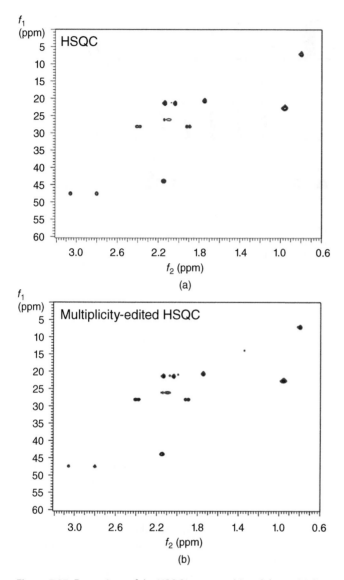

Figure 7.27 Expansions of the HSQC spectrum (a) and the multiplicity-edited spectrum (b) of T-2 toxin. The colors red and blue correspond to positive and negative peaks.

differentiate two-bond from three-bond (and occasionally greater) X—H couplings. This is an unfortunate shortcoming, especially with regard to problems involving structural elucidation (Chapter 8).

Sorensen [16, 17], however, developed a variation of the HMBC experiment, H2BC, shown in Figure 7.28, which can distinguish, with limitations, between two-bond and three-bond (or greater) X—H couplings for *protonated* carbons or nitrogens. The basis of the two- and three-bond differentiation lies in the modulation of two-bond X—H couplings by three-bond H—H couplings and is illustrated for C—H couplings in the

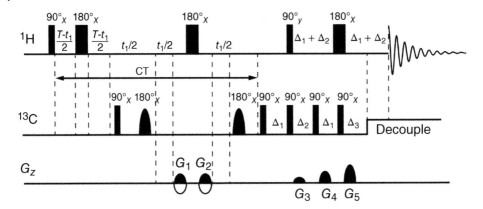

Figure 7.28 The H2BC pulse sequence. Time T is adjusted to make the constant time, CT, around 20 ms. Delays Δ_1–Δ_3 are part of a low-pass J filter and are set to the inverse of an average, a low and a high value of $^1J_{CH}$, respectively. Gradients G_1 and G_2 alternate polarity for *echo/anti-echo* selection and gradients G_3–G_5 are set for coherence selection. Gradient strengths for G_1–G_5 are 10, 10, 1.51, 3.52, and 10.1 G cm^{-1}, respectively.

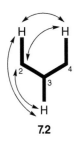

7.2

C-ring of T-2 toxin, **7.2**. If we examine the coupling between H-2 and H-3, on the one hand, and H-2 and H-4, on the other, we see that H-2 can be three-bond (vicinally) coupled to H-3, but only weakly four-bond coupled ($^4J_{HH} \sim 0$), if at all, to H-4 (Chapter 4). If we next consider the two-bond coupling between C-2 and H-3 and the three-bond coupling between C-2 and H-4, we conclude that H-2 can modulate the coupling of C-2 to H-3, but not that of C-3 to H-4 because of the above strong coupling of H-2 and H-3 and the very weak, to nil, coupling of H-2 and H-4. A correlation is thus observed in the H2BC spectrum between C-2 and H-3 but not between C-2 and H-4. Clearly, the corresponding situation applies in which H-4 modulates the coupling of C-4 to H-3, and a correlation is found between C-4 and H-3.

The experimental consequence of this H—H modulation of C—H couplings is that cross peaks corresponding to two-bond correlations in HMBC spectra are present while those corresponding to three-bond correlations are absent. A note of caution, however, is most important for situations like those observed for the very C$_2$—C$_4$ fragment described above. If the coupling between H-3 and H-4 is small because (i) the H-3/H-4 dihedral angle is approaching 90° or (ii) the H-3/H-4 coupling is reduced because of presence of electronegative elements such as oxygen, on C-3 and C-4, it may not be possible to differentiate the coupling between H-2 and H-4 to C-3. Comparison H2BC and HMBC spectra of T-2 toxin are given in Figure 7.29.

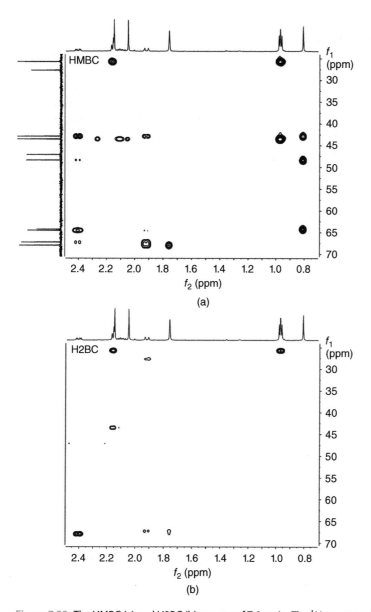

Figure 7.29 The HMBC (a) and H2BC (b) spectra of T-2 toxin. The ^{1}H spectrum is shown on top and the carbon spectrum is shown vertically on the left.

Several features can be observed in the comparison spectra of Figure 7.29. First, the two-bond, C—H cross peaks appearing at the coordinates δ 26.0/δ 2.17 (C-3′, H-2′), δ 26.0/δ 0.98 and δ 0.96 (C-3′, CH_3-4′, and CH_3-5′), and δ 68.3/δ 2.41 (C-8, H-7A) are clear. Second, the lower sensitivity of the H2BC experiment becomes evident when it is realized that, while both experiments were performed with the same number of (i) increments and (ii) transients per increment, the H2BC spectrum is shown at a vertical scale three times that of the HMBC spectrum.

When applying the H2BC experiment, the spectroscopist must be mindful of several things. First, the experiment is applicable only to *protonated* nuclei, and H2BC correlations are, therefore, not observed for quaternary carbons 5 and 6 and geminally coupled protons: δ 48.8/δ 0.81 (C-5, CH_3-14) and δ 43.3/2.41 or δ 1.91 (C-6, H-7A or H-7B). Like COSY and DQF-COSY experiments, HMBC and H2BC are complementary, not supplementary, and thus should be performed in addition to the usual HMBC experiments. Second, discrimination between two- and three-bond X—H couplings relies on the presence of adequate vicinal (three-bond) H—H coupling. When these coupling constants are too small, the resulting cross peaks arising from two-bond X—H couplings also may be absent. An example is observed in Figure 7.29, in which a strong two-bond correlation is observed for H-7A (δ 2.41) and C-8 (δ 68.3), but none for H-7B (δ 1.91) and C-8. Inspection of Table 8.6 at the end of Chapter 8 reveals that $^3J_{7A,8} = 5.5$ Hz while $^3J_{7B,8} = 0$.

In addition, longer-range C—H couplings can be observed when molecular geometry is conducive to long-range H—H couplings. Figure 7.29 also shows very weak correlations between C-11 (δ 67.6) and both H-7B (δ 1.91) and CH_3-16 (δ 1.75). These, respective, weak three-bond and four-bond C—H correlations result from equally weak (i) four-bond, W-coupling and (ii) five-bond, homoallylic coupling, respectively, between H-11 (δ 4.34) and H-7B (δ 1.91) and H-11 and CH_3-16 (δ 1.75).

Nevertheless, if properly used, the H2BC experiment is a helpful technique for those who determine molecular structures. The following parameters are appropriate for H2BC experiments:

1) $f_2 = -0.5$ to 9.5 ppm
2) $f_1 = -5$ to 220 ppm
3) SS = 16
4) ns/i = 16–64
5) DT = 0.5–1.0 second
6) $^1J_{CH} = 145$ Hz (or 130 Hz and 165 Hz if a two-step J-filter is used)
7) Δ_{CT} (fixed time) = 16–22 ms
8) f_2 apodization: Gaussian
9) f_1 apodization: Gaussian
10) np = 2048, $ZF(f_2) = 4096$, ni = 256, LP = 1024, $ZF(f_1) = 2048$.

7.3.5.5 Nonuniform Sampling

We saw in Section 7.3.5.4 that sensitivity vs time issues are especially critical in the t_1 dimension of 2D experiments, as there generally are twice as many data points in the f_2 dimension as there are in the f_1 domain even when employing LP and zero filling. Another factor that reduces f_1 resolution in heteronuclear 2D experiments is the, generally, much wider f_1 dimension compared with the f_2 domain. Now, already scarce data points must be spread over even greater frequency ranges. This disparity in resolution between the f_1 and the f_2 dimensions can be seen easily in Figure 7.9, in which the cross peaks are appreciably narrower in the horizontal (f_2) dimension than in the vertical (f_1) dimension.

A key factor that limits resolution along the indirect (f_1) dimension is the requirement that data sampling be performed at uniform intervals, a requirement that is imposed by the discrete Fourier transform. Several decades ago, a technique known as NUS [18–21] was proposed, in which the equidistant time increments (ni) used during the evolution periods of 2D NMR experiments could instead be replaced by a series of *irregularly spaced intervals*. The frequency spectrum is no longer obtainable by the discrete Fourier transform but can be provided by several types of algorithms, such as maximum entropy reconstruction (MaxEnt), iterative soft thresholding (IST), and compressed sensing (CS). The primary requirement in current practice is that the actual evolution times used for the nonuniform acquisition must be a subset of the uniformly spaced evolution times that would have been used to acquire the 2D spectrum with the same f_1 spectral window and maximum evolution time.

NUS is then carried out in the following manner at the beginning of the 2D data collection process. The NMR data for a 2D NMR spectrum are obtained for only a *selected subset* of evolution times that would have been acquired uniformly, sometimes called the Nyquist grid, and generally including the last increment. Sampling coverage by NUS refers to the fraction of the set of uniform increments that is selected via NUS, and is often of the order of 25–33% of the Nyquist grid. This requirement, therefore, reduces the data acquisition time from 67% to 75%. Figure 7.30 presents two interferograms in which the blue trace on top (a) is a 50% sampled interferogram that shows "zero-intensity" points in those places for which data were not acquired, which line up along the horizontal line marked (a), while the bottom (b) trace shows the resulting interferogram after NUS reconstruction.

The practice of NUS has become much more common, especially for obtaining 2D NMR data of small molecules in solution. Several criteria should be considered in choosing to acquire data nonuniformly in the indirect evolution period. It is critical in the NUS process that the smaller set of collected data points have adequate signal-to-noise. As a general rule, if the S/N ratio in the indirectly acquired dimension (f_1) is deemed to be too low to acquire the 2D NMR data by conventional uniform sampling, then the use of

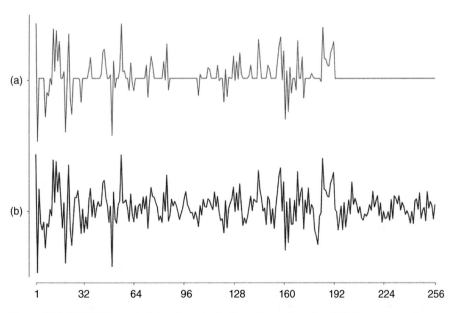

Figure 7.30 50% NUS sampled data from the indirect dimension of an HSQC experiment is shown in (a). The same data reconstructed to the full 256 point grid is in (b).

NUS will not be successful either. Also, the spectroscopist must pay attention that the number of samples not be too small. The number of intervals in the t_1 dimension to be sampled nonuniformly should be at least twice the number of signals expected in any given f_1 slice. The higher the S/N in the direct dimension (f_2), the more the coverage can be reduced such that a very concentrated solution should be treated more carefully with 33% or 50% coverage. Furthermore, if the degree of coverage is too small, then large gaps in the nonuniformly spaced intervals lead to noise-like artifacts in the final reconstructed spectrum, regardless of which method is used to reconstruct the data.

NUS can be used in one of two ways: to decrease experimental acquisition times or to increase spectral resolution, and representative spectra are shown in Figure 7.31. A conventional HSQC spectrum for T-2 toxin is shown in Figure 7.31a with ni = 128 increments and an acquisition time of six minutes. Figure 7.31b has a NUS density of 25% (ni = 36), but, due to the increased spectral resolution of NUS data processing, its data-point resolution is essentially identical to that in Figure 7.31a but acquired in one third of the time. Figure 7.31c has an acquisition time of 22 minutes, which is nearly four times longer than that in Figure 7.31a. With 512 increments, however, its spectral

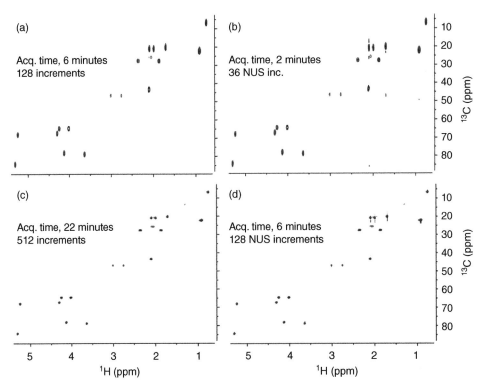

Figure 7.31 The NUS HSQC spectra of T-2 toxin. The spectra are multiplicity edited so that CH and CH_3 correlations are red and CH_2 peaks are blue. (a) Conventional sampling, ni = 128, t = 6 minutes, (b) 25% sampled with 36 NUS increments on a 128 point grid, t = 2 minutes, (c) conventional sampling, ni = 512, t = 22 minutes, (d) 25% sampled 128 NUS increments from a 512 point grid, t = 6 minutes. Both (a) and (d) were acquired in 6 minutes but (d) has the same resolution as (c), which took 22 minutes to acquire. The S/N is roughly the same in all. There are some artifacts that appear close to the noise level in the NUS experiments.

resolution is considerably better. Figure 7.31d achieves the same increased resolution of the 22-minute experiment in just 6 minutes, because it has the same number of increments as Figure 7.31a (ni = 128). It also is evident that, while resolution has been considerably increased in Figure 7.31c,d, it has been at the expense of sensitivity. The reason for this effect is that the nonuniformly sampled data points are now dispersed over a larger t_1 dimension, and the later data points are reduced in intensity as the interferograms decay to zero intensity. If spectral crowding is a problem, however, the extra time spent in spectral acquisition is certainly well worth the resulting excellent data-point resolution.

Mestrelab Research offers software that processes NUS experiments rapidly and with data from any of the major spectrometer manufacturers.

7.3.5.6 Pure Shift NMR

Pure shift NMR [22–24] is the second of three recent developments (NUS being the first and covariance NMR the third (Section 7.3.5.7)), in the field of NMR spectroscopy. Here, ^{1}H NMR spectra, either 1D or 2D, are ^{1}H decoupled so that all of the proton multiplets are collapsed to single lines, in the absence, of course, of NMR-active nuclei such as ^{31}P or ^{19}F that would continue to exert spin-coupling effects. So-called broadband proton decoupling is based on the idea of refocusing the evolution of proton–proton coupling while permitting the evolution of proton chemical shifts. Proton homonuclear couplings thus are removed from the resulting ^{1}H NMR spectrum, and only ^{1}H chemical shifts remain. Standard 1D and ^{1}H pure shift NMR spectra of menthol (**7.3**) are shown in Figure 7.32.

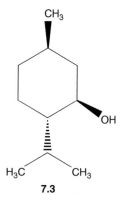

7.3

In the spectra of menthol pictured in Figure 7.32, a control spectrum is shown in the top trace (a). The spectral region between δ 0.9–1.0 is especially crowded and that between δ 0.8 and 1.1 is sufficiently so that the chemical shifts of protons 4B, 6B, 7, and 9 are uncertain. The middle trace employs the older Zangger–Sterk refocusing element and shows considerable simplification of the conventional ^{1}H spectrum. The signals of H-2 (δ 1.10), H-3B (δ 0.83), and H-4B (δ 0.98), however, exhibit considerable distortions. The bottom trace employs the Pure Shift Yielded by CHirp Excitation (PSYCHE) refocusing element, which decouples the signal of every ^{1}H nucleus in menthol from its proton neighbors. A ^{1}H spectrum now is produced that consists of 13 single lines, and the chemical shifts of once obscured H-4B, H-6B, Me-7, and Me-9 are easily discernible.

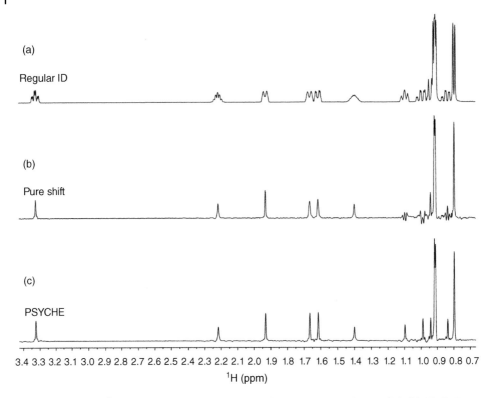

Figure 7.32 The 1D ^{1}H spectra of menthol. Conventional (a), Zangger–Sterk pure shift (b), PSYCHE pure shift (c). The vertical scale of the pure-shift spectra were increased by ~600 times compared with the normal 1D.

Standard and pure shift HSQC spectra of menthol appear in Figure 7.33. The difference between the standard and ^{1}H-decoupled HSQC spectra is not so dramatic as that for the previous, crowded 1D spectra primarily because the HSQC signals are now dispersed into two dimensions. An advantage here is that there is no sensitivity loss. Decoupling is not done by spatial separation with gradients. The HSQC procedure already edits out the protons bound to C-12, so that the only coupling that remains after ^{1}H—^{1}H decoupling is that of the CH$_2$ protons to their geminal partners. These protons cannot be decoupling by this method, as they share the same carbon. The C—H correlations are seen to be narrower along the horizontal, f_2 axis due to the removal of ^{1}H—^{1}H coupling for all except the CH$_2$ geminal partners.

Pure shift NMR may be understood by examining the behavior of a pair of spin-coupled nuclei, H$_A$ and H$_X$, and following the evolution of their magnetization over a time period τ. During time τ a 180° refocusing pulse is applied *selectively* only to H$_X$ at the midpoint ($\tau/2$), which creates a spin-echo (Section 5.5, Figure 5.13). If we examine the behavior of H$_A$, we see that its chemical shift evolves during time τ because H$_A$ is not affected by the refocusing pulse. Conversely, J_{AX} spin-coupling evolution is refocused at the end of time τ because only H$_X$ was inverted by the selective, refocusing pulse. This selective inversion has the effect of reversing the direction of the spin coupling evolution, thus eliminating spin coupling between H$_A$ and H$_X$.

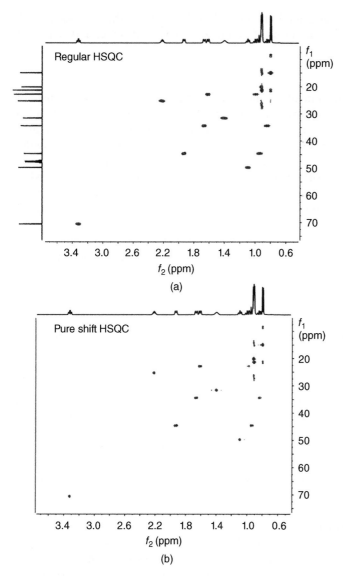

Figure 7.33 The multiplicity-edited HSQC spectra of menthol. Conventional (a) and pure shift (b). The geminal protons in blue are not ^{1}H-decoupled from each other and so the resolution enhancement along f_2 is not so large as for the CH and CH_3 protons in red.

Pure shift experiments require that for the generation of broadband-decoupled proton NMR spectra, such as for spin decoupling of each proton in a molecule from its spin-coupled partners to occur, the latter must be inverted at the same time that the observed proton is unaffected. This is often an especially severe requirement when one considers that the chemical shifts of many geminal protons are very similar and thus exhibit very strong spin coupling. Selectively irradiating any proton without affecting a close-lying coupling partner is not an easy task. There are several methods for accomplishing this imposing requirement, and the PSYCHE method of Morris et al. [22, 23] is especially appealing in terms of minimizing sensitivity losses and achieving full homonuclear decoupling.

7.3.5.7 Covariance NMR

Covariance NMR is certainly one of the most noteworthy developments in the field of NMR spectroscopy, as proposed by Brüschweiler [25–28]. In this technique, two 2D NMR experiments are interwoven in such a manner that composite 2D spectra are produced with greatly enhanced resolution and sensitivity. We already have touched on the subject of f_1 vs f_2 resolution in Sections 7.2.1.2 and 7.3.5.5, in which we saw that f_2 resolution almost always is greater than that of f_1 because the number of points used to describe the f_2 dimension should, especially with today's very large data systems, almost always be larger than the number of increments that are used to describe f_1. While LP and NUS can somewhat alleviate this mismatch, signal broadening of correlation peaks in the f_1 dimension is a problem that NMR spectroscopists and chemists have had to live with, and as NMR spectra become more congested, the problem has become more acute.

Covariance NMR combines individual 2D spectra to produce a third 2D spectrum. An absolute requirement is that the 2D spectra that are to be combined must have a common axis (although not exactly identical spectral widths), and the f_1 dimensions, which have lower digital resolution than their f_2 counterparts, are perfect candidates. In homonuclear experiments such as COSY, two 2D spectra can be combined along their f_1 domains to produce COSY spectra in which f_2 resolution in both dimensions is identical.

In the case of heteronuclear combination, such as HSQC plus COSY or TOCSY, the resulting (HSQC + COSY) or especially (HSQC + TOCSY) 2D spectra could be more difficult, if not impossible, to observe directly if one were sample limited. Note that we use the following terms, (HSQC + COSY) or (HSQC + TOCSY), to differentiate these covariance experiments from their directly determined HSQC–COSY or HSQC–TOCSY procedures. Two examples of the most common covariance spectra are discussed in the next subsections.

7.3.5.7.1 Direct Covariance NMR

Direct covariance NMR concerns homonuclear experiments such as COSY, TOCSY, NOESY, and ROESY [25, 27, 28]. In a conventional 2D NMR experiment, a (t_2, t_1) time matrix is Fourier transformed along the direct time dimension (t_2) to yield a mixed time–frequency matrix (f_2, t_1), which is then Fourier transformed along the indirect time dimension (t_1) to yield the familiar, symmetrical (f_2, f_1) frequency matrix plots (Section 6.1). By comparison, when covariance processing is employed, the (t_2, t_1) time matrix is again Fourier transformed along the direct time dimension (t_2) to yield a mixed time–frequency matrix (f_2, t_1), denoted by **S**. In the second step, however, a covariance matrix $(\mathbf{C}^2)$ is computed by multiplying the transpose of the above mixed time–frequency matrix by itself such that $\mathbf{C}^2 = (f_2, t_1)^T (f_2, t_1)$ or $\mathbf{S}^T\mathbf{S}$.

If we examine the two procedures described above, we see that, for instance, the conventional 2D spectrum may have 1024 data points along f_2 and 256 data points along f_1. Conversely, the covariance 2D spectrum, in which the resulting

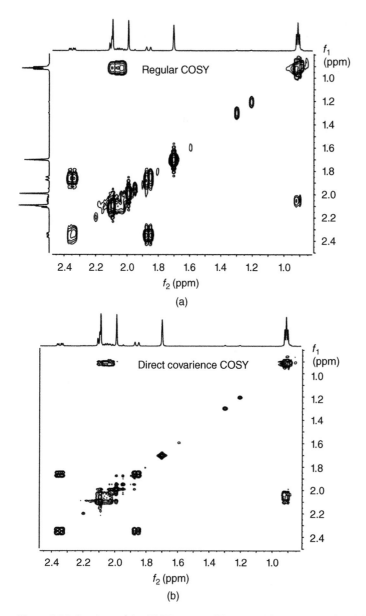

Figure 7.34 Sections of the COSY spectra of T2-toxin: (a) conventional and (b) direct covariance. The resolution enhancement in the covariance processed spectrum is apparent along f_1. The regular ^{1}H spectrum is attached on the top and left side.

data matrix ($\mathbf{C}^2$) has four times as many data points in the t_1 dimension, has 1024 data points along both dimensions. In this manner, resolution is effectively transferred from the direct dimension (f_2) to the indirect dimension (f_1). Figure 7.34 illustrates the greatly enhanced f_1 resolution that can be achieved with direct covariance NMR.

7.3.5.7.2 Generalized Indirect Covariance NMR

The direct covariance experiment, presented in the above section, produces *symmetrical* spectral matrices, from which square roots can be calculated, and symmetrical ^{1}H–^{1}H spectra thus derived. Unsymmetrical matrices, e.g. ^{1}H–^{13}C, however, are not amenable to square root operations. The generalized indirect covariance experiment was developed by Brüschweiler and Snyder [26, 28] to produce "intermediate symmetric covariance matrices," from which the square roots of what would have been otherwise unsymmetrical matrices could be calculated.

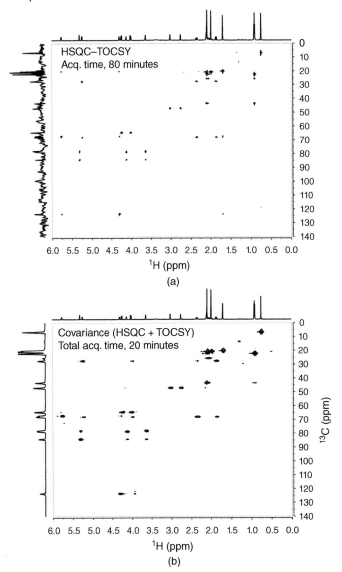

Figure 7.35 The HSQC–TOCSY spectra of T-2 toxin: (a) conventional and (b) covariance (HSQC + TOCSY). The vertical traces on the left of each spectrum show the f_1 projections that are compared in Figure 7.36. The regular ^{1}H spectrum is on the top.

For example, an HSQC (H) and a TOCSY (T) can be combined to produce a new stacked spectrum **S**:

$$\mathbf{S} = \begin{bmatrix} H \\ C \end{bmatrix}$$

In the next step, **S** is multiplied by its transpose to produce a generalized covariance matrix **C**:

$$\mathbf{C} = \mathbf{SS}^{\mathrm{T}} = \begin{bmatrix} H \\ C \end{bmatrix} \begin{bmatrix} H^{\mathrm{T}} C^{\mathrm{T}} \end{bmatrix}$$

The critical point is that matrix **C** is now symmetric and thus amenable to square root operations. This procedure has led to the combination of HSQC and either COSY or TOCSY spectra to produce (HSQC + COSY) and (HSQC + TOCSY) spectra in far less time than would be required to perform the direct HSQC–COSY and, especially, HSQC–TOCSY experiments [29–31].

Martin has demonstrated that for the direct HSQC–TOCSY vs the covariance (HSQC + TOCSY) procedure, the experimental times are 4 hours vs <20 minutes and S/N is c. 77 : 1 vs 8 : 1 [29]. Comparison spectra for T-2 toxin are shown in Figure 7.35.

The great sensitivity advantages of general indirect covariance experiments over their directly observed analogs are vividly illustrated for the HSQC–TOCSY experiment for T-2 toxin. Figure 7.35a shows a directly observed HSQC–TOCSY spectrum with an acquisition time of 80 minutes. The total acquisition times for the HSQC and TOCSY spectra of the corresponding (HSQC + TOCSY) covariance spectrum in Figure 7.35b were just 20 minutes. Inspection of the two spectra shows that the sensitivity of the latter is considerably better than the former. The sensitivity advantage of the covariance experiment over that of the directly observed experiment, however, is dramatically illustrated in Figure 7.36 for the corresponding projections along the f_1 axis for the above experiments. The general indirect covariance (HSQC + TOCSY) spectrum now is seen to exhibit very much greater sensitivity than that of the directly observed HSQC–TOCSY spectrum.

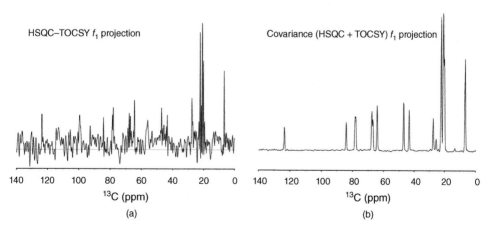

Figure 7.36 The HSQC–TOCSY f_1 projections of T-2 toxin: (a) conventional and (b) covariance (HSQC + TOCSY). The S/N is approximately 10 times larger in the covariance-processed spectrum.

7.3.6 Pure Shift-Covariance NMR

A more recent development in covariance NMR spectroscopy concerns the combination of covariance and pure shift methods to produce pure shift-covariance NMR experiments, in which ^{1}H signals appear as single lines [32, 33]. These experiments should be very useful in situations in which the ^{1}H spectra are unusually complex due to severe signal overlapping. Mestrelab Research is one of several widely available data-processing platforms that offers covariance NMR processing rapidly and with very little input from user.

References

7.1 Bildsoe, H., Donstrup, S., Jakobsen, H.S., and Sorensen, O.W. (1983). *J. Magn. Reson.* 53: 154–162.
7.2 Doddrell, D.M., Pegg, D.T., and Bendall, M.R. (1982). *J. Magn. Reson.* 48: 323–327.
7.3 Sanders, J.K.M. and Mersh, J.D. (1982). *Prog. Nucl. Magn. Reson. Spectrosc.* 15: 353–400.
7.4 Hwang, T.L. and Shaka, A.J. (1995). *J. Magn. Reson. A* 112: 275–279.
7.5 Reynolds, W.F. and Enriquez, R.G. (2002). *J. Nat. Prod.* 65: 221.
7.6 Hoch, J.C. and Stern, A.S. (1996). *NMR Data Processing*. New York: Wiley-Liss.
7.7 Bax, A. and Freeman, R. (1981). *J. Magn. Reson.* 44: 542.
7.8 Bax, A. and Davis, D.G. (1985). *J. Magn. Reson.* 65: 355.
7.9 Bax, A. and Subramanian, S. (1986). *J. Magn. Reson.* 67: 565.
7.10 Bodenhausen, G. and Rubin, D.J. (1980). *Chem. Phys. Lett.* 69: 185.
7.11 Bax, A. and Morris, G.A. (1981). *J. Magn. Reson.* 42: 501.
7.12 Bax, A. and Summers, M.F. (1986). *J. Am. Chem. Soc.* 108: 2093.
7.13 Reynolds, W.F., McLean, S., Perpick-Dumont, M., and Enriquez, R.G. (1989). *Magn. Reson. Chem.* 27: 162.
7.14 Macura, S., Huang, Y., Suter, D., and Ernst, R.R. (1981). *J. Magn. Reson.* 43: 259–281.
7.15 Bothner-By, A.A., Stephens, R.L., Lee, J. et al. (1984). *J Am. Chem. Soc.* 106: 811–813.
7.16 Nyberg, N.T., Duus, J.O., and Sorensen, O.W. (2005). *J. Am. Chem. Soc.* 127: 6154–6155.
7.17 Nyberg, N.T., Duus, J.O., and Sorensen, O.W. (2005). *Magn. Reson. Chem.* 43: 971–974.
7.18 Barna, J.C.J., Laue, E.D., Mayger, M.R.S. et al. (1987). *J. Magn. Reson.* 73: 69.
7.19 Barna, J.C.J. and Laue, E.D. (1987). *J. Magn. Reson.* 75: 384.
7.20 Mobli, M. and Hoch, J.C. (2008). *Concepts Magn. Reson. Part A* 32: 436–448.
7.21 Mobli, M. and Hoch, J.C. (2014). *Prog. Nucl. Magn. Reson. Spectrosc.* 83: 21–41.
7.22 Foroozandeh, M., Adams, R.W., Nilsson, M., and Morris, G.A. (2014). *J. Am. Chem. Soc.* 136: 11867–11869.
7.23 Foroozandeh, M., Adams, R.W., Nilsson, M., and Morris, G.A. (2014). *Angew. Chem. Int. Ed.* 53: 6990–6992.
7.24 Adams, R.W. (2014). *eMagRes* 2014 (3): 1–15.
7.25 Brüschweiler, R. and Zhang, F. (2004). *J. Chem. Phys.* 120: 5253–5260.
7.26 Snyder, D.A. and Brüschweiler, R. (2009). *J. Phys. Chem. A* 113: 12898–12903.
7.27 Snyder, D.A. and Brüschweiler, R. (2010). *Multidimensional Correlation Spectroscopy by Covariance NMR*, 97–105. Wiley.

7.28 Snyder, D.A. and Brüschweiler, R. (2016). Multidimensional spin correlations by covariance NMR. In: *Modern NMR Approaches to the Structure Elucidation of Natural Products*, 244–258. Cambridge: RSC Books.
7.29 Blinov, K.A., Larin, N.I., Williams, A.J. et al. (2006). *J. Heterocycl. Chem.* 43: 163–166.
7.30 Blinov, K.A., Larin, N.I., Williams, A.J. et al. (2006). *Magn. Reson. Chem.* 44: 107–109.
7.31 Martin, G.E., Hilton, B.D., Irish, P.A. et al. (2007). *J. Nat. Prod.* 70: 1393–1396.
7.32 Fredi, A., Nolis, P., Cobas, C. et al. (2016). *J. Magn. Reson.* 266: 16–22.
7.33 Fredi, A., Nolis, P., Cobas, C., and Parella, T. (2016). *J. Magn. Reson.* 270: 161–168.

8

Structural Elucidation: Two Methods

8.1 Part A: Spectral Analysis

A considerable number of NMR experiments have been introduced and discussed in this book. In order to illustrate how a series of such experiments is used in practice, we will take the readers through the complete structural elucidation exercise for the compound T-2 toxin, many of whose 2D spectra were encountered as examples in Chapter 7. If we had no knowledge of the compound's structure, such a study would begin with the determination of its infrared, ^{1}H and ^{13}C NMR spectra, and high-resolution mass spectrum. This compound is a white solid that is soluble in chloroform.

The high-resolution mass spectrum (MW = 466.5573) gives us several pieces of information about the compound. The even molecular weight (466) indicates that the unknown contains an *even* number of nitrogen atoms and probably none (zero, as in the attached proton test (APT) experiment, is considered to be an even number). If we make the preliminary, and usually justified, assumption that the unknown compound contains only carbon, hydrogen, and oxygen, the high-resolution, mass spectra-determined molecular weight supports a molecular formula of $C_{24}H_{34}O_9$. This formula, in turn, indicates the presence of eight units of unsaturation. While this unsaturation number is compatible with the presence of an aromatic compound, the H-to-C ratio appreciably greater than unity suggests that the unknown compound likely is composed primarily of aliphatic carbons. The presumptive molecular formula also indicates that T-2 toxin is heavily oxygenated.

Strong infrared absorptions centered at 1740 and 1100–1300 cm^{-1} suggest the presence of two or more carbonyl groups, which more likely are ester functionalities than ketone groups. Moreover, the occurrence of nine oxygens in the molecular formula supports the inference that the carbonyl absorptions are probably due to ester, rather than ketone, groups.

8.1.1 ^{1}H NMR Data

The ^{1}H NMR spectrum of a 15-mg sample of the unknown compound is shown in Figure 8.1.

The ^{1}H signals are designated by letters (skipping "H" to avoid confusion) in decreasing chemical shift order (A—V). Their chemical shifts, any coupling data, and integrations are given in Table 8.1. Chemical shifts, coupling constants, and integrals can be measured with considerable accuracy from expansions of Figure 8.1.

Nuclear Magnetic Resonance Spectroscopy: An Introduction to Principles, Applications, and Experimental Methods,
Second Edition. Joseph B. Lambert, Eugene P. Mazzola, and Clark D. Ridge.
© 2019 John Wiley & Sons Ltd. Published 2019 by John Wiley & Sons Ltd.

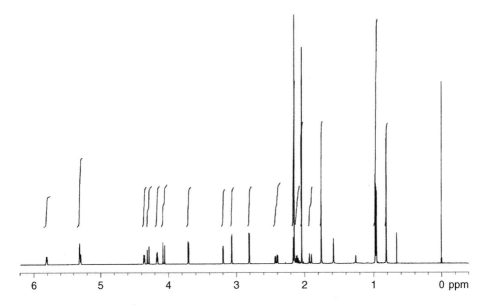

Figure 8.1 The 500-MHz ^{1}H NMR spectrum of T-2 toxin.

Integration of ^{1}H spectra is especially important for unknown materials. The presence of six signals, each of which integrates to three protons, indicates that this unknown contains six methyl groups. The only real difficulty in integration arises for the signals centered at δ 2.17, 2.16, and 2.11. Those at δ 2.17 and 2.16 cannot be integrated separately and total five protons. Since the signal at δ 2.16 (O) is a sharp singlet, however, it must in all likelihood be a methyl group. Two protons remain for the signal (N) at δ 2.17, which appears to be a methylene doublet whose low-frequency (high field) signal is obscured by the sharp, methyl singlet (O) at δ 2.16. The fact that these methylene protons exhibit just one coupling (6.3 Hz) means that they are enantiotopic and likely are located on a side chain. If they were incorporated in a ring, they would almost certainly be diastereotopic.

The signal at δ 2.11 (between the methyl signals at δ 2.16 and 2.03) is a complex multiplet but integrates cleanly to one proton. In addition, the integrals sum to 34, which is the number of protons required by the high-resolution mass spectrum. Moreover, according to the tabulated coupling constant data, the protons whose signals appear at δ 4.28 and 4.06 (both with $J = 12.6$ Hz) and at δ 2.41 and 1.91 (both with $J = 14$ Hz) appear to be geminal pairs. Other couplings that are similar and even identical are observed, but the establishment of additional spin systems is better deferred until the correlation spectroscopy (COSY) spectrum is recorded.

A quick look at the data in Table 8.1 confirms the earlier inference that few of the protons in the unknown compound can be attached to alkenic or aromatic carbons (Section 3.2 and Figure 3.9).

8.1.2 ^{13}C NMR Data

The ^{1}H-decoupled ^{13}C NMR spectrum of the 50-mg sample of the unknown compound is shown in Figure 8.2.

Table 8.1 ^{1}H NMR data.

Designation	Chemical shift[a]	Coupling constants (Hz)	Integration (H)
A	5.81 d	5.0	1
B	5.31 d	2.4	1
C	5.28 d	5.5	1
D	4.34 d	5.0	1
E	4.28 d	12.6	1
F	4.16 ddd	5.0, 2.4, 2.4	1
G	4.06 d	12.6	1
I	3.70 d	5.0	1
J	3.20 d	2.4	1
K	3.06 d	4.0	1
L	2.80 d	4.0	1
M	2.41 dd	14.0, 5.5	1
N	2.17 d	6.3	2
O	2.16 s	—	3
P	2.11 m	6.4, 6.3	1
Q	2.03 s	—	3
R	1.91 d	14.0	1
S	1.75 s	—	3
T	0.98 d	6.4	3
U	0.96 d	6.4	3
V	0.81 s	—	3

a) Multiplicities: d = doublet, m = multiplet, s = singlet.

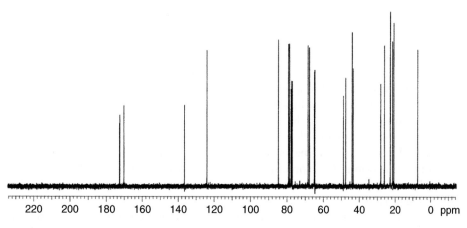

Figure 8.2 The 125 MHz ^{13}C NMR spectrum of T-2 toxin.

Table 8.2 ^{13}C and HSQC NMR Data.

Designation	Chemical shift	n[a]	Directly attached hydrogen(s)
1	173.0	0	—
2	172.8	0	—
3	170.5	0	—
4	136.6	0	—
5	124.2	1	A
6	84.5	1	B
7	79.2	1	I
8	78.5	1	F
9	68.3	1	C
10	67.6	1	D
11	64.9	2	E, G
12	64.6	0	—
13	48.8	0	—
14	47.4	2	K, L
15	43.8	2	N
16	43.3	0	—
17	27.9	2	M, R
18	26.0	1	P
19	22.5	3	T
20	22.4	3	U
21	21.1	3	O
22	21.0	3	Q
23	20.3	3	S
24	7.0	3	V

a) Number of attached protons.

The ^{13}C signals are designated by numbers in decreasing chemical shift order (1–24). Their chemical shifts and the number of attached protons are given in Table 8.2. A list of chemical shifts is printed by the spectrometer for those ^{13}C signals that exceed a threshold value that is selected to include likely signals and exceed the height of the noise. Quaternary carbons give rise to the smallest signals in the ^{13}C spectrum, and the intensities of the highest frequency (lowest field) resonances are generally what determine the threshold level. Since very small signals (those with small nuclear Overhauser effects (NOEs) or unusually long T_1's) sometimes barely exceed the noise, threshold levels must be set with caution and with the realization that real resonances may be missed in the signal-listing process.

Especially small resonances can sometimes be differentiated from noise spikes by their line widths, the latter tending to be very thin. Missing signals must always be a concern with unknown compounds if the number of observed ^{13}C signals cannot be confirmed by a high-resolution mass spectrum. While molecular weight data from such mass spectra can reveal the occurrence of absent resonances, the spectroscopist has to be cautious because molecular symmetry can result in fewer observed ^{13}C (and ^{1}H) signals than required by the mass spectral data. Keep in mind that the ^{13}C peak intensities (in contrast to ^{1}H intensities) are not necessarily a good measure of the relative numbers of contributing carbon atoms. In the present case of T-2 toxin, the number of carbon resonances is 24, the same number of carbons as required by the high-resolution mass spectrum.

A cursory examination of the data in Table 8.2 also supports the assumption that the unknown compound is essentially aliphatic in nature (Section 3.5 and Figure 3.11). The three highest frequency (downfield) signals (carbons 1–3) are in the ester carbonyl range (Section 3.5.3), while carbons 4 and 5 appear to be alkenic and require the presence of a monosubstituted, carbon–carbon double bond (Section 3.5.2). The resonances corresponding to carbons 6–12 are in the chemical shift range for aliphatic carbons that are attached to oxygen (Section 3.5.1).

8.1.3 The DEPT Experiment

For unknown compounds, the number of attached protons per carbon is best determined by the DEPT experiment (Sections 5.6 and 7.1.2.2). The DEPT spectrum of the 50-mg sample of T-2 toxin is shown in Figure 8.3.

It reveals the presence of six methyl, four methylene, and seven methine carbons. Subtracting these 17 protonated carbons from 24 identifies 7 quaternary carbons. The DEPT data are listed in the third column of Table 8.2. In addition, summing the 18 methyl protons, 8 methylene, and 7 methine accounts for 33 protons. The molecular formula and the integrals in Table 8.1 require 34 so that the final proton must be attached to oxygen.

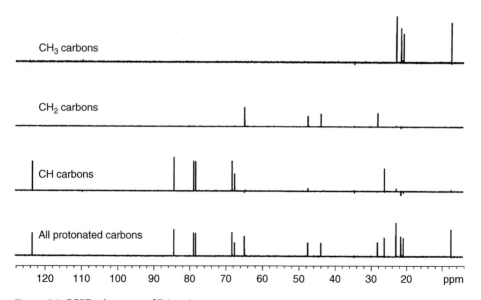

Figure 8.3 DEPT subspectra of T-2 toxin.

8.1.4 The HSQC Experiment

At this point in a structural elucidation exercise, some NMR spectroscopists prefer to begin assembling ^{1}H NMR spin systems by means of COSY experiments. COSY correlations, however, can be difficult to interpret in unknown compounds, since protons that are several bonds removed may exhibit spin coupling. In addition, vicinal couplings cannot always be distinguished from geminal couplings in congested spectra even if COSY-45 experiments are performed (Section 6.1). It is useful, therefore, to determine initially which protons are directly attached to specific carbons by an heteronuclear single quantum correlation (HSQC) experiment (Sections 6.2 and 7.3.2.2). Standard and gradient HSQC spectra of the 15-mg sample of the unknown compound are shown in Figure 7.15 for all of the protonated carbons except the signal at δ 124.2. Both of the full spectra, moreover, reveal that this carbon is directly bonded to the proton at δ 5.81.

The HSQC data are presented in the fourth column of Table 8.2. Thus we see that H_A is attached to C-5, H_B to C-6, H_E and H_G to C-11, H_S to C-23, and H_V to C-24. The seven quaternary carbons, of course, exhibit no direct proton connectivities, while H_J displays no carbon connectivity. Thus H_J must be the proton attached to oxygen. The HSQC data confirm that protons E and G and protons M and R are geminal pairs, as expected from their coupling constants (12.6 and 14 Hz, respectively). These data, however, also demonstrate that protons K and L constitute a third geminal system. They might easily have been mistaken for a vicinal pair because of their unusually small, 4-Hz coupling (Section 4.6). In addition, the doublet at δ 2.17 (N) is, indeed, found to represent a fourth geminal pair of protons. The partial redundancy found in the DEPT and HSQC experiments is very helpful with unknowns because it serves as a built-in check of the consistency of spectral interpretations.

8.1.5 The COSY Experiment

The construction of proton spin coupling networks is achieved by a variety of COSY experiments (Sections 6.1 and 7.3.1.1). Standard COSY and double-quantum-filtered—COSY (DQF—COSY) spectra of the 15-mg sample of the unknown compound are shown in Figure 7.11, and the data from these contour plots are summarized in Table 8.3. Coupling constants, which have been measured in the ^{1}H NMR spectrum, correspond to cross peaks and are included in parentheses in the table.

Four ^{1}H spin systems are observed in the COSY contour plots, three of which are illustrated in Figure 8.4.

The first, and simplest, is the K,L pair, which also was identified in the ^{1}H spectrum. The second group comprises H_P, protons N, and the methyl groups T and U. H_P is coupled to protons N and to methyls T and U, suggesting an isopropyl array with H_P in the middle. The third system is composed of protons B, F, I, and J. H_F is coupled to the other three members of this group, but there is no coupling among protons B, I, and J. The fourth and largest network includes protons A, C, D, E, G, M, R, and methyl S. In addition to the geminal pairs of protons already identified in this system (E and G, M and R), H_A and H_D constitute a vicinal pair, as do H_C and H_M.

Weaker cross peaks are observed between (i) methyl S and protons A, C, and D, (ii) H_D and protons G and R, (iii) H_E and H_M, and (iv) H_C and H_R. Since H_A is attached to an sp^2-hybridized carbon (5), and methyl S, from its chemical shift, might be likewise (Section 3.2.2), couplings between H_A and its vicinal partner (H_D) to methyl S (allylic and homoallylic, respectively, Section 4.7) are not unreasonable. Moreover, protons

Table 8.3 COSY NMR data.

Designation	Spin coupling partners[a]
A	C, D (5.0), S
B	F (2.4)
C	A, M (5.5), R, S
D	A (5.0), G, R, S
E	G (12.6), M
F	B (2.4), I (5.0), J (2.4)
G	D, E (12.6)
I	F (5)
J	F (2.4)
K	L (4.0)
L	K (4.0)
M	C (5.5), E, R (14.0)
N	P (6.3)
P	N (6.3), T (6.4), U (6.4)
R	C, D, M (14.0)
S	A, C, D
T	P (6.4)
U	P (6.4)

[a] Coupling constants from the ^{1}H spectrum in parentheses.

C, M, and R form a three-spin subsystem with $^3J_{CM} = 5.5$ Hz. Since $^3J_{CR}$ is present as a small cross peak in the COSY spectrum but does not appear in the ^{1}H NMR spectrum, the H_C—C_9—C_{17}—H_R dihedral angle must be close to 90°. This subsystem is an excellent example of the way in which small vicinal and long-range couplings can give rise to COSY cross peaks that are very similar in appearance. Weak cross peaks between H_C and methyl S and between H_D and H_R serve to link the H_C—H_M—H_R and the H_D—H_A—methyl S subsystems.

A total correlation spectroscopy (TOCSY) experiment is not employed in this structural determination exercise despite the fact that the extended proton spin network A, C, D, E, G, M, R, and methyl S would appear to take advantage of relayed coupling. As a general rule, however, only those spin systems in which there is *appreciable* coupling between each member of the system and both the preceding and following protons are appropriate for TOCSY experiments. If there are very small couplings at certain points in the chain, relaying will tend to stop there. It can be readily seen in Figure 7.11a and Table 8.1 that 7 of the 11 correlations in this extended system are so small that they appear only in the COSY contour plots. A TOCSY experiment would, in all likelihood, add little information to that inferred from one of the COSY experiments.

In addition, based on their chemical shifts (δ 67.6–84.5), carbons 6, 7, 8, 9, and 10 are almost certainly attached to oxygen atoms (Section 3.5.1). The three molecular fragments shown in Figure 8.4 can be constructed tentatively from the COSY correlations described above. If they are verified, these fragments would account for over half of the carbons and hydrogens in the unknown compound and two units of unsaturation.

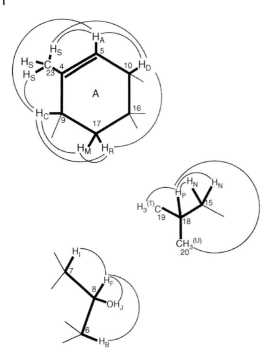

Figure 8.4 Molecular fragments established by COSY correlations.

8.1.6 The HMBC Experiment

Since the ^{13}C NMR spectrum of the unknown compound is not more congested than its ^{1}H spectrum, heteronuclear multiple bond correlation (HMBC) (Sections 6.2 and 7.3.3.1) is the experiment of choice to establish the longer range, C—H correlation networks. Standard and gradient HMBC spectra of the 15-mg sample of the unknown compound are shown in Figure 7.18, and the data from these contour plots are summarized in Table 8.4.

8.1.7 General Molecular Assembly Strategy

Of the many techniques available to the NMR spectroscopist in structural elucidations, none is so valuable as the indirect, chemical shift correlation experiments like HMBC, TOCSY (both homo- and heteronuclear varieties), and FLOCK. Once molecular fragments have been identified by the COSY and HSQC experiments, attaching these fragments together is attempted by means of these techniques. As indispensable as these methods have become to NMR spectroscopists, they suffer a common limitation: two-bond, C—H couplings cannot generally be distinguished from three-bond, C—H coupling constants. The *occasional* observation of four- and five-bond, C—H couplings further complicates matters. Unfortunately, there is no overall solution to this uncertainty concerning the number of bonds over which C—H coupling is being detected.

The process of molecular assembly can be approached in the following manner. If possible, a carbon atom is selected from which the remainder of the molecular skeleton can be built in just *one* direction. Methyl groups are, of course, excellent starting points. Adjacent protons, if any, can be identified from a COSY contour plot, and the carbon

Table 8.4 HMBC NMR data.

Designation	Chemical shift	Longer-range attached hydrogens
1	173.0	C, N
2	172.8	B, C(H$_O$)$_3$
3	170.5	E, G, C(H$_Q$)$_3$
4	136.6	C, D, R, C(H$_S$)$_3$
5	124.2	C, D, C(H$_S$)$_3$
6	84.5	F, I, C(H$_V$)$_3$
7	79.2	B, J, K
8	78.5	B, I, J
9	68.3	A, D, R, C(H$_S$)$_3$
10	67.6	A, E, G, I, R, C(H$_S$)$_3$
11	64.9	M, R
12	64.6	B, I, K, L, C(H$_V$)$_3$
13	48.8	B, D, E, G, I, C(H$_V$)$_3$
14	47.4	None
15	43.8	C(H$_T$)$_3$, C(H$_U$)$_3$
16	43.3	B, C, D, E, G, M, R, C(H$_V$)$_3$
17	27.9	C, D, E, G
18	26.0	N, C(H$_T$)$_3$, C(H$_U$)$_3$
19	22.5	N, C(H$_U$)$_3$
20	22.4	N, C(H$_T$)$_3$
21	21.1	None
22	21.0	None
23	20.3	A
24	7.0	None

to which they are directly attached from an HSQC or HMQC plot. The HMBC or FLOCK spectra can then be scanned using either the contour plot (for uncongested spectra) or methyl-proton or methyl-carbon traces (whichever is less congested) if spectral congestion or weak cross peaks are a problem. Cross peaks may be found for (i) the adjacent carbon (if methyl-proton traces are viewed) or protons (if methyl-carbon traces are observed), which represent two-bond couplings in either case and (ii) any other carbons or protons, indicative of three-bond couplings (but always being mindful that one, or more, members of the second group could possibly be due to $^nJ_{CH} > 3$). The fortunate redundancy of these 2D experiments is seen, whereby an adjacent carbon may be identified by a combination of COSY ($^3J_{HH}$) and HSQC ($^1J_{CH}$) connectivities and also by HMBC/FLOCK ($^2J_{CH}$) correlations.

The second carbon atom in the fragment (adjacent to the original methyl group) likely shows two-bond couplings with directly attached proton(s) if present, both backward to the original methyl carbon and forward to the third carbon in the series.

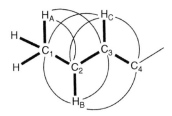

Figure 8.5 Generalized two- and three-bond C—H couplings.

Carbon atom connectivities thus can be built up using two-bond, C—H couplings to confirm previously determined C—C connectivities generally and then employing three-bond, H—H and C—H couplings to extend the developing molecular structure. These forward- and backward-looking two- and three-bond, C—H coupling networks are illustrated in Figure 8.5. To avoid confusion from too many lines, the corresponding H—H couplings are not shown.

Alternatively, when a methyl ^{1}H signal is nothing more than a broadened singlet, the COSY spectrum (either the contour plot or the methyl trace) can be scanned for cross peaks due to longer range coupling. As before, either the contour plot or the methyl ^{1}H trace of the HMBC or FLOCK spectrum can be scanned for cross peaks due to (i) two-bond coupling to adjacent (quaternary) carbons and (ii) three-bond coupling to farther-removed carbons (with the usual $^nJ_{CH} > 3$ caveat). Finally, a note of caution. Like NOEs, three-bond, C—H correlations are not necessarily symmetrical, e.g. in the four-carbon fragment pictured in Figure 8.5, a strong cross peak may, in fact, be observed between H_A and C_3 while a weak one, or none at all, is seen between H_C and C_1. The main reason for these weak or missing correlations is the dependence of vicinal couplings on the $H_A-C_1-C_2-C_3$ and $H_C-C_3-C_2-C_1$ dihedral angles (Karplus relationship, Section 4.6), which are seldom identical. Since similar considerations are largely absent for two-bond couplings, cross peaks should be detected between H_A and C_2 and also between H_B and C_1. Extensive redundancy of the type described above, however, is in fact observed for vicinal C—H correlations and is invaluable in the construction of molecular structures.

The same factors that influence H—H couplings, e.g. dihedral angles, substituent electronegativity, bond length, and bond order (Chapter 4), apply to C—H couplings. As a general rule, C—H coupling constants are approximately $2/3$ the value of the corresponding H—H couplings. In alkenes, for example, average *cis* and *trans* H—H couplings are c. 11 and 18 Hz, respectively, while the corresponding C—H coupling constants are c. 7 and 12 Hz, respectively.

8.1.8 A Specific Molecular Assembly Procedure

On an examination of the three fragments of Figure 8.4 as possible starting points for the structural elucidation of the unknown compound, the isobutyl group is an attractive choice if, of course, additional connections can be made to carbons and protons elsewhere in the molecule. If such connections cannot be made, the methyl protons (S) of the putative cyclohexene system provide another possible starting point.

The HMBC data (Table 8.4) provide confirmation of the isobutyl structure. Examination of the methyl group protons (T and U) shows HMBC correlations to both C_{18} and C_{15}. In addition, methyl protons T exhibit a connectivity to C_{20} while the methyl protons U do likewise to C_{19}. Correlations due to H_P are obscured because of its proximity to methyls O and Q. Connectivities, however, are found from the methylene protons N to carbons 18, 19, and 20, thus confirming the presence of the isobutyl group.

A fourth HMBC correlation is detected between the methylene protons N and the most deshielded carbon. This carbon (C_1) appears to be an ester carbonyl from its chemical shift (δ 173.0, Section 3.5.3) and exhibits a second connectivity to H_C. This correlation is crucial because it is the only one that connects C_1 indirectly to C_9 (via an oxygen as an isovaleryl ester group) of the presumptive cyclohexene fragment. If confirmed, the molecular fragment **8.1**

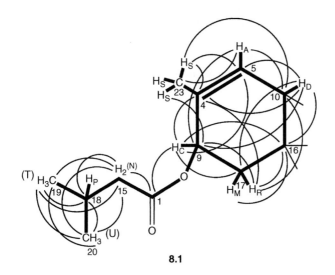

8.1

(with HMBC connectivities shown) accounts for three units of unsaturation and 12 carbons (including the previously surmised, carbon–carbon double bond), 17 hydrogens, and 2 oxygens.

The next step in this structural determination is confirmation of the proposed cyclohexene (A) ring. Several COSY and HMBC correlations provide this confirmation. The $H_C C_9$ fragment is central to the issue, and vicinal coupling between H_C and H_M previously established that C_9 is attached to the $C_{17} H_M H_R$ fragment. Confirming HMBC connectivities between H_C and C_{17} and between H_R and C_9 also are observed as well as COSY cross peaks between H_C and H_R. In addition, C_9 displays HMBC connectivities to H_A, H_D, and the methyl protons S, while H_C exhibits HMBC correlations to C_4, C_5, and C_{16}.

Since H_A and H_D also display vicinal coupling, their directly attached carbons (5 and 10) must be adjacent, and confirming HMBC connectivities between H_A and C_{10} and between H_D and C_5 are observed. Further, inspection of the ^{13}C NMR data indicates that there are only two non carbonyl, sp^2 carbons, and it is logical to pair them as C_4 and C_5. H_C and H_D each exhibit HMBC connectivities to both carbons 4 and 5, confirming the 4,5-double bond. These data demonstrate also that C_5 is located between C_4 and C_{10}.

Several COSY and HMBC correlations suggest that methyl-23 be positioned at C_4 and that it, in turn, be attached to C_9. First, COSY cross peaks between the methyl protons S and H_A and H_C and HMBC connectivities between (i) the methyl protons S and C_4, C_5, and C_9, and (ii) H_A and methyl C_{23} argue that methyl-23 be located at C_4. Second, the above correlations plus (i) (from COSY) that between H_A and H_C and (ii) (from HMBC) those between H_A and C_9 and between H_C and C_4 require that C_4 be joined to C_9.

Finally, the HMBC connectivities that C_{16} displays to H_C, H_D, H_M, and H_R suggest that it be situated between C_{10} and C_{17} to complete the six-membered A-ring. HMBC correlations between H_D and C_{17} and H_R and C_{10} support this placement. The 4,5-double bond and A-ring account for the second and third units of unsaturation.

Two allylic (four-bond), C—H connectivities are thus observed in the A-ring (H_D to C_9 and the methyl protons S to C_{10}), underscoring the warning that HMBC cross peaks are not restricted just to two- and three-bond correlations. Unsaturated ring systems pose special problems for structural analysis via HMBC experiments because of long-range connectivities from allylic and homoallylic C—H couplings that can be mistaken for low-intensity cross peaks from small vicinal coupling constants. This situation can be illustrated by alternatively assembling the A-ring as shown in structure **8.2**.

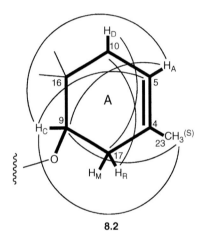

8.2

Most of the HMBC correlations used to derive the A-ring are *possible* for the rearranged ring. A problem, however, arises with their intensities. Where there are two allylic couplings in **8.1** (H_D to C_9 and methyl-S to C_{10}), there must now be three allylic couplings (H_A to C_9, H_D to C_{17}, and H_R to C_{10}), one W-type coupling (Section 4.7) (methyl-S to C_9), and one nondescript, four-bond coupling between H_C and C_5 in **8.2**. Even if these couplings did exist, their cross-peak intensities in the HMBC spectrum would not be expected to be anywhere nearly so great as those of the corresponding couplings in **8.1**.

Another problem with the arrangement of **8.2** concerns the observed COSY cross peaks between H_A and H_C and between H_C and the methyl protons S. These represent allylic and W-type couplings, respectively, in **8.1** but would have to be nondescript, five-bond couplings in **8.2**. Substructure **8.2** is, therefore, rejected as a likely possibility.

In addition to the connectivities noted, C_{16} displays two more from the geminal protons E and G, and this methylene pair also shows correlations to both C_{10} and C_{17}. Furthermore, three-bond connectivities are seen from C_{11} to protons M and R. These correlations are illustrated in the lower right-side portion of substructure **8.3** and necessitate that C_{11} be attached directly to C_{16}. Note that as fragments are added to **8.1** to generate **8.3** and succeeding substructures, these additions are shown in red.

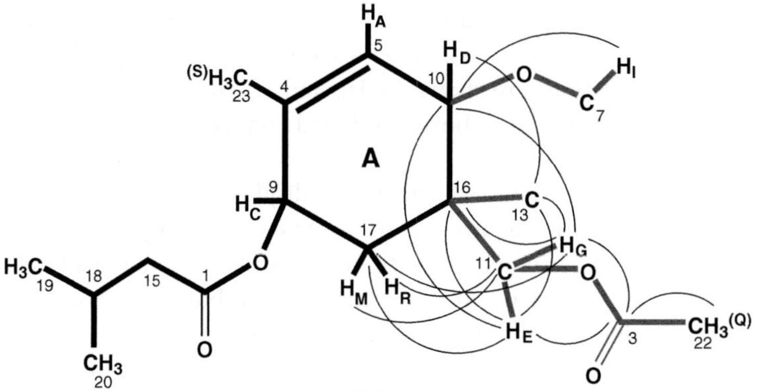

8.3

The considerably high-frequency (low field) chemical shifts of H_E (δ 4.28) and H_G (δ 4.06) suggest both that they are carbinol protons and that the oxygen, to which their directly bonded carbon is attached, is part of an ester group (Section 3.2.1). Protons E and G also exhibit connectivities to C_3, which, in turn, displays correlations to methyl protons Q. These connectivities require that the acetate group, $(H_Q)_3C_{22}(O)O$, be attached to C_{11}. H_D and methylene protons E and G also show a three-bond correlation to C_{13} and thus necessitate that it be attached to C_{16}.

The chemical shift of C_{10} (δ 67.6) indicates that it is bonded to an oxygen atom (Section 3.5.1). In addition, it exhibits what must be a three-bond connectivity from H_I. The chemical shift of C_7 (δ 79.2), to which H_I is directly attached, suggests that it too is bonded to oxygen. Connection of C_{10} and C_7 by an oxygen atom is consistent with the chemical shift and HMBC data and completes substructure **8.3**, which accounts for a fourth unit of unsaturation and a total of 17 carbons, 23 hydrogens, and 5 oxygens.

Extension of the developing structure to C_7 is especially important because it unites substructure **8.3** with the third molecular fragment (comprising C_7—C_8—C_6) shown in Figure 8.4, to produce substructure **8.4**.

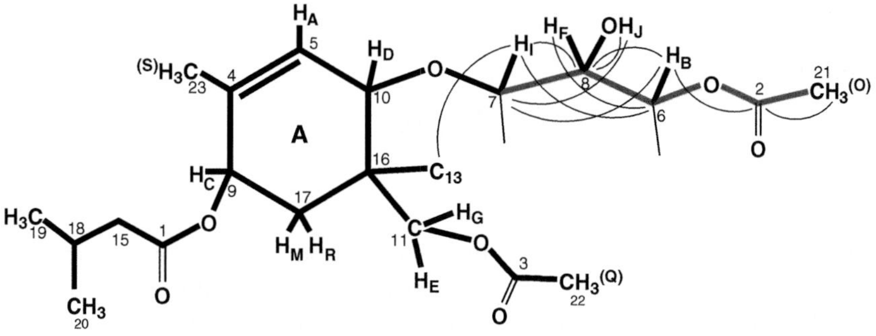

8.4

Similar to our approach to the cyclohexene system, the next step in this structural determination is to confirm the structure of the third fragment. Unlike the former subunit, the latter is well defined by its COSY and 1D spin coupling data. Nevertheless, the HMBC data are checked for consistency. H_I displays two- and three-bond correlations to C_8 and C_6, respectively, while C_7 exhibits connectivities to H_B and the hydroxyl proton J. In addition, two-bond correlations are shown by H_B and H_F to C_8 and C_6, respectively. The middle carbon (8) exhibits the abovementioned connectivities and another (two-bond) from OH_J.

H_B displays an HMBC correlation to C_2, which, in turn, displays connectivities to the methyl protons O. These correlations require that the acetate group, $(H_O)_3C_{22}(O)O$, be attached to C_6. Substructure **8.4** accounts for a fifth unit of unsaturation and a total of 21 carbons, 29 hydrogens, and 8 oxygens. At this point, three carbons, five hydrogens, one oxygen, and three units of unsaturation remain undetermined.

The carbons remaining to be added to substructure **8.4** are the following: quaternary C_{12}, the methylene fragment $C_{14}H_KH_L$, and the methyl group $C_{24}(H_V)_3$. In addition, carbons 6 and 7 each needs to be connected to one more atom and quaternary C_{13} requires three. Furthermore, since only one oxygen and no sp^2 carbons remain unaccounted for, the remaining three units of unsaturation must be due to ring systems rather than carbon–carbon or carbon–oxygen double bonds. The methyl protons V exhibit connectivities to C_{16} and C_{13}, which have been placed in the structure, and C_6 and C_{12}, which have not. These correlations necessitate that C_{12} and $C_{24}(H_V)_3$ also be attached to C_{13}. They account for the methyl group and are illustrated in structure **8.5**.

8.5

In addition to its already noted connectivities to C_7 and C_8, H_B shows correlations to C_{12}, C_{13}, and C_{16}. The last one to C_{16} is particularly strong and suggests that it is a three-bond, rather than some type of longer range, coupling. If this assumption is correct, then C_6 must be connected to C_{13} as part of the three-bond coupling pathways of H_B to C_{12} and C_{16}.

The chemical shift of C_{12} (δ 64.6) indicates that it is bonded to the remaining oxygen. Moreover, the chemical shift of C_7 (δ 79.2) requires that it be attached to only one oxygen (Section 3.5.1) as illustrated in substructure **8.4**, and that its last unassigned bond be to another carbon (12 or 14). With C_{12} already connected to C_{13}, its remaining two bonds must, therefore, be to C_7 and C_{14}. This arrangement necessitates that the last unassigned bond of C_{14} be to the same oxygen that is bonded to C_{12} to form an epoxide. At this point all of the carbons, hydrogens, and oxygens are accounted for. In addition, the complete structure of the unknown compound, shown in structure **8.5**, satisfies the required eight units of unsaturation.

The final HMBC connectivities support structure **8.5**. C_{12}, for example, displays correlations to H_B, H_I, H_K, and H_L, while C_7 shows one to H_K. The directly attached carbon (14) of H_K and H_L is somewhat isolated in that it displays no proton connectivities. These last correlations illustrate the difficulties in determining the substructures of ring compounds in which junction carbons display many HMBC connectivities that cannot be differentiated as two- or three-bond correlations.

8.1.9 The NOESY Experiment

With completion of the two-dimensional structural elucidation of the unknown compound, questions arise concerning its three-dimensional shape, i.e. the relative orientations of atoms or groups such as H_M or methyl-24 at various carbons. For a molecule of MW = 466, the nuclear Overhauser effect spectroscopy (NOESY) experiment (Sections 6.3 and 7.3.4.1) can provide a wealth of such stereochemical information. The NOESY spectrum of the 15-mg sample of the compound is shown in Figure 7.23a. The data from these contour plots are summarized in Table 8.5 and illustrated, in part, in structure **8.6**.

8.6

A number of NOESY cross peaks are critical in determining the overall stereochemistry of T-2 toxin. NOEs, for example, demonstrate that protons B, D, E, G, and J are located in one region of the molecule, protons C, L, and M in another, and protons F, I, and K in a third. In addition, $C_{24}(H_V)_3$ and H_L are spatially proximate.

Table 8.5 NOESY NMR data.

Designation	NOE enhancements[a]
A	s D, S
B	s D, E, w F, J, O, V
C	M, w Q, w R, S
D	A, s B, E, w G, w J
E	B, D, s G, w O, w V
F	w B, s I, J, w O
G	w D, s E, R, s V
I	s F, w K
J	B, w D, F
K	w I, s L
L	s K, w M, s V
M	C, w L, s R
N	w Q, T, U (P obscured)
O	B, w E, w F
P	T, U (N obscured)
Q	w C, w N
R	w C, G, s M, s V
S	A, C
T	P
U	P
V	B, w E, s G, s L, s R

a) s = strong, w = weak.

The A-ring must exist in a half-chair conformation because of the 4,5-double bond. Moreover, COSY and ^{1}H NMR coupling data require that H_C unequally bisect the H_M—C_{17}—H_R angle, i.e. the H_C—C_9—C_{17}—H_M dihedral angle must be c. 20° while that between H_C and H_R be c. 90° (Section 4.6). This arrangement can be achieved by having C-17 puckered up while C-16 is puckered down. A reversed orientation keeps the c. 20° dihedral angle between H_C and H_M, but the H_C—C_9—C_{17}—H_R dihedral angle would open up to c. 180°. This arrangement would result in three large and easily detected couplings between H_C, H_M, and H_R, contrary to observation.

The NOESY correlations between H_D, H_E, and H_G are crucial in establishing how the six-membered ring B is attached to ring A. A trans AB-ring junction is eliminated from consideration because it requires that H_D and $C_{11}H_EH_G$ be on *opposite* sides of the AB-ring plane, and NOEs between H_D and both H_E and H_G are impossible in this orientation.

Two *cis* AB-ring systems can be fashioned in the following way: (i) C_{13} of ring B can be attached at the equatorial position of C_{16}, and the ether oxygen at the pseudo-axial site of C_{10} (both with respect to ring A), shown in substructure **8.6** or (ii) C_{13} can occupy the axial position of C_{16} and the ether oxygen the pseudo-equatorial site of C_{10}, given in substructure **8.7**.

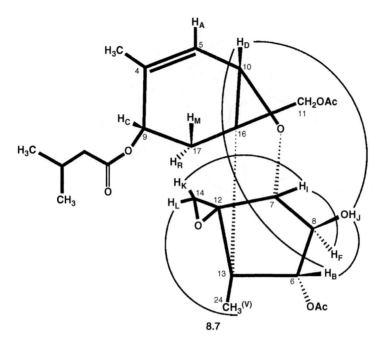

8.7

While both substructures accommodate the NOESY connectivities between H_L and $C_{24}(H_V)_3$, H_K and H_I, H_I and H_F, and H_B, H_D, and OH_J, that between H_M and H_L is reconciled only by **8.6**. Substructure **8.7** is, therefore, rejected because it places H_M and the epoxy group on *opposite* sides of ring A (this arrangement may be hard to visualize in the drawing but can be seen easily with a Dreiding model), for which the strong NOESY enhancement exhibited by H_M and H_L is impossible. Ring B thus exists in a chair conformation in which C_{13} is puckered down, the ether oxygen is puckered up, and carbons 7, 12, 10, and 16 lie in a common plane.

Furthermore, in this representation, H_D is pseudo-equatorial and the $C_{11}H_EH_GOAc$ group is axial, (again, both with respect to ring A). The five-membered ring C is nearly perpendicular to ring B with C_7, C_8, C_6, and C_{13} residing in a common plane and C_{12} puckered up (these puckerings cannot all be illustrated in any one drawing). In addition, if ring A is positioned in the plane of the paper, the epoxide group at C_{12} projects outward toward the reader as in substructure **8.6**. Note that the epoxy oxygen points toward the acetoxy group at C_6, and H_L is oriented toward H_M to accommodate the H_L/H_M NOESY correlation.

The $C_{11}H_EH_GOAc$ group is interesting because it can, in principle, rotate to some degree about the C_{11}–C_{16} bond. In addition to the NOESY connectivities that H_E and H_G display with each other and their abovementioned correlation to H_D, they both exhibit a NOESY connectivity to $C_{24}(H_V)_3$. The NOE of H_E, however, is much stronger to H_D while that of H_G to $C_{24}(H_V)_3$ is more intense. Further, H_E exhibits NOESY correlations to H_B and to $C_{21}(H_O)_3$ while H_G does so to H_R. Moreover, $(H_Q)_3C_{22}C_3(O)O$ displays connectivities to H_C and to the methylene protons N. When ring A is placed in the plane of the paper, these NOESY correlations (shown in structure **8.8**)

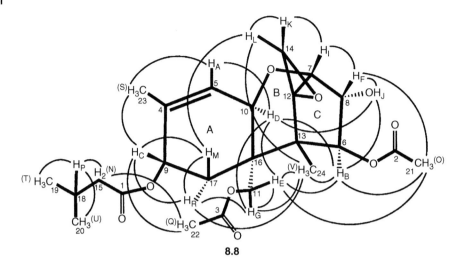

8.8

suggest that the $C_{11}H_EH_GOC_3(O)C_{22}(H_Q)_3$ group orients itself so that H_G is pointing toward the reader, H_E is pointing back toward C_8 and somewhat parallel to the C_{16}–C_{17} bond, and $(H_Q)_3C_{22}C_3(O)O$ is positioned back toward methyl-23 where it participates in relaxation of the protons N and H_C.

Structure **8.8** is, to a reasonable approximation, a complete three-dimensional representation of T-2 toxin. It can readily be seen that there is no *molecular plane* in this molecule. What is more, the meaning of stereochemical descriptors like α and β is, to a certain degree, indefinite. Nevertheless, with ring A positioned horizontally, a final description of the relative orientation of the remaining substituents is as follows: the isovaleryl ester group at C_9 is α and axial, H_M is β and axial, and H_R is α and equatorial. Substituents at the AB-ring junction cannot be described easily as axial or equatorial because these descriptors depend on whether ring A or ring B is used as a reference point. A better choice is to label both H_D and the $C_{11}H_EH_G$OAc moieties as α. At the BC-ring junction, both H_I and methyl-24 are β.

In ring C, several lines of reasoning support the inference that H_B and OH_J are α. First, this arrangement puts both hydrogens in close proximity to H_D, and NOESY enhancements are observed between all three nuclei. Both groups would be oriented away from H_D if they were β. Second, OH_J must be α (and H_F necessarily β) due to the 5-Hz coupling observed between H_I and H_F. Since the H_I–C_7–C_8–$R_β$ dihedral angle is c. 20° and the H_I–C_7–C_8–$R_α$ dihedral angle is c. 90°, if H_F were not β, it would be only slightly, if at all, coupled to H_I (Section 4.6).

In the epoxide group, NOESY connectivities between H_I and H_K and between H_L and both H_M and $C_{24}(H_V)_3$ indicate that H_K is β and pointed toward H_I, while H_L is α and oriented toward methyl-24. With the interpretation of these final NOESY correlations, the two- and three-dimensional structural elucidation of T-2 toxin is complete.

8.2 Part B: Computer-Assisted Structure Elucidation

NMR spectroscopy was recognized as a potentially powerful tool for the determination of structures of organic compounds in the late 1950s. Interests soon arose concerning whether structural determination processes might be amenable to computer analysis, and the first computer-assisted structure elucidation (CASE) programs were developed for small organic compounds in the late 1960s. Early structure elucidation programs, however, were considerably limited in many ways, particularly the very limited capability of computers, especially by today's standards. Moreover, available NMR experiments then were restricted to one-dimensional ^{1}H NMR spectroscopy, as practical one-dimensional ^{13}C and two-dimensional NMR experiments had not yet been developed.

CASE capabilities improved markedly with the advent of (i) ^{13}C NMR spectroscopy, (ii) 2D NMR experiments capable of determining both direct and indirect homonuclear, e.g. H—H COSY (Section 6.1) and TOCSY (Section 6.1), and heteronuclear, e.g. C—H HSQC (Section 6.2) and HMBC (Section 6.2) correlation spectra, and (iii) the development of powerful computer systems. One of the great strengths of CASE programs is that they avoided the "bias of familiarity." This situation occurs when a number of similar compounds have been isolated in the past, and a current unknown presents similar NMR spectra and thus appears to be yet another member of the same group.

CASE programs have no memory and are immune to this bias. While some of their proposed structures may seem highly unlikely or chemically impossible, they more than compensate for this deficiency by suggesting structures and classes of compounds that may have not even been considered. More powerful structural elucidation programs are now available, and one of these, "ACD/Structure Elucidator" of Advanced Chemistry Development, Inc. is described below.

Section 8.1 presented a systematic procedure for the elucidation of an unknown organic chemical structure. Section 8.2 now concerns the computerized determination of structures using the ACD/Labs program, and the following methodology is largely taken from its manual.

8.2.1 CASE Procedures

1) Spectral data requirements:
   (a) ^{1}H NMR: critical for signal multiplicity identification and integral information.
   (b) ^{1}H/^{13}C HSQC, ^{1}H/^{1}H COSY, and ^{1}H/^{13}C HMBC: essential for establishing atomic connectivities.
   (c) ^{1}H/^{1}H TOCSY: useful when extended spin systems are present.
   (d) ^{13}C NMR data: essential for quaternary carbons that do not exhibit HMBC connectivities.
   (e) High-resolution mass spectral data to provide a molecular formula.
   (f) IR and UV/visible spectral data to furnish information on functional groups.
2) NMR data are submitted in one of the following two formats:
   (a) As one- and two-dimensional spectral data (free induction decays (FIDs) (Section 1.4) together with their processing parameters), in which 1D and 2D NMR cross-peaks (Section 6.1) are selected by the software. This approach is preferable because data entry is rapid, direct, and avoids transcription errors.
   (b) As text (TXT) files in which the chemist/spectroscopist has analyzed the NMR spectra to establish various homo- and heteronuclear connectivities. The latter

approach is less favored because data entry is both laborious and introduces the possibility of transcription errors.

3) NMR data processing can involve numerous conditions, but the following are generally applied:
   (a) HMBC correlations are assigned as follows:
      (i) Strong: two to three bond C—H couplings
      (ii) Weak: two to four bond C—H couplings
   (b) Direct heteroatom-to-heteroatom connectivities are disallowed.
   (c) Triple bonds within three, four, and five-membered rings are disallowed.

4) A molecular-connectivity map is then generated, which shows all of the NMR data in one place and is correlated to the molecular formula. The NMR data include heteronuclear (HMBC) and homonuclear (COSY and TOCSY) connectivities. These maps are informative but generally far too complicated to be used alone to generate possible structures.

5) Potential structures, numbering in the tens or hundreds, are generated, and ^{13}C chemical shifts are calculated for each candidate structure. Differences between *predicted* and *experimental* data are initially reported as a "fast-deviation" statistic, $d_F(^{13}C)$, and possible structures are then ranked in order of *increasing* $d_F(^{13}C)$.

6) More accurate ^{13}C chemical shifts calculations are next performed on the smaller of either (i) all structures with $d_F(^{13}C) \leq 4$ ppm/carbon or (ii) the first 50 ranked structures. Differences between the more accurately predicted and experimental data are reported as an "accurate-deviation" statistic, using neural net $(d_N(^{13}C))$ values or HOSE (Hierarchically Ordered Spherical Description of Environment) $(d_A(^{13}C))$. The two methods give somewhat similar results. HOSE tends to yield better results when compounds that are *similar* to the unknown are contained in the ACD/Labs spectral library while Neural Net tends to work better for those compounds that are not. Potential structures are re-ranked in order of increasing $d_{A/N}(^{13}C)$ numbers. Structures having $d_{A/N}(^{13}C)$ values >4 ppm/carbon are generally discarded. ACD/Labs' experience is that correct structures are usually identified at this point.

7) In situations for which the smallest calculated $d_{A/N}(^{13}C)$ values are very close, a best structure may be arrived at by calculating $d_{A/N}(^{1}H)$, and, if good MS data are available, $d(MS)$ can be calculated as well.

8.2.2 T-2 Toxin

T-2 toxin is a sesquiterpene and a member of the trichothecene family of mycotoxins whose structure (**7.1**) appears in Section 7.1.3.1. NMR spectra of T-2 have been employed extensively in this book to illustrate many types of NMR experiments.

Recently, one- and two-dimensional NMR raw spectral data, viz, FIDs and their processing parameters, were processed by the ACD/Labs' Structural Elucidator program as an exercise. These data are given in Section 8.1 in the following tables: ^{1}H NMR (Table 8.1), ^{13}C and HSQC (Table 8.2), COSY (Table 8.3), and HMBC (Table 8.4). The Structure Elucidator program generated four structures with respective $d_A(^{13}C)$ values of 0.816, 2.187, 4.095, and 4.459 (**8.9a**). Only the first two structures, which contain epoxy groups (numbers **1** and **2**), passed filtering, i.e. their $d_A(^{13}C)$ values were less than 4 ppm and more discriminating than the corresponding $d_N(^{13}C)$ values. Experimental ^{13}C and ^1H chemical shifts are shown for the two highest-ranked structures **1** (**8.9b**) and **2** (**8.9c**). The two structures are very similar and differ in the placement of the methylene carbon at δ 27.86 ppm and the quaternary carbon at δ 43.25, which are switched in the two structures. Structure **1**, however, has a considerably better $d_A(^{13}C)$ value than **2** (0.816 vs 2.187 ppm). Moreover, as we saw in Section 8.1, **1** proved to be correct. What is especially noteworthy in these two structural elucidation processes is that the data entry and data analysis in the CASE procedure took only one hour of spectral processing time!

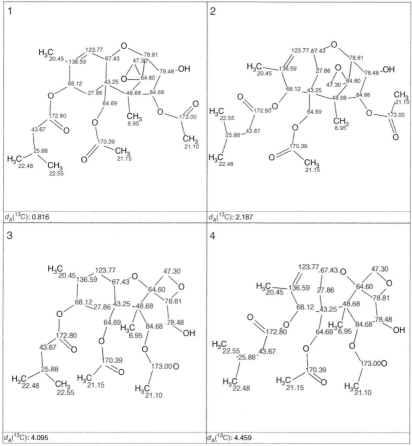

8.9a The best four structures

8.9b Correct structure (top ranked above)

8.9c Incorrect structure (second ranked above)

The difference between the correct (**1**) and incorrect (**2**) structures is rather minute and involves the interchange of adjacent methylene (7) and quaternary (6) carbons. With the 12 HMBC correlations that are observed in both structures (8 for the central quaternary carbon and 4 for the methylene carbon), it seems almost inconceivable that the HMBC experiment cannot distinguish between these two structures. Careful inspection of structures **1** and **2**, however, demonstrates that all of the HMBC connectivities to these adjacent carbons are such that each one is two-bond in one structure and three-bond in the other. Moreover, we have seen in Section 6.2, two- and three-bond C—H couplings cannot usually be distinguished by the HMBC experiment. The presence of a key, 3-bond, 5.5-Hz coupling between the vicinal protons at δ 2.41 (H-7A) and 5.28 (H-8), however, permits identification of **1** as the correct structure. In structure **2**, this H-7A/H-8 coupling would have to occur over four bonds. Since the largest four-bond couplings are no greater than 2 Hz in rigid, planar systems, structure **2** is inconsistent with the observed 5.5-Hz coupling. Table 8.6 contains ^{13}C and ^{1}H chemical shift data for T-2 toxin that are sorted by actual position number.

Table 8.6 T-2 toxin NMR data.

Position	^{13}C	^{1}H
2	79.2	3.70 d (5.0)
3	78.5	4.16 ddd (5.0, 2.4, 2.4)
4	84.5	5.31 d (2.4)
5	48.8	—
6	43.3	—
7	27.9	2.41 dd (14.0, 5.5), 1.91 d (14.0)
8	68.3	5.28 d (5.5)
9	136.6	—
10	124.2	5.81 d (5.0)
11	67.6	4.34 d (5.0)
12	64.6	—
13	47.4	3.06 d (4.0), 2.80 d (4.0)
14	7.0	0.81 s
15	64.9	4.28 d (12.6), 4.06 d (12.6)
16	20.3	1.75 s
1'	172.8	—
2'	43.8	2.17 d (6.3)
3'	26.0	2.11 m (6.4, 6.3)
4'	22.5	0.98 d (6.4)
5'	22.4	0.96 d (6.4)
4-OAc (CH$_3$)	21.1	2.03 s
4-OAc (CO)	173.0	—
15-OAc (CH$_3$)	21.0	2.16 s
15-OAc (CO)	170.5	—
3-OH	—	3.19

Appendix A

Derivation of the NMR Equation

The magnitude of the magnetic moment μ_N of a nucleus N is directly proportional to its spin angular momentum $\boldsymbol{J}$, with the constant of proportionality defined as the gyromagnetic ratio γ_N (Eq. (A.1)).

$$\boldsymbol{\mu}_N = \gamma_N \boldsymbol{J} \tag{A.1}$$

Vector quantities, having both magnitude and direction, are indicated by boldfaced fonts. Scalar quantities, having only magnitude, are indicated by simple Roman fonts. Physicists and chemists have preferred to discuss these concepts in terms of a dimensionless spin $\boldsymbol{I}$ ($\boldsymbol{J} = \hbar \boldsymbol{I}$, in which $\hbar$ is Planck's constant divided by 2π). Eq. (A.1) then becomes Eq. (A.2).

$$\boldsymbol{\mu}_N = \gamma_N \hbar \boldsymbol{I} \tag{A.2}$$

Spin is quantized, i.e. it takes on only certain values. The magnitude I has values of 0 (^{12}C, ^{16}O), ½ (^{1}H, ^{13}C), 1 (^{2}H, ^{14}N), 3/2 (^{7}Li, ^{35}Cl), and so on in increments of ½. The spin in the direction (z) of the external magnetic field (B_0) can have $2I + 1$ values, varying from $-I$ to $+I$ in increments of one, e.g. $I_z = -½$, and $+½$, for $I = ½$; $I_z = -1$, 0, and $+1$ for $I = 1$, etc.

In the absence of B_0, all orientations of the nuclear magnets in space have the same energy. The B_0 field, however, interacts with the nuclear magnets to alter their energies as a function of I_z. Because energy is a scalar quantity, it is obtained by the scalar or dot product of the magnetic moment and the external magnetic field (Eq. (A.3)).

$$\text{energy} = \boldsymbol{\mu} \cdot \boldsymbol{B} \tag{A.3}$$

The expression for the nuclear magnet in Eq. (A.2) may be substituted into Eq. (A.3) to give Eq. (A.4),

$$\text{energy} = \gamma \hbar \boldsymbol{I} \cdot \boldsymbol{B} \tag{A.4}$$

in which the subscript N has been dropped. Expansion of this expression in the usual fashion ($\boldsymbol{I} \cdot \boldsymbol{B} = I_x B_x + I_y B_y + I_z B_z$) is simplified because there are no x or y components of the magnetic field and B_z is represented by B_0. The energy thus is expressed by Eq. (A.5),

$$\text{energy} = \gamma \hbar I_z B_0 \tag{A.5}$$

Nuclear Magnetic Resonance Spectroscopy: An Introduction to Principles, Applications, and Experimental Methods,
Second Edition. Joseph B. Lambert, Eugene P. Mazzola, and Clark D. Ridge.
© 2019 John Wiley & Sons Ltd. Published 2019 by John Wiley & Sons Ltd.

in which I_z can assume only the two values ($-\tfrac{1}{2}$ and $+\tfrac{1}{2}$). Two states result, whose energies are given, respectively, by Eqs. (A.5a) and (A.5b).

$$E_1(I_z = -\tfrac{1}{2}) = -\tfrac{1}{2}\gamma\hbar B_0 \tag{A.5a}$$

$$E_2(I_z = \tfrac{1}{2}) = \tfrac{1}{2}\gamma\hbar B_0 \tag{A.5b}$$

This splitting into multiple states has been called the *Zeeman effect*. By convention, we shall assign the lower energy E_1 to the state with $I_z = -\tfrac{1}{2}$. The difference in energy between the two spin states then is given by Eq. (A.6).

$$\Delta E = E_2 - E_1 = \gamma\hbar B_0 \tag{A.6}$$

By the Planck relationship ($\Delta E = h\nu_0$), this quantity corresponds to the energy of an electromagnetic wave with frequency ν_0. Eq. (A.7)

$$\Delta E = h\nu_0 = \gamma\hbar B_0 \tag{A.7}$$

indicates the equivalence of the two energy expressions. Cancellation of Planck's constant in the two expressions in the rightmost equality of Eq. (A.7) leads to Eq. (A.8)

$$\nu_0 = \frac{\gamma B_0}{2\pi} \tag{A.8}$$

in linear frequency or to Eq. (A.9)

$$\omega_0 = \gamma B_0 \tag{A.9}$$

in angular frequency ($\omega = 2\pi\nu$). The quantities ν_0 and ω_0 (in different units) are called the *Larmor frequency* and depend only on the laboratory field B_0 and the gyromagnetic ratio γ.

Appendix B

The Bloch Equations

The Bloch equations provide a mathematical expression for the magnetization based on classical vector mechanics. The overall magnetization M is the sum of three Cartesian components (Eq. (B.1)).

$$M = iM_x + jM_y + kM_z \tag{B.1}$$

The magnetic field B_0 that gives rise to the Zeeman splitting is in the z direction, and the applied magnetic field B_1 operates in the xy plane.

When perturbed from equilibrium in the z direction, for example by the NMR experiment, the system returns to equilibrium in a first-order rate process with a time constant T_1 (the spin-lattice, or longitudinal, relaxation time). Consequently, the rate of return from a perturbed value M_z to the equilibrium value M_0 is given by Eq. (B.2).

$$\frac{dM_z}{dt} = \frac{1}{T_1}(M_0 - M_z) \tag{B.2}$$

Similarly, relaxation processes in the xy plane are controlled by the spin–spin, or transverse, relaxation time T_2 so that Eqs. (B.3) and (B.4)

$$\frac{dM_x}{dt} = -\frac{M_x}{T_2} \tag{B.3}$$

$$\frac{dM_y}{dt} = -\frac{M_y}{T_2} \tag{B.4}$$

describe the rates of return from perturbed values M_x and M_y to the equilibrium values of zero.

The various magnetization components are influenced by the magnetic fields B_0 and B_1. The forces between the M and B vectors provide the mechanism to move the magnetization out of equilibrium in the NMR experiment. Equations (B.2)–(B.4) describe only the return of magnetization to equilibrium. To include the interaction of the magnetization with the magnetic fields, Bloch followed a purely phenomenological approach rather than one based on theory. The overall effect of all magnetic fields B on a single nuclear magnet μ is given by the vector cross product, as is the case for the force (or moment or torque) between any two vectors. In classical mechanics, the moment is the same thing as the rate of change of angular momentum J with time (Eq. B.5).

$$\frac{dJ}{dt} = \mu \times B \tag{B.5}$$

Nuclear Magnetic Resonance Spectroscopy: An Introduction to Principles, Applications, and Experimental Methods,
Second Edition. Joseph B. Lambert, Eugene P. Mazzola, and Clark D. Ridge.
© 2019 John Wiley & Sons Ltd. Published 2019 by John Wiley & Sons Ltd.

(For example, see *The Feynman Lectures of Physics*, Vol. 1, Eqs. (20.11–20.13).) From Eq. (A.2) ($\mu = \gamma \hbar I$) and from the definition of dimensionless spin ($J = \hbar I$), Eq. (B.5) can be recast as Eq. (B.6).

$$\frac{dI}{dt} = \gamma I \times B \tag{B.6}$$

Since M is the sum of all spins ($M = \Sigma I$), the rate of change of magnetization, perturbed by the magnetic fields, is given by Eq. (B.7),

$$\frac{dM}{dt} = \gamma M \times B = \gamma \begin{vmatrix} i & j & k \\ M_x & M_y & M_z \\ B_x & B_y & B_z \end{vmatrix} \tag{B.7}$$

which is expanded into the standard matrix form for the vector cross product.

Our eventual aim is to obtain simple, closed expressions for the three magnetization components in the middle row of the matrix in Eq. (B.7). We can evaluate the values for the rate of change of each component of M from the force of B on M in Eq. (B.7) to give the results in Eqs. (B.8)–(B.10).

$$\frac{dM_z}{dt} = \gamma M_x B_y - \gamma M_y B_x \tag{B.8}$$

$$\frac{dM_x}{dt} = \gamma M_y B_z - \gamma M_z B_y \tag{B.9}$$

$$\frac{dM_y}{dt} = -\gamma M_x B_z + \gamma M_z B_x \tag{B.10}$$

Although we can substitute $B_0 = B_z$, we still need expressions for B_x and B_y in terms of B_1. As depicted in Figure B.1, the linear oscillation of B_1 along the x direction is equivalent to a circular sweep in the xy plane. More rigorously, the linear vector is equivalent to two vectors rotating circularly in opposite directions. The x components of two circular vectors sum to give the linear vector along the x direction, whereas the y components (directed up and down) cancel out. At any time t, the components of the clockwise vector are $B_x = B_1 \cos \omega t$ and $B_y = -B_1 \sin \omega t$, and those of the counterclockwise vector are $B_x = B_1 \cos \omega t$ and $B_y = B_1 \sin \omega t$ (ω is the Larmor frequency). By convention, the clockwise vector is used to describe the B_1 field.

Substituting these values into Eqs. (B.8)–(B.10) and combining those equations with Eqs. (B.2)–(B.4) leads to Eqs. (B.11)–(B.13).

$$\frac{dM_z}{dt} = -\gamma M_x B_1 \sin \omega t - \gamma M_y B_1 \cos \omega t + \frac{(M_0 - M_z)}{T_1} \tag{B.11}$$

$$\frac{dM_x}{dt} = -\gamma M_y B_0 + \gamma M_z B_1 \sin \omega t - \frac{M_x}{T_2} \tag{B.12}$$

$$\frac{dM_y}{dt} = -\gamma M_x B_0 + \gamma M_z B_1 \cos \omega t - \frac{M_y}{T_2} \tag{B.13}$$

These expressions may be simplified before they are solved by a change in coordinates that corresponds to the use of a *rotating coordinate system*. There is no change along the z direction, but a set of coordinates rotating around z at rate ω replaces x and y. The new coordinate x' rotates along (in phase) with B_1, and the new coordinate y' rotates

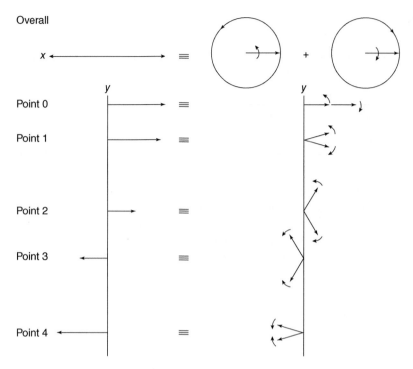

Figure B.1 The equivalence of a linearly oscillating frequency to two oppositely circulating oscillating frequencies.

90° behind x' (out of phase). Thus, the magnetization in the xy plane can be expressed as real u and imaginary v components of a complex number $M_{xy} = u + iv$ (u is the same as $M_{x'}$ and v as $M_{y'}$). Equations (B.14) and (B.15)

$$M_x = u \cos \omega t - v \sin \omega t \tag{B.14}$$

$$M_y = -u \sin \omega t - v \cos \omega t \tag{B.15}$$

fully define the change in coordinates. Substitution of these quantities into Eqs. (B.11)–(B.13), leads, after considerable algebra and trigonometry, to new differential Eqs. (B.16)–(B.18)

$$\frac{dM_z}{dt} = \gamma B_1 v + \frac{(M_0 - M_z)}{T_1} \tag{B.16}$$

$$\frac{du}{dt} = -(\omega_0 - \omega)v - \frac{u}{T_2} \tag{B.17}$$

$$\frac{dv}{dt} = (\omega_0 - \omega)u - \frac{v}{T_2} - \gamma B_1 M_z \tag{B.18}$$

(with the additional substitution of ω_0 for γB_0, the resonance condition).

At the equilibrium condition of slow adiabatic passage, all three of these time variations are small and may be set equal to zero. We then have three equations in three unknowns (u, v, and M_z). These equations may be solved algebraically to give

Eqs. (B.19)–(B.21).

$$u = M_0 \frac{\gamma B_1 T_2^2(\omega_0 - \omega)}{1 + T_2^2(\omega_0 - \omega)^2 + \gamma^2 B_1^2 T_1 T_2} \tag{B.19}$$

$$v = -M_0 \frac{\gamma B_1 T_2}{1 + T_2^2(\omega_0 - \omega)^2 + \gamma^2 B_1^2 T_1 T_2} \tag{B.20}$$

$$M_z = M_0 \frac{1 + T_2^2(\omega_0 - \omega)^2}{1 + T_2^2(\omega_0 - \omega)^2 + \gamma^2 B_1^2 T_1 T_2} \tag{B.21}$$

Although either u or v could be detected, normally the component $(v \text{ or } M_{y'})$ that is out of phase with respect to the B_1 field is detected and observed. This component is called the *absorption mode*, and the component u or $M_{x'}$ that is in phase with B_1 is called the *dispersion mode*. Whereas the dispersion-mode signal goes to zero at resonance, when $\omega = \omega_0$ (and is used in electron spin resonance), the absorption mode goes to $-M_0 \gamma B_1 T_2 / (1 + \gamma^2 B_1^2 T_1 T_2)$, which has this familiar maximum value at $\omega = \omega_0$.

The phase relationship between u and v is best seen from the Eqs. (B.22) and (B.23)

$$u = M_x \cos \omega t - M_y \sin \omega t \tag{B.22}$$

$$v = -M_x \sin \omega t - M_y \cos \omega t \tag{B.23}$$

for the reverse coordinate transformation (compare Eqs. (B.14) and (B.15)). If there is no x magnetization, then u is a sine function, v is a cosine function, and the two are 90° out of phase. Similarly, if there is no y magnetization, then u is a cosine function, v is a cosine

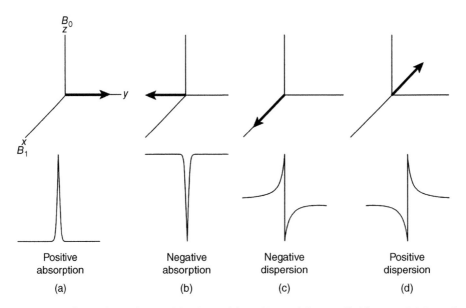

Figure B.2 Phase relationships and the slope of the NMR signal. Source: Claridge 1999 [1]. Reproduced with permission of Elsevier.

function, and, again, the two are 90° out of phase. The linear sums of x and y magnetization then also are 90° out of phase. The phase relationships are illustrated in Figure B.2. When the magnetization is lined up along the y-axis of the rotating coordinate system (called y'), the resulting signal is *positive absorption* (Figure B.2a). When the magnetization is lined up along the $-y$-axis, the signal is *negative absorption* (Figure B.2b). When the signal is lined up along the x-axis (and hence is in phase with B_1), the signal is *negative dispersion* (Figure B.2c). Finally, when the signal is lined up along the $-x$-axis, the signal is *positive dispersion* (Figure B.2d).

Reference

B.1 Claridge, T.D.W. (1999). *High-Resolution NMR Techniques in Organic Chemistry*, 19. Amsterdam: Pergamon Press.

Appendix C

Quantum Mechanical Treatment of the Two-Spin System

Modern spectral analysis has largely lost track of the quantum mechanical foundation of NMR spectroscopy. As a result, many common NMR phenomena must be taken on faith. Without knowledge of the quantum mechanical treatment, however, one cannot answer common questions such as the following:

- Where is the cause of the scalar coupling constant J?
- Why do equivalent or isochronous nuclei not split each other, even when they couple?
- What do the terms *closely coupled* and *second order* mean in terms of spin states?
- How are second-order intensities determined?
- What does the sign of the coupling constant signify?
- How is the chemical-shift difference measured in an AB spectrum?

To answer these and other questions, we describe in this appendix the quantum mechanical treatment for two spins. The description provides one of the few examples of an exact (closed) solution of the Schrödinger equation.

Schrödinger's wave equation (Eq. (C.1))

$$\mathcal{H}\Psi = E\Psi \tag{C.1}$$

provides a process whereby the energies of wave functions may be evaluated. Whereas, in molecular science, wave functions are used to describe molecules, as regards NMR they describe spin states $+\frac{1}{2}, -\frac{1}{2}$. The Hamiltonian operator $\mathcal{H}$ provides the instruction to be executed on the wave function whereby the energy is obtained according to Eq. (C.1). Our objective in this presentation is fourfold: (i) to describe wave functions for spin systems, (ii) to obtain the energies of the wave functions from the Schrödinger equation, (iii) to derive the frequencies of individual transitions between spin states, and (iv) to evaluate the intensities of the resulting NMR absorptions. This information provides all the parameters needed to define the NMR spectrum, except the relaxation time, which controls the spectral line width.

The Hamiltonian operator for NMR is composed of a series of terms, the most important of which define the chemical shift and the coupling constant. Since we are dealing with the scalar energies of the interactions between vectors (between the magnetic moment $\boldsymbol{\mu}$ and the magnetic field $\boldsymbol{B}$ for the chemical shift and between two spins

I for the coupling constant), the Hamiltonian operators are vector dot products, as in Eqs. (C.2a) and (C.2b).

$$\text{chemical shift:} \quad \mathcal{H}^0 = \sum_i \mu_i \cdot B \tag{C.2a}$$

$$\text{coupling constant:} \quad \mathcal{H}^1 = \sum_i \sum_j J_{ij} I(i) \cdot I(j) \tag{C.2b}$$

The superscript on the Hamiltonian signifies that the chemical shift is considered to be the zeroth-order Hamiltonian ($\mathcal{H}^0$) and the coupling constant the first-order Hamiltonian ($\mathcal{H}^1$). The familiar coupling constant J is seen to be the proportionality constant between the Hamiltonian operator and the dot product of the spins and signifies quantitatively how strongly the spins interact. The summation over i in $\mathcal{H}^0$ allows the inclusion of all nuclei, and the double summation over i and j ($i<j$) in $\mathcal{H}^1$ ensures that interactions between all spins are included, but only once.

In Appendix A, we saw how the expression $\mu \cdot B$ can be expanded. Substitution into Eq. (C.2a) gives $\mathcal{H}^0 = \gamma \hbar I \cdot B$, which corresponds to $\gamma \hbar I_z B_0$ (Eq. (A.5)) in the presence of only B_0. By Eqs. (A.6) and (A.7), the Hamiltonian operator for the chemical shift becomes $h\nu_0 I_z$, or, in units of frequency (Hz), $\nu_0 I_z$. The full Hamiltonian operator then is given by Eq. (C.3).

$$\mathcal{H} = \mathcal{H}^0 + \mathcal{H}^1 = \sum_i \nu_i I_z(i) + \sum_i \sum_j J_{ij} I(i) \cdot I(j) \tag{C.3}$$

In order to apply the Schrödinger equation, we need to define spin wave functions. For a single spin with $I = 1/2$, the wave function for the lower-energy spin state with $I_z = -1/2$ is simply called β and that for the higher-energy spin state with $I_z = +1/2$ is called α. To construct wave functions for systems containing multiple spins, products of these wave functions are used as a basis, or beginning, set. Thus, for two spins, the basis set comprises the products β(1)β(2), α(1)β(2), β(1)α(2), and α(1)α(2). It is always understood that the first character represents nucleus one, and so on, so the parentheses may be dropped, giving ββ, αβ, βα, and αα. For larger spin systems, this process is carried out in logical fashion (for three spins, βββ, ββα, βαβ, αββ, and so on). The basis wave functions are orthonormal; that is, the projection of a wave function onto itself gives unity (normality) and that of one wave function onto another gives zero (orthogonality). In Eqs. (C.4a) and (C.4b),

$$\langle \alpha | \alpha \rangle = 1 \quad \langle \beta | \beta \rangle = 1 \tag{C.4a}$$

$$\langle \alpha | \beta \rangle = 0 \quad \langle \beta | \alpha \rangle = 0 \tag{C.4b}$$

the shorthand Dirac quantum mechanical notation for integration is used. Thus, $\langle \alpha | \alpha \rangle = \int \alpha \alpha dv$, in which the integration is over a spatial element v. Extention of these equations to two spins thus implies that $\langle \alpha\alpha | \alpha\alpha \rangle$, and so on, are unity and $\langle \alpha\alpha | \alpha\beta \rangle$, and so on, are zero.

The complete wave function for a spin system is constructed by linearly combining the various basis-set wave functions, as in the formula of Eq. (C.5),

$$\Phi_q = \sum_m a_{qm} \Psi_m \tag{C.5}$$

in which Φ_q is the complete function for the qth spin state, Ψ_m is the set of m basis functions, and a_{qm} is the mixing coefficient for each of the m basis functions. In a first-order, two-spin system (AX rather than AB), by definition, there is no mixing between the spin states, so the basis functions also are the final, or stationary-state, wave functions. This is the quantum mechanical meaning of first order. Table C.1 provides a list of the first-order wave functions Φ_q. The quantity S_z is the sum of the constituent I_z states. Thus for αα, $S_z = (½ + ½) = 1$; for αβ and βα, $S_z = (-½ + ½) = 0$; and for ββ, $S_z = (-½ - ½) = -1$. The energies of a given state are obtained from the Hamiltonian matrix, whose component values are $H_{mn} = \langle \Phi_m | \mathcal{H} | \Phi_n \rangle$. With orthonormal wave functions, this matrix is diagonalized in the first-order case, so $H_{mn}(m \neq n) = 0$ and H_{nn} is given by Eq. (C.6).

$$H_{nn} = \langle \Phi_n | \mathcal{H} | \Phi_n \rangle = E_n \langle \Phi_n | \Phi_n \rangle = E_n \qquad (C.6)$$

Thus, the energies of the spin state are the diagonal elements of the diagonalized Hamiltonian matrix. In the first-order case, the off-diagonal elements are zero (there is no mixing), so the energies are calculated directly from the diagonal elements (Eq. (C.6)).

When $J = 0$, the Hamiltonian operator is given by Eq. (C.7).

$$\mathcal{H} = \mathcal{H}^0 = \sum_i v_i I_z(i) \qquad (C.7)$$

For example, for αα, I_z is $+½$ for both nuclei, so $\mathcal{H} = \frac{1}{2}v_A + \frac{1}{2}v_X$, as recorded in the fourth column of Table C.1. The other three values are calculated analogously. Figure C.1 illustrates the order of spin states in what is called an *energy-level diagram*. Resonance corresponds to transitions between any two of the levels shown. Only transitions between adjacent values of S_z, however, are allowed; that is, there is a selection rule of $\Delta S_z = 1$. The forbidden transitions are $4 \rightarrow 1$ and $3 \rightarrow 2$, which, respectively, are called double ($\Delta S_z = 2$) or zero ($\Delta S_z = 0$) quantum transitions and occasionally can be seen as very weak peaks. There are two allowed transitions of A nuclei ("A-type transitions"): from 3 to 1 (13) and from 4 to 2 (24). There are also two allowed transitions of X nuclei ("X-type transactions"): from 2 to 1 (12) and from 4 to 3 (34). Table C.1 shows that, when the subtractions are carried out, $\Delta E_{13} = \Delta E_{24} = v_A$ and

Table C.1 Two-spin, first-order (AX) parameters without coupling ($J=0$) and with coupling ($J \neq 0$).

q	Φ_q	S_z	$E_q (J=0)$	$E_q (J \neq 0)$
1	αα	+1	$\frac{1}{2}(v_A + v_X)$	$\frac{1}{2}(v_A + v_X) + \frac{1}{4}J$
2	αβ	0	$\frac{1}{2}(v_A - v_X)$	$\frac{1}{2}(v_A - v_X) - \frac{1}{4}J$
3	βα	0	$-\frac{1}{2}(v_A - v_X)$	$-\frac{1}{2}(v_A - v_X) - \frac{1}{4}J$
4	ββ	-1	$-\frac{1}{2}(v_A + v_X)$	$-\frac{1}{2}(v_A + v_X) + \frac{1}{4}J$

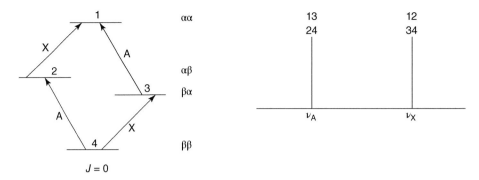

Figure C.1 The energy level diagram and the transitions for a first order, two spin system (AX) with $J=0$.

$\Delta E_{12} = \Delta E_{24} = v_X$. These are the resonance frequencies of the four allowed transitions. The spectrum, illustrated diagrammatically at the right of Figure C.1, thus is composed of two singlets, respectively, at v_A and v_X ($v_A > v_X$, arbitrarily), each composed of two coincident transitions. The position of the double quantum transition can be obtained by equating $2h\nu$ to ΔE_{14}. The resulting frequency occurs at $\frac{1}{2}(v_A + v_X)$, the midpoint of the spectrum.

In order to describe the coupled AX spectrum ($J \neq 0$), it is necessary to evaluate the $\mathcal{H}^1$ part of the Hamiltonian matrix and add it to $\mathcal{H}^0$. The simple results already are given in the last column of Table C.1, where it is seen that the term from $\mathcal{H}^0$ is modulated by a term related to the coupling constant J from $\mathcal{H}^1$ (Eq. (C.3)). The integral to be evaluated is of the type $\langle \Phi_n | \mathcal{H}^1 | \Phi_m \rangle = J_{ij} \langle \Phi_n | I(i) \cdot I(j) | \Phi_m \rangle$. Evaluation of this integral requires a specification of the effects of the operator I on the wave functions, as in Eq. (C.8), in which $i = \sqrt{-1}$.

$$I_x \alpha = \frac{1}{2}\beta$$
$$I_y \alpha = \frac{1}{2}i\beta$$
$$I_z \alpha = \frac{1}{2}\alpha \qquad (C.8)$$
$$I_x \beta = \frac{1}{2}\alpha$$
$$I_y \beta = -\frac{1}{2}i\alpha$$
$$I_z \beta = -\frac{1}{2}\beta$$

Since the system is first order, the H_{nm} terms are zero and the energy is given by H_{nn} in Eq. (C.6). Figure C.2 illustrates the evaluation of $\mathcal{H}^1$ for E_1. In this way, the energies are obtained for $J \neq 0$ in Table C.1. The energies illustrated in Figure C.1 thus are nudged up (for $\alpha\alpha$ and $\beta\beta$) and down (for $\alpha\beta$ and $\beta\alpha$) by $1/4 J$, as indicated by the arrows beside the energy levels in the leftmost diagram of Figure C.3. When the subtractions are carried out between the energies E_q ($J \neq 0$) in Table C.1, the transition energies, which are the differences, are found to be $\Delta E_{13} = v_A + 1/2 J$, $\Delta E_{24} = v_A - 1/2 J$, $\Delta E_{12} = v_X + 1/2 J$, and $\Delta E_{34} = v_X - 1/2 J$. Now each allowed transition has a different

Figure C.2 Evaluation of the diagonal elements of the Hamiltonian for scalar coupling in a first-order, two-spin system (AX).

$$E_1^1 = H_{11}^1 = \langle\alpha\alpha|\mathcal{H}^1|\alpha\alpha\rangle$$
$$= J\langle\alpha\alpha|I(1)\cdot I(2)|\alpha\alpha\rangle$$
$$= J\langle\alpha\alpha|(I_xI_x + I_yI_y + I_zI_z)|\alpha\alpha\rangle$$
$$= J(\langle\alpha\alpha|I_xI_x|\alpha\alpha\rangle + \langle\alpha\alpha|I_yI_y|\alpha\alpha\rangle + \langle\alpha\alpha|I_zI_z|\alpha\alpha\rangle)$$

Separate into terms for the individual spins:

$$E_1^1 = J(\langle\alpha(1)|I_x(1)|\alpha(1)\rangle\langle\alpha(2)|I_x(2)|\alpha(2)\rangle +$$
$$\langle\alpha(1)|I_y(1)|\alpha(1)\rangle\langle\alpha(2)|I_y(2)|\alpha(2)\rangle +$$
$$\langle\alpha(1)|I_z(1)|\alpha(1)\rangle\langle\alpha(2)|I_z(2)|\alpha(2)\rangle)$$

Spins 1 and 2 are equivalent:

$$E_1^1 = J(\langle\alpha|I_x|\alpha\rangle^2 + \langle\alpha|I_y|\alpha\rangle^2 + \langle\alpha|I_z|\alpha\rangle^2)$$

The integrals are evaluated by Eq. (C.8)

$$E_1^1 = J(\langle\alpha|\tfrac{1}{2}\beta\rangle^2 + \langle\alpha|\tfrac{1}{2}i\beta\rangle^2 + \langle\alpha|\tfrac{1}{2}\alpha\rangle^2)$$
$$= J(0 + 0 + \tfrac{1}{4})$$
$$= \tfrac{1}{4}J$$

Similarly,

$$E_2^1 = J\langle\alpha\beta|I\cdot I|\alpha\beta\rangle = -\tfrac{1}{4}J$$
$$E_3^1 = J\langle\beta\alpha|I\cdot I|\beta\alpha\rangle = -\tfrac{1}{4}J$$

and

$$E_4^1 = J\langle\beta\beta|I\cdot I|\beta\beta\rangle = \tfrac{1}{4}J$$

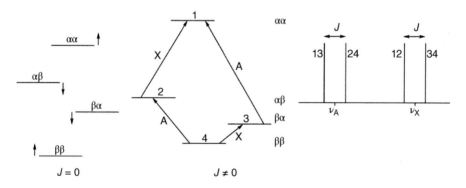

Figure C.3 The energy-level diagram and the transitions for a first-order, two-spin system (AX) with $J \neq 0$.

energy, and the spectrum is composed of four peaks. The pairs of peaks are centered, respectively, about v_A and v_X. The spacings within each pair are the coupling constant J: $[(v_A + \tfrac{1}{2}J) - (v_A - \tfrac{1}{2}J) = J]$ and $[(v_X + \tfrac{1}{2}J) - (v_X - \tfrac{1}{2}J) = J]$.

As the A and X chemical shifts move closer together, mixing occurs between some of the basic wave functions, the off-diagonal matrix elements can no longer be ignored, and the spectrum is said to be second order. Only wave functions with the same S_z can mix. Thus, $\alpha\alpha$ and $\beta\beta$, the sole representatives for $S_z = +1$ and $S_z = -1$, respectively, do not mix with any other wave functions and continue to be stationary-state wave functions. The $S_z = 0$ states ($\alpha\beta$ and $\beta\alpha$), however, do mix.

It is easiest first to examine the extreme at which the chemical shifts coincide, that is, the case of A_2. We can construct the stationary-state wave functions with the knowledge that contributions from each spin must be equivalent, a condition that is possible only with equal mixing coefficients: $(\alpha\beta + \beta\alpha)$ and $(\alpha\beta - \beta\alpha)$. In order to make the wave functions orthonormal, mixing coefficients of $1/\sqrt{2}$ must be used, that is, $(\alpha\beta + \beta\alpha)/\sqrt{2}$ and $(\alpha\beta - \beta\alpha)/\sqrt{2}$, as can be seen by evaluating the integrals $\langle(\alpha\beta + \beta\alpha)/\sqrt{2}|(\alpha\beta + \beta\alpha)/\sqrt{2}\rangle$ and $\langle(\alpha\beta - \beta\alpha)/\sqrt{2}|(\alpha\beta - \beta\alpha)/\sqrt{2}\rangle$.

These wave functions have important symmetry properties. For $\alpha\alpha$, $\beta\beta$, and $(\alpha\beta + \beta\alpha)/\sqrt{2}$, the two A nuclei may be interchanged to give the identical initial wave functions back. Such wave functions are said to be *symmetric*. In contrast, interchanging nuclei in $(\alpha\beta - \beta\alpha)/\sqrt{2}$ gives $(\beta\alpha - \alpha\beta)/\sqrt{2}$, which is the negative of the original wave function. Such wave functions are said to be *antisymmetric*. There is a second selection rule when wave functions have these properties, namely, that transitions are allowed only between wave functions of the same symmetry. Thus, transitions are forbidden between symmetric and antisymmetric wave functions.

This second selection rule can be rationalized by introducing two new operators: the raising operator F_+, which can raise β to α, but cannot operate on α, and the lowering operator, which can lower α to β, but cannot operate on β. (We will need to use only the raising operator.) The operation rules are given in Eq. (C.9).

$$F_+|\beta\rangle = |\alpha\rangle \quad F_+|\alpha\rangle = 0$$
$$F_-|\beta\rangle = 0 \quad F_-|\alpha\rangle = |\beta\rangle \tag{C.9}$$

A transition probability is the squared integral $\langle\Phi_n|F_+|\Phi_m\rangle^2$, which evaluates the probability of raising Φ_m to Φ_n. Figure C.4 evaluates two of these integrals and confirms

Probability of transition from $\beta\beta$ to $(\alpha\beta + \beta\alpha)/\sqrt{2}$
$$= \langle(\alpha\beta + \beta\alpha)/\sqrt{2}|F_+|\beta\beta\rangle^2$$
$$= \tfrac{1}{2}\langle(\alpha\beta + \beta\alpha)|(\alpha\beta + \beta\alpha)\rangle^2$$
$$= \tfrac{1}{2}(\langle\alpha\beta|\alpha\beta\rangle + \langle\alpha\beta|\beta\alpha\rangle + \langle\beta\alpha|\alpha\beta\rangle + \langle\beta\alpha|\beta\alpha\rangle)^2$$
$$= \tfrac{1}{2}(1 + 0 + 0 + 1)^2$$
$$= 2$$

Probability of transition from $\beta\beta$ to $(\alpha\beta - \beta\alpha)/\sqrt{2}$
$$= \langle(\alpha\beta - \beta\alpha)/\sqrt{2}|F_+|\beta\beta\rangle^2$$
$$= \tfrac{1}{2}\langle(\alpha\beta - \beta\alpha)|(\alpha\beta + \beta\alpha)\rangle^2$$
$$= \tfrac{1}{2}(\langle\alpha\beta|\alpha\beta\rangle + \langle\alpha\beta|\beta\alpha\rangle - \langle\beta\alpha|\alpha\beta\rangle - \langle\beta\alpha|\beta\alpha\rangle)^2$$
$$= \tfrac{1}{2}(1 + 0 - 0 - 1)$$
$$= 0$$

Figure C.4 Evaluation of the transition probabilities in a two-spin system with identical chemical shifts (A_2).

the symmetry selection rule. Note that the operation of F_+ on $\beta\beta$ gives $(\alpha\beta + \beta\alpha)$, because F_+ must operate separately on each spin (i.e. $F_+ = F_{1+} + F_{2+}$, in which F_{1+} operates on the first spin ($\beta\beta \to \alpha\beta$) and F_{2+} operates on the second spin ($\beta\beta \to \beta\alpha$)). The result is zero for transitions between states of unlike symmetry and 2 for transitions between states of like symmetry. The significance of the value 2 is noted presently.

Table C.2 contains the parameters for the case of A_2. The wave functions have been sorted according to symmetry. Let us first consider the case for which $J = 0$. The E^0 term is evaluated from $\sum_i v_i I_z(i)$ (Eq. (C.7)) and the Hamiltonian component H_{nm} (Eq. (C.6)). The results for $\alpha\alpha$ and $\beta\beta$ must be the same as in Table C.1 when $v_A = v_X$. The results for $(\alpha\beta \pm \beta\alpha)/\sqrt{2}$ are interesting in that the resonance frequency cancels out, as in Eq. (C.10a) and (C.10b)).

$$E^0(S_0) = \frac{[E^0(\alpha\beta) + E^0(\beta\alpha)]}{\sqrt{2}} = \frac{\left[\frac{1}{2}(v_A - v_A) + \frac{1}{2}(-v_A + v_A)\right]}{\sqrt{2}} = 0 \quad \text{(C.10a)}$$

$$E^0(A_0) = \frac{[E^0(\alpha\beta) - E^0(\beta\alpha)]}{\sqrt{2}} = \frac{\left[\frac{1}{2}(v_A - v_A) - \frac{1}{2}(-v_A + v_A)\right]}{\sqrt{2}} = 0 \quad \text{(C.10b)}$$

Thus, the E^0 term is zero for the S_0 and A_0 states, which is the same thing as saying that the B_0 field does not interact with those states.

Figure C.5 shows the energy-level diagram. Subtraction of the energies for allowed transitions between spin states indicates that resonances occur at $\Delta E(S_0 \to S_1) = v_A$ and at $\Delta E(S_{-1} \to S_0) = v_A$. Thus, two transitions occur at the same position, and the result is a singlet at v_A. Although there are four spin states, only two of these can make upward transitions. These transitions, however, have a probability of 2 (Figure C.4) so that the average probability for all four states is $\frac{1}{4}(0 + 0 + 2 + 2)$, or unity.

When $J \neq 0$, the values for $\mathcal{H}^1$ are evaluated as in Figure C.2. The symmetric states are raised in energy by $\frac{1}{4}J$, and the antisymmetric state is lowered by $-\frac{1}{4}J$, as illustrated in Figure C.6. Again, the arrows beside the energy levels in the leftmost diagram indicate the direction of the change in energy. Because the symmetric states are perturbed by identical amounts, the transition energies remain unchanged: $\Delta E(S_0 \to S_1) = \Delta E(S_{-1} \to S_0) = v_A$. Again, the two transitions occur at the same position, and the result is a singlet at v_A. Hence, the explanation as to why equivalent nuclei do

Table C.2 Parameters for two equivalent spin (A_2) without coupling ($J=0$) and with coupling ($J \neq 0$).

q	Φ_q	S_z	$E_q (J=0)$	$E_q (J \neq 0)$
S_1	$\alpha\alpha$	$+1$	v_A	$v_A + \frac{1}{4}J$
S_0	$(\alpha\beta + \beta\alpha)/\sqrt{2}$	0	0	$\frac{1}{4}J$
S_{-1}	$\beta\alpha$	-1	$-v_A$	$-v_A + \frac{1}{4}J$
A_0	$(\alpha\beta - \beta\alpha)/\sqrt{2}$	0	0	$-\frac{3}{4}J$

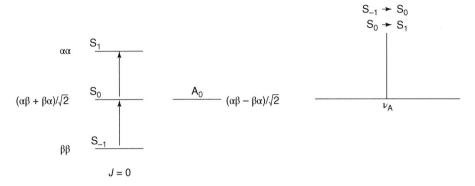

Figure C.5 The energy-level diagram and the transitions for the two-spin system with identical chemical shifts (A_2) with $J = 0$.

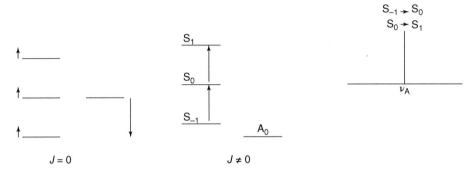

Figure C.6 The energy-level diagram and the transitions for the two-spin system with identical chemical shifts (A_2) with $J \neq 0$.

not split each other is subtle. First, transitions between symmetric and antisymmetric states are forbidden, so different perturbations of these families are irrelevant. Second, each symmetric state is perturbed by the same amount, so the coupling drops out of the difference between states. It even drops out of the forbidden double quantum transition from S_{-1} to S_1, which would occur with low intensity at $v_A(2h\nu) = 2v_A$.

When two nuclei have similar, but nonidentical, chemical shifts ($\Delta v/J \lesssim 10$), the AX system becomes AB. The $\alpha\alpha$ and $\beta\beta$ states are still stationary-state wave functions, but now $\alpha\beta$ and $\beta\alpha$ are mixed to an unknown, unequal extent. The general wave function is $[a(\alpha\beta) + b(\beta\alpha)]$, with two pairs of values for a and b to generate two wave functions. We need to obtain these values to evaluate the energies of the spin states, to calculate the frequencies of the resonances, and to determine the intensities of the resonances.

The energies are determined from the Hamiltonian matrix, and those for $\alpha\alpha$ and $\beta\beta$ are the same as in Table C.1. The Hamiltonian matrix is then reduced to a 2×2 determinant. The off-diagonal elements (H_{23} and H_{32}) represent the extent of mixing. This determinant must be diagonalized by subtracting the variable from the diagonal elements, setting the results equal to zero, and working out the algebra. The process is carried out in Figure C.7. The diagonal elements, H_{22} and H_{33}, are the same as in Table C.1. The off-diagonal elements—for example, H_{23}—are calculated from $\langle \Psi_m | H | \Psi_n \rangle$. The E^0

Figure C.7 Diagonalization of the $S_z = 0$ Hamiltonian matrix for the second-order, two-spin system (AB) and evaluation of the spin state energies.

$$H_{mn} = \begin{vmatrix} H_{22} & H_{32} \\ H_{23} & H_{33} \end{vmatrix}$$

$$\begin{vmatrix} H_{22} - E & H_{32} \\ H_{23} & H_{33} - E \end{vmatrix} = 0$$

$$\begin{vmatrix} \tfrac{1}{2}(v_A - v_B) - \tfrac{1}{4}J - E & \tfrac{1}{2}J \\ \tfrac{1}{2}J & -\tfrac{1}{2}(v_A - v_B) - \tfrac{1}{4}J - E \end{vmatrix} = 0$$

$$\left[\left(\tfrac{1}{2}(v_A - v_B) - \tfrac{1}{4}J - E\right)\left(-\tfrac{1}{2}(v_A - v_B) - \tfrac{1}{4}J - E\right) - \tfrac{1}{4}J^2\right] = 0$$

$$-\tfrac{1}{4}(v_A - v_B)^2 + \tfrac{1}{16}J^2 + \tfrac{1}{2}JE + E^2 - \tfrac{1}{4}J^2 = 0$$

$$\left(E + \tfrac{1}{4}J\right)^2 = \tfrac{1}{4}[(v_A - v_B)^2 + J^2]$$

$$E_\pm = \pm\tfrac{1}{2}[(v_A - v_B)^2 + J^2]^{\tfrac{1}{2}} - \tfrac{1}{4}J$$

$$E_\pm = \pm C - \tfrac{1}{4}J$$

Note:

$$2C = [(v_A - v_B)^2 + J^2]^{\tfrac{1}{2}}$$

$$4C^2 = (v_A - v_B)^2 + J^2$$

$$v_A - v_B = (4C^2 - J^2)^{\tfrac{1}{2}}$$

components are calculated from $\langle\alpha\beta|\mathcal{H}^1|\beta\alpha\rangle$ and $\langle\beta\alpha|\mathcal{H}^1|\alpha\beta\rangle$ and are found to be $\tfrac{1}{2}J$, by the method illustrated in Figure C.2.

A new parameter, C, is defined in Figure C.7. The chemical-shift differences $(v_A - v_B)$ can be calculated if J and C are known. Table C.3 summarizes the results. After C is inserted, and when the chemical-shift difference is much greater than the coupling constant $[(v_A - v_B) \gg J]$, the values of E_q are the same as those for the first-order case in Table C.1.

Figure C.8 illustrates the energy levels and transitions for the AB case. Subtraction of the energies leads to four transition frequencies. By convention, the center of the spectrum is given zero frequency, so that $\tfrac{1}{2}(v_A + v_B) = 0$. Consequently, the four transition frequencies are $(-C + \tfrac{1}{2}J)$ for $2 \to 1$ (12), $(-C - \tfrac{1}{2}J)$ for $4 \to 3$ (34), $(C + \tfrac{1}{2}J)$ for $3 \to 1$ (13), and $(C - \tfrac{1}{2}J)$ for $2 \to 4$ (24) with respect to the center point. The value of J thus corresponds to the difference between the first (13) and second (24) or between the third (12) and fourth (34) peaks, and the value of C corresponds to one-half the

Table C.3 Parameters for two-spins, second-order (AB) with coupling ($J \neq 0$).

q	Φ_q	S_z	E_q
1	$\alpha\alpha$	$+1$	$\tfrac{1}{2}(v_A + v_B) + \tfrac{1}{4}J$
2	$a(\alpha\beta) + b(\beta\alpha)$	0	$C - \tfrac{1}{4}J$
3	$a'(\alpha\beta) + b'(\beta\alpha)$	0	$-C - \tfrac{1}{4}J$
4	$\beta\beta$	-1	$-\tfrac{1}{2}(v_A - v_B) + \tfrac{1}{4}J$

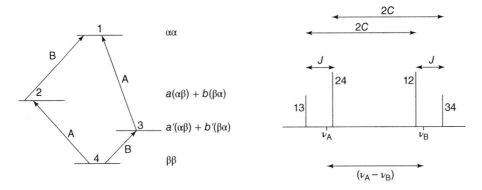

Figure C.8 The energy-level diagram and the transitions for the second-order, two-spin system (AB).

difference between the first and third or between the second and fourth peaks, as illustrated in the figure. The chemical-shift difference comes from these two quantities: $(v_A - v_B) = (4C^2 - J^2)^{1/2}$. When $J = 0$, the spectrum collapses to the two-line form seen in Figure C.1.

What remains is to evaluate the coefficients in the stationary-state wave function for $S_z = 0$ and to determine the relative intensities of the inner and outer peaks, which are seen to be unequal in Figure C.8. One method of determining the coefficients is to recognize that their ratio corresponds to the ratio of their matrix cofactors. (The cofactor of a particular matrix element is obtained by deleting the top row and the column in which it resides.) With reference to Figure C.7, the cofactor for $\alpha\beta$ (H_{22}) is $\left[-\frac{1}{2}(v_A - v_B) - \frac{1}{4}J - E\right]$ and the cofactor for $\beta\alpha$ (H_{33}) is $\frac{1}{2}J$. The formula of Eq. (C.11)

$$\frac{a}{b} = \frac{1}{2}J[-(v_A - v_B) \pm 2C] \tag{C.11}$$

gives the coefficient ratio when the values of E_+ and E_- from Figure C.7 are substituted for E in the cofactor for $\alpha\beta$. In order to obtain normalized wave functions, we also require that $(a^2 + b^2) = 1$. The coefficients a and b are thus two unknowns to be obtained from two equations. A simplification results when one defines an angle (with no physical significance) from the expression $\sin 2\theta = J/2C$. When this substitution is made, it is found that $a = \cos\theta$ when $b = \sin\theta$ and $a = \sin\theta$ when $b = -\cos\theta$. (Note that $(\sin^2\theta + \cos^2\theta) = 1$, corresponding to $(a^2 + b^2) = 1$.) The stationary-state wave functions then are given by the Eq. (C.12a) and (C.12b).

$$\phi_2 = \cos\theta(\alpha\beta) + \sin\theta(\beta\alpha) \tag{C.12a}$$

$$\phi_3 = \sin\theta(\alpha\beta) - \cos\theta(\beta\alpha) \tag{C.12b}$$

When $\theta = 0$, ϕ_2 becomes $\alpha\beta$ and ϕ_3 becomes $\beta\alpha$, which are the stationary-state wave functions for the first-order (AX) case. When $\theta = 45°$, ϕ_2 becomes $(\alpha\beta + \beta\alpha)/\sqrt{2}$ and ϕ_3 becomes $(\alpha\beta - \beta\alpha)/\sqrt{2}$, which are the stationary-state wave functions for the case of A_2.

The intensities or transition probabilities are obtained with the use of the raising operators, as shown in Figure C.9. The intensity ratio (inner/outer) thus is seen to be $(1 + 2ab)/(1 - 2ab)$, which may be manipulated algebraically to $(a + b)^2/(a - b)^2$, or

Figure C.9 Evaluation of the transition probabilities in the second-order, two-spin system (AB).

Probability of $4 \rightarrow 2$:

$$\{\langle [a(\alpha\beta) + b(\beta\alpha)]|F_+|\beta\beta\rangle\}^2$$
$$= \{\langle [a(\alpha\beta) + b(\beta\alpha)]|(\alpha\beta + \beta\alpha)\rangle\}^2$$
$$= \{\langle [a(\alpha\beta|\alpha\beta) + b(\beta\alpha|\alpha\beta) + a(\alpha\beta|\beta\alpha) + b(\beta\alpha|\beta\alpha)]\rangle\}^2$$
$$= [a(1) + b(0) + a(0) + b(1)]^2$$
$$= (a + b)^2$$
$$= 1 + 2ab$$

Probability of $3 \rightarrow 1$:

$$\{\langle [\alpha\alpha|F_+|[a'(\alpha\beta)] + b'(\beta\alpha)]\rangle\}^2$$
$$= \{\langle \alpha\alpha|F_+|[b(\alpha\beta) - a(\beta\alpha)]\rangle\}^2 \quad \text{(From Eq. (C.12))}$$
$$= (b - a)^2$$
$$= 1 - 2ab$$

$(1+a/b)^2/(1-a/b)^2$. The ratio a/b from Eq. (C.11) is inserted into the latter expression, yielding the intensity ratio from J, C, and $(v_A - v_B)$. The intensity ratio may be obtained more directly by inserting the definition $2ab = J/2C = \sin 2\theta = 2 \sin \theta \cos \theta$ into the probability expressions in Figure C.9. The intensities of transitions 24 and 12 are thus $1+J/2C$ and of transitions 13 and 34 are $1-J/2C$. Their ratio becomes $(1+J/2C)/(1-J/2C)$, or $(2C+J)/(2C-J)$, which is the same as $(v_{13} - v_{34})/(v_{24} - v_{12})$. The intensity ratio thus corresponds to the ratio of the distances between the outer and the inner peaks.

Expressions for three-spin systems or higher can be developed in the same fashion, but with increasingly larger matrix elements to evaluate.

Appendix D

Analysis of Second-Order, Three- and Four-Spin Systems by Inspection

The analysis of the second-order, two-spin (AB) system by inspection is outlined in Section 4.7 and is explained in Appendix C. A few other second-order systems may be analyzed without recourse to computer methods. Analyzing a three-spin system ranges from the trivial (AX_2, AMX) to the impossible (many ABC systems). As the AX chemical-shift difference in the AX_2 system decreases, degeneracies are lifted, intensities change, and a new peak can appear (Figure D.1). In the AB_2 extreme, a total of nine peaks can be observed. Four of these peaks result from spin flips of A protons, four from spin flips of B protons, and one from simultaneous spin flips of both A and B protons. The ninth peak, called a combination line, is forbidden in the first-order case (AX_2) and is rarely observed even in the AB_2 extreme. The combination line is seen only in the most closely coupled case at the top, as a very low intensity peak (no. 9) at the far right.

When eight peaks are observed, the AB_2 spectrum may be analyzed by inspection. By convention, the peaks are numbered from the A to the B nuclei, from left to right in Figure D.1. The spectrum in Figure D.1e is nearly first order, and peaks 2/3, 5/6, and 7/8 are degenerate pairs giving the triplet and doublet (five peaks) expected of AX_2. In Figure D.1b,c, all degeneracies are lifted to give the full complement of eight peaks. In Figure D.1a, peaks 5/6 are degenerate again, and the combination line 9 is barely visible. Intensities increase toward the middle of the spectrum as the chemical-shift difference approaches zero for the A_3 extreme. The chemical shift of the A proton always corresponds to the position of the third peak (Eq. D.1).

$$\nu_A = \nu_3 \tag{D.1}$$

The chemical shift of the B proton occurs at the average of the fifth and seventh peaks (Eq. D.2),

$$\nu_B = \frac{1}{2}(\nu_5 + \nu_7) \tag{D.2}$$

and the AB coupling constant is obtained from a linear combination of four peak positions (Eq. D.3).

$$J_{AB} = \frac{1}{3}(\nu_1 - \nu_4 + \nu_6 - \nu_8) \tag{D.3}$$

Thus, the entire spectrum may be analyzed without reference to the combination line. When not all eight of the remaining peaks are distinct, as in Figure D.1a,d, care must

Nuclear Magnetic Resonance Spectroscopy: An Introduction to Principles, Applications, and Experimental Methods,
Second Edition. Joseph B. Lambert, Eugene P. Mazzola, and Clark D. Ridge.
© 2019 John Wiley & Sons Ltd. Published 2019 by John Wiley & Sons Ltd.

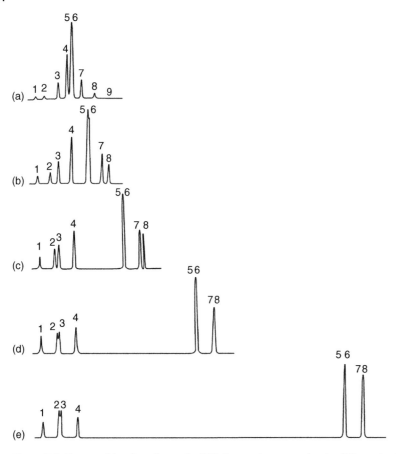

Figure D.1 The transition from first order (AX$_2$, bottom) to second order (AB$_2$, top), with J_{AB} held at 10 Hz and Δv_{AB} (Δv_{AX}) varied: 10 (top), 20, 40, 80, and 160 Hz.

be taken to recognize which peaks overlap. Recourse to computer methods then may be necessary.

The next level of spectral complexity is the case in which all three nuclei have different chemical shifts, but one of the resonances is well removed from the other two (ABX). This spectrum is determined by six parameters: v_A, v_B, v_X, J_{AB}, J_{AX}, and J_{BX}. The X resonance of an ABX spectrum generally has six lines (two overlapping triplets), with v_X at the midpoint (Figure D.2). The distance between the two tallest peaks is equal to the sum of J_{AX} and J_{BX}. The separation between the outermost lines is twice the sum of a pair of important quantities, which have been designated D_+ and D_-, and the separation of the innermost peaks is twice the difference of these two quantities. Thus, D_+ and D_-, which are defined by the formula of Eq. (D.4),

$$D_\pm = \frac{1}{2}\left\{\left[(v_A - v_B) \pm \frac{1}{2}(J_{AX} - J_{BX})\right]^2 + J_{AB}^2\right\}^{1/2} \tag{D.4}$$

may be derived from the X resonance. The value of J_{AB} is obtained from the AB part of the spectrum, as described shortly. With the knowledge of D_+, D_-, and J_{AB}, the pair of

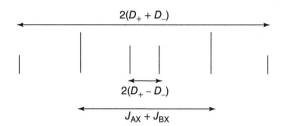

Figure D.2 The X portion of an ABX spectrum.

Eq. (D.4) gives values for $(\nu_A - \nu_B)$ and $(J_{AX} - J_{BX})$. Since $J_{AX} + J_{BX}$ is known from the X part, the individual values for J_{AX} and J_{BX} may be calculated. Two solutions, however, are possible, depending on whether these two coupling constants have the same sign (either both plus or both minus) or opposite signs (one plus and one minus). The two possibilities may be distinguished by examining the AB portion of the spectrum.

The AB part of the ABX spectrum is made up of two overlapping quartets (Figure D.3). The doublet separation between the first and second and between the third and fourth peaks in both quartets gives four independent measurements of J_{AB}. (No absolute or relative information about the sign is obtained.) The separation of alternate peaks (first and third or second and fourth) provides redundant determinations of D_+ and D_-. The separation of the midpoints of the two quartets gives an additional measurement of $\frac{1}{2}(J_{AX} + J_{BX})$. All of these spacings are illustrated in the figure.

If the arithmetic for a particular ABX spectrum is followed through, the exact values of ν_X, $(\nu_A - \nu_B)$, and J_{AB} may be obtained, but two sets of J_{AX} and J_{BX} are produced, with either like or unlike signs. Although both of these solutions correctly reproduce all the peak positions, only one gives the X portion with the proper intensities. The most straightforward method of differentiating the two sets, therefore, is to use computer methods to calculate the spectrum corresponding to each set of signs. Most modern spectrometers have simple programs for calculating spectra from an input of all the chemical shifts and coupling constants. The correct set of parameters reproduces the experimental X intensities.

The ABC spectrum, in which all three nuclei are closely coupled, contains up to 15 lines (four A-type, four B-type, four C-type, and three combination lines), but analysis by inspection is very difficult. Accordingly, recourse must be made to a general computer method, such as LAOCN3 or DAVINS, or to Castellano and Waugh's EXAN

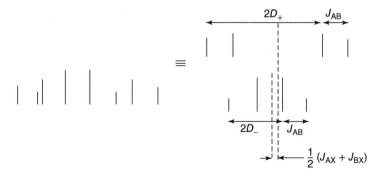

Figure D.3 The AB portion of an ABX spectrum.

II program, which was designed specifically for ABC spectra and is successful in more than 90% of cases.

Arithmetic analysis of four spin systems is very limited. First-order spectra (AX_3, A_2X_2) provide no difficulty. Of the second-order spectra, however (AA′XX′, AA′BB′, ABXY, ABCX, ABCD, etc.), only the AA′XX′ can be analyzed readily. This common pattern, which is second order only because of magnetic nonequivalence, is determined by six parameters: ν_A, ν_X, $J_{AA'}, J_{XX'}, J_{AX}$ ($= J_{A'X'}$), and $J_{AX'}$ ($= J_{A'X}$). Note that such a spectrum is dependent on the coupling between chemically equivalent nuclei ($J_{AA'}$ and $J_{XX'}$). The A and X parts of the spectrum are identical, and the chemical shifts (ν_A and ν_X) are obtained readily from their respective midpoints. Figure D.4 gives a diagrammatic representation of the A part of an AA′XX′ spectrum with mirror symmetry about ν_A. As ($\nu_A - \nu_X$) gets smaller, this symmetry is lost, and the spectrum becomes AA′BB′.

As Eq. (D.5)

$$K = J_{AA'} + J_{XX'}; L = J_{AX} - J_{AX'}; M = J_{AA'} - J_{XX'}; N = J_{AX} + J_{AX'} \tag{D.5}$$

and the following discussion show, the AA′XX′ spectrum is dependent primarily on the sums and differences of coupling constants, rather than on the couplings themselves. The distance between the two dominant peaks ($\nu_{1,2} - \nu_{3,4}$), each of which is a degenerate pair of peaks, corresponds to N, the sum of J_{AX} and $J_{AX'}$. The remainder of the spectrum corresponds to two AB-like quartets (ν_{5-8} and ν_{9-12}). The separation of adjacent outer peaks in a given quartet corresponds to K [($\nu_5 - \nu_6$) and ($\nu_7 - \nu_8$)] and M [($\nu_9 - \nu_{10}$) and ($\nu_{11} - \nu_{12}$)]. Finally, L may be obtained from K, M, and the separation between alternate peaks in either quartet via the relationships in Eqs. (D.6a) and (D.6b).

$$(\nu_5 - \nu_7) = (K^2 + L^2)^{1/2} \tag{D.6a}$$

$$(\nu_9 - \nu_{11}) = (M^2 + L^2)^{1/2} \tag{D.6b}$$

If all 10 peaks are observed, the analysis of the AA′XX′ spectrum is quite straightforward, as could be done, for example, with the spectrum of 1,1-difluoroethene in Figure 4.2. When fewer peaks are observed, as in the spectrum of 1,2-dichlorobenzene (which is AA′BB′ rather than AA′XX′) (Figure 4.3), the analysis is more difficult and normally requires the use of computer programs. Often, such spectra are termed *deceptively simple*, and the analysis does not distinguish $J_{AA'}$ from $J_{XX'}$ or J_{AX} from $J_{AX'}$.

A common example of the AA′XX′ spectrum occurs in the bismethylene fragment ($-CH_2 - CH_2-$) of both acyclic and cyclic compounds. Both $J_{AA'}$ and $J_{XX'}$ are geminal couplings, and J_{AX} and $J_{AX'}$ are vicinal couplings. Because geminal couplings are usually

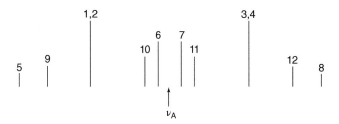

Figure D.4 The A part of an AA′XX′ spectrum.

Figure D.5 Two examples of the special case of AA'XX' spectra (A part) for —CH$_2$CH$_2$— fragment in a cyclic compound: (a) when $M = 0$ and (b) when M is small.

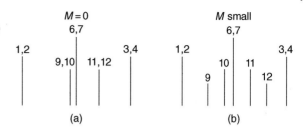

about −12 Hz in such situations, K is quite large in absolute value. As a result, v_5 and v_8 are well removed and of very small intensity, and v_6 and v_7 become degenerate. Because $J_{AA'}$ and $J_{XX'}$ usually have similar values and often are nearly identical, the separation (M) between v_9 and v_{10} and between v_{11} and v_{12} is small or zero, so that v_9 and v_{12} move inside the two large peaks ($v_{1,2}$ and $v_{3,4}$) that give the sum of the vicinal couplings (N). If M is zero, then the difference between the vicinal couplings (L) is given by the separation between $v_{9,10}$ and $v_{11,12}$ (Figure D.5a). Otherwise, L is obtained from M and Eq. (D.6b) (Figure D.5b). This spectrum thus readily gives J_{AX}, $J_{AX'}$, and $(J_{AA'} - J_{XX'})$, but not $(J_{AA'} + J_{XX'})$. As a result, the geminal couplings often are not available. The spectrum of 2-chloroethanol in Figure 4.4 is AA'BB', but provides an example close to that in Figure D.5a, in which peaks 9/10 fall on top of 6/7 to give the large central peak in each half and peaks 11/12 appear as a small peak on the side of the central one.

Appendix E

Relaxation

Spin-lattice relaxation occurs by the interaction of an excited nucleus with a magnetic field fluctuating in the approximate frequency range 10^8-10^9 s^{-1}. Energy passes from the excited nucleus to the source of the magnetic field in the surrounding lattice. Since molecular tumbling is the primary cause of the fluctuating magnetic field, the frequency is most appropriately expressed in the angular units of ω (radians per second). The time constant for this process (the average time to rotate through one radian) is called the effective correlation time τ_c, with units of seconds. Whereas ω is an instantaneous rate, τ_c is an average period. Thus, τ_c is only roughly equivalent to the inverse of the tumbling rate. The effective correlation time depends on a number of factors, including the size of the molecule, the viscosity of the solution, the temperature of the experiment, hydrogen bonding interactions of both solvent and solute, and even the pH. It is not the same for every atom in the molecule, as segmental motion can cause some groups in a molecule (methyl, in particular) to rotate more rapidly than other components. Relaxation is most efficient when $1/\tau_c$ and the resonance (Larmor) frequency ω_0 are comparable in magnitude, i.e. $\omega_0 \tau_c \approx 1$.

The most common source of the fluctuating field is tumbling of a nearby magnetic dipole. Such a dipole is supplied by a magnetic nucleus, usually a proton (called the S nucleus, for "sensitive"; protons have nearly the highest magnetic sensitivity as expressed by the gyromagnetic ratio). Relaxation of a resonating proton (called the I nucleus, for "insensitive"; when the resonating nucleus is not a proton, e.g. ^{13}C or ^{15}N, it invariably is less sensitive than the proton) occurs by interaction with fields from nearby tumbling protons, which are geminal in methylene (CH$_2$) or methyl groups (CH$_3$) and vicinal or more distant for methinyl protons (CH). Carbon or nitrogen (I) relaxation usually occurs by interaction with fields from attached protons (S), if present, or more distant protons, if not. Equation (5.1) (in Chapter 5) shows the distance dependence of this dipolar interaction.

Tumbling of a very small molecule such as methanol can occur at a rate ω as high as 10^{12} s^{-1}. Not all molecules in a given sample are tumbling at this rate, however. Molecules collide with each other and with the walls of the vessel so that their tumbling rate can drop to zero and then accelerate back to the maximum value of 10^{12} s^{-1}. The existence of a range of tumbling rates within a sample means that there also is a range of magnetic fields that they generate. The concentration of fields at a particular frequency ω (or the fraction of time the molecule tumbles at frequency ω) is called the *spectral density*, $J(\omega)$. The value of the spectral density at the resonance frequency, $J(\omega_0)$, is critical for relaxation to be efficient. Because small molecules tumble at rates around

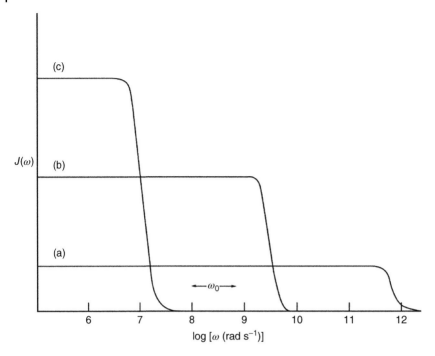

Figure E.1 Plot of spectral density vs logarithm of tumbling frequency.

10^{12} s^{-1}, their spectral density at the desirable range, 10^8–10^9 s^{-1}, is low ($1/\tau_c \gg \omega_0$) and falls to zero above 10^{12}. This situation is illustrated in Figure E.1a. Relaxation rates increase with larger molecules, as the tumbling rate slows ($\omega_0 \tau_c \approx 1$), as in Figure E.1b. The spectral density now is sufficiently high at the Larmor frequency to provide efficient relaxation. For very large, slowly tumbling molecules, the plot (Figure E.1c) backs off further and has a very low value at the Larmor frequency ($1/\tau_c \gg \omega_0$).

$$J(\omega) = 2\tau_c / (1 + \omega^2 \tau_c^2) \tag{E.1}$$

Equation (E.1) describes the spectral density in terms of the tumbling frequency and the correlation time. It is noteworthy that the spectral density is constant from zero up to the maximum rate and then rapidly drops to zero (Figure E.1). The long, flat portion of the curve corresponds to the region in which $1/\tau_c$ is much greater than ω, so that $\omega^2 \tau_c^2 \ll 1$ and $J(\omega) = 2\tau_c$ (a constant). This is the *extreme narrowing condition*. Under this condition, any dipolar line broadening is completely averaged out, so narrowing is at its extreme. For liquid samples of molecules with molecular weights up to a few thousand, the extreme-narrowing condition normally holds.

The size of the dipolar interaction depends primarily on the distance and orientation between the two dipoles, not on the correlation time. By contrast, the *rate of change* of the dipolar interactions depends on τ_c and hence is relevant to the efficiency of relaxation. The total amount of fluctuating fields is independent of τ_c, although τ_c determines the upper limit of the frequencies of the fields. The three curves in Figure E.1 must enclose the same area, but their upper limits vary. In curve (a), molecular tumbling is very rapid, and the spectral density is low. In curves (b) and (c), the upper limit of

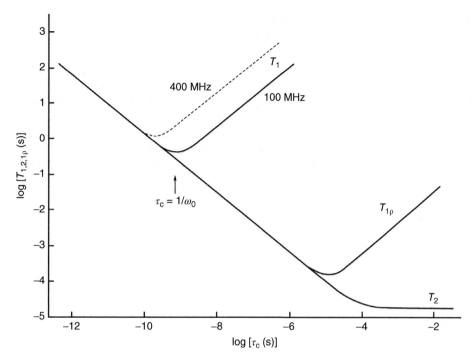

Figure E.2 Logarithmic plot of relaxation times vs the effective correlation time.

frequencies decreases with the lengthening of τ_c, so the spectral density increases proportionately to maintain a constant area.

Since relaxation is most efficient when $\omega_0 \tau_c \approx 1$ (when $1/\tau_c$ is on the order of ω_0), the relaxation time goes through a minimum when plotted against the correlation time (Figure E.2). Equation (E.1) can be transformed into Eq. (E.2)

$$T_1^{-1} = \gamma^2 \left[B_{xL}^0\right]^2 \left[\frac{2\tau_c}{(1+\omega_0^2 \tau_c^2)}\right] \tag{E.2}$$

for the relaxation rate (the reciprocal of the relaxation time). In this expression, γ is the gyromagnetic ratio, and B_{xL}^0 is the root-mean-square average of the x component of the fluctuating local field. Note that ω_0 replaces ω in the expression for the relaxation rate, which, under the extreme-narrowing condition, becomes $2\gamma^2[B_{xL}^0]^2\tau_c$. Thus, the relaxation time T_1 decreases as τ_c increases, i.e. as mobility decreases, for example, with the larger size of molecules or with lower temperatures. This regime is depicted for T_1 at the left of Figure E.2; note that T_1 is independent of the resonance frequency ω_0 here.

When $\tau_c = 1/\omega_0$, the relaxation time attains its minimum (the relaxation rate achieves its maximum), at $1/T_1 = \gamma^2[B_{xL}^0]^2/\omega_0$. The minimum is dependent on the resonance frequency. Since T_1 is directly proportional to ω_0, the minimum occurs at longer T_1 for a higher frequency. This situation is illustrated in Figure E.2 by the dashed line. (For example, the solid line represents 100 MHz and the dashed line 400 MHz.) The T_1 minimum occurs as the spectral density moves off the long, flat region and quickly drops to zero. The minimum for spin-lock relaxation, or relaxation in the rotating frame, $T_{1\rho}$, occurs at a much shorter T_1 because the effective resonance frequency (the spin-lock frequency) is very low—approximately 40 kHz.

When $1/T_c$ is even shorter, as with macromolecules, $\omega_0^2 \tau_c^2$ is much greater than unity. The denominator of Eq. (E.2) then becomes $\omega_0^2 \tau_c^2$, and $1/T_1 = \gamma^2 [B_{xL}^0]^2 [2/(\omega_0^2 \tau_c)]$. In this regime, T_1 and the correlation time are directly proportional. Thus, T_1 increases as τ_c increases, i.e. as the mobility decreases, the opposite to the extreme-narrowing condition. Such a situation is illustrated with the continuations of the solid and dashed lines in the figure upwards and toward the right. This segment is clearly dependent on the resonance frequency, so the dashed and solid lines do not coincide.

Transverse, or spin–spin, relaxation (T_2) is influenced not only by the effects of fluctuating magnetic fields but also by direct interactions between spins. The mutual exchange of two spins occurs without energy transfer to the lattice (adiabatically) and without altering the net magnetization in the z direction. Thus spin–spin interactions alter T_2 without affecting T_1. In the absence of the adiabatic term, the two relaxation times are approximately equal. As T_1 increases (signifying a slower relaxation rate) with decreased mobility outside the extreme-narrowing limit in Figure E.2, the adiabatic term does not respond to reduced mobility in the same way, but instead continues to decrease monotonically until the lattice becomes rigid at very long τ_c. (Spins dephase without relaxation.) When τ_c is longer than T_2, T_2 reaches an asymptotic limit (the rigid-lattice value).

Spin-lattice relaxation can occur by several mechanisms, in addition to dipolar interactions:

1) *Chemical shielding anisotropy.* Because shielding at the nucleus varies with the orientation of the molecule with respect to the B_0 field, the magnetic field acting on the nucleus varies (except for extremely symmetrical molecules). Molecular tumbling, therefore, modulates the local field and provides a mechanism for relaxation. This tumbling process is identical to that for dipolar relaxation, so both mechanisms have the same correlation time τ_c. Consequently, in the extreme-narrowing limit, both mechanisms respond in the same way to a change in temperature, i.e. they decrease at lower temperatures. This mechanism becomes more important at higher fields. Its magnitude varies with the square of B_0. Because many instruments operate at extremely high fields today, line broadening can arise when relaxation by chemical shielding anisotropy causes very short relaxation times.

2) *Spin rotation.* Coherent molecular rotations (or rotations of molecular segments) generate a magnetic field. These rotations are interrupted by molecular collisions, causing the magnetic field to fluctuate. The correlation time for this mechanism, τ_{sr} ("SR" standing for spin rotation) depends on the time between collisions and therefore is different from τ_c for relaxation through dipole–dipole interactions and chemical shielding anisotropy. Because τ_{sr} decreases with increased temperature, relaxation by spin rotation is faster at higher temperatures [$T_1(SR)$ decreases], the opposite of relaxation by dipolar interactions [$T_1(DD)$] in the extreme-narrowing regime. This mechanism is important for very small molecules for which τ_c is too short for effective dipolar relaxation (Figure E.1a). It also is important for rapidly rotating groups within a molecule so that a terminal methyl carbon may have an important component from spin rotation, whereas the internal methylene carbons, rotating more slowly, do not. The trifluoromethyl carbon (CF_3) almost

always relaxes by spin rotation. A plot of T_1 vs $1/T$ exhibits a maximum as dipolar relaxation at lower temperatures changes over to relaxation by spin rotation at higher temperatures. At the maximum of the plot, the contributions from the two mechanisms are equal. Curved temperature dependencies are an indication of a mixture of these mechanisms.

3) *Scalar coupling.* The J coupling constant can be varied by two mechanisms: (i) If the molecule is engaged in molecular exchange between two forms by a dynamic process, differences between magnitudes of the J values in the two forms can provide a time dependence of J, which in turn can modulate the local magnetic field and cause relaxation. This mechanism has been called *scalar relaxation of the first kind* and is relatively rare. Its correlation time is the correlation time for the exchange process. (ii) When the excited nucleus is coupled to a quadrupolar nucleus, the value of the coupling can depend on the relaxation rate of the latter. For example, we usually think of bromine as being a nonmagnetic nucleus, but, of course, the spin of both ^{79}Br and ^{81}Br is and is $3/2$. Rapid quadrupolar relaxation interconverts the bromine spin states so that a neighboring ^{1}H or ^{13}C is affected only by the average and appears uncoupled. The interaction, however, is dependent on the rate of quadrupolar relaxation of the excited nucleus. Modulation of the value of J hence can provide a fluctuating magnetic field and a mechanism for relaxation. For example, carbons directly bonded to bromine may undergo additional relaxation from scalar coupling and hence exhibit broadened peaks. This effect is largest when the Larmor frequencies of the coupled nuclei are similar, as is the case for ^{13}C and ^{79}Br. The time constant for the mechanism is the time constant for relaxation of the excited nucleus.

4) *Quadrupole.* This mechanism of relaxation applies only to quadrupolar nuclei and has been discussed in detail in Section 5.1.

5) *Unpaired electrons.* The electron has a magnetic moment more than three orders of magnitude greater than that of the proton. Consequently, dipolar interactions between an excited nucleus and an electron are extremely strong, and the resulting fluctuating magnetic fields are quite effective at producing relaxation. Thus, paramagnetic impurities can shorten relaxation times to the point that uncertain broadening occurs and line widths are increased, sometimes significantly. For the measurement of T_1, samples routinely are degassed so that unwanted dissolved oxygen, with its unpaired electrons, does not contribute to the relaxation time. Unpaired spins within a molecule can have the same effect, resulting in contact shifts as well as the broadening of peaks. Relaxation agents containing unpaired spins sometimes are introduced intentionally at trace levels in order to shorten relaxation times and allow shorter times between pulses.

When multiple mechanisms contribute to relaxation, the rate is the sum of all contributions (Eq. (E.3))

$$T_1^{-1} = T_1^{-1}(\text{DD}) + T_1^{-1}(\text{SR}) + \ldots \tag{E.3}$$

In addition to utilizing these intramolecular mechanisms, relaxation can occur intermolecularly, as has been noted already for dissolved oxygen. Intermolecular dipolar relaxation can occur, particularly with solvent, but the contributions usually are small.

The nuclear Overhauser effect (NOE) requires dipolar relaxation of the insensitive (I) nucleus by the sensitive (S) nucleus. As illustrated in Figure 5.10, W_2 relaxation provides a mechanism whereby the intensity of the I nucleus is enhanced. This mechanism applies to relatively small molecules in the extreme-narrowing limit, since motional frequencies are fast enough to correspond to W_2. The maximum intensity change is +50% when I and S both are protons (and the terms do not really apply) and is $100\gamma_S/2\gamma_I$ (in %) when the nuclei are different. When $\omega_0^2\tau_c^2 \gg 1$, as is the case with extremely large molecules, motional frequencies correspond to W_0, and relaxation by W_1 and W_2 is inefficient. In this regime, the maximum intensity change is −100% when I and S are both protons. Thus, for macromolecules, the NOE can be expected to be negative and twice as large as for molecules in the extreme-narrowing condition. The W_0 mechanism of relaxation occurs without any change in the net z magnetization, as the I and S nuclei interchange their spins. In this fashion, spin disturbances can propagate beyond the nucleus that originally received the perturbation, in a process called *spin diffusion*. The regime in which $\omega_0^2\tau_c^2 \gg 1$ sometimes is called the *spin-diffusion limit*. For example, spin saturation can propagate throughout the sample and suppress macromolecular signals. With spin diffusion, the distance specificity of the NOE is lost. When the I nucleus has a low gyromagnetic ratio, as with ^{13}C or ^{15}N, there is essentially no change in the signal for slowly tumbling molecules, in contrast to the very important enhancement of the signal obtained for ^{13}C in the extreme-narrowing limit. This result follows from the small perturbation on I spin populations by W_0 relaxation. The situation can be advantageous when one is examining samples containing both small and large molecules, e.g. in cells, as carbons in the small molecules will experience the normal, large NOE. The effects of spin diffusion can be limited by truncating the NOE experiment. If the irradiation time is shortened, for example, distance specificity can be restored.

Appendix F

Product-Operator Formalism and Coherence-Level Diagrams

The Bloch equations (Appendix B) describe the magnetization that results from the effects of the external magnetic field B_0 and the applied magnetic field B_1. In order to include the effects of coupling between nuclei, it was necessary to apply quantum mechanical methods (Appendix C), which permitted the calculation of the energies of spin states and the probabilities of transitions between them. These approaches were developed during the age of continuous-wave experiments. The use of pulses and pulse sequences, however, emphasizes the importance of the evolution of magnetic phenomena over time. A new formalism, therefore, became necessary to understand the effects of pulses, relaxation during pulse sequences, interactions between spins, and multiple time domains. The *product-operator formalism* was introduced by Ernst and others to provide a more straightforward theory than that offered by the more complete density matrix theory of quantum mechanics.

Appendix C used spin operators to calculate the effects of coupling. The overall Hamiltonian operator for the NMR experiment (Eq. (C.3)) is repeated here as Eq. (F.1),

$$\mathcal{H} = \mathcal{H}^0 + \mathcal{H}^1 = \sum_i v_i I_z(i) + \sum_i \sum_j J_{ij} \mathbf{I}(i) \cdot \mathbf{I}(j) \tag{F.1}$$

in which the frequency quantities (v_i, J_{ij}) are expressed in units of Hertz. It is necessary at this point to move into angular frequencies ($\omega = 2\pi v$) through multiplication of the linear frequency v by 2π. Accordingly, Eq. (F.1) may be converted to Eq. (F.2),

$$\mathcal{H} = \mathcal{H}^0 + \mathcal{H}^1 = \sum_i \omega_i I_z(i) + \sum_i \sum_j 2\pi J_{ij}(i) I_z(j) \tag{F.2}$$

in which the effects of coupling are expressed only for the z-axis, as initially there is no magnetization in the *xy* plane. In Appendix C, the effects of the spin operator I were examined and evaluated along each axis (Eq. (C.8)). The product-operator formalism expands upon this approach by exploring the effects of adding a time factor to the Hamiltonian. In this fashion, we can evaluate the evolution of magnetization over time.

The effects of three general phenomena over time must be considered: (i) pulses, (ii) chemical shifts, and (iii) spin–spin coupling. To express the effect of a 90° ($\pi/2$) pulse along the *x*-axis on *z* magnetization, we use the formalism of Eq. (F.3),

$$I_z \xrightarrow{90° I_x} I_y \tag{F.3}$$

which indicates that a 90° I_x pulse (over the arrow) operates on I_z (z magnetization) to transform it into I_y (y magnetization), as illustrated in Figure F.1 by the familiar vector

---

Nuclear Magnetic Resonance Spectroscopy: An Introduction to Principles, Applications, and Experimental Methods,
Second Edition. Joseph B. Lambert, Eugene P. Mazzola, and Clark D. Ridge.
© 2019 John Wiley & Sons Ltd. Published 2019 by John Wiley & Sons Ltd.

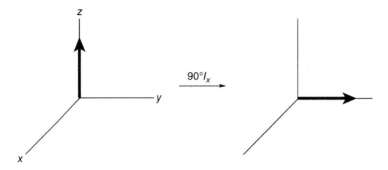

Figure F.1 The vector representation of a 90° pulse in the x direction.

picture. Equation (F.3) and Figure F.1 represent the same phenomenon. Other pulses can be similarly represented. For example, Eq. (6.4)

$$I_z \xrightarrow{90°I_y} -I_x \qquad (F.4)$$

represents the effect of a 90° pulse along the y axis on z magnetization, Eq. (F.5)

$$I_z \xrightarrow{90°(-I_x)} -I_y \qquad (F.5)$$

represents the effect of a 90° pulse along the −x axis on z magnetization, Eq. (F.6)

$$I_y \xrightarrow{90°I_x} -I_z \qquad (F.6)$$

represents the effect of a 90° pulse along the x axis on y magnetization, and, finally, Eq. (F.7)

$$I_x \xrightarrow{90°I_x} I_x \qquad (F.7)$$

represents the effect of a 90° pulse along the x-axis on x magnetization. Note that in the last case a parallel pulse has no effect on magnetization, as was mentioned earlier. These equations introduce an operator formalism that is equivalent to the vector diagrams. For a general pulse angle θ, the effect is described, for example for z magnetization, by Eq. (F.8).

$$I_z \xrightarrow{\theta I_x} I_z \cos\theta + I_y \sin\theta \qquad (F.8)$$

In the vector formalism, the effect of chemical shifts (differences in Larmor frequencies) is represented by the rotation of a vector in the xy plane from a position on the y-axis to one ahead of it by an angle ωt (Figure F.2). The coordinate system is rotating around the z-axis at the reference frequency ω_r so that the frequency of the magnetization vector moves away from the axis at a rate $\Delta\omega = (\omega - \omega_r)$, subtending an angle of $(\Delta\omega)t$ with the y-axis. For simplicity, we will drop the Δ and refer both to the frequency of the nucleus and its difference from the reference frequency as ω, as in Eq. (F.2) (corresponding actually to the special case of $\omega_r = 0$). In the product operator formalism, the effect of chemical shifts is represented by the operation of the Hamiltonian term H^0 on magnetization, illustrated for the three Cartesian coordinates by Eqs. (F.9)–(F.11).

$$I_x \xrightarrow{\omega t I_z} I_x \cos\omega t - I_y \sin\omega t \qquad (F.9)$$

Figure F.2 Evolution of an uncoupled ($J = 0$) spin vector with frequency ω over time t in the xy plane.

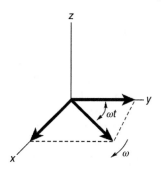

$$I_y \xrightarrow{\omega t I_z} I_y \cos \omega t + I_x \sin \omega t \tag{F.10}$$

$$I_z \xrightarrow{\omega t I_z} I_z \tag{F.11}$$

As mentioned earlier, the operator has been multiplied by time t in order to include the evolution of the chemical shift with time and to convert the operator to the unit of angles (radians).

There is no evolution of the chemical shift along the z-axis (Eq. (F.11)), and its evolution in the xy plane occurs as the sum of sine and cosine components (Eqs. (F.9) and (F.10)). For example, in Figure F.2, magnetization begins ($t = 0$) aligned along the y-axis (represented by the left side of Eq. (F.10)), but then, under the influence of the chemical shift (the symbols over the arrow), evolves into a cosine component along the y-axis (starting at unity) and into a sine component along the x-axis (starting at zero). The result is represented by the right side of Eq. (F.10). Fourier transformation of the trigonometric components yields the Larmor frequency as $\pm\omega$. The experiment cannot distinguish between signals that are equally spaced on either side of the reference frequency. Quadrature detection then performs this selection. In the absence of coupling, Eq. (F.12)

$$I_z \xrightarrow{90° I_x} I_y \xrightarrow{\omega t I_z} I_y \cos \omega t + I_x \sin \omega t \tag{F.12}$$

describes a typical NMR experiment in the product-operator formalism: initial condition with only longitudinal (z) magnetization, application of a 90° x pulse to create transverse (y) magnetization, and evolution of the chemical shift during the detection period, corresponding to the vector diagrams illustrated, for example, in Figures F.1 and F.2.

During the continuous-wave era, scalar coupling was primarily a tool for assigning multiplicities of neighboring protons and for determining stereochemistry through Karplus considerations. In complex pulse sequences, the coupling constant often is the engine that produces spectral modification resulting from structural connectivities. In terms of the product-operator formalism, the coupling-constant term of the Hamiltonian (Eq. (F.2)), multiplied by time t is given, for nucleus i coupled to nucleus j, by Eqs. (F.13) and (F.14)

$$I_x(i) \xrightarrow{2\pi J_{ij} I_z(i) I_z(j) t} I_x(i) \cos(\pi J_{ij} t) - 2I_y(i) I_z(j) \sin(\pi J_{ij} t) \tag{F.13}$$

$$I_y(i) \xrightarrow{2\pi J_{ij} I_z(i) I_z(j) t} I_y(i) \cos(\pi J_{ij} t) + 2I_x(i) I_z(j) \sin(\pi J_{ij} t) \tag{F.14}$$

for magnetization along the x- and y-axes, respectively. The vector description of the evolution of coupling for one member of a two-spin system is illustrated in Figure 5.9a–c. The reference frequency of the rotating coordinate system is set at the Larmor frequency so that the two vectors of nucleus i that result from coupling to nucleus j move in the plus and minus directions away from the y-axis at equal velocities ($J/2$ from the axis, or J from each other). Figure F.3a picks up from Figure 5.9c. The two vectors are projected onto the xy plane in Figure F.3b, one vector representing coupling to the α vector of the j nucleus and the other to the β vector. Each of these vectors can be resolved into x and y components, as depicted in Figure F.3c. The in-phase component along the y-axis starts off large and decreases as the vectors fan out, so it is modulated by a cosine function and is given mathematically by the first term in Eq. (F.14): $I_y(i) \cos(\pi J_{ij} t)$. This is the signal that is detected and Fourier transformed in the experiment to produce the spectrum.

The out-of-phase (or *antiphase*) x component is made up of mutually canceling terms along the +x- and −x-axes (Figure F.3c; the term moving counterclockwise in the xy plane is exactly canceled by the term moving clockwise). The antiphase x component is not detected, as the cancelation process leads to zero macroscopic magnetization. This component can be described as $I_x(i)[I_\beta(j) - I_\alpha(j)]$ if one says that the β spin of nucleus j points to +x and the α spin to −x. The convention is arbitrary and depends on the sign of the coupling constant between i and j. The difference in population $[I_\beta(j) - I_\alpha(j)]$ between the α and β levels corresponds to $2I_z$, so that the x component is expressed mathematically as $I_x(i)[2I_z(j)]$. This quantity is modulated by a sine function over time and becomes the second term in Eq. (F.14), which, then, describes the evolution of the two vectors illustrated in Figures 5.9c, F.3a, and elsewhere, expressed as distinct components, respectively, along the y- and x-axes.

The second (antiphase) term in Eq. (F.14) contains the product of *two* operators, $I_x(i)I_z(j)$, for the first time. Coupling thus can be considered to create a new operator through the process $I_y \to 2I_x I_z$, whereby one operator is transformed into the product of two. It is worth considering a particular case when the amount of time t is set to a constant value, as has been done in numerous examples in Chapters 5 and 6. If t is set equal to $(2J)^{-1}$, then the transformation of Eq. (F.14) takes the form of Eq. (F.15)

$$I_y(i) \xrightarrow{\pi I_z(i) I_z(j)} 2I_x(i)I_z(j) \tag{F.15}$$

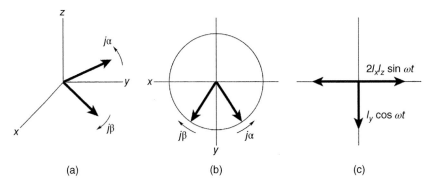

Figure F.3 Evolution of the two vectors from scalar coupling (J) in the xy plane from two perspectives (a) and (b) and resolved into components along the x- and y-axes (c).

since $\cos(\pi/2) = 0$ and $\sin(\pi/2) = 1$. At this time, there is no in-phase component, and the only magnetization is the antiphase component. Selection for antiphase magnetization occurs in this way, for example, in the INEPT sequence (Section 5.6). If one allows precession to continue under the influence of the same coupling constant, the antiphase component is transformed in a fashion parallel to that of Eq. (F.14), with a reversal of roles so that the antiphase component is multiplied by a cosine term and the in-phase component by a sine term, giving Eq. (F.16).

$$2I_x(i)I_z(j) \xrightarrow{2\pi J_{ij}I_z(i)I_z(j)t} 2I_x(i)I_z(j)\cos(\pi J_{ij}t) - I_y(i)\sin(\pi J_{ij}t) \tag{F.16}$$

When another period $t = (2J)^{-1}$ passes, this expression becomes Eq. (F.17).

$$2I_x(i)I_z(j) \xrightarrow{\pi I_z(i)I_z(j)} -I_y(i) \tag{F.17}$$

Magnetization thus is refocused as an echo along the negative y-axis, as we saw in Section 5.6. The second stage of this sequence involves the annihilation of an operator through the process $2I_xI_z \to I_y$. Thus, the operation of coupling can serve both to create and annihilate operators. In this way, coherence is said to be transferred. It is useful to include Eq. (F.18)

$$2I_y(i)I_z(j) \xrightarrow{2\pi J_{ij}I_z(i)I_z(j)t} 2I_y(i)I_z(j)\cos(\pi J_{ij}t) + I_x(i)\sin(\pi J_{ij}t) \tag{F.18}$$

for the transformation of antiphase magnetization along the y-axis as a complement to Eq. (F.16).

A two-spin system has a total of 18 product operators, three each for the Cartesian coordinates of the two spins: $I_x(i)$, $I_y(i)$, $I_z(i)$, $I_x(j)$, $I_y(j)$, and $I_z(j)$. By Eqs. (F.3)–(F.14), these six operators transform according to the various effects of pulses, chemical shifts, and coupling constants. The effects of chemical shifts and coupling constants occur simultaneously. In the absence of strong coupling (the first-order condition), they may be depicted as occurring sequentially, since the operators commute. Not all possible equations are given, but the others may be derived by analogy. There also is an identity operator or unity operator that always operates without changing the function, and there are nine product operators that involve two spins. We have already seen the four antiphase spin magnetizations—$2I_x(i)I_z(j)$, $2I_y(i)I_z(j)$, $2I_z(i)I_x(j)$, and $2I_z(i)I_y(j)$—whose transformations through coupling are given by Eqs. (F.16) and (F.18) or analogous ones. The operator $2I_z(i)I_z(j)$ is said to have longitudinal two-spin order and refers to certain spin perturbations that do not lead to any net spin polarization. The remaining four operators—$2I_x(i)I_x(j)$, $2I_y(i)I_y(j)$, $2I_x(i)I_y(j)$, and $2I_y(i)I_x(j)$—describe two-spin coherence, which requires further explanation.

Coherence refers to a connection between the states of two spins via a spin flip. The initial condition of the NMR experiment, at which all spins are precessing randomly about the z-axis, is incoherent with respect to phase (Figure 1.10). As magnetization moves into the xy plane through simple spin flips, z magnetization decreases, xy magnetization appears, and phases become coherent rather than random (Figure 1.11). The term *coherence* has taken on more complex meanings within the realm of pulse NMR phenomena. A simple spin flip, involving a β spin becoming an α spin in a one-spin system, implies a connection between two adjacent spin levels. The difference in their spin quantum numbers, $\Delta m = [\frac{1}{2} - (-\frac{1}{2})]$, is unity. The sign of Δm corresponds to absorption or emission and is not important at this stage. The number Δm now is defined as the *coherence order p*, so that a simple spin flip is said to have unity order ($p = \pm 1$) and constitute *single quantum coherence*. In more complex systems, with coupling, the phenomenon carries over, so that, for example, the transition from $\beta\beta$ to $\alpha\beta$ involves a single spin flip and constitutes single quantum coherence.

Previously, we described transitions that involve two simultaneous spin flips as being forbidden and hence inherently of low probability. Although this statement is true, the phenomena can be realized by pulse processes. When two spins flip at the same time, the net change in magnetization, Δm, can be either 0 or 2. For example, the transition from $\beta\beta$ to $\alpha\alpha$, $\Delta m = 2$, is referred to as *double quantum coherence* ($p = \pm 2$), and the transition from $\beta\alpha$ to $\alpha\beta$, $\Delta m = 0$, is referred to as *zero quantum coherence* ($p = 0$). During pulse sequences, such coherences may be created, exploited, or annihilated so that it no longer is a simple question of looking for low-intensity double or zero quantum transitions. We will not be concerned with coherences involving more than two states.

Only single quantum coherences can be observed. This assertion is equivalent to saying that only single spin flips are observable. In the product-operator formalism, these are represented by symbols such as $I_y(i)$, denoting the observation of nucleus i along the y-axis, as in Eq. (F.3). Two-spin coherences (double spin flips), such as $2I_x(i)I_x(j)$ and $2I_y(i)I_x(j)$, are not observable. Antiphase coherences, such as $2I_x(i)I_z(j)$, may be transformed into single quantum coherence and can be observed when coupling is present between i and j by processes analogous to those represented in Eq. (F.17). Antiphase coherences are recognized as having exactly one transverse component (I_x or I_y) (corresponding to a single spin flip) and any number of longitudinal components (I_z). Longitudinal magnetization by itself, of course, is not observable. Thus, an ensemble of nuclei can pass through multiple or zero quantum states during a pulse sequence, but must end up as either I_x or I_y for detection.

Transformations can occur between any of the various operators. For example, Eqs. (F.19) and (F.20)

$$2I_y(i)I_z(j) \xrightarrow{90°[I_x(i)+I_x(j)]} -2I_z(i)I_y(j) \tag{F.19}$$

$$2I_x(i)I_z(j) \xrightarrow{90°[I_x(i)+I_x(j)]} 2I_x(i)I_y(j) \tag{F.20}$$

illustrate, respectively, the transformation of antiphase i magnetization into antiphase j magnetization and the transformation of antiphase i magnetization into two-spin coherence. These relationships require that product operators carry out their transformations independently. There are sign conventions that we will not go into but are discussed, for example, in Ref. [1], which also presents a complete table of all possible operations in a two-spin system.

Pulse sequences have been portrayed herein in a number of different ways: (i) as a listing of pulse angles and time periods, e.g. 90°(acquire) for the single pulse and 180°–τ–90°(acquire) for the inversion recovery sequence; (ii) as a series of blocks connected by continuous lines, as in Figure F.1 for the single pulse and for the COSY experiment; and (iii) as vectors moving in a three-dimensional Cartesian coordinate system, as in Figure 1.15 for the single pulse and Figure F.2 for the COSY experiment. These representations successfully treat the movement of incoherent magnetization

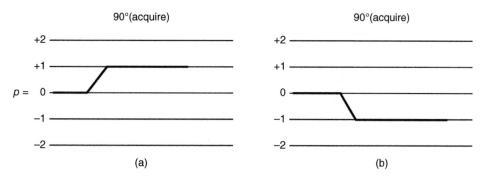

Figure F.4 Coherence-level diagram for a 90° pulse.

(I_z) into single quantum coherences (I_y) or into antiphase arrangements $[2I_x(i)I_z(j)]$, but there is no way for them to depict zero or multiple quantum coherences, such as $2I_x(i)I_x(j)$ or $2I_y(i)I_x(j)$. To till in this gap, Ernst, Bodenhausen, Bain, and others developed a fourth method of depicting NMR experiments, which emphasizes coherence orders.

Figure F.4 illustrates the changes in coherence for the single 1D 90° pulse, followed by acquisition. The experiment begins with spins randomly precessing about the z-axis, depicted by the thick horizontal line with zero coherence ($p = 0$). When the 90° pulse is applied, the vector moves to the y-axis in the rotating coordinate system, and single quantum coherence develops, represented by the movement of the thick horizontal line to $p = 1$. This signal may be thought of either as a frequency oscillating along the y-axis or as two signals rotating respectively clockwise and counterclockwise in the xy plane (analogous to Figures 1.12 and B.1). The signals with either +ω or −ω are, respectively, labeled $p = +1$ and -1 (the difference in sign is an arbitrary distinction that is not used consistently in the literature), and these two possibilities are represented in parts (a) and (b) of the figure. Quadrature detection is tuned to receive one or the other of the two signals, which are 90° out of phase from each other. This sort of depiction is called a *coherence-level diagram*.

The effect of a 180° pulse in such a diagram depends on what has gone before it. In the inversion recovery experiment, the 180° pulse inverts incoherent spins. There is no change in coherence order, so the pulse does not alter the horizontal line. The coherence-level diagram for the inversion recovery experiment [180°–τ–90°(acquire)] differs from those in Figure F.4 only in that the change in the thick line at 90°(acquire) is preceded by the constant period τ. If a 90° pulse, however, has initially moved the coherence order to +1, the effect of the 180° pulse is to change it to −1, as in the simple spin echo experiment, 90°–τ–180°–τ(acquire), with no coupling. Figure 5.13 illustrates the vector diagram and Figure F.5 the coherence-level diagram for the inversion recovery pulse sequence, in which only the $p = +1$ component is shown. The refocusing of chemical shifts is not evident in the latter depiction.

Now let us examine the most fundamental 2D experiment, COSY, in terms of the product-operator formalism. Most of the relationships developed in earlier equations will be utilized. The pulse sequence (Figure F.1) involves two 90° pulses separated by a period t_1 and followed by a second period t_2: 90°–t_1–90°–t_2(acquire). The effects of chemical shifts and coupling constants are expressed during both periods. The theory

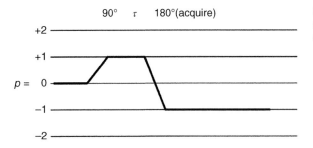

Figure F.5 Coherence-level diagram for the inversion recovery experiment.

may be understood by examining a two-spin coupled system (nuclei i and j). The expression in Eq. (F.21a)

$$I_z(i) \xrightarrow{90°I_x} I_y(i) \xrightarrow{\omega_i I_z(i) t_1} I_y \cos\omega_i t_1 + I_x \sin\omega_i t_1 \tag{F.21a}$$

includes the product operator for the effects of the initial 90° pulse and the evolution of the chemical shift on nucleus i. An analogous equation would apply for nucleus j. Magnetization moves onto the y-axis and evolves during period t_1 according to the Larmor frequency ω_i. This result follows from Eqs. (F.3) and (F.10). Equation (F.21b)

$$\xrightarrow{2\pi J_{ij} I_z(i) I_z(j) t_1} \left[I_y(i) \cos\omega_i t_1 + I_x(i) \sin\omega_i t_1\right] \cos(\pi J_{ij} t_1)$$
$$+ \left[2I_x(i)I_z(j) \cos\omega_i t_1 - 2I_y(i)I_z(j) \sin\omega_i t_1\right] \sin(\pi J_{ij} t_1) \tag{F.21b}$$

then gives the transformation arising from the effects of the coupling constant J_{ij}. There are two terms in this transformation: the cosine term, which corresponds to the in-phase signal, and the sine term, which represents the antiphase signal. This result follows from Eqs. (F.13) and (F.14).

At this stage, the second 90° pulse is applied along the x-axis. Each of the terms in Eq. (F.21b) is transformed appropriately: the first in-phase term according to Eq. (F.6), the second in-phase term by Eq. (F.7), the first antiphase term by Eq. (F.20) (the transformation of antiphase i into multiple quantum coherence), and the second antiphase term in accordance with Eq. (F.19) (the transformation of antiphase i magnetization into antiphase j magnetization). The overall result is given by Eq. (F.21c).

$$\xrightarrow{90°[I_x(i)+I_x(j)]} \left[-I_z(i) \cos\omega_i t_1 + I_x(i) \sin\omega_i t_1\right] \cos(\pi J_{ij} t_1)$$
$$+ \left[2I_x(i)I_y(j) \cos\omega_i t_1 + 2I_z(i)I_y(j) \sin\omega_i t_1\right] \sin(\pi J_{ij} t_1) \tag{F.21c}$$

During the new period t_2, magnetization evolves according to the values of the chemical shifts and coupling constants during the detection period. Only the second and fourth terms of Eq. (F.21c) result in detectable signals, as the first term is longitudinal and the third term contains two transverse components. Thus, Eq. (F.20) simplifies to Eq. (F.21d)

$$\xrightarrow{90°[I_x(i)+I_x(j)]} I_x(i) \sin\omega_i t_1 \cos(\pi J_{ij} t_1) + 2I_z(i)I_y(j) \sin\omega_i t_1 \sin(\pi J_{ij} t_1) \tag{F.21d}$$

during the detection period t_2.

Now consider the two terms of this equation separately for their evolution during t_2. The operation of the chemical shift of nucleus i on the in-phase term of Eq. (F.21d) gives Eq. (F.21e)

$$I_x(i) \sin \omega_i t_1 \cos(\pi J_{ij} t_1) \xrightarrow{\omega_i I_z(i) t_2} I_x(i) \sin \omega_i t_1 \cos(\pi J_{ij} t_1) \cos \omega_i t_2$$
$$- I_y(i) \sin \omega_i t_1 \cos(\pi J_{ij} t_1) \sin \omega_i t_2 \tag{F.21e}$$

according to the result of Eq. (F.9). The operation of coupling (Eqs. (F.13) and (F.14)) on Eq. (F.21e) in turn leads to Eq. (F.21f),

$$\xrightarrow[\text{(in-phase terms)}]{2\pi J_{ij} I_z(i) I_z(j) t_2} I_x(i) \sin \omega_i t_1 \cos(\pi J_{ij} t_1) \cos \omega_i t_2 \cos(\pi J_{ij} t_2)$$
$$- I_y(i) \sin \omega_i t_1 \cos(\pi J_{ij} t_1) \sin \omega_i t_2 \cos(\pi J_{ij} t_2) \tag{F.21f}$$

which is simplified because we need include only the in-phase terms of Eqs. (F.13) and (F.14). The antiphase terms introduce unobservable multiple quantum states, and it must be kept in mind that t_2 is the detection period. The observable quantity along the y-axis in Eq. (F.21f) (the second term) has the frequency of nucleus i (ω_i) during both the t_2 evolution and the t_2 detection periods, so that in a two-dimensional plot this detected magnetization falls on the diagonal. Because the signal is modulated by the cosine of the coupling term in t_2, it is in phase. (Recall the definition of *in phase* from Figure F.3c.) The first term of Eq. (F.21d) therefore leads to the diagonal signals in the 2D experiment.

From Eq. (F.10), the operation of the chemical shift on the second term of Eq. (F.21d) leads to Eq. (F.21g).

$$2I_z(i)I_y(j) \sin \omega_i t_1 \sin(\pi J_{ij} t_1) \xrightarrow{\omega_i I_z(i) t_2} 2I_z(i)I_y(j) \sin \omega_i t_1 \sin(\pi J_{ij} t_1) \cos \omega_j t_2$$
$$+ 2I_z(i)I_x(j) \sin \omega_i t_1 \sin(\pi J_{ij} t_1) \sin \omega_j t_2 \tag{F.21g}$$

and, from Eqs. (F.13) and (F.14), the operation of coupling leads to Eq. (F.21h).

$$\xrightarrow{2\pi J_{ij} I_z(i) I_z(j) t_2} 2I_z(i)I_y(j) \sin \omega_i t_1 \sin(\pi J_{ij} t_1) \cos \omega_j t_2 \cos(\pi J_{ij} t_2)$$
$$+ I_x(j) \sin \omega_i t_1 \sin(\pi J_{ij} t_1) \cos \omega_j t_2 \sin(\pi J_{ij} t_2)$$
$$+ 2I_z(i)I_x(j) \sin \omega_i t_1 \sin(\pi J_{ij} t_1) \sin \omega_j t_2 \cos(\pi J_{ij} t_2)$$
$$- I_y(j) \sin \omega_i t_1 \sin(\pi J_{ij} t_1) \sin \omega_j t_2 \sin(\pi J_{ij} t_2) \tag{F.21h}$$

In this case, we must include both the in-phase and the antiphase terms of Eqs. (F.13) and (F.14). The first and third terms derive from the in-phase term of the coupling operation, whereby the two terms in Eq. (F.21g) are modulated by the cosine term of Eqs. (F.13) and (F.14). In the final expression, these terms are antiphase and hence undetectable, so they may be ignored. The second and fourth terms began as the antiphase term of the coupling operation. The second application of the coupling operator, however, annihilates an operator. The operations are best taken from the table in Freeman. (The operation of $I_z(i)I_z(j)$ on $2I_z(i)I_y(j)$ gives $I_x(j)$, and the operation of the same operator on $2I_z(i)I_x(j)$ gives $-I_y(j)$.) The operations also may be worked out, with the additional fact that $I_z(i)I_z(i) = \frac{1}{4}$, which cancels the factor of 2. The second and fourth terms in Eq. (F.21h) are the detectable signals. Note that, for the signals in these terms, the chemical shift during t_1 was ω_i, but during t_2 it was ω_j. Thus, the signals constitute cross peaks at different frequencies in the two frequency domains after Fourier transformation. These terms are modulated by the sine of the coupling term in t_2 and hence differ in phase from the diagonal peaks by 90°. When the diagonal peaks are dispersive, for example, the cross peaks are absorptive, and vice versa.

It is worthwhile to streamline the analysis of these equations, in order to avoid getting lost in the trigonometry. Equation (F.22)

$$I_z(i) \xrightarrow{90°I_x} I_y(i) \xrightarrow{\omega} I_x(i)\sin\omega_i t_1 \xrightarrow[\text{(in-phase term)}]{J} I_x(i)\sin\omega_i t_1 \cos(\pi J_{ij}t_1)$$

$$\xrightarrow[\text{(in-phase term)}]{90°I_x} I_x(i)\sin\omega_i t_1 \cos(\pi J_{ij}t_1) \xrightarrow{\omega} I_y(i)\sin\omega_i t_1 \cos(\pi J_{ij}t_1)\sin\omega_i t_2$$

$$\xrightarrow{J} I_y(i)\sin\omega_i t_1 \cos(\pi J_{ij}t_1)\sin\omega_i t_2 \cos(\pi J_{ij}t_2) \qquad (F.22)$$

gives each of the steps in the 2D experiment that leads to the observed diagonal signal for nucleus i. The initial longitudinal magnetization of nucleus i moves to the y-axis when the 90° pulse is applied along the x-axis and, during the period t_1, is subject to the effects of nucleus i's chemical shift and coupling to nucleus j. In this abbreviated version, only magnetization along the x-axis has been retained, as it ultimately leads to the detected y magnetization. The second 90° pulse has no effect on the in-phase x magnetization, which then evolves during the period t_2 according to the aforesaid chemical shift and coupling constant to give the term describing the observed magnetization. This term, which contains the same frequency ω_i in both time domains in order that the observed signal falls on the diagonal of the 2D plot, is modulated by the cosine of the coupling term during the detection period. Hence, the signal is in phase.

Equation (F.23)

$$I_z(i) \xrightarrow{90°I_x} I_y(i) \xrightarrow{\omega} I_x(i)\sin\omega_i t_1 \xrightarrow[\text{(antiphase term)}]{J} 2I_y(i)I_z(j)\sin\omega_i t_1 \sin(\pi J_{ij}t_1)$$

$$\xrightarrow[\text{(antiphase term)}]{90°I_x} 2I_z(i)I_y(j)\sin\omega_i t_1 \sin(\pi J_{ij}t_1) \xrightarrow{\omega} 2I_z(i)I_x(j)\sin\omega_i t_1 \sin(\pi J_{ij}t_1)\sin\omega_j t_2$$

$$\xrightarrow{J} I_y(j)\sin\omega_i t_1 \sin(\pi J_{ij}t_1)\sin\omega_j t_2 \sin(\pi J_{ij}t_2) \qquad (F.23)$$

gives each of the steps in the 2D experiment that leads to the observed off-diagonal peaks for nucleus i. The initial steps are the same as in Eq. (F.22) for the diagonal signal, until the antiphase term is selected during the evolution of coupling in t_1. After the second 90° pulse, the term that came from antiphase i magnetization evolves during the second period t_2 according to the relevant chemical shift and coupling constant. We select the term from the chemical-shift operation that eventually produces y magnetization. The coupling operation now serves to annihilate an operator, producing single quantum coherence along the y-axis. Signs have been omitted for simplicity in this streamlined depiction. The final signal exhibits ω_i during t_1 and ω_j during t_2, so it appears as a cross peak. Because it is modulated by the sine of the coupling term during the period t_2, this signal is 90° out of phase with the signals along the diagonal.

The coherence-level diagram for the COSY experiment is illustrated in Figure F.6. The initial 90° pulse generates y magnetization, which may precess either clockwise or counterclockwise, corresponding to the coherence modes $p = \pm 1$ in Figure F.4. Phase cycling selects the type of coherence to be detected at the end of the experiment, for example, $p = -1$. When the initial and detected coherences have the same coherence

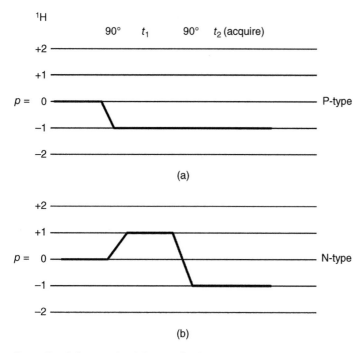

Figure F.6 Coherence-level diagram for the COSY experiment.

number, as in Figure F.6a, the experiment is called a P-type (or antiecho) experiment, and when the initial and detected coherences have opposite coherence numbers, as in Figure F.6b, the experiment is called an N-type (or echo) experiment. They differ only in the sense of precession ($\pm\omega$) of the original signal. The second 90° pulse changes the coherence number p from +1 to −1 in the N-type experiment, but has no effect on the P-type experiment (as the result of the antiphase shift caused by the coupling operator). In the phase-sensitive COSY experiment, both ±1 pathways must be retained during t_1.

As discussed in Section 6.3, the NOESY and EXSY experiments exploit the residual longitudinal magnetization left after the second 90° pulse in the COSY experiment. This longitudinal magnetization is allowed to develop during the mixing time τ_m, when intensities are modulated either through dipolar relaxation that generates the NOE or through chemical exchange that alters nuclear identities. A third 90° pulse then moves the altered longitudinal magnetization into the xy plane for detection. Figure F.7 shows the coherence-level diagram for the NOESY and EXSY experiments. In terms of coherence orders, the initial 90° pulse generates single quantum coherence (both $+\omega$ and $-\omega$ are shown), which evolves according to chemical shifts during t_1. The magnetization that is returned to the z-axis by the second 90° pulse (which generates zero quantum coherence) then is perturbed by dipolar relaxation or chemical exchange during τ_m. Since the COSY signal, which includes both single and double quantum coherence, also is present, the experiment must select the zero quantum signal and convert it to $p = -1$ by the third 90° pulse for observation. This process was described in Section 6.3 as a simple 90° pulse that left unwanted signals along the z-axis. In practice, coherence numbers are selected either by phase cycles or by pulsed field gradients. The process of phase

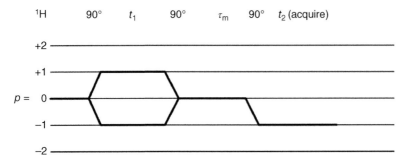

Figure F.7 Coherence-level diagram for the NOESY or EXSY experiment.

cycling, described in Section 5.8, can select a particular phase coherence by exploiting differences between phases. Pulsed field gradients generally perform the operation more efficiently (see Section 6.6).

The double-quantum-filtered COSY experiment (DQF-COSY) was described in Section 6.1 as a method that produces a COSY spectrum lacking signals from nuclei with no homonuclear coupling partner. The COSY experiment generates double quantum coherences (the third term in Eq. (F.21c), which we ignored as unobservable. Figure F.8 shows the development of these coherences during the DQF-COSY experiment. Zero (the first term in Eq. (F.21c)) and single quantum coherences (the second and fourth terms) also are present, but are not shown in the figure. Phase cycling or pulse field gradients are used to select the double quantum coherences and convert them to observable single quantum coherences. A comparison of Figures F.7 and F.8 indicates a strong similarity between the NOESY and the DQF-COSY experiments, which differ only in the length of the mixing time that follows the second 90° pulse. This time is selected for dipolar relaxation, on the one hand, or for homonuclear coupling, on the other. In addition, phase cycling of the third 90° pulse differs, depending upon the coherence selection.

Finally, let us reexamine the 2D INADEQUATE experiment (Section 6.4) in terms of its coherence-level diagram (Figure F.9). This experiment utilizes pulses only at the frequency of the insensitive nucleus, such as ^{13}C. The initial 90° pulse generates single quantum coherence, and the periods τ are selected to produce an antiphase disposition of ^{13}C nuclei with ^{13}C coupling partners. The center-band signal from ^{13}C nuclei

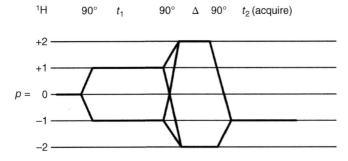

Figure F.8 Coherence-level diagram for the DQF-COSY experiment.

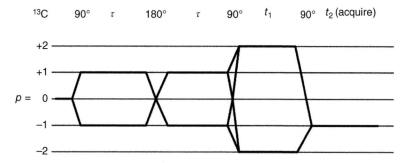

Figure F.9 Coherence-level diagram for the 2D INADEQUATE experiment.

lacking adjacent ^{13}C nuclei is left on the z-axis. During the period t_1, double quantum coherences develop by processes similar to the DQF-COSY experiment. These coherences evolve as the sums of the Larmor frequencies $\omega_i + \omega_j$ in the notation of Eq. (F.21a) and (F.21b) and are selected by phase cycling of the final 90° pulse. The signals that are observed during t_2 appear as cross peaks at $\omega_i + \omega_j$ on the v_1 axis and at the positions of ω_i and ω_j on the v_2 axis, as, for example, in Figure 6.33. There are no diagonal peaks, because they have been selected out. This experiment is usually accompanied by heteronuclear decoupling to suppress ^{1}H–^{13}C coupling and to exploit possible heteronuclear NOE's.

Reference

F.1 Freeman, R. (1987). *A Handbook of Nuclear Magnetic Resonance*. Essex: Longman Scientific & Technical.

Appendix G

Stereochemical Considerations

The terms isochrony, equivalence, and topicity long ago were introduced to describe nuclei of interest in NMR spectroscopy. *Isochronous* nuclei or groups were seen to be chemically equivalent. *Magnetic equivalence*, however, was found to be a stricter requirement than *chemical equivalence*, as it is determined by the coupling constant(s) of each nucleus in a group of potentially magnetically equivalent nuclei (Section 4.2). Finally, *topicity* was seen to be dependent on the nature of symmetry operations that interchange chemically equivalent nuclei or groups.

The particular type of organization that is concerned with *prochiral* groups (the "a" ligands in CaabCxyz systems) remains a source of confusion in NMR spectroscopy. Typical examples are the R—CH_2—R' and R—$C(CH_3)_2$—R' arrangements of, for example, ethylbenzene and isopropylbenzene, respectively. With even the presence of a chiral center in either the R or R' group, a common misconception is that the geminal protons or methyl groups of these systems are equivalent (this term usually is not defined but, from context, refers to magnetically equivalence by the chemical shift criterion, Section 4.2) under suitable conditions. Moreover, it is often expected that the methylene protons are magnetically equivalent by the coupling constant criterion and act in concert to split adjacent protons into a triplet (the spin coupled protons, of course, may be further split by additional neighboring protons or heteronuclei). Observation of nonequivalence in prochiral groups or failure of prochiral protons to spin couple in the expected manner frequently is ascribed to slow rotation of that part of the molecule containing the prochiral moiety.

Enantiotopic nuclei, or groups, are capable of fulfilling all, or most of the above symmetry-related expectations. Their chemical shifts depend, in addition, on both the medium in which the NMR experiment is conducted and the spectral resolution of the spectrometer. The latter is influenced by, for example, the magnetic field strength. Enantiotopic groups are isochronous in achiral or racemic media and constitute A_2, X_2, etc. systems. Moreover, they are potentially anisochronous in chiral media.

The vast majority of prochiral groups, however, cannot be interchanged by any type of symmetry operation due to the presence of one or more chiral centers in the molecule. They are *diastereotopic* and anisochronous, and they constitute AB, XY, etc. systems. For this reason, when encountering prochiral groups, chemists should expect them to be diastereotopic and be, perhaps, pleasantly surprised when they are not.

Before further investigating the matter of chemical shift nonequivalence of diastereotopic, prochiral groups in the presence of a chiral center, let us examine examples of more familiar homotopic and enantiotopic groups to see how their NMR behavior differs critically from that of their diastereotopic counterparts. In particular, we will see why methyl protons are magnetically equivalent and their chemical shifts are, indeed, averaged by fast rotation, while diastereotopic, prochiral protons are magnetically nonequivalent and their chemical shifts are not averaged by rapid rotation.

G.1 Homotopics Groups

Let us first consider the methyl protons of the molecule $H_aH_bH_cC$-Cxyz illustrated in Figure G.1. It is well known that, even in the presence of an adjacent chiral center, methyl protons are chemically equivalent (and thus isochronous). The isochrony of these methyl protons can be demonstrated in the following manner. For the sake of simplicity, only the three staggered conformations of one enantiomer (**A**, **B**, and **C** in Figure G.1) are considered, and the populations of all other conformers are assumed to be negligible.

Furthermore, the chemical shift of any particular proton, say H_a, is a function not only of its neighboring groups (including H_b and H_c in Figure G.1) but also of the *geometrical* relationships that exist among these groups. For the purpose of their contribution to nearby chemical shifts (like those pictured in Figures G.1–G.4), three types of neighboring groups can be identified from the Newman projections: (i) those that are immediately gauche to a proton in question (α), (ii) those that are, in turn, gauche to the first group (β), and (iii) that group that is anti to the subject proton (γ). The orientations of these substituents result in differential effects on the chemical shifts of their neighbors because of interactions *between* these groups, e.g. steric inhibition of resonance, which are specific to their particular geometric arrangements in the molecule.

The chemical shift of, for example, H_a in conformers **A**, **B**, and **C** can be expressed as a function of $\delta_{yH(b)z/xH(c)z}$, $\delta_{xH(b)y/zH(c)y}$, and $\delta_{zH(b)x/yH(c)x}$, respectively, where the substituents are given in the order α, β, γ for clockwise/counterclockwise viewing of the conformers in Figure G.1. In addition, the mole fractions of conformers **A**, **B**, and **C** are n_A, n_B, and n_C, respectively. The average chemical shifts of H_a, H_b, and H_c are then given by Eqs. (G.1)–(G.3). For chemical shift purposes,

$$\delta_{H(a)} = n_A \delta_{yH(b)z/xH(c)z} + n_B \delta_{xH(b)y/zH(c)y} + n_C \delta_{zH(b)x/yH(c)x} \qquad (G.1)$$

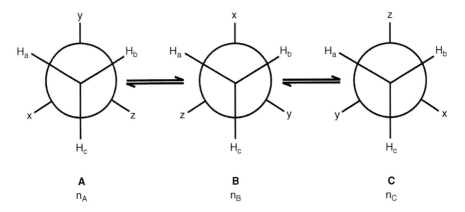

Figure G.1 Homotopic ligands (H_a, H_b, H_c).

$$\delta_{H(b)} = n_A \delta_{zH(c)x/yH(a)x} + n_B \delta_{yH(c)z/xH(a)z} + n_C \delta_{xH(c)y/zH(a)y} \qquad (G.2)$$

$$\delta_{H(c)} = n_A \delta_{xH(a)y/zH(b)y} + n_B \delta_{zH(a)x/yH(b)x} + n_C \delta_{yH(a)z/xH(b)z} \qquad (G.3)$$

H_a, H_b, and H_c are, of course, indistinguishable, and $n_A = n_B = n_C = 1/3$. The average chemical shifts of the methyl protons can then be rewritten as Eqs. (G.4)–(G.6), and the average chemical shifts

$$\delta_{H(a)} = 1/3\delta_{yHz/xHz} + 1/3\delta_{xHy/zHy} + 1/3\delta_{zHx/yHx} \qquad (G.4)$$

$$\delta_{H(b)} = 1/3\delta_{zHx/yHx} + 1/3\delta_{yHz/xHz} + 1/3\delta_{xHy/zHy} \qquad (G.5)$$

$$\delta_{H(c)} = 1/3\delta_{xHy/zHy} + 1/3\delta_{zHx/yHx} + 1/3\delta_{yHz/xHz} \qquad (G.6)$$

of H_a, H_b, and H_c can be expressed finally as Eq. (G.7).

$$\delta_{H(a)} = \delta_{H(b)} = \delta_{H(c)} = 1/3\delta_{yHz/xHz} + \delta 1/3_{xHy/zHy} + \delta 1/3_{zHx/yHx} \qquad (G.7)$$

We can see from Eq. (G.7) that each methyl proton has the same three contributions to its overall chemical shift, and thus the three protons have *potentially* identical chemical shifts. If rotation about the $H_a H_b H_c C$-$Cxyz$ carbon–carbon bond were very slow, three equal-intensity, methyl-proton signals would be observed. Since only one methyl-proton NMR signal is actually detected, rotation about the subject carbon–carbon bond must be sufficiently rapid, on the NMR time scale, to average the three chemical shift contributions. On the time average, the methyl protons are, therefore, isochronous. It is important to recognize, however, that the signal of, for instance, H_a can be averaged with those of H_b and H_c only because the summed chemical shift contributions of all three protons in the three, equally populated conformers are identical. The three signals *are*, in fact, averaged because, in addition, the rotational speed about the carbon–carbon bond of $H_a H_b H_c C$-$Cxyz$ is very fast relative to the time scale of the NMR experiment.

Methyl protons also are known to be magnetically equivalent under conditions of rapid rotation. This magnetic equivalence can be demonstrated, to an approximation, by first substituting H_1 and H_2 for substituents y and z in Figure G.1 to produce Figure G.2

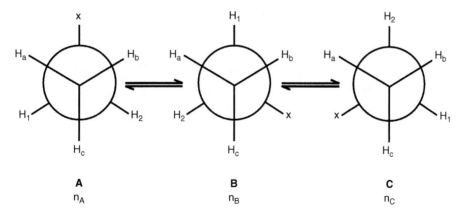

Figure G.2 Coupling of homotopic ligands (H_a, H_b, H_c) with H_1 and H_2.

and then by determining whether couplings between H_a, H_b, H_c and H_1, H_2, are gauche or anti. The coupling constants between, for example, H_a and H_1 are gauche in conformers **A** and **B** while that in **C** is anti. Examination of the three rotamers in Figure G.2 shows that the average coupling constants between H_a, H_b, H_c and H_1, H_2 are given by Eqs. (G.8)–(G.13).

$$J_{H(a)H(1)} = n_A J_{gauche} + n_B J_{gauche} + n_C J_{anti} = 1/3 J_{anti} + 2/3 J_{gauche} \tag{G.8}$$

$$J_{H(b)H(1)} = n_A J_{anti} + n_B J_{gauche} + n_C J_{gauche} = 1/3 J_{anti} + 2/3 J_{gauche} \tag{G.9}$$

$$J_{H(c)H(1)} = n_A J_{gauche} + n_B J_{anti} + n_C J_{gauche} = 1/3 J_{anti} + 2/3 J_{gauche} \tag{G.10}$$

$$J_{H(a)H(2)} = n_A J_{anti} + n_B J_{gauche} + n_C J_{gauche} = 1/3 J_{anti} + 2/3 J_{gauche} \tag{G.11}$$

$$J_{H(b)H(2)} = n_A J_{gauche} + n_B J_{anti} + n_C J_{gauche} = 1/3 J_{anti} + 2/3 J_{gauche} \tag{G.12}$$

$$J_{H(c)H(2)} = n_A J_{gauche} + n_B J_{gauche} + n_C J_{anti} = 1/3 J_{anti} + 2/3 J_{gauche} \tag{G.13}$$

If the various gauche couplings are essentially identical, and likewise for the different anti couplings, then we can easily determine from Eqs. (G.8)–(G.13) that H_a, H_b, and H_c each has the same *averaged* coupling constant to both H_1 and H_2. By definition, the three protons then are magnetically equivalent. The same arguments that were invoked for the averaging of chemical shift contributions into one detected signal, similarly, are employed here for the averaging of spin coupling contributions into a single, observed coupling constant.

G.2 Enantiotopic Groups

Let us next consider the methylene protons of the molecule $H_a H_b YC\text{-}CR_1 R_2 X$, shown in Figure G.3, and focus primarily on the prochiral group comprising H_a and H_b. If $R_1 = R_2$, these two protons are chemically equivalent and enantiotopic by virtue of a σ plane that

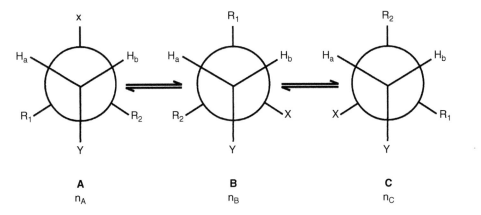

Figure G.3 Enantiotopic ligands (H_a, H_b).

bisects the X and Y groups in conformer **A**. They also are isochronous in achiral and racemic media. The more limited isochrony of H_a and H_b (and, of course, R_1 and R_2) can be established in the following way.

The same considerations and assumptions concerning three primary conformations, the insignificance of other conformers, and the geometric effects of neighboring groups on chemical shifts that were applied in Section G.1 are employed for the rotamers of Figure G.3. Employing identical neighboring-group clockwise/counterclockwise substituent notation as before, the average chemical shifts of H_a and H_b are given by Eqs. (G.14) and (G.15). Again, for chemical shift

$$\delta_{H(a)} = n_A \delta_{XH(b)R(2)/R(1)YR(2)} + n_B \delta_{R(1)H(b)X/R(2)YX} + n_C \delta_{R(2)H(b)R(1)/XYR(1)} \quad (G.14)$$

$$\delta_{H(b)} = n_A \delta_{R(2)YR(1)/XH(a)R(1)} + n_B \delta_{XYR(2)/R(1)H(a)R(2)} + n_C \delta_{R(1)YX/R(2)H(a)} \quad (G.15)$$

purposes, both R_1 and R_2 and H_a and H_b are indistinguishable. While $n_B = n_C$, n_A may not, however, be equal to these two mole fractions. The chemical shifts of H_a and H_b can then and then be rewritten as Eqs. (G.16) and (G.17). Since $\delta_{XHR/RYR} = \delta_{RYR/XHR}$, $\delta_{RHX/RYX} = \delta_{RYX/RHX}$, and $\delta_{RHR/XYR} =$

$$\delta_{H(a)} = n_A \delta_{XHR/RYR} + n_B \delta_{RHX/RYX} + n_C \delta_{RHR/XYR} \quad (G.16)$$

$$\delta_{H(b)} = n_A \delta_{RYR/XHR} + n_B \delta_{XYR/RHR} + n_C \delta_{RYX/RHX} \quad (G.17)$$

$\delta_{XYR/RHR}$, the chemical shifts of H_a and H_b can be expressed further as Eq. (G.18). We can see that

$$\delta_{H(a)} = \delta_{H(b)} = n_A \delta_{XHR/RYR} + n_B \delta_{RHX/RYX} + n_C \delta_{RHR/XYR} \quad (G.18)$$

H_a and H_b each has the same three contributions to its overall chemical shift, and, therefore, both protons have potentially identical chemical shifts. The same argument of very slow rotation, which was made for the *possible* observation of separate methyl-proton signals, applies also to H_a and H_b (and R_1 and R_2). As we saw for the methyl protons, rapid rotation about the carbon–carbon bond of $H_a H_b YC\text{-}CR_1 R_2 X$ averages the three chemical shift contributions, and the H_a/H_b (and R_1/R_2) pair is isochronous in achiral and racemic media. If groups X and Y, however, are very large, and bond rotation commensurately is slow, separate signals could, in principle, be detected.

If $R_1 = R_2 = H$, the appearance of H_a, H_b, H_1, and H_2 depends upon (i) the relative populations of the three conformers (n_A, n_B, and n_C), (ii) the four coupling constants [$J_{H(a)H(1)}$, $J_{H(a)H(2)}$, $J_{H(b)H(1)}$, and $J_{H(b)H(2)}$], and (iii) the rate of rotation about the carbon–carbon bond of $H_a H_b YC\text{-}CH_1 H_2 X$. Since the bond rotational speed is almost always sufficiently rapid to average chemical shift contributions (see above), the third criterion is not considered further with respect to the averaging of coupling constant contributions.

The mole fractions of conformers **A**, **B**, and **C** and the above four coupling constants dictate whether H_a and H_b (and H_1 and H_2) are magnetically equivalent or nonequivalent. Overall couplings can be determined by a gauche/anti coupling constant analysis similar to that carried out previously for the methyl protons. Examination of the three rotamers in Figure G.3 (where $n_B = n_C$) shows that the average coupling constants between H_a, H_b, H_1, and H_2 are given by Eqs. (G.19)–(G.22).

$$J_{H(a)H(1)} = n_A J_{gauche} + n_B J_{gauche} + n_C J_{anti} \quad (G.19)$$

$$J_{H(b)H(1)} = n_A J_{anti} + n_B J_{gauche} + n_C J_{gauche} \tag{G.20}$$

$$J_{H(a)H(2)} = n_A J_{anti} + n_B J_{gauche} + n_C J_{gauche} \tag{G.21}$$

$$J_{H(b)H(2)} = n_A J_{gauche} + n_B J_{anti} + n_C J_{gauche} \tag{G.22}$$

If $n_A = n_B = n_C$ and both the various gauche and different anti couplings are essentially identical, then, from Eqs. (G.19)–(G.22), H_a has the same coupling constants to H_1 and H_2 as H_b. By definition, both H_a and H_b and H_1 and H_2 are magnetically equivalent, constitute an A_2B_2 or A_2X_2 spin system, and appear as two triplets (excluding coupling to other nuclei). When approximate anti and gauche coupling of 13 and 4 Hz, respectively, are substituted into Eqs. (G.19)–(G.22) (and for equally populated rotamers), an average of 7 Hz is obtained for all four coupling constants. This value is in good agreement with those typically observed for ethylene fragments. Such behavior is common for $H_a H_b YC\text{-}CH_1 H_2 X$ systems in which the X and Y groups are relatively small, e.g. $BrCH_2 CH_2 OH$.

Conversely, if the X and Y groups are moderate in size, the population of conformer A can increase relative to those of B and C, where $n_A > n_B = n_C$. In this situation, $J_{H(a)H(2)} > J_{H(a)H(1)}$ and $J_{H(b)H(1)} > J_{H(b)H(2)}$. Due to the coupling constant inequalities, both H_a and H_b and also H_1 and H_2 are magnetically nonequivalent, by definition, and constitute an AA′BB′ or AA′XX′ spin system. The spectral appearance of such systems depends not only on the relative conformer populations and coupling constants but also on the chemical shift separation between the AA′ and BB′ protons.

G.3 Diastereotopic Groups

Finally, let us consider a compound of the type CaabCxyz where the "a" ligands are methylene protons, and the molecule is rewritten as $H_a H_b mC\text{-}Cxyz$ shown in Figure G.4. H_a and H_b constitute a prochiral group and cannot be interchanged by any type of symmetry operation due to the presence of the Cxyz chiral center. In compounds of this type, Gutowsky elegantly demonstrated that the H_a and H_b ligands

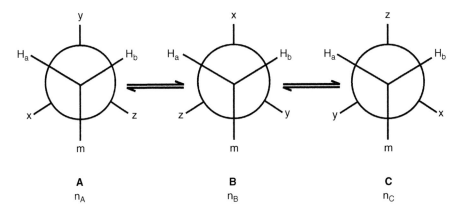

Figure G.4 Diastereotopic ligands (H_a, H_b).

Table G.1 Summary of stereochemical relationships.

Topicity	Symmetry type	Ligand type	NMR behavior
Homotopic	Rotation (C_n)	Equivalent	Isochronous
Enantiotopic	Plane/center	Equivalent	Isochronous in achiral or racemic media but potentially anisochromous in chiral media
Diastereotopic	(none)	Nonequivalent	Potentially anisochronous in achiral or racemic media

are inherently anisochronous (see Ref. [1]). Moreover, as pointed out by Eliel and Wilen, this intrinsic anisochrony is independent of the rate of rotation of the prochiral groups about the H_aH_bmC–Cxyz carbon–carbon bond (see Ref. [2]). The fundamental anisochrony of these diastereotopic groups can be developed in the following manner.

Once more, the same considerations and assumptions concerning three primary conformations, the insignificance of other conformers, and the geometric effects of neighboring groups on chemical shifts that were applied in Sections G.1 and G.2 are employed for the rotamers of Figure G.4. The average chemical shifts of H_a and H_b are given by Eqs. (G.23) and (G.24).

$$\delta_{H(a)} = n_A \delta_{yH(b)z/xmz} + n_B \delta_{xH(b)y/zmy} + n_C \delta_{zH(b)x/ymx} \tag{G.23}$$

$$\delta_{H(b)} = n_A \delta_{zmx/yH(a)x} + n_B \delta_{ymz/xH(a)z} + n_C \delta_{xmy/zH(a)y} \tag{G.24}$$

Again, for chemical shift purposes, H_a and H_b are indistinguishable, and the chemical shifts of H_a and H_b then can be rewritten as Eqs. (G.25) and (G.26). We can see from these expressions that the

$$\delta_{H(a)} = n_A \delta_{yHz/xmz} + n_B \delta_{xHy/zmy} + n_C \delta_{zHx/ymx} \tag{G.25}$$

$$\delta_{H(b)} = n_A \delta_{zmx/yHx} + n_B \delta_{ymz/xHz} + n_C \delta_{xmy/zHy} \tag{G.26}$$

chemical shift terms (expressed in the $n\delta_{abc/xyz}$ format) of H_a and H_b are similar in appearance but not *identical*. For example, the x, m, and z groups are located in the α, β, and γ positions (with respect to H_a) in conformer **A**, but their relative positions are reversed with respect to H_b in this rotamer. Moreover, these three substituents are clustered in this manner only in conformer **A**.

H_a and H_b in this system are, therefore, chemically distinct, diastereotopic, and anisochronous in either achiral or racemic media. Since they constitute an AB or AX system and are magnetically nonequivalent by the chemical shift criterion, any considerations of magnetic equivalence by the coupling constant criterion are, therefore, not applicable. The divisions of ligand types are summarized in Table G.1.

References

G.1 Gutowsky, H.S. (1962). *J. Chem. Phys.* 37: 2196.
G.2 Eliel, E.L. and Wilen, S.H. (1994). *Stereochemistry of Organic Compounds*. New York: Wiley.

Index

a

acquisition time   48, 49, 52, 53, 60, 61, 317, 318, 353–354
active coupling   248
aldehydes   83, 89, 105
aliphatics
   saturated
      alkanes   85–86
      functionalized alkanes   86–87
      structural units   85, 86
   unsaturated
      aldehydes   89
      alkenes   88–89
      alkynes   87
alkanes
   acyclic
      α, β, and γ effects   98–99
      butane fragment, anti and gauche geometries   101
      carbon-13 chemical shifts   100, 101
      methylene carbons   100
   cyclic   101
   cyclopropane   85–86
   functionalized
      carbon substituent parameters   102, 103
      chemical shifts   101–102
      in 1,3-dichloropropane   102, 103
      methyl groups   86–87
alkenes   88–89, 103–104
alkynes   87, 104
apodization   61, 62
APT. *See* attached proton test (APT)
aromatics
   anisole and pyridine   90
   nitrobenzene   89
   nitro group   89–90
   pyridine and pyrrole   104–105
attached proton test (APT)
   C—H couplings   311
   double-resonance procedure   201
   drawbacks   312
   homonuclear decoupling experiment   202
   methyl and methine carbons   312
   proton irradiation   202, 203
   pulse sequence   311
   spectral editing experiment   202–203
   spin vectors   201–202

b

BB decoupling   43, 57
bilinear rotation decoupling (BIRD) sequence   257, 331
BIRD. *See* bilinear rotation decoupling (BIRD) sequence
BIRD-HMQC
   delay time (DT)   259
   pulsed field gradients   257
   selection of signals from protons   258, 259
Bloch equations
   absorption and dispersion mode   394
   Cartesian components   391
   linearly oscillating frequency   393
   magnetization components   391–392
   phase relationship   394
   positive and negative absorption   394, 395

Bloch equations (*contd.*)
    positive and negative dispersion   394, 395
    rotating coordinate system   392
Bloch–Siegert shift   191
Boltzmann's law   3
BURP (Band selective, Uniform Response, Pure phase)   217

C

calibrations
    decoupler field strength   72–73
    pulse width (flip angle)   70–72
CAMELSPIN   266
carbon–carbon correlation
    C,C–COSY   269
    1D INADEQUATE experiment   268
    2D INADEQUATE spectrum of menthol   268–270
    INEPT-INADEQUATE   270
carbon–carbon single bond   81
carbon chemical shifts and structure
    carbonyl groups   105
    electronegativity of groups   97–98
    empirical calculations   105
    factors influencing   96–98
    heavy atom effect   98
    multiple bonding   98
    paramagnetic shielding   96–97
    saturated aliphatics
        acyclic alkanes   98–101
        cyclic alkanes   101
        functionalized alkanes   101–103
    unsaturated compounds   103–105
        alkenes   103–104
        alkynes and nitriles   104
        aromatics   104–105
carbon connectivity
    INADEQUATE spectrum   212–213
    one-bond ^{13}C–^{13}C coupling   212
carboxyl protons   29
CASE. *See* computer-assisted structure elucidation (CASE)
chemical and magnetic equivalence   435
    AA′XX′ system   128, 130
    alkenic protons in cyclopropene   127, 128
    bromochloromethane, protons in   127
    2-chloroethanol   130
    deuterium   127
    diastereotopic protons   131–132
    1,2-dichlorobenzene   129
    difluoroethene or difluoromethane, protons in   126–127, 128, 129
    effect of methyl rotation   131
    magnetic nonequivalence   129, 130
    3-methylcyclopropene   127
    Newman projections   130
    NMR spectrum, role of symmetry   126
chemical shielding anisotropy   31, 418
chemical shift
    alkenes   108–109
    aromatics   109
    carbon (*See* carbon chemical shifts and structure)
    carbonyl compounds   110
    CW field sweep   9
    empirical calculations   105
    isotope effects
        1,4-dioxane   95–96
        in undeuterated dioxane   96
    medium effects
        aromatic solvent-induced shifts   94, 95
        electric-field effect   94
        intermolecular shielding   92
        methyl chemical shifts   95
        in *N*,*N*-dimethylformamide   95
        solute and solvent   92–93
        solute methane   94
    methyl acetate
        ^{1}H spectrum and ^{13}C spectrum   6, 7
        resonances   7, 8
    methyl and methylene groups   106–107
    of nucleus   8–9
    proton (*See* proton chemical shifts and structure)
    resonance frequency   6
    saturated ring systems   107–108
    shielding   6
    spectral conventions   9–10
    tetramethylsilane   8, 9
^{13}C NMR data

Index | 445

chemical shifts and number of attached protons 368
high-resolution mass spectrum 368–369
spectrum of T-2 toxin 366, 367
coherence-level diagrams
  COSY experiment 427–428, 430–431
  2D INADEQUATE experiment 432–433
  double quantum coherence 426
  DQF-COSY 432
  in-phase and the antiphase terms 428–430
  inversion recovery experiment 428
  NOESY and EXSY experiments 431–432
  pulse sequences 426–427
  single quantum coherences 426
  transformation of antiphase magnetization 426
  two-spin coherence 425–426, 427
  zero quantum coherence 426
COLOC. *See* correlation spectroscopy *via* long-range coupling (COLOC)
combination line 147
composite pulses 215
computer-assisted structure elucidation (CASE) 382–383
  procedures 383–384
  T-2 toxin
    correct and incorrect structures 386
    NMR data 387
    one- and two-dimensional NMR raw spectral data 384–385
continuous-wave (CW) field sweep 9, 39
correlation spectroscopy (COSY)
  basic COSY 326–327
  contour representation of 242
  COSY 45 experiment 247–248, 327
  COSY 90 experiment 247
  DQF COSY 328–329
  J coupling, proton-proton correlation 237, 238
  LR COSY 328
  molecular fragments 371–372
  NMR data 370–371
  for other nuclides 254
  stacked representation 241–242
  TOCSY experiment 371
correlation spectroscopy *via* long-range coupling (COLOC)
  FLOCK sequence 260
  of vanillin 260, 261
COSY. *See* correlation spectroscopy (COSY)
coupling constants
  carbon 155–156
  chemical and magnetic equivalence 126–132
  in 1-chloro-4-nitrobenzene 16, 17
  couplings over one bond 134–135
  diethyl ether 18, 20
  double resonance 23
  first-order spectra 16–17, 18–19, 20, 21, 125–126
  geminal couplings 136–138
  geminal proton–proton (H—C—H) 153–154
  3-hydroxybutyric acid 22, 23
  indirect spin–spin coupling 17–18
  isotope satellites 150–151
  long-range couplings 143–146
  nitrogen-15 156
  one-bond 152–153
  Pascal's triangle 19, 20–21
  proton decoupling 22, 23
  second-order spectra 21, 125–126, 147–148
  shift reagents 150
    paramagnetic 150
  signs and mechanisms
    direct coupling 133
    Fermi contact mechanism 132, 133
    Pauli Exclusion Principle 132–133
  spectral analysis 146–147
  1,1,2-trichloroethane 18, 19
  vicinal couplings 139–142
  vicinal proton–proton (H—C—C—H) 154–155
  virtual coupling 149–150
covariance NMR 358
  direct 358–359
  generalized indirect
    advantages 361
    HSQC–TOCSY spectra of T-2 toxin 360–361

CP. *See* cross polarization (CP)
cross polarization (CP)   32
cross polarization and magic angle spinning (CP/MAS)   33
^{13}C spectral editing experiments
   APT experiment   311–312
   DEPT experiment   312–313
CW field sweep. *See* continuous-wave (CW) field sweep

d

1D and advanced 2D experiments
   covariance NMR
      direct   358–359
      generalized indirect   360–361
   1D NOESY and ROESY experiments   347
   1D TOCSY experiment
      comparison spectra   346
      four-spin system of T-2 toxin   345–346
      parameters   347
   H2BC experiment   348–352
   multiplicity-edited HSQC experiment
      expansions of   348, 349
      pulse sequence   347–348
   nonuniform sampling   352–355
   pure shift NMR   355–358
data-acquisition parameters, 2D
   acquisition time   317
   flip angle   318
   number of data points   316–317
   number of scans per time increment   319
   number of time increments   317
   receiver gain   318–319
   relaxation delay   318
   spectral widths   317
   transmitter offset   318
data display, 2D
   phasing and zero referencing   324–325
   symmetrization   325
   use of cross sections in analysis   325
data-processing parameters, 2D
   digital resolution   321–322
   linear prediction   322–324
   weighting functions   319–321
   zero filling   321
deceptive simplicity   147–148
decoupler field strength   72–73
decoupler modulation frequency   73
decoupling   23
delayed COSY   248–249, 279
delays alternating with nutation for tailored excitation (DANTE) experiment   217
density functional theory (DFT)   92
DEPT. *See* distortionless enhancement by polarization transfer (DEPT) experiment
depth gauge   42
deshielding   78, 79
deuterium   44
deuterium lock system   44
DFT. *See* density functional theory (DFT)
2D HMQC spectrum   271
diamagnetic   75
diamagnetic anisotropy   78
diamagnetic shielding   96
diastereotopic groups
   chemical shifts   441
   coupling constant criterion   441
   diastereotopic ligands   440
difference decoupling spectrum   190
diffusion ordered spectroscopy (DOSY)
   molecular diffusion   277–278
   three-component system   278–279
   transformation   278
digital filtration   56
digital resolution (DR)   59, 321–322
digital signal filtration technique   56
digitization noise   56
dioctyl phthalate (DOP)   278
dipolar coupling
   homonuclear chemical-shift correlation
      NOESY experiment   342–343
      ROESY experiment   343–344
dipole–dipole, dipolar, direct, or D-coupling   30–31
dipole–dipole relaxation (T_1(DD))   174
distortionless enhancement by polarization transfer (DEPT) experiment   56, 311, 312–313, 369

carbon types, θ values   312
disadvantage   313
revisited
  edited spectra   211
  MQC and HMQC   211
  one-bond $^{13}C-^{1}H$ couplings   211
  protonated carbon resonances   210–211
  single quantum coherence   211
sequence
  carbon substitution patterns   204
  trisaccharide gentamycin   204, 205
spectral parameters   312
subspectra of T-2 toxin   369
1D NOESY and ROESY experiments   347
3D NOESY/HMQC experiment   271–272, 273
2D NOESY spectrum   270–271
DOSY. See diffusion ordered spectroscopy (DOSY)
double PFG spin echo (DPFGSE) experiment   275
double-pulse, field-gradient, spin-echo NOE experiment (DPFGSE-NOE)   315–316
double quantum coherence   426
double quantum filtered COSY (DQF-COSY)   279
  phase sensitive experiments   328–329
  T-2 toxin   329
double resonance or double irradiation   23, 188
doublet of doublets   24
DQF-COSY. See double quantum filtered COSY (DQF-COSY)
1D TOCSY experiment   345–347
dwell time   49
dynamic effects
  cyclohexane   29, 30
  of methanol   28

e

enantiotopic groups
  chemical shifts   439
  coupling constants   439–440
  enantiotopic ligands   438–439
  methylene protons   438
equation, NMR   389–390
exchange spectroscopy (EXSY)   188, 264, 280
excitation
  absorption of energy   11
  collection of nuclei   10
  linearly and circularly oscillating fields   11–12
  magnetization (M)   10–11
  rotating coordinate system   12
excitation sculpting   275, 276, 277
exponential weighting
  free-induction decay   59, 61
  resolution enhancement   59, 60
  sensitivity enhancement   59, 60
EXSY. See exchange spectroscopy (EXSY)

f

Fermi contact mechanism   132, 133
FID. See free induction decay (FID)
filter bandwidth   52
first-order spectra
  characteristics   125
  correction   65
  left-phase   64
  paramagnetic effect   96
  phase correction   64
  spin–spin splitting patterns   20, 21
  three-spin system (A_2X)   18–19
  two-spin system (AX)   16–17, 125, 126
  A_2X_3 spectrum   18, 20
flip angle   52–54, 318
flip-flop mechanism   13
FLOCK experiment   260, 325
  BIRD pulses   338–339
  COLOC sequence   338
  fixed delay times   339
  resolution enhancement   339–340
  WALTZ decoupling   339
Fourier transformation (FT)   15
free induction decay (FID)   14, 15–16, 237, 238, 239
  acquisition times   60–61
  apodization   61, 62

free induction decay (FID) (*contd.*)
　lock signal　45
　truncation artifacts　61–62
full width at half maximum (FWHM)　63

g

GARP or WURST sequences　331
geminal couplings
　for alkanes　136
　effect of π withdrawal　137
　H—C—F couplings　138
　σ effects (induction)　136
　two-bond couplings　138
　vicinal H—C—C—H coupling constant　139
gHMBC. *See* gradient HMBC (gHMBC) pulse sequence
gradient echo　274
gradient HMBC (gHMBC) pulse sequence　336, 337, 338
gradient pulse. *See* pulsed field gradients (PFG)
gradient shimming　48
gyromagnetic or the magnetogyric ratio　2, 5

h

Hartmann–Hahn condition　32
H2BC. *See* heteronuclear two-bond correlation (H2BC) experiment
heavy atom effect　98
HETCOR. *See* heteronuclear chemical-shift correlation (HETCOR)
heteronuclear chemical-shift correlation (HETCOR)　317
　adamantane derivative　256
　advantages　256–257
　COSY spectra　257
　decoupling　255–256
　delay times　335
　WALTZ decoupling　335
　X-nucleus-detected experiment　334–335
heteronuclear double resonance experiment　191
heteronuclear multiple bond correlation (HMBC) experiment　325, 372

gHMBC pulse sequence　336, 337, 338
　H—C couplings　260–261
　mixed-mode processing　336
　pulse sequence　261–262
　spectrum of heterocycle　262–263
heteronuclear multiple quantum coherence (HMQC) experiment　211
　for camphor　257, 258
　carbon decoupling　331
　gradient-selected HMQC (absolute-value) experiments　332
　inverse detection　257
　LP and NUS methods　332
　pulse sequence　257
heteronuclear relay coherence transfer
　dimethyl acetal of acrolein　263–264
　H–H–C RCT　264
heteronuclear shift correlation (HSC)　256
heteronuclear single quantum correlation (HSQC) experiment　260, 370
　double-INEPT pulse sequence　332, 333
　expansion spectra　332, 333
　GARP or WURST sequences　333
　gradient-selected　334
　LP and NUS　333
heteronuclear two-bond correlation (H2BC) experiment
　comparison spectra　349, 350
　HMBC experiment　348–349
　longer-range C—H couplings　352
　parameters　352
　pulse sequence　349–350
HMBC. *See* heteronuclear multiple bond correlation (HMBC) experiment
HMQC. *See* heteronuclear multiple quantum coherence (HMQC) experiment
homonuclear double resonance experiments　190
HOmonuclear HArtmann-HAhn or HOHAHA, experiment　252
homotopics groups
　chemical shifts　436, 437
　gauche couplings　438
　homotopic ligands　436, 437

methyl protons 436, 437–438
Newman projections 436
HSQC. *See* heteronuclear single quantum correlation (HSQC) experiment
HSQC–TOCSY experiment
 comparison 341–342
 pulse sequence 341
 spectral dispersion 340–341
Hückel rule for aromaticity 80

i

indirect coupling 18
INEPT. *See* insensitive nuclei enhanced by polarization transfer (INEPT) sequence
INEPT-INADEQUATE 270, 280
insensitive nuclei enhanced by polarization transfer (INEPT) sequence 206
 antiphase 206
 carbon transitions 207
 of pyridine 207–208
 refocused (*See* refocused INEPT)
 spin vectors 206
 two-spin system 206–207
integrals 68
integration 68
interferograms 319
inversion-recovery experiment 175, 310–311
inversion-recovery-Fourier transformation (IR-FT) method 309–310
isochronous nuclei or groups 435
isotopes 2, 95–96
 satellites 150–151

j

J coupling, proton-proton correlation
 for annulene 242–243
 axial peaks 246
 AX spin system 240
 Fourier transformation 239, 240
 free-induction decay (FID) 237, 238, 239
 J-resolved spectroscopy 252–254
 LRCOSY or delayed COSY 248–249
 magnetization or population transfer 240–241
 multiple quantum filtration 250–252
 phase-sensitive COSY (φ-COSY) 249–250
 Pro–Leu–Gly 245, 246
 relayed COSY 252
 symmetrization 246
 TOCSY 252
 tripeptide Pro–Leu–Gly in DMSO 244–245
J-filter 336
J modulation. *See* attached proton test (APT)
J-resolved spectroscopy
 glucose derivative 253
 proton–proton decoupled proton spectrum 253, 254
 spin echo experiment 252–254

k

Karplus equation 139

l

Larmor frequency 3, 4, 390
linear prediction (LP) 317
 coefficients 322
 data-processing method 323–324
 2D experiments 322
 expanded HSQC spectra 324
 FIDs 322–323
line broadening functions 59
lock phase 44
lone-pair anisotropy 83
long-range COSY (LR-COSY) 248–249, 279, 328
long-range couplings
 lone-pair-mediated, through-space couplings 145–146
 σ–π overlap
  alkynic and allenic systems 144
  benzylic couplings 144
  five-bond doubly allylic coupling (homoallylic) 143
  four-bond allylic coupling 143
 zigzag pathways
  aromatic meta couplings 145
  percaudal interaction 145
LP. *See* linear prediction (LP)

LR-COSY. *See* long-range COSY (LR-COSY)

m

magic angle spinning (MAS)   31–32
magnetic equivalence   435
magnetic field homogeneity   63
magnetic resonance imaging (MRI)   273
magnetization (M)   10
medium effects   92–95
methyl acetate, resonances   7, 8
MLEV-16 (Malcolm LEVitt)   194
modern spectrometers   40
molecular assembly procedure
  allylic and W-type couplings   376
  C—H couplings   376
  chemical shifts   379
  COSY and HMBC correlations   375–376
  cyclohexene fragment   375
  four- and five-bond, C—H couplings   372
  HMBC   374–375
  three-bond correlation   377, 378
  two- and three-bond C—H couplings   373–374
  two-bond correlations   378
  vicinal coupling   375
MQC. *See* multiple quantum coherence (MQC)
multinuclear spectrometers   40
multiple irradiation. *See* multiple resonance
multiple quantum coherence (MQC)   211
multiple quantum filtration
  DQF-COSY experiment   250–251
  TOCSY spectra of lysine   251
  TQF-COSY experiment   252
multiple resonance
  difference decoupling   190
  experiments, classes of   190–191
  off-resonance decoupling   191–194
  spin decoupling   188–190

n

nitriles   104
NOE. *See* nuclear Overhauser effect (NOE)

NOE spectroscopy (NOESY) experiment   264–265, 280
  AB-ring systems   380–381
  delay (DT) times   339–340
  Dreiding model   381
  EXSY experiments   342, 343
  NMR data   379–380
  phase-sensitive experiments, parameters   343
  three-dimensional representation of T-2 toxin   381–382
noise decoupling   191
non-selective irradiation or broadband decoupling   191
nonuniform sampling
  conventional uniform sampling   353–354
  heteronuclear 2D experiments   352
  NUS   352–353, 354–355
nonuniform sampling (NUS) method   332
nuclear Overhauser effect (NOE)
  applications
    heteronuclear examples   199
    on internuclear distances   199–200
    spin–lattice relaxation   199
  difference experiment   314–315
    of progesterone   198
    three-spin effect   199
  double-pulse, field-gradient, spin-echo NOE experiment   315–316
  enhancements   313
  limitations   200
  observation
    dipolar mechanism   196–197
    double irradiation   195–196, 197
    nondipolar relaxation mechanisms   196
  structural determination   313
  two-spin (AX) system   194–195
nuclei, magnetic properties
  benzene   4, 5
  classes of   1, 2
  energy between spin states   3–6
  external magnetic field   2–3
  gyromagnetic ratio   5
  magnetic moment   1, 2
  NMR properties of   26

nonmagnetic (nonspinning) nuclei 1–2
precessional motions 3
resonance frequency 4
spinning nucleus 1, 2
spin quantum number 1–2
Zeeman effect 2
nuclides 2, 26
natural abundance 27
natural sensitivity 27
receptivity 27
spin 26–27
number of scans (ns) 55
NUS. *See* nonuniform sampling (NUS) method

o

off-resonance decoupling procedure 72
composite pulses and phase cycling 194
heteronuclear decoupling 193
irradiation frequency 192–193
spectral editing 201
of vinyl acetate 192
one bond couplings
carbon-13 and protons 134
CH couplings 134
INADEQUATE technique 135
nitrogen and hydrogen 135
one-dimensional NMR spectroscopy
carbon connectivity 212–213
composite pulses 215
^{13}C spectral editing experiments 311–313
multiple resonance 188–194
NOE experiments 194–200, 313–316
phase cycling 213–215
sensitivity enhancement 205–211
shaped pulses 215–217
spectral editing 200–205
spin–lattice and spin–spin relaxation 173–180
time scale, reactions on 180–188
T_1 measurements 309–311
oversampling 56

p

PANACEA 280
parallel transition 240
paramagnetic shielding 96
parameters, NMR
acquisition parameters 69
chemical shifts and coupling constants 66–68
peak-picking programs 66, 67–68
processing parameters 69
spectral display 69
Pascal's triangle 19, 21
Pauli exclusion principle 132–133
peak-picking programs 66–68
peak suppression or solvent suppression 177
PFG. *See* pulsed field gradients (PFG)
phase cycling
broadband heteronuclear decoupling 214
inversion recovery experiment 213–214
quadrature detection 215
reference frequency 214–215
selection of coherence pathways 215
phase-sensitive COSY (φ-COSY)
dispersion-mode and absorption-mode spectra 249
2D phase quadrants 250
magnitude, or absolute-value, spectrum 249–250
phasing and zero referencing 324–325
planar W. *See* zigzag pathways
Planck's constant 390
polar or inductive effects 75
polymer polyvinyl chloride (PVC) 278
precession 3
prochiral groups 435
product-operator formalism
chemical shifts 422–423
out-of-phase (or antiphase) component 424
pulses 421–422
scalar coupling 421–422
second (antiphase) term 424–425
spin-spin coupling 421
progressive transition 240
proton chemical shifts and structure
aromatics 89–90
empirical calculations 91–92
carbon, hybridization of 77

proton chemical shifts and structure (*contd.*)
    electron density   76
    methyl resonances   75–76
    polar or inductive effects   75
    unshielded nucleus   76
  nonlocal fields
    benzene ring, shielding geometry   79
    carbon–carbon single bond   81
    in 1-chloro-2-fluorobenzene   84–85
    diamagnetic anisotropic properties   78, 82–83, 84
    electron withdrawal or donation   83–84
    ^{19}F spectroscopy   84, 85
    Hückel rule for aromaticity   80
    methano[10]annulene   79
    methyl protons   81–82
    in *N*-methylpiperidine   82
    nonspherical substituents   84
    oblate ellipsoid, shielding   78
    in perfluorocyclohexane   85
    polar bonds   84
    prolate ellipsoid   80–81
    spherical (isotropic) group, shielding   77–78
    van der Waals effect   84, 85
  on oxygen and nitrogen   90–91
  saturated aliphatics   85–87
  unsaturated aliphatics   87–89
proton decoupling   23
proton–heteronucleus correlation
  BIRD-HMQC   257–260
  carbon-13   254
  COLOC   260
  HETCOR   255–257
  heteronuclear relay coherence transfer   263–264
  HMBC   260–263
  HMQC   257
  HSQC   260
pulsed experiments
  FID   14, 15–16
  magnetization vector M   13–14
  y-axis, induced magnetization   14–15
pulsed field gradients (PFG)
  for brucine   274–275
  DPFGSE experiment   275–276
  excitation sculpting   275, 276, 277
  INADEQUATE   274
  NOE experiment   275
  phase cycling   274
  rephasing process   274
  transverse magnetization   273
  WATERGATE   274
pulse Fourier transform   39
pulse width (flip angle)   57
  ^{13}C spectra   72
  magnetization vectors   70
  one-scan spectrum   71
pure shift-covariance NMR   362
pure shift NMR
  broadband proton decoupling   355, 357–358
  multiplicity-edited HSQC spectra of menthol   356, 357
  PSYCHE   355
  Zangger–Sterk refocusing element   355–356
pure shift yielded by chirp excitation (PSYCHE)   355

q

quadrupolar nuclei   2
quantitation and complex splitting
  of ethyl *trans*-crotonate   23–24
  resonance, overlapping peaks   24–25
quantization process   3
quantum mechanical treatment, two-spin system
  energy-level diagram   399, 400, 403, 404, 405, 406
  first-order wave functions   399
  Hamiltonian matrix   400, 401, 404, 405
  NMR, Hamiltonian operator for   397–398
  Schrödinger's wave equation   397
  second-order (AB) with coupling   405, 407
  spin wave functions   398
  stationary-state wave function   406
  three-spin systems   407
  transition probabilities   402

two equivalent spin, parameters for 403–404
wave functions 401–402, 403
quartet 24

r

radio frequency (RF) coils 43
RCT. *See* relayed coherence transfer (RCT)
recovered magnetization 310
refocused INEPT 208
   ^{13}C spectrum of chloroform 208, 209
   spectral editing
      carbon resonances 209, 210
      for methylene and methyl groups 208–209
      signal intensities 209, 210
   spin echo 208
regressive transition 240
relaxation
   correlation time 417
   dipolar interaction 416–417
   extreme narrowing condition 416
   I nucleus 415
   Larmor frequency 416
   nuclear Overhauser effect (NOE) 419–420
   spin–lattice or longitudinal 12–13, 415
      chemical shielding anisotropy 418
      quadrupole 419
      scalar coupling 419
      spin rotation 418–419
      unpaired electrons 419
   spin-lock relaxation 417–418
   spin–spin or transverse 12, 13, 418
   tumbling frequency 415–416
relayed coherence transfer (RCT) 252
relayed COSY
   COSY and RCT 252, 253
   three-spin systems (AMX and A′M′X′) 252, 253
resolution enhancement function 59
resonance 4
resonance frequency 6
ROESY. *See* rotating-frame NOESY experiment (ROESY) experiment
rotating coordinate system 12
rotating-frame NOESY experiment (ROESY) experiment 266–267, 280
   enhancement factors 343
   NOESY experiments 344
   TOCSY artifacts 343–344

s

sample tube placement 42–43
saturation or magnetization transfer 187–188
scalar coupling
   direct heteronuclear chemical-shift correlation
      HMQC experiment 331–332
      HSQC experiment 332–335
      X-nucleus-detected experiments 331
   homonuclear chemical-shift correlation experiments
      COSY family 326–329
      TOCSY experiment 330–331
   indirect heteronuclear chemical-shift correlation
      FLOCK experiment 338–340
      HMBC experiment 336–338
      HSQC–TOCSY experiment 340–342
second-order spectra
   AX_2 system 409–410
   paramagnetic effect 96
   three- and four-spin systems
      AA′XX′ spectrum, four spin systems 412–413
      ABC spectrum 141, 411–412
      AB_2 spectrum 147, 409–410
      ABX spectrum 134, 148, 149, 410–411
selective irradiation or selective decoupling 191
sensitivity enhancement
   DEPT revisited 210–211
   INEPT sequence 206–208
   refocused INEPT 208
      spectral editing with 208–210
shaped pulses
   DANTE pulses 217
   Gaussian shape 216–217
   hard pulses 215–216
   soft pulses 216

shielding 6, 75, 78, 79
shimming process 45
  gradient shimming 48
  homogeneity requirement 44–45
  maximum lock signal level 45
  misadjusted shim settings, effects of
    46–47
  superconducting magnets 46
sine bell function
  pseudo-echo 320
  shifted 320, 321
  squared 320, 321
single quantum coherences 426
space or chemical exchange, proton-proton
    correlation
  COSY signals 264, 265
  1D NOE experiment 265–266
  EXSY experiment 267–268
  NOE and chemical exchange 264
  NOESY experiment 264–265, 266
  ROESY experiment 266–267
  spin diffusion 266
spectral-acquisition parameters
  acquisition time 52
  dwell time 49
  experiments 57–58
  filter bandwidth 52
  flip angle 52–54
  number of data points 50
  number of scans 55
  oversampling and digital filtration 56
  pulse sequence 58
  receiver gain 54
  sinusoidal signals 48–49
  spectral resolution 48
  spectral width 50–51, 52
  steady-state scans 55–56
  transmitter offset 52
  X nuclei, decoupling 56–57
spectral analysis
  ^{13}C NMR data 366–369
  COSY experiment 370–371
  DEPT experiment 369
  HMBC experiment 372
  ^{1}H NMR data 365–366, 367
  HSQC experiment 370
  molecular assembly strategy

  general 372–374
  specific 374–379
  NOESY experiment 379–382
  second-order, two-spin (AB) system
    146, 147
  three-spin systems 147
  trial-and-error procedure 147
spectral editing
  attached proton test 201–204
  DEPT sequence 204–205
  off-resonance decoupling procedure
    200–201
  spin–echo experiment 201
spectral-processing parameters
  exponential weighting 59
  FID truncation and spectral artifacts
    60–62
  resolution 62–63
  zero filling 59–60
spectral resolution (SR) 48
spectral width (sw) 50–51, 317
spectra of solids
  chemical shielding anisotropy 31
  CP/MAS 32–33
  dipole–dipole, dipolar, direct, or
    D-coupling 30–31
  J-coupling 31
  MAS 31–32
  polycrystalline β-quinol methanol
    clathrate 32
  relaxation times 32
spectra: spectral presentation
  baseline correction 65–66
  NMR parameters 66–69
  signal phasing
    absorption and dispersion signal 64
    first-order or left-phase control
      64–65
    zero-order and first-order phase
      correction 64
    zero-order or right-phase control 64
  signal-truncation effects 65
  zero referencing 66, 67
spectrometer
  components of 39–41
  field/frequency locking 43–44
  NMR instrumentation 39–40

probe tuning   43
sample preparation   41–42
sample tube placement   42–43
shimming (*See* shimming process)
spectroscopy   4, 5
spin decoupling   188–190
spin diffusion   420
spin diffusion limit   420
spin–echo experiment   201
spin–lattice and spin–spin relaxation   12, 13, 309
   anisotropic motion   177–178
   causes of relaxation
      carbon relaxation   174
      dipole–dipole relaxation   174
      fluctuating magnetic fields   174–175
   measurement of relaxation time
      inversion recovery experiment   175–176
   partially relaxed spectra   178
   quadrupolar relaxation
      ^{14}N decoupling   180
      nitromethane   179
      spin states for nucleus   178–179
   segmental motion   178
   structural ramifications   177
   transverse relaxation
      mechanisms of *xy* relaxation   176–177
      spin diffusion   176–177
spin locking   32, 188
spinner turbine   42
spin-orbit coupling   98
spin–spin splitting, indirect coupling, or J-coupling   17
steady-state, or dummy, scans   55–56
stereochemical considerations
   diastereotopic groups   440–441
   enantiotopic groups   438–440
   homotopics groups   436–438
subtraction artifacts   275
symmetrization   325

t

tetrahydrofuran (THF)   278
tetramethylsilane (TMS)   8, 9, 66
thermal noise   56

THF. *See* tetrahydrofuran (THF)
time scale, reactions
   atomic inversion   183–184
   fast and slow exchange   181
   hindered rotation   181–182
   laboratory time scale   180
   magnetization transfer and spin locking   187–188
   quantification   187
   ring reversal   183
   valence tautomerizations and bond shifts   185–187
T_1 measurements
   inversion-recovery experiment   310–311
   IR-FT method   309–310
   recovered magnetization   310
TMS. *See* tetramethylsilane (TMS)
TOCSY. *See* total correlation spectroscopy (TOCSY) experiment
TOCSY–HMQC   272
topicity   435
total correlation spectroscopy (TOCSY) experiment   252, 326
   HOHAHA experiment   252
   pulse sequence   330
   spectra of lysine   251, 252
   and Z-TOCSY (phase-sensitive) experiments, parameters   330–331
TQF-COSY. *See* triple quantum filtered COSY (TQF-COSY) experiment
triple quantum filtered COSY (TQF-COSY) experiment   252
triple-resonance experiment   188
triplet–quartet pattern   18
T-2 toxin   327, 384–387
two-bond couplings   138
two-dimensional NMR spectroscopy
   carbon–carbon correlation   268–270
   diffusion-ordered spectroscopy   277–279
   higher dimensions   270–273
   proton–heteronucleus correlation   254–264
   proton–proton correlation
      through *J* coupling   237–246

two-dimensional NMR spectroscopy (*contd.*)
  through space or chemical exchange   264–268
  pulsed field gradients   273–277
two-dimensional techniques
  data-acquisition parameters   316–319
  data display   324–325
  data-processing parameters   319–324
  dipolar coupling   342–344
  experiments   345–361
  pure shift-covariance NMR   362
  scalar coupling constants (*See* scalar coupling)

v

valence tautomerizations
  and bond shifts   185–186
  cyclooctatetraene, fluxional behavior   185
  fluxional organometallic species   186
  3,4-homotropilidine, Cope rearrangement   185
  1,5-sigmatropic shifts   186–187
van derWaals effect   84
vicinal couplings
  acrylonitrile   141, 142
  in alkenes   141
  benzene derivatives   142
  cyclohexanes   139–140
  substituent electronegativity   141–142

vicinal H—C—C—H coupling constant   139
virtual coupling
  β-methylglutaric acid   149–150
  dimethylbenzoquinones   150

w

WALTZ-16   194
WALTZ decoupling   331
  scheme   57, 70
WATER suppression by gradient-tailored excitation (WATERGATE)   274
weighting functions
  absolute-value data   319–320
  interferograms   319
  modern spectrometers   320–321
  phase-sensitive data   321

x

X nuclei, decoupling for   56–57

z

Zangger–Sterk refocusing element   355–356
Zeeman effect   2, 390
zero filling   59–60, 321
zero-order or right-phase control   64
zero quantum coherence   426
zero referencing
  chemical shift data   66, 67
  tetramethylsilane   66
zigzag pathways   144–145

Printed and bound by CPI Group (UK) Ltd, Croydon, CR0 4YY
09/06/2025

14685657-0001